MONOGRAPHS AND TEXTBOOKS IN PURE AND APPLIED MATHEMATICS

T0172861

1. *K. Yano,* Integral Formulas in Riemannian Geometry (1970)
2. *S. Kobayashi,* Hyperbolic Manifolds and Holomorphic Mappings (1970)
3. *V. S. Vladimirov,* Equations of Mathematical Physics (A. Jeffrey, ed.; A. Littlewood, trans.) (1970)
4. *B. N. Pshenichnyi,* Necessary Conditions for an Extremum (L. Neustadt, translation ed.; K. Makowski, trans.) (1971)
5. *L. Narici et al.,* Functional Analysis and Valuation Theory (1971)
6. *S. S. Passman,* Infinite Group Rings (1971)
7. *L. Domhoff,* Group Representation Theory. Part A: Ordinary Representation Theory. Part B: Modular Representation Theory (1971, 1972)
8. *W. Boothby and G. L. Weiss, eds.,* Symmetric Spaces (1972)
9. *Y. Matsushima,* Differentiable Manifolds (E. T. Kobayashi, trans.) (1972)
10. *L. E. Ward, Jr.,* Topology (1972)
11. *A. Babakhanian,* Cohomological Methods in Group Theory (1972)
12. *R. Gilmer,* Multiplicative Ideal Theory (1972)
13. *J. Yeh,* Stochastic Processes and the Wiener Integral (1973)
14. *J. Barros-Neto,* Introduction to the Theory of Distributions (1973)
15. *R. Larsen,* Functional Analysis (1973)
16. *K. Yano and S. Ishihara,* Tangent and Cotangent Bundles (1973)
17. *C. Procesi,* Rings with Polynomial Identities (1973)
18. *R. Hermann,* Geometry, Physics, and Systems (1973)
19. *N. R. Wallach,* Harmonic Analysis on Homogeneous Spaces (1973)
20. *J. Dieudonné,* Introduction to the Theory of Formal Groups (1973)
21. *I. Vaisman,* Cohomology and Differential Forms (1973)
22. *B.-Y. Chen,* Geometry of Submanifolds (1973)
23. *M. Marcus,* Finite Dimensional Multilinear Algebra (in two parts) (1973, 1975)
24. *R. Larsen,* Banach Algebras (1973)
25. *R. O. Kujala and A. L. Vitter, eds.,* Value Distribution Theory: Part A; Part B: Deficit and Bezout Estimates by Wilhelm Stoll (1973)
26. *K. B. Stolarsky,* Algebraic Numbers and Diophantine Approximation (1974)
27. *A. R. Magid,* The Separable Galois Theory of Commutative Rings (1974)
28. *B. R. McDonald,* Finite Rings with Identity (1974)
29. *J. Satake,* Linear Algebra (S. Koh et al., trans.) (1975)
30. *J. S. Golan,* Localization of Noncommutative Rings (1975)
31. *G. Klambauer,* Mathematical Analysis (1975)
32. *M. K. Agoston,* Algebraic Topology (1976)
33. *K. R. Goodearl,* Ring Theory (1976)
34. *L. E. Mansfield,* Linear Algebra with Geometric Applications (1976)
35. *N. J. Pullman,* Matrix Theory and Its Applications (1976)
36. *B. R. McDonald,* Geometric Algebra Over Local Rings (1976)
37. *C. W. Groetsch,* Generalized Inverses of Linear Operators (1977)
38. *J. E. Kuczkowski and J. L. Gersting,* Abstract Algebra (1977)
39. *C. O. Christenson and W. L. Voxman,* Aspects of Topology (1977)
40. *M. Nagata,* Field Theory (1977)
41. *R. L. Long,* Algebraic Number Theory (1977)
42. *W. F. Pfeffer,* Integrals and Measures (1977)
43. *R. L. Wheeden and A. Zygmund,* Measure and Integral (1977)
44. *J. H. Curtiss,* Introduction to Functions of a Complex Variable (1978)
45. *K. Hrbacek and T. Jech,* Introduction to Set Theory (1978)
46. *W. S. Massey,* Homology and Cohomology Theory (1978)
47. *M. Marcus,* Introduction to Modern Algebra (1978)
48. *E. C. Young,* Vector and Tensor Analysis (1978)
49. *S. B. Nadler, Jr.,* Hyperspaces of Sets (1978)
50. *S. K. Segal,* Topics in Group Kings (1978)
51. *A. C. M. van Rooij,* Non-Archimedean Functional Analysis (1978)
52. *L. Corwin and R. Szczarba,* Calculus in Vector Spaces (1979)
53. *C. Sadosky,* Interpolation of Operators and Singular Integrals (1979)
54. *J. Cronin,* Differential Equations (1980)
55. *C. W. Groetsch,* Elements of Applicable Functional Analysis (1980)

56. *I. Vaisman*, Foundations of Three-Dimensional Euclidean Geometry (1980)
57. *H. I. Freedan*, Deterministic Mathematical Models in Population Ecology (1980)
58. *S. B. Chae*, Lebesgue Integration (1980)
59. *C. S. Rees et al.*, Theory and Applications of Fourier Analysis (1981)
60. *L. Nachbin*, Introduction to Functional Analysis (R. M. Aron, trans.) (1981)
61. *G. Orzech and M. Orzech*, Plane Algebraic Curves (1981)
62. *R. Johnsonbaugh and W. E. Pfaffenberger*, Foundations of Mathematical Analysis (1981)
63. *W. L. Voxman and R. H. Goetschel*, Advanced Calculus (1981)
64. *L. J. Corwin and R. H. Szczarba*, Multivariable Calculus (1982)
65. *V. I. Istrătescu*, Introduction to Linear Operator Theory (1981)
66. *R. D. Järvinen*, Finite and Infinite Dimensional Linear Spaces (1981)
67. *J. K. Beem and P. E. Ehrlich*, Global Lorentzian Geometry (1981)
68. *D. L. Armacost*, The Structure of Locally Compact Abelian Groups (1981)
69. *J. W. Brewer and M. K. Smith, eds.*, Emmy Noether: A Tribute (1981)
70. *K. H. Kim*, Boolean Matrix Theory and Applications (1982)
71. *T. W. Wieting*, The Mathematical Theory of Chromatic Plane Ornaments (1982)
72. *D. B.Gauld*, Differential Topology (1982)
73. *R. L. Faber*, Foundations of Euclidean and Non-Euclidean Geometry (1983)
74. *M. Carmeli*, Statistical Theory and Random Matrices (1983)
75. *J. H. Carruth et al.*, The Theory of Topological Semigroups (1983)
76. *R. L. Faber*, Differential Geometry and Relativity Theory (1983)
77. *S. Barnett*, Polynomials and Linear Control Systems (1983)
78. *G. Karpilovsky*, Commutative Group Algebras (1983)
79. *F. Van Oystaeyen and A. Verschoren*, Relative Invariants of Rings (1983)
80. *I. Vaisman*, A First Course in Differential Geometry (1984)
81. *G. W. Swan*, Applications of Optimal Control Theory in Biomedicine (1984)
82. *T. Petrie and J. D. Randall*, Transformation Groups on Manifolds (1984)
83. *K. Goebel and S. Reich*, Uniform Convexity, Hyperbolic Geometry, and Nonexpansive Mappings (1984)
84. *T. Albu and C. Năstăsescu*, Relative Finiteness in Module Theory (1984)
85. *K. Hrbacek and T. Jech*, Introduction to Set Theory: Second Edition (1984)
86. *F. Van Oystaeyen and A. Verschoren*, Relative Invariants of Rings (1984)
87. *B. R. McDonald*, Linear Algebra Over Commutative Rings (1984)
88. *M. Namba*, Geometry of Projective Algebraic Curves (1984)
89. *G. F. Webb*, Theory of Nonlinear Age-Dependent Population Dynamics (1985)
90. *M. R. Bremner et al.*, Tables of Dominant Weight Multiplicities for Representations of Simple Lie Algebras (1985)
91. *A. E. Fekete*, Real Linear Algebra (1985)
92. *S. B. Chae*, Holomorphy and Calculus in Normed Spaces (1985)
93. *A. J. Jerri*, Introduction to Integral Equations with Applications (1985)
94. *G. Karpilovsky*, Projective Representations of Finite Groups (1985)
95. *L. Narici and E. Beckenstein*, Topological Vector Spaces (1985)
96. *J. Weeks*, The Shape of Space (1985)
97. *P. R. Gribik and K. O. Kortanek*, Extremal Methods of Operations Research (1985)
98. *J.-A. Chao and W. A. Woyczynski, eds.*, Probability Theory and Harmonic Analysis (1986)
99. *G. D. Crown et al.*, Abstract Algebra (1986)
100. *J. H. Carruth et al.*, The Theory of Topological Semigroups, Volume 2 (1986)
101. *R. S. Doran and V. A. Belfi*, Characterizations of C*-Algebras (1986)
102. *M. W. Jeter*, Mathematical Programming (1986)
103. *M. Altman*, A Unified Theory of Nonlinear Operator and Evolution Equations with Applications (1986)
104. *A. Verschoren*, Relative Invariants of Sheaves (1987)
105. *R. A. Usmani*, Applied Linear Algebra (1987)
106. *P. Blass and J. Lang*, Zariski Surfaces and Differential Equations in Characteristic $p >$ 0 (1987)
107. *J. A. Reneke et al.*, Structured Hereditary Systems (1987)
108. *H. Busemann and B. B. Phadke*, Spaces with Distinguished Geodesics (1987)
109. *R. Harte*, Invertibility and Singularity for Bounded Linear Operators (1988)
110. *G. S. Ladde et al.*, Oscillation Theory of Differential Equations with Deviating Arguments (1987)
111. *L. Dudkin et al.*, Iterative Aggregation Theory (1987)
112. *T. Okubo*, Differential Geometry (1987)

113. *D. L. Stancl and M. L. Stancl*, Real Analysis with Point-Set Topology (1987)
114. *T. C. Gard*, Introduction to Stochastic Differential Equations (1988)
115. *S. S. Abhyankar*, Enumerative Combinatorics of Young Tableaux (1988)
116. *H. Strade and R. Farnsteiner*, Modular Lie Algebras and Their Representations (1988)
117. *J. A. Huckaba*, Commutative Rings with Zero Divisors (1988)
118. *W. D. Wallis*, Combinatorial Designs (1988)
119. *W. Więsław*, Topological Fields (1988)
120. *G. Karpilovsky*, Field Theory (1988)
121. *S. Caenepeel and F. Van Oystaeyen*, Brauer Groups and the Cohomology of Graded Rings (1989)
122. *W. Kozlowski*, Modular Function Spaces (1988)
123. *E. Lowen-Colebunders*, Function Classes of Cauchy Continuous Maps (1989)
124. *M. Pavel*, Fundamentals of Pattern Recognition (1989)
125. *V. Lakshmikantham et al.*, Stability Analysis of Nonlinear Systems (1989)
126. *R. Sivaramakrishnan*, The Classical Theory of Arithmetic Functions (1989)
127. *N. A. Watson*, Parabolic Equations on an Infinite Strip (1989)
128. *K. J. Hastings*, Introduction to the Mathematics of Operations Research (1989)
129. *B. Fine*, Algebraic Theory of the Bianchi Groups (1989)
130. *D. N. Dikranjan et al.*, Topological Groups (1989)
131. *J. C. Morgan II*, Point Set Theory (1990)
132. *P. Biler and A. Witkowski*, Problems in Mathematical Analysis (1990)
133. *H. J. Sussmann*, Nonlinear Controllability and Optimal Control (1990)
134. *J.-P. Florens et al.*, Elements of Bayesian Statistics (1990)
135. *N. Shell*, Topological Fields and Near Valuations (1990)
136. *B. F. Doolin and C. F. Martin*, Introduction to Differential Geometry for Engineers (1990)
137. *S. S. Holland, Jr.*, Applied Analysis by the Hilbert Space Method (1990)
138. *J. Okniński*, Semigroup Algebras (1990)
139. *K. Zhu*, Operator Theory in Function Spaces (1990)
140. *G. B. Price*, An Introduction to Multicomplex Spaces and Functions (1991)
141. *R. B. Darst*, Introduction to Linear Programming (1991)
142. *P. L. Sachdev*, Nonlinear Ordinary Differential Equations and Their Applications (1991)
143. *T. Husain*, Orthogonal Schauder Bases (1991)
144. *J. Foran*, Fundamentals of Real Analysis (1991)
145. *W. C. Brown*, Matrices and Vector Spaces (1991)
146. *M. M. Rao and Z. D. Ren*, Theory of Orlicz Spaces (1991)
147. *J. S. Golan and T. Head*, Modules and the Structures of Rings (1991)
148. *C. Small*, Arithmetic of Finite Fields (1991)
149. *K. Yang*, Complex Algebraic Geometry (1991)
150. *D. G. Hoffman et al.*, Coding Theory (1991)
151. *M. O. González*, Classical Complex Analysis (1992)
152. *M. O. González*, Complex Analysis (1992)
153. *L. W. Baggett*, Functional Analysis (1992)
154. *M. Sniedovich*, Dynamic Programming (1992)
155. *R. P. Agarwal*, Difference Equations and Inequalities (1992)
156. *C. Brezinski*, Biorthogonality and Its Applications to Numerical Analysis (1992)
157. *C. Swartz*, An Introduction to Functional Analysis (1992)
158. *S. B. Nadler, Jr.*, Continuum Theory (1992)
159. *M. A. Al-Gwaiz*, Theory of Distributions (1992)
160. *E. Perry*, Geometry: Axiomatic Developments with Problem Solving (1992)
161. *E. Castillo and M. R. Ruiz-Cobo*, Functional Equations and Modelling in Science and Engineering (1992)
162. *A. J. Jerri*, Integral and Discrete Transforms with Applications and Error Analysis (1992)
163. *A. Charlier et al.*, Tensors and the Clifford Algebra (1992)
164. *P. Biler and T. Nadzieja*, Problems and Examples in Differential Equations (1992)
165. *E. Hansen*, Global Optimization Using Interval Analysis (1992)
166. *S. Guerre-Delabrière*, Classical Sequences in Banach Spaces (1992)
167. *Y. C. Wong*, Introductory Theory of Topological Vector Spaces (1992)
168. *S. H. Kulkarni and B. V. Limaye*, Real Function Algebras (1992)
169. *W. C. Brown*, Matrices Over Commutative Rings (1993)
170. *J. Loustau and M. Dillon*, Linear Geometry with Computer Graphics (1993)
171. *W. V. Petryshyn*, Approximation-Solvability of Nonlinear Functional and Differential Equations (1993)

172. *E. C. Young*, Vector and Tensor Analysis: Second Edition (1993)
173. *T. A. Bick*, Elementary Boundary Value Problems (1993)
174. *M. Pavel*, Fundamentals of Pattern Recognition: Second Edition (1993)
175. *S. A. Albeverio et al.*, Noncommutative Distributions (1993)
176. *W. Fulks*, Complex Variables (1993)
177. *M. M. Rao*, Conditional Measures and Applications (1993)
178. *A. Janicki and A. Weron*, Simulation and Chaotic Behavior of α-Stable Stochastic Processes (1994)
179. *P. Neittaanmäki and D. Tiba*, Optimal Control of Nonlinear Parabolic Systems (1994)
180. *J. Cronin*, Differential Equations: Introduction and Qualitative Theory, Second Edition (1994)
181. *S. Heikkilä and V. Lakshmikantham*, Monotone Iterative Techniques for Discontinuous Nonlinear Differential Equations (1994)
182. *X. Mao*, Exponential Stability of Stochastic Differential Equations (1994)
183. *B. S. Thomson*, Symmetric Properties of Real Functions (1994)
184. *J. E. Rubio*, Optimization and Nonstandard Analysis (1994)
185. *J. L. Bueso et al.*, Compatibility, Stability, and Sheaves (1995)
186. *A. N. Michel and K. Wang*, Qualitative Theory of Dynamical Systems (1995)
187. *M. R. Darnel*, Theory of Lattice-Ordered Groups (1995)
188. *Z. Naniewicz and P. D. Panagiotopoulos*, Mathematical Theory of Hemivariational Inequalities and Applications (1995)
189. *L. J. Corwin and R. H. Szczarba*, Calculus in Vector Spaces: Second Edition (1995)
190. *L. H. Erbe et al.*, Oscillation Theory for Functional Differential Equations (1995)
191. *S. Agaian et al.*, Binary Polynomial Transforms and Nonlinear Digital Filters (1995)
192. *M. I. Gil'*, Norm Estimations for Operation-Valued Functions and Applications (1995)
193. *P. A. Grillet*, Semigroups: An Introduction to the Structure Theory (1995)
194. *S. Kichenassamy*, Nonlinear Wave Equations (1996)
195. *V. F. Krotov*, Global Methods in Optimal Control Theory (1996)
196. *K. I. Beidar et al.*, Rings with Generalized Identities (1996)
197. *V. I. Arnautov et al.*, Introduction to the Theory of Topological Rings and Modules (1996)
198. *G. Sierksma*, Linear and Integer Programming (1996)
199. *R. Lasser*, Introduction to Fourier Series (1996)
200. *V. Sima*, Algorithms for Linear-Quadratic Optimization (1996)
201. *D. Redmond*, Number Theory (1996)
202. *J. K. Beem et al.*, Global Lorentzian Geometry: Second Edition (1996)
203. *M. Fontana et al.*, Prüfer Domains (1997)
204. *H. Tanabe*, Functional Analytic Methods for Partial Differential Equations (1997)
205. *C. Q. Zhang*, Integer Flows and Cycle Covers of Graphs (1997)
206. *E. Spiegel and C. J. O'Donnell*, Incidence Algebras (1997)
207. *B. Jakubczyk and W. Respondek*, Geometry of Feedback and Optimal Control (1998)
208. *T. W. Haynes et al.*, Fundamentals of Domination in Graphs (1998)
209. *T. W. Haynes et al., eds.*, Domination in Graphs: Advanced Topics (1998)
210. *L. A. D'Alotto et al.*, A Unified Signal Algebra Approach to Two-Dimensional Parallel Digital Signal Processing (1998)
211. *F. Halter-Koch*, Ideal Systems (1998)
212. *N. K. Govil et al., eds.*, Approximation Theory (1998)
213. *R. Cross*, Multivalued Linear Operators (1998)
214. *A. A. Martynyuk*, Stability by Liapunov's Matrix Function Method with Applications (1998)
215. *A. Favini and A. Yagi*, Degenerate Differential Equations in Banach Spaces (1999)
216. *A. Illanes and S. Nadler, Jr.*, Hyperspaces: Fundamentals and Recent Advances (1999)
217. *G. Kato and D. Struppa*, Fundamentals of Algebraic Microlocal Analysis (1999)
218. *G. X.-Z. Yuan*, KKM Theory and Applications in Nonlinear Analysis (1999)
219. *D. Motreanu and N. H. Pavel*, Tangency, Flow Invariance for Differential Equations, and Optimization Problems (1999)
220. *K. Hrbacek and T. Jech*, Introduction to Set Theory, Third Edition (1999)
221. *G. E. Kolosov*, Optimal Design of Control Systems (1999)
222. *N. L. Johnson*, Subplane Covered Nets (2000)
223. *B. Fine and G. Rosenberger*, Algebraic Generalizations of Discrete Groups (1999)
224. *M. Väth*, Volterra and Integral Equations of Vector Functions (2000)
225. *S. S. Miller and P. T. Mocanu*, Differential Subordinations (2000)

226. R. Li et al., Generalized Difference Methods for Differential Equations: Numerical Analysis of Finite Volume Methods (2000)
227. H. Li and F. Van Oystaeyen, A Primer of Algebraic Geometry (2000)
228. R. P. Agarwal, Difference Equations and Inequalities: Theory, Methods, and Applications, Second Edition (2000)
229. A. B. Kharazishvili, Strange Functions in Real Analysis (2000)
230. J. M. Appell et al., Partial Integral Operators and Integro-Differential Equations (2000)
231. A. I. Prilepko et al., Methods for Solving Inverse Problems in Mathematical Physics (2000)
232. F. Van Oystaeyen, Algebraic Geometry for Associative Algebras (2000)
233. D. L. Jagerman, Difference Equations with Applications to Queues (2000)
234. D. R. Hankerson et al., Coding Theory and Cryptography: The Essentials, Second Edition, Revised and Expanded (2000)
235. S. Dăscălescu et al., Hopf Algebras: An Introduction (2001)
236. R. Hagen et al., C*-Algebras and Numerical Analysis (2001)
237. Y. Talpaert, Differential Geometry: With Applications to Mechanics and Physics (2001)
238. R. H. Villarreal, Monomial Algebras (2001)
239. A. N. Michel et al., Qualitative Theory of Dynamical Systems: Second Edition (2001)
240. A. A. Samarskii, The Theory of Difference Schemes (2001)
241. J. Knopfmacher and W.-B. Zhang, Number Theory Arising from Finite Fields (2001)
242. S. Leader, The Kurzweil-Henstock Integral and Its Differentials (2001)
243. M. Biliotti et al., Foundations of Translation Planes (2001)
244. A. N. Kochubei, Pseudo-Differential Equations and Stochastics over Non-Archimedean Fields (2001)
245. G. Sierksma, Linear and Integer Programming: Second Edition (2002)
246. A. A. Martynyuk, Qualitative Methods in Nonlinear Dynamics: Novel Approaches to Liapunov's Matrix Functions (2002)
247. B. G. Pachpatte, Inequalities for Finite Difference Equations (2002)
248. A. N. Michel and D. Liu, Qualitative Analysis and Synthesis of Recurrent Neural Networks (2002)
249. J. R. Weeks, The Shape of Space: Second Edition (2002)
250. M. M. Rao and Z. D. Ren, Applications of Orlicz Spaces (2002)
251. V. Lakshmikantham and D. Trigiante, Theory of Difference Equations: Numerical Methods and Applications, Second Edition (2002)
252. T. Albu, Cogalois Theory (2003)
253. A. Bezdek, Discrete Geometry (2003)
254. M. J. Corless and A. E. Frazho, Linear Systems and Control: An Operator Perspective (2003)
255. I. Graham and G. Kohr, Geometric Function Theory in One and Higher Dimensions (2003)

Additional Volumes in Preparation

LINEAR SYSTEMS AND CONTROL

An Operator Perspective

MARTIN J. CORLESS
ARTHUR E. FRAZHO
Purdue University
West Lafayette, Indiana, U.S.A.

MARCEL DEKKER, INC.   NEW YORK · BASEL

Library of Congress Cataloging-in-Publication Data
A catalog record for this book is available from the Library of Congress.

ISBN: 0-8247-0729-X

This book is printed on acid-free paper.

Headquarters
Marcel Dekker, Inc.
270 Madison Avenue, New York, NY 10016
tel: 212-696-9000; fax: 212-685-4540

Eastern Hemisphere Distribution
Marcel Dekker AG
Hutgasse 4, Postfach 812, CH-4001 Basel, Switzerland
tel: 41-61-260-6300; fax: 41-61-260-6333

World Wide Web
http://www.dekker.com

The publisher offers discounts on this book when ordered in bulk quantities. For more information, write to Special Sales/Professional Marketing at the headquarters address above.

To our parents

Preface

This monograph is intended for engineers, scientists, and mathematicians interested in linear systems and control. It is concerned with the analysis and control of systems whose behavior is governed by linear ordinary differential equations. Our approach is based mainly on state space models. These models arise naturally in the description of physical systems. Many of the classical results in linear systems, including the linear quadratic regulator problem, are presented. An introduction to H^∞ analysis and control is also given. The major prerequisite is a standard graduate-level course in linear algebra. However, on occasion we use some elementary results from least squares optimization theory and operator theory. These results are reviewed in the appendix. It is believed that an operator perspective provides a deep understanding of certain aspects of linear systems, along with a set of tools which are useful in system analysis and control design.

The monograph begins by presenting the basic state space model used in linear systems and control. Some physical examples from mechanics are given to motivate the state space model. Then we study both state space and input-output stability of linear systems. Lyapunov techniques are developed to analyze state space stability. Finally, we use these results to present some stability results for mechanical systems.

Next we study controllability and observability for state space systems. All the classical results on controllability and observability are presented. Standard least squares optimization techniques are used to solve a controllability optimization problem and an observability optimization problem. This naturally leads to the controllability and observability Gramians. The connections between stability and these Gramians are studied. The duality between controllability and observability is given. Using invariant subspaces and matrix representations of linear operators, we present the controllable and observable decomposition for a linear system. The singular value decomposition provides an efficient algorithm for computing the controllable and observable decomposition.

The backward shift operator plays a fundamental role in operator theory. Here we use the backward shift operator to develop realization theory. The backward shift approach to realization theory provides a natural geometric setting for realization theory and simplifies some of the classical proofs. All the standard results in realization theory are given. For example, we show that a realization is controllable and observable if and only if it is minimal. Then we show that all minimal realizations of the same transfer function are similar. The classical Kalman-Ho algorithm is presented.

Using standard linear algebra techniques, we present the classical solutions to the eigenvalue placement problem in state feedback control of controllable linear systems. First we study the single input-single output case. Ackermann's formula is presented. Then we solve the eigenvalue placement problem for multivariable systems.

State feedback results are used to develop state estimators for detectable systems. By combining the stabilization results with the state estimators, we present the classical observer-based output feedback controllers. Finally, we present a short chapter on the zeros of a state space system.

The solution to the linear quadratic regulator problem plays a fundamental role in designing feedback controllers. Here we present a solution to the linear quadratic regulator problem by using standard least squares optimization techniques. For example, the adjoint of a certain operator readily yields the corresponding two-point boundary value problem, the feedback solution and a closed form solution for the Riccati differential equation. We also provide a simple derivation of the solution to the tracking problem. Many of the classical results associated with the quadratic regulator problem are also presented. For instance, we obtain stabilizing controllers from the solution of the algebraic Riccati equation. A spectral factorization of a certain positive operator is presented. Then this factorization is used to develop the classical root locus interpretation of the feedback controllers obtained from the solution of the linear quadratic regulator problem.

Motivated by the linear quadratic regulator problem, we present a separate chapter on the Hamiltonian matrix and its role in computing the stabilizing solution to the algebraic Riccati equation. The Hamiltonian matrix plays a fundamental role in linear system theory and is also used in studying H^∞ analysis and control.

The last two chapters present an introduction to H^∞ analysis and control. First we study the analysis problem. Our approach is based on an abstract optimization problem. Then we introduce a max-min optimization problem to study the full state feedback H^∞ control problem. This max-min optimization problem is essentially a combination of the linear quadratic regular problem and the disturbance attenuation problem. The solutions to these optimization problems are obtained by completing the squares of certain operators. This yields the appropriate Riccati equation and the corresponding H^∞ controller. Finally, we also present a controller which is a trade-off between the optimal H^∞ and H^2 controller.

The authors are indebted to George Leitmann and Bob Skelton for their encouragement and inspiration.

Martin J. Corless
Arthur E. Frazho

Contents

Preface   **v**

1   Systems and Control   **1**
  1.1   Notation   1
  1.2   State space description   3
  1.3   Transfer functions   6
  1.3.1   Proper rational functions   7
  1.3.2   Power series expansions   9
  1.4   State space realizations of transfer functions   13
  1.5   Notes   17

2   Stability   **19**
  2.1   State space stability   19
  2.2   Mechanical systems   20
  2.3   Input-output stability   23
  2.4   The H^∞ norm   25
  2.5   Notes   27

3   Lyapunov Theory   **29**
  3.1   Basic Lyapunov theory   29
  3.1.1   Lyapunov functions   33
  3.2   Lyapunov functions and related bounds   34
  3.2.1   Bounds on e^{At}   34
  3.2.2   Some system bounds   36
  3.3   Lyapunov functions for mechanical systems   38
  3.4   Notes   39

4   Observability   **41**
  4.1   Observability   41
  4.2   Unobservable eigenvalues and the PBH test   45
  4.3   An observability least squares problem   46
  4.4   Stability and observability   50
  4.5   Notes   53

5 Controllability   55
    5.1 Controllability   55
    5.2 Uncontrollable eigenvalues and the PBH test   58
    5.3 A controllability least squares problem   59
    5.4 Stability and controllability   61
    5.5 Notes   64

6 Controllable and Observable Realizations   67
    6.1 Invariant subspaces   67
    6.2 The controllable and observable decomposition   69
         6.2.1 A minimal realization procedure   75
    6.3 The minimal polynomial and realizations   77
    6.4 The inverse of a transfer function   80
    6.5 Notes   82

7 More Realization Theory   83
    7.1 The restricted backward shift realization   83
    7.2 System Hankel operators   89
    7.3 Realizations and factoring Hankel matrices   91
    7.4 Partial realizations and the Kalman-Ho Algorithm   96
         7.4.1 The Kalman-Ho Algorithm   102
    7.5 Matrix representation of operators   102
    7.6 Shift realizations for proper rational functions   106
    7.7 Jordan form realizations   113
    7.8 Notes   116

8 State Feedback and Stabilizability   119
    8.1 State feedback and stabilizability   119
    8.2 Simple stabilizing controllers   120
    8.3 Stabilizability and uncontrollable eigenvalues   122
    8.4 Eigenvalue placement   123
         8.4.1 An eigenvalue placement procedure   123
         8.4.2 Ackermann's Formula   125
    8.5 Controllable canonical form   126
         8.5.1 Transformation to controllable canonical form   128
         8.5.2 Eigenvalue placement by state feedback   130
    8.6 Multivariable eigenvalue placement   130
    8.7 Two canonical forms   135
    8.8 Transfer functions and feedback   141
    8.9 Notes   143

9 State Estimators and Detectability   145
    9.1 Detectability   145
    9.2 State estimators   148
    9.3 Eigenvalue placement for estimation error   149

9.4 Notes 151

10 Output Feedback Controllers   **153**
10.1 Static output feedback 153
      10.1.1 Transfer function considerations 154
10.2 Dynamic output feedback 156
10.3 Observer based controllers 158
10.4 Notes 162

11 Zeros of Transfer Functions   **163**
11.1 Zeros 163
11.2 The system matrix 165
11.3 Notes 170

12 Linear Quadratic Regulators   **171**
12.1 The finite horizon problem 171
      12.1.1 Problems with control weights 174
12.2 An operator approach 175
      12.2.1 An operator based solution 175
      12.2.2 The adjoint system 177
      12.2.3 The Riccati equation 179
12.3 An operator quadratic regulator problem 180
12.4 A linear quadratic tracking problem 182
12.5 A spectral factorization 184
      12.5.1 A general tracking problem 186
12.6 The infinite horizon problem 189
12.7 The algebraic Riccati equation 190
12.8 Solution to the infinite horizon problem 193
      12.8.1 Problems with control weights 195
12.9 An outer spectral factorization 196
12.10 The root locus and the quadratic regulator 198
      12.10.1 Some comments on the outer spectral factor . . . 202
12.11 Notes 204

13 The Hamiltonian Matrix and Riccati Equations   **205**
13.1 The Hamiltonian matrix and stabilizing solutions 205
      13.1.1 Computation of the stabilizing solution 210
13.2 Characteristic polynomial of the Hamiltonian matrix . . . 211
13.3 Some special cases 212
13.4 The linear quadratic regulator 214
13.5 H^∞ analysis and control 216
13.6 The Riccati differential equation 219
      13.6.1 A two point boundary value problem 219
      13.6.2 Some properties 221
13.7 Notes 224

14 H^∞ Analysis   **227**

14.1 A disturbance attenuation problem 227

14.2 A Riccati equation 229

14.3 An abstract optimization problem 234

14.4 An operator disturbance attenuation problem 235

14.5 The disturbance attenuation problem revisited 236

    14.5.1 The adjoint system 237

    14.5.2 The Riccati equation 239

14.6 A spectral factorization 241

    14.6.1 A general disturbance attenuation problem 242

14.7 The infinite horizon problem 244

    14.7.1 Stabilizing solutions to the algebraic Riccati equation 247

    14.7.2 An outer spectral factor 249

14.8 The root locus and the H^∞ norm 250

14.9 Notes 254

15 H^∞ Control   **255**

15.1 A H^∞ control problem 255

    15.1.1 Problem solution 256

    15.1.2 The central controller 261

15.2 Some abstract max-min problems 263

15.3 The Riccati differential equation and norms 268

15.4 The infimal achievable gain 270

15.5 A two point boundary value problem 271

    15.5.1 The Riccati differential equation 273

15.6 The infinite horizon problem 274

    15.6.1 The stabilizing solution 277

    15.6.2 The scalar valued case 281

15.7 The central controller 282

15.8 An operator perspective 284

15.9 A tradeoff between norms 288

    15.9.1 The L^2 norm of an operator 288

    15.9.2 The L^2 norm of a system 289

    15.9.3 The L^2 optimal cost 291

    15.9.4 A tradeoff between d_∞ and d_2 293

15.10 A tradeoff between the H^2 and H^∞ norms. 295

15.11 Notes 297

16 Appendix: Least Squares   **299**

16.1 The Projection Theorem 299

16.2 A general least squares optimization problem 302

    16.2.1 Computation of orthogonal projections 305

16.3 The Gram matrix 309

16.4 An application to curve fitting 312

16.5 Minimum norm problems 313

16.6 The singular value decomposition 317
16.7 Schmidt pairs 319
16.8 A control example 323
    16.8.1 State space 324
    16.8.2 The L^2-L^∞ gain 325
16.9 Notes 327

Bibliography   **329**

Index   **337**

Chapter 1

Systems and Control

In this chapter we introduce the basic state space model used throughout this monograph. Some elementary examples from mechanical systems are given to motivate this model. Then we introduce the transfer function for a state space model. Some basic state space realization results are presented.

1.1 Notation

In this section we introduce some notation used throughout this monograph. The reader can choose to go directly to the next section and refer back to this section as the need arises. The set of all complex numbers is denoted by $\mathbb{C}$. Throughout, all linear spaces are complex vector spaces unless stated otherwise. The *inner product* on a linear space is denoted by $(\cdot, \cdot)$. To be precise, the inner product on a linear space $\mathcal{H}$ is a complex valued function mapping $\mathcal{H} \times \mathcal{H}$ into $\mathbb{C}$ with the following properties

(i) $(f, h) = \overline{(h, f)}$;

(ii) $(\alpha f + \beta g, h) = \alpha(f, h) + \beta(g, h)$;

(iii) $(h, h) \geq 0$;

(iv) $(h, h) = 0$ if and only if $h = 0$.

Here f, g and h are arbitrary vectors in $\mathcal{H}$ while α and β are arbitrary complex numbers. The inner product $(\cdot, \cdot)$ is linear in the first variable and conjugate linear in the second variable. An *inner product space* is simply a linear space with an inner product. If $\mathcal{H}$ is an inner product space, then the norm of the vector h is the non-negative real number defined by $\|h\| = \sqrt{(h, h)}$. The Cauchy-Schwartz inequality states that $|(f, h)| \leq \|f\| \, \|h\|$ for all f and h in $\mathcal{H}$. Moreover, we have equality $|(f, h)| = \|f\| \, \|h\|$ if and only if f and h are linearly dependent. Finally, it is noted that if $\mathcal{H}$ is an inner product space, then we have the triangle inequality, namely, $\|f + h\| \leq \|f\| + \|h\|$ for all f and h in $\mathcal{H}$.

Let $\mathcal{H}$ be an inner product space. Then we say that a sequence $\{h_i\}_0^\infty$ is a *Cauchy sequence* in $\mathcal{H}$ if each h_i is in $\mathcal{H}$ and $\|h_i - h_j\|$ approaches zero as i and j tend to infinity. An inner product space is *complete* if every Cauchy sequence converges to a vector in $\mathcal{H}$, that is, if $\{h_i\}_0^\infty$ is any Cauchy sequence in $\mathcal{H}$, then $\{h_i\}_0^\infty$ converges to a vector h in $\mathcal{H}$. A *Hilbert space* is simply a complete inner product space. A Hilbert space is separable if it is the closure of a countable set. Throughout we only consider separable Hilbert spaces. If $\mathcal{H}$ is a (separable) Hilbert space, then there exists an orthonormal basis $\{\varphi_j\}_1^m$ for $\mathcal{H}$ where m is possibly infinite. In fact, m is the dimension of $\mathcal{H}$. In this case, every h in $\mathcal{H}$ admits a Fourier series expansion of the form

$$h = \sum_{j=1}^{m} (h, \varphi_j)\varphi_j \qquad (h \in \mathcal{H}). \qquad (1.1)$$

Moreover, we also have Parseval's relation

$$(h, g) = \sum_{j=1}^{m} (h, \varphi_j)(\varphi_j, g) \qquad (h, g \in \mathcal{H}). \qquad (1.2)$$

In particular, $\|h\|^2 = \sum_1^m |(h, \varphi_j)|^2$.

If $\mathcal{F}$ is a subset of a linear space $\mathcal{H}$, then $\bigvee \mathcal{F}$ denotes the linear span of $\mathcal{F}$, that is the space of all linear combinations of elements of $\mathcal{F}$. If $\mathcal{H}$ is a Hilbert space, then $\bigvee \mathcal{F}$ is the closed linear span of $\mathcal{F}$.

Throughout $\mathbb{C}^n$ is the Hilbert space determined by the set of all complex n-tuples of the form $[x_1, x_2, \cdots, x_n]^{tr}$ where x_j is in $\mathbb{C}$ for all $j = 1, 2, \cdots, n$ and tr denotes the transpose. The inner product on $\mathbb{C}^n$ is given by

$$(x, y) = \sum_{j=1}^{n} x_j \bar{y}_j$$

where $x = [x_1, x_2, \cdots, x_n]^{tr}$ and $y = [y_1, y_2, \cdots, y_n]^{tr}$.

Furthermore, $L^2[a, b]$ is the Hilbert space formed by the set of all square integrable Lebesgue measurable functions over the interval $[a, b]$, that is, f is in $L^2[a, b]$ if and only if f is Lebesgue measurable on $[a, b]$ and

$$\int_a^b |f(t)|^2 \, dt < \infty.$$

The inner product on $L^2[a, b]$ is given by

$$(f, g) = \int_a^b f(t)\overline{g(t)} \, dt.$$

Now let $\mathcal{U}$ be a Hilbert space. Then $L^2([a, b], \mathcal{U})$ denotes the Hilbert space formed by the set of all square integrable Lebesgue measurable functions on $[a, b]$ with values in $\mathcal{U}$. The inner product on $L^2([a, b], \mathcal{U})$ is given by

$$(f, g) = \int_a^b (f(t), g(t))_\mathcal{U} \, dt$$

where $(f(t), g(t))_{\mathcal{U}}$ is the inner product on $\mathcal{U}$. For example, $L^2([a, b], \mathbb{C}^n)$ is the Hilbert space formed by the set of n-tuples of the form $f = [f_1, f_2, \cdots, f_n]^{tr}$ where f_j is a vector in $L^2[a, b]$ for all $j = 1, 2, \cdots, n$. The inner product on $L^2([a, b], \mathbb{C}^n)$ is given by

$$(f, g) = \sum_{j=1}^{n} \int_a^b f_j(t) \overline{g_j(t)} \, dt$$

where $g = [g_1, g_2, \cdots, g_n]^{tr}$ is in $L^2([a, b], \mathbb{C}^n)$.

We say that T is an *operator* if T is a linear map from a linear space $\mathcal{H}$ into a linear space $\mathcal{K}$. The kernel or null space of T is denoted by $\ker T$, and the range of T is denoted by $\operatorname{ran} T$. Now let $\mathcal{H}$ and $\mathcal{K}$ be Hilbert spaces and T be an operator from $\mathcal{H}$ into $\mathcal{K}$. The norm of T is defined by

$$\|T\| := \sup\{\|Th\| : h \in \mathcal{H} \text{ and } \|h\| \le 1\}.$$

We say that T is a *bounded operator* if $\|T\|$ is finite. Throughout this monograph we deal mainly with bounded operators. So, all of our operators acting between Hilbert spaces are bounded unless stated otherwise. The set of all bounded operators from $\mathcal{H}$ into $\mathcal{K}$ is denoted by $\mathcal{L}(\mathcal{H}, \mathcal{K})$. Notice that if T, R and S are operators acting between the appropriate Hilbert spaces, then $\|TR\| \le \|T\| \|R\|$ and $\|T + S\| \le \|T\| + \|S\|$.

As before, let T be a bounded operator mapping $\mathcal{H}$ into $\mathcal{K}$. Then the adjoint T^* of T is the linear operator from $\mathcal{K}$ into $\mathcal{H}$ uniquely determined by $(Th, k) = (h, T^*k)$ for all h in $\mathcal{H}$ and k in $\mathcal{K}$. It is well known that T and T^* have the same norm, that is, $\|T\| = \|T^*\|$. If $\mathcal{H}$ is finite dimensional, then $\|T\|^2$ equals the largest eigenvalue of T^*T. Finally, it is noted that if T is a matrix from $\mathbb{C}^n$ into $\mathbb{C}^m$, then T^* is the conjugate transpose of T, that is, $T^* = \bar{T}^{tr}$. On several occasions we will use some elementary results from Hilbert spaces. For some references on Hilbert spaces see Akhiezer-Glazman [2], Balakrishnan [8], Conway [30], Gohberg-Goldberg [53], Halmos [59] and Taylor-Lay [117].

1.2 State space description

A common model for describing a dynamic system is given by the following state space representation

$$\begin{aligned} \dot{x} &= Ax + Bu \\ y &= Cx + Du \end{aligned} \tag{1.3}$$

where the *state* $x(t)$ belongs to $\mathcal{X}$ while the *input* $u(t)$ lives in $\mathcal{U}$ and the *output* $y(t)$ lives in $\mathcal{Y}$ for all time $t \ge 0$. Unless otherwise noted, we always assume that the *state space* $\mathcal{X}$, the *input space* $\mathcal{U}$ and the *output space* $\mathcal{Y}$ are finite dimensional Hilbert spaces. The *system operator* A acts on $\mathcal{X}$ and B maps $\mathcal{U}$ into $\mathcal{X}$, while C maps $\mathcal{X}$ into $\mathcal{Y}$ and D maps $\mathcal{U}$ into $\mathcal{Y}$. The state space system in (1.3) is denoted by $\{A, B, C, D\}$. In many applications A, B, C and D are matrices acting between the appropriate Euclidean spaces. If an initial condition $x_0 = x(0)$ is specified and the input u is given, then there is a unique solution to the differential equation in (1.3). In this case, the output y is uniquely determined by the initial state x_0 and the input u.

Example 1.2.1 *The forced linear oscillator.* Probably one of the simplest models in studying mechanical systems is the forced linear oscillator shown in Figure 1.1. It consists of a small body of mass $m > 0$, a linear spring with spring constant $k > 0$, and a linear dashpot with damping coefficient $c > 0$. At each instant of time t, the body is subject to a force $u(t)$. Applying Newton's second law, we arrive at the following differential equation

$$m\ddot{y} + c\dot{y} + ky = u \tag{1.4}$$

where $y(t)$ is the displacement of the body from its equilibrium position. Now let $x_1 = y$ and $x_2 = \dot{y}$. Using $\dot{x}_1 = x_2$ and $\dot{x}_2 = \ddot{y}$, it follows that (1.4) admits the state space form

$$\begin{bmatrix} \dot{x}_1 \\ \dot{x}_2 \end{bmatrix} = \begin{bmatrix} 0 & 1 \\ -k/m & -c/m \end{bmatrix} \begin{bmatrix} x_1 \\ x_2 \end{bmatrix} + \begin{bmatrix} 0 \\ 1/m \end{bmatrix} u$$

$$y = \begin{bmatrix} 1 & 0 \end{bmatrix} \begin{bmatrix} x_1 \\ x_2 \end{bmatrix}.$$

In this case the displacement $y = x_1$ is the output. Notice that in this example $D = 0$. However, if we choose the acceleration $\ddot{y}$ of the mass as the output, then the corresponding D matrix is nonzero.

Figure 1.1: The forced linear oscillator

Example 1.2.2 By following the technique in the previous example, one can convert any linear differential equation to a state space system of the form (1.3). To see this, consider any scalar input-output system, with input $u(t)$ and output $y(t)$, described by an n-th order differential equation of the form

$$y^{(n)} + a_{n-1}y^{(n-1)} + \cdots + a_1\dot{y} + a_0 y = u \tag{1.5}$$

where $y^{(j)}$ denotes the j-th derivative of y and a_j is a scalar for $j = 0, 1, \cdots, n-1$. To convert this system to state space form, let $x_1 = y$, $x_2 = \dot{y}, \cdots, x_n = y^{(n-1)}$. Then using $\dot{x}_j = x_{j+1}$ for $j = 1, 2, \cdots, n-1$ along with $\dot{x}_n = y^{(n)}$, we obtain

$$\begin{bmatrix} \dot{x}_1 \\ \dot{x}_2 \\ \vdots \\ \vdots \\ \dot{x}_{n-1} \\ \dot{x}_n \end{bmatrix} = \begin{bmatrix} 0 & 1 & 0 & \cdots & 0 & 0 \\ 0 & 0 & 1 & \cdots & 0 & 0 \\ \vdots & \vdots & & \ddots & & \vdots \\ \vdots & \vdots & & & \ddots & \vdots \\ 0 & 0 & 0 & \cdots & 0 & 1 \\ -a_0 & -a_1 & -a_2 & \cdots & -a_{n-2} & -a_{n-1} \end{bmatrix} \begin{bmatrix} x_1 \\ x_2 \\ \vdots \\ \vdots \\ x_{n-1} \\ x_n \end{bmatrix} + \begin{bmatrix} 0 \\ 0 \\ \vdots \\ \vdots \\ 0 \\ 1 \end{bmatrix} u \tag{1.6}$$

$$y = \begin{bmatrix} 1 & 0 & 0 & \cdots & 0 & 0 \end{bmatrix} x.$$

The initial state is given by $x(0) = [y(0), y^{(1)}(0), \cdots, y^{(n-1)}(0)]^{tr}$ where tr denotes the transpose. Therefore, one can readily obtain a state space representation for any system described by (1.5).

Example 1.2.3 *Axisymmetric spacecraft.*

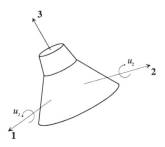

Figure 1.2: The axisymmetric spacecraft

We consider here the rotational motion of an axisymmetric spacecraft which is subject to two body-fixed input torques u_1 and u_2 about axes which are perpendicular to the symmetry axis. This is illustrated in Figure 1.2. If we choose a body-fixed axes system with origin at the center of mass, with axes corresponding to u_1, u_2 and whose 3-axis corresponds to the axis of symmetry, the rotational motion of the axisymmetric spacecraft can be described by

$$
\begin{aligned}
I_1\dot{\omega}_1 &= (I_2 - I_3)\omega_2\omega_3 + u_1 \\
I_2\dot{\omega}_2 &= (I_3 - I_1)\omega_3\omega_1 + u_2 \\
I_3\dot{\omega}_3 &= 0
\end{aligned}
$$

where $\omega_1, \omega_2, \omega_3$ denote the components of the spacecraft angular velocity vector with respect to the body-fixed axes; the positive scalars I_1, I_2, I_3 are the principal moments of inertia of the spacecraft with respect to the body-fixed axes and $I_1 = I_2$. From the third equation above ω_3 is constant. If we introduce the constants $\alpha := (I_1 - I_3)\omega_3/I_1$ and $\beta := 1/I_1$, define state variables $x_1 := \omega_1$, $x_2 := \omega_2$ and output variables $y_1 := x_1$, $y_2 := x_2$, we obtain

$$
\begin{aligned}
\dot{x}_1 &= \alpha x_2 + \beta u_1 \\
\dot{x}_2 &= -\alpha x_1 + \beta u_2 \\
y_1 &= x_1 \\
y_2 &= x_2 \quad .
\end{aligned}
$$

This readily yields the state space form in (1.3) where

$$
A = \begin{bmatrix} 0 & \alpha \\ -\alpha & 0 \end{bmatrix}, \qquad
B = \begin{bmatrix} \beta & 0 \\ 0 & \beta \end{bmatrix}, \qquad
C = \begin{bmatrix} 1 & 0 \\ 0 & 1 \end{bmatrix}, \qquad
D = \begin{bmatrix} 0 & 0 \\ 0 & 0 \end{bmatrix}.
$$

Now let us return to the state space description (1.3). For a specified initial condition $x(0) = x_0$ the state $x(t)$ at time t for the system in (1.3) is uniquely given by

$$
x(t) = e^{At}x_0 + \int_0^t e^{A(t-\tau)}Bu(\tau)\,d\tau. \tag{1.7}
$$

The output $y(t)$ of the state space system $\{A, B, C, D\}$ in (1.3) is determined by

$$y(t) = Ce^{At}x_0 + \int_0^t Ce^{A(t-\tau)}Bu(\tau)d\tau + Du(t) \tag{1.8}$$

where the initial state $x_0 = x(0)$ is specified. Clearly,

$$y(t) = Ce^{At}x_0 + \int_0^t G(t-\tau)u(\tau)d\tau \tag{1.9}$$

where G is the operator valued function defined by

$$G(t) = Ce^{At}B + \delta(t)D. \tag{1.10}$$

Here $\delta(t)$ is the unit Dirac delta function. So, if the initial state $x_0 = 0$, then (1.9) defines a linear time invariant map of the form

$$y(t) = \int_0^t G(t-\tau)u(\tau)d\tau. \tag{1.11}$$

In particular, if $u(t) = \delta(t)u_0$ for some constant u_0 in $\mathcal{U}$ and $x_0 = 0$, then $y(t) = G(t)u_0$. Motivated by this, we call the function G in (1.10) the *impulse response* for $\{A, B, C, D\}$. Clearly, if $\mathcal{U} = \mathbb{C}^1$ and $u_0 = 1$, then $G(t)$ is precisely the output $y(t)$ for $u(t) = \delta(t)$. Finally, it is worth emphasizing that, under zero initial conditions, the input-output response of $\{A, B, C, D\}$ is completely determined by its impulse response via (1.11).

1.3   Transfer functions

In this section we introduce the notion of a transfer function for a linear system. Recall that the Laplace transform of a measurable function f on $[0, \infty)$ with values in a finite dimensional space $\mathcal{E}$ is defined by

$$\mathbf{f}(s) = \mathcal{L}(f)(s) = \int_0^\infty e^{-st}f(t)\, dt \tag{1.12}$$

for appropriate values of s in $\mathbb{C}$. (The set of all complex numbers is denoted by $\mathbb{C}$.) The Laplace transform of the function f is denoted by $\mathbf{f}$ and has values in $\mathcal{E}$. Notice that a boldface $\mathbf{f}$ is used to represent the Laplace transform of f. If f is exponentially bounded, that is, if $\|f(t)\| \leq Me^{\gamma t}$ for some finite scalars M and γ, then $\mathbf{f}$ is an analytic function in the region of $\mathbb{C}$ defined by $\Re(s) > \gamma$. Clearly, the Laplace transform operator $\mathcal{L}$ is linear. Finally, by convention, $\mathcal{L}(\delta) = 1$ where δ is the unit Dirac delta function.

Recall that $\mathcal{L}(\dot{f})(s) = s\mathbf{f}(s) - f(0)$. By taking the Laplace transform of the state space equations in (1.3), we obtain

$$\begin{aligned} s\mathbf{x}(s) - x_0 &= A\mathbf{x}(s) + B\mathbf{u}(s) \\ \mathbf{y}(s) &= C\mathbf{x}(s) + D\mathbf{u}(s) \end{aligned}$$

where $\mathbf{x}, \mathbf{y}, \mathbf{u}$ are the Laplace transforms of x, y, u, respectively and $x_0 = x(0)$. Solving these equations for $\mathbf{y}$ in terms of x_0 and $\mathbf{u}$ yields

$$\mathbf{y}(s) = C(sI - A)^{-1}x_0 + \left[C(sI - A)^{-1}B + D \right] \mathbf{u}(s). \qquad (1.13)$$

The *transfer function* $\mathbf{G}$ for system $\{A, B, C, D\}$ is defined by

$$\mathbf{G}(s) = C(sI - A)^{-1}B + D. \qquad (1.14)$$

Recall that the impulse response of system $\{A, B, C, D\}$ given by

$$G(t) = Ce^{At}B + \delta(t)D.$$

Using the following Laplace transform result:

$$\mathcal{L}(e^{At})(s) = (sI - A)^{-1}$$

we have $\mathbf{G} = \mathcal{L}(G)$, that is, the transfer function $\mathbf{G}$ is the Laplace transform of the impulse response G. Clearly, $\mathbf{G}$ is analytic everywhere except possibly at the eigenvalues of A.

When $x(0) = 0$, the relationship between the Laplace transform of the input $\mathbf{u}$ and the Laplace transform of the output $\mathbf{y}$ is given by

$$\mathbf{y}(s) = \mathbf{G}(s)\mathbf{u}(s). \qquad (1.15)$$

In other words, when the initial conditions are zero, the transfer function is the multiplication operator which maps $\mathbf{u}$ into $\mathbf{y}$. Recall that convolution in the time domain corresponds to multiplication in the s-domain. Hence,

$$y(t) = \int_0^t G(t-\tau)u(\tau)\,d\tau.$$

Obviously, $\mathbf{G}$ and G uniquely determine each other. So, for zero initial conditions, the input-output response of the system $\{A, B, C, D\}$ is completely determined by its transfer function.

1.3.1   Proper rational functions

A operator valued polynomial N is a function from $\mathbb{C}$ into some space $\mathcal{L}(\mathcal{U}, \mathcal{Y})$ of operators which can be expressed as

$$N(s) = N_0 + sN_1 + \cdots + s^n N_n$$

where $N_0, \ldots, N_n$ are in $\mathcal{L}(\mathcal{U}, \mathcal{Y})$. If N is nonzero, then the degree of N (which we denote by $\deg N$) is the largest integer m for which N_m is non-zero. We say that an operator valued function $\mathbf{G}$ of a complex variable is a *proper rational* function if $\mathbf{G}(s) = N(s)/d(s)$ where N is an operator valued polynomial, d is a scalar valued polynomial and $\deg N \leq \deg d$. If $\deg N < \deg d$, then we say that $\mathbf{G}$ is a *strictly proper rational* function. Notice that $\mathbf{G}$ is a proper rational function with values in $\mathcal{L}(\mathcal{U}, \mathcal{Y})$ if and only if $\mathbf{G}$ admits a decomposition of the form $\mathbf{G}(s) = N(s)/d(s) + D$ where N is an operator valued polynomial, d is a scalar valued polynomial satisfying $\deg N < \deg d$, and D is a constant in $\mathcal{L}(\mathcal{U}, \mathcal{Y})$. In other words, $\mathbf{G}$ is a proper rational function if and only if $\mathbf{G}$ admits a decomposition of the form $\mathbf{G}(s) = \mathbf{G}_o(s) + D$ where $\mathbf{G}_o$ is a strictly proper rational function and D is a constant.

We claim that the transfer function $\mathbf{G}$ of a finite dimensional system $\{A$ on $\mathcal{X}, B, C, D\}$ is a proper rational function. To see this, recall that $(sI - A)^{-1}$ can be expressed as

$$(sI - A)^{-1} = \operatorname{adj}(sI - A)/\det[sI - A] \tag{1.16}$$

where $\operatorname{adj}(sI - A)$ is the algebraic adjoint (or adjugate) of $sI - A$ and $\det[sI - A]$ is the characteristic polynomial of A. Since $\operatorname{adj}(sI - A)$ is an operator valued polynomial whose degree is strictly less than the degree of $\det[sI - A]$, it follows that $(sI - A)^{-1}$ is a strictly proper rational function. Hence, $C(sI - A)^{-1}B$ is also a strictly proper rational function. Recalling from (1.14) that $\mathbf{G}(s) = C(sI - A)^{-1}B + D$, it follows that the transfer function $\mathbf{G}$ is a proper rational function. Hence, $\mathbf{G}$ is a proper rational function of the form $\mathbf{G} = N/d$ where d is the characteristic polynomial of A. Therefore, the poles of $\mathbf{G}$ are contained in the eigenvalues of A. We say that A is *stable* if all the eigenvalues of A are contained in the open left half plane $\{s : \Re s < 0\}$. So, if A is stable, then $\mathbf{G}$ is also stable, that is, all the poles of $\mathbf{G}$ are contained in the open left hand plane. Finally, it is noted that $\mathbf{G}$ is a strictly proper rational function if and only if D equals zero.

Suppose $\mathbf{G} = N/d$ is a proper rational function. Then, we define the *relative degree* of $\mathbf{G}$ to be $\deg d - \deg N$. If $\mathbf{G}$ is a transfer function for $\{A, B, C, D\}$, then $\mathbf{G}$ has relative degree zero if and only if D is non-zero. When D is zero, the following result provides a state space characterization of relative degree.

Lemma 1.3.1 *Suppose $\mathbf{G}$ is a transfer function for a system $\{A, B, C, 0\}$ and*

$$CA^j B = 0 \quad for \quad j = 0, 1, \cdots, r - 2, \tag{1.17a}$$
$$CA^{r-1} B \neq 0. \tag{1.17b}$$

Then

$$CA^j (sI - A)^{-1} B = s^j \mathbf{G}(s) \quad for \quad j = 0, 1, \cdots, r - 1 \tag{1.18a}$$
$$CA^r (sI - A)^{-1} B = s^r \mathbf{G}(s) - CA^{r-1}B. \tag{1.18b}$$

Furthermore, $\mathbf{G} = N/d$ where d is the characteristic polynomial of A while N is a polynomial of degree $\deg(d) - r$ and whose highest order coefficient is $CA^{r-1}B$. In particular, $\mathbf{G}$ has relative degree r. Finally, if r equals the dimension of the state space, then $\mathbf{G}(s) = CA^{r-1}B/d(s)$.

PROOF. We first demonstrate (1.18a) by induction. Since,

$$\mathbf{G}(s) = C(sI - A)^{-1}B,$$

the equality in (1.18a) clearly holds for $j = 0$. Suppose now the equality (1.18a) holds for some $j = k \leq r - 2$. Then for $j = k + 1$,

$$\begin{aligned} CA^{k+1}(sI - A)^{-1}B &= C(A - sI + sI)A^k(sI - A)^{-1}B \\ &= -CA^k B + sCA^k(sI - A)^{-1}B \\ &= s^{k+1}\mathbf{G}(s). \end{aligned}$$

Hence, (1.18a) holds for $j = k + 1$. So, by induction it holds for $j = 0, 1, \cdots, r - 1$. We also obtain that

$$CA^r(sI - A)^{-1}B = -CA^{r-1}B + sCA^{r-1}(sI - A)^{-1}B = -CA^{r-1}B + s^r\mathbf{G}(s);$$

hence (1.18b) also holds.

Recall that $\mathbf{G} = N/d$ where d is the characteristic polynomial of A and N is a polynomial of degree strictly less than the degree of d. Hence, $CA^r(sI - A)^{-1}B = M(s)/d(s)$ where M is a polynomial of degree strictly less that the degree of d. It now follows from (1.18b) that

$$\frac{M(s)}{d(s)} = -CA^{r-1}B + s^r\frac{N(s)}{d(s)}$$

that is,

$$s^r N(s) = CA^{r-1}Bd(s) + M(s).$$

The properties of d and M imply that $s^r N(s)$ is a polynomial of degree $\deg(d)$ and whose highest order coefficient is $CA^{r-1}B$. Hence, N is a polynomial of degree $\deg(d) - r$ and the coefficient of its highest order term is $CA^{r-1}B$. ∎

1.3.2 Power series expansions

We now claim that if $\mathbf{G}$ is a proper rational function with values in $\mathcal{L}(\mathcal{U}, \mathcal{Y})$, then for sufficiently large s, the function $\mathbf{G}$ admits a power series expansion of the form

$$\mathbf{G}(s) = \sum_{i=0}^{\infty} s^{-i}G_i \tag{1.19}$$

where each G_i is in $\mathcal{L}(\mathcal{U}, \mathcal{Y})$. To see this let $\mathbf{G} = N/d$ where N is a polynomial with values in $\mathcal{L}(\mathcal{U}, \mathcal{Y})$ and d is a scalar valued polynomial satisfying $\deg N \le \deg d$. Considering $s = 1/\lambda$ and letting $\tilde{G}(\lambda) = \mathbf{G}(1/\lambda)$, we obtain

$$\tilde{G}(\lambda) = \frac{N(1/\lambda)}{d(1/\lambda)} = \frac{\lambda^n N(1/\lambda)}{\lambda^n d(1/\lambda)} = \frac{\tilde{N}(\lambda)}{\tilde{d}(\lambda)}$$

where n is the degree of d. Here $\tilde{N}$ and $\tilde{d}$ are the polynomials of degree at most n defined by $\tilde{N}(\lambda) = \lambda^n N(1/\lambda)$ and $\tilde{d}(\lambda) = \lambda^n d(1/\lambda)$, respectively. Since $\tilde{d}(0)$ equals the coefficient of s^n in the polynomial d, it is nonzero. Hence, $\tilde{G}$ is analytic at zero. So, for sufficiently small λ, the function $\tilde{G}$ has a power series expansion of the form $\tilde{G}(\lambda) = \sum_0^\infty \lambda^i G_i$ where each G_i is in $\mathcal{L}(\mathcal{U}, \mathcal{Y})$. Therefore, for s sufficiently large, $\mathbf{G}$ admits a power series expansion of the form (1.19). Moreover, the infinite sequence $\{G_i\}_0^\infty$ and $\mathbf{G}$ uniquely determine each other.

Now assume that $\mathbf{G}$ is a proper rational function of the form $\mathbf{G} = N/d$ where d is a scalar valued polynomial and N is an operator valued polynomial defined by

$$\begin{aligned} d(s) &= d_0 + sd_1 + \cdots + s^{n-1}d_{n-1} + s^n d_n & (d_n \ne 0) \tag{1.20} \\ N(s) &= N_0 + sN_1 + \cdots + s^{n-1}N_{n-1} + s^n N_n. \tag{1.21} \end{aligned}$$

Hence,

$$N(s) = d(s)\mathbf{G}(s) = \sum_{i=0}^{\infty} d(s)s^{-i}G_i \tag{1.22}$$

where G_i is the coefficient of s^{-i} in the power series expansion of $\mathbf{G}$. Equating the coefficients of s^j in (1.22) for $j = n, n-1, \cdots, 0$, yields

$$
\begin{aligned}
N_n &= d_n G_0 \\
N_{n-1} &= d_{n-1} G_0 + d_n G_1 \\
&\ \ \vdots \\
N_j &= d_j G_0 + d_{j+1} G_1 + \cdots + d_{n-1} G_{n-j-1} + d_n G_{n-j} \\
&\ \ \vdots \\
N_0 &= d_0 G_0 + d_1 G_1 + \cdots + d_{n-1} G_{n-1} + d_n G_n .
\end{aligned}
\tag{1.23}
$$

These equations can be written in the following compact form:

$$
\begin{bmatrix}
N_0 \\
N_1 \\
N_2 \\
\vdots \\
N_{n-1} \\
N_n
\end{bmatrix}
=
\begin{bmatrix}
d_0 I & d_1 I & \cdots & d_{n-2} I & d_{n-1} I & d_n I \\
d_1 I & d_2 I & \cdots & d_{n-1} I & d_n I & 0 \\
d_2 I & d_3 I & \cdots & d_n I & 0 & 0 \\
\vdots & \vdots & & \vdots & \vdots & \vdots \\
d_{n-1} I & d_n I & \cdots & 0 & 0 & 0 \\
d_n I & 0 & \cdots & 0 & 0 & 0
\end{bmatrix}
\begin{bmatrix}
G_0 \\
G_1 \\
G_2 \\
\vdots \\
G_{n-1} \\
G_n
\end{bmatrix}.
\tag{1.24}
$$

(The block matrix in the above expression is a Hankel matrix.) Since N is a polynomial, the coefficients of s^{-j} in $\sum_0^{\infty} d(s)s^{-i}G_i$ are zero for all integers $j \geq 1$. For $j = 1$, we obtain

$$0 = d_0 G_1 + d_1 G_2 \cdots + d_{n-1} G_n + d_n G_{n+1} . \tag{1.25}$$

In fact, for any integer $k \geq 1$, we have

$$0 = d_0 G_k + d_1 G_{k+1} + \cdots + d_{n-1} G_{k+n-1} + d_n G_{k+n} \qquad (k \geq 1) . \tag{1.26}$$

The above considerations lead to the following result.

Lemma 1.3.2 *Let $\mathbf{G} = \sum_0^{\infty} s^{-i} G_i$ be the power series expansion for a proper rational function $\mathbf{G}$ and suppose d is a scalar valued polynomial of the form (1.20). Then $\mathbf{G} = N/d$ where N is an operator valued polynomial if and only if (1.26) holds. Moreover, in this case, N is given by (1.21) and (1.23).*

When $\mathbf{G} = N/d$ and N and d are known, the above results allow us to recursively compute the operators in the sequence $\{G_i\}_0^{\infty}$ directly from the polynomials N and d. Specifically, $\{G_i\}_0^n$ can be recursively computed from (1.23) or (1.24), while $\{G_i\}_{n+1}^{\infty}$ is recursively computed from

$$G_k = -\frac{d_{n-1} G_{k-1} + d_{n-2} G_{k-2} + \cdots + d_0 G_{k-n}}{d_n} \qquad (k \geq n+1) . \tag{1.27}$$

The following result uses state space methods to compute the coefficients in the power series expansion for a transfer function.

Proposition 1.3.3 *Let $\sum_0^\infty s^{-1} G_i$ be the power series expansion for a proper rational function G. Then G is the transfer function for a system $\{A, B, C, D\}$ if and only if*

$$G_0 = D \quad \text{and} \quad G_i = CA^{i-1}B \quad \text{(for all } i \geq 1 \text{).} \tag{1.28}$$

The proof of this result depends upon the following classical result.

Lemma 1.3.4 *Let T be an operator on $\mathcal{X}$ satisfying $\|T\| < 1$. Then $I - T$ is invertible. Moreover, in this case,*

$$(I - T)^{-1} = \sum_{k=0}^{\infty} T^k . \tag{1.29}$$

Finally, $\|(I - T)^{-1}\| \leq 1/(1 - \|T\|)$.

PROOF. Let R_n be the operator on $\mathcal{X}$ defined by $R_n = \sum_{k=0}^n T^k$. We claim that R_n converges to a bounded operator $R = \sum_{k=0}^\infty T^k$ (in the operator topology) as n tends to infinity. To see this, let $r = \|T\|$. Obviously, $\|T^k\| \leq r^k$ for all integers $k \geq 0$. If n and m are any two positive integers satisfying $n > m$, then

$$\|R_n - R_m\| = \| \sum_{k=m+1}^{n} T^k\| \leq \sum_{k=m+1}^{n} \|T^k\| \leq \sum_{k=m+1}^{\infty} r^k = \frac{r^{m+1}}{1-r} \to 0$$

as m tends to infinity. The last equality follows from the geometric series $\sum_0^\infty r^k = 1/(1-r)$. So, R_n converges to the operator $R = \sum_0^\infty T^k$ as n approaches infinity. To verify that R is a bounded operator simply notice that

$$\|R\| \leq \sum_{k=0}^{\infty} \|T^k\| \leq \sum_{k=0}^{\infty} r^k = \frac{1}{1-r} = \frac{1}{1 - \|T\|} .$$

To show that R is the inverse of $I - T$, observe that

$$(I - T)R_n = (I - T)(I + T + T^2 + T^3 + \cdots + T^n) = I - T^{n+1} .$$

Because T^{n+1} converges to zero, the sequence $\{(I - T)R_n\}_0^\infty$ converges to I. Since $\{R_n\}_0^\infty$ converges to R, this implies that $(I - T)R = I$. Because R_n and $I - T$ commute, we also have $R(I - T) = I$. Therefore, $I - T$ is invertible and $\sum_0^\infty T_k$ is the inverse of $I - T$. ∎

PROOF OF PROPOSITION 1.3.3. By definition, $\mathbf{G}$ is the transfer function for a system $\{A, B, C, D\}$ if and only if $\mathbf{G}(s) = C(sI - A)^{-1}B + D$. Notice that

$$(sI - A)^{-1} = \frac{1}{s}\left(I - \frac{A}{s}\right)^{-1} .$$

Whenever $|s| > \|A\|$, it follows that $\|A/s\| < 1$. According to Lemma 1.3.4, we have

$$(sI - A)^{-1} = \sum_{i=0}^{\infty} \frac{A^i}{s^{i+1}} \quad \text{(if } |s| > \|A\| \text{).} \tag{1.30}$$

Therefore, $\mathbf{G}$ is the transfer function for a system $\{A, B, C, D\}$ if and only if it has a power series expansion of the form

$$\mathbf{G}(s) = D + \sum_{i=1}^{\infty} \frac{CA^{i-1}B}{s^i}. \tag{1.31}$$

By matching like coefficients of $1/s^i$, it follows that $\mathbf{G}(s) = \sum_0^{\infty} G_i/s^i$ is the transfer function for a system $\{A, B, C, D\}$ if and only if (1.28) hold. ∎

We say that $\{A, B, C, D\}$ is a *state space realization* of $\mathbf{G}$ if $\mathbf{G}$ is the transfer function for $\{A, B, C, D\}$. Consider any proper rational function $\mathbf{G}$ with a power series expansion of the form $\sum_0^{\infty} G_i/s^i$. Then the above analysis shows that $\{A, B, C, D\}$ is a realization of $\mathbf{G}$ if and only if (1.28) holds. So, with out loss of generality we say that $\{A, B, C, D\}$ is a *realization of a sequence of operators* $\{G_i\}_0^{\infty}$ if (1.28) holds.

Now suppose that $\{A, B, C, D\}$ is a state space realization of $\mathbf{G}$ and d is a scalar valued polynomial defined by (1.20). Then, according to Lemma 1.3.2 and Proposition 1.3.3, we can express $\mathbf{G}$ as $\mathbf{G} = N/d$ where N is an operator valued polynomial if and only if

$$CA^{k-1}d(A)B = 0 \qquad (k \geq 1) \tag{1.32}$$

where $d(A) = d_0 I + d_1 A + \cdots d_{n-1}A_{n-1} + d_n A^n$. Since A and $d(A)$ commute, the above condition is equivalent to

$$CA^i d(A)A^j B = 0 \quad (i, j \geq 0). \tag{1.33}$$

Moreover, in this case, N is given by (1.21) and

$$\begin{aligned}
N_n &= d_n D \\
N_{n-1} &= d_{n-1}D + d_n CB \\
&\vdots \\
N_j &= d_j D + d_{j+1}CB + \cdots + d_{n-1}CA^{n-j-2}B + d_n CA^{n-j-1}B \\
&\vdots \\
N_0 &= d_0 D + d_1 CB + \cdots + d_{n-1}CA^{n-2}B + d_n CA^{n-1}B.
\end{aligned} \tag{1.34}$$

When $D = 0$, we obtain that $N_n = 0$ and

$$\begin{bmatrix} N_0 & N_1 & \cdots & N_{n-1} \end{bmatrix} = \begin{bmatrix} CB & CAB & \cdots & CA^{n-1}B \end{bmatrix} \Upsilon \tag{1.35}$$

or

$$\begin{bmatrix} N_0 \\ N_1 \\ \vdots \\ N_{n-1} \end{bmatrix} = \Upsilon \begin{bmatrix} CB \\ CAB \\ \vdots \\ CA^{n-1}B \end{bmatrix} \tag{1.36}$$

where Υ is the invertible Hankel matrix given by

$$\Upsilon = \begin{bmatrix}
d_1 I & d_2 I & \cdots & d_{n-2}I & d_{n-1}I & d_n I \\
d_2 I & d_3 I & \cdots & d_{n-1}I & d_n I & 0 \\
d_4 I & d_4 I & \cdots & d_n I & 0 & 0 \\
\vdots & \vdots & & \vdots & \vdots & \vdots \\
d_{n-1}I & d_n I & \cdots & 0 & 0 & 0 \\
d_n I & 0 & \cdots & 0 & 0 & 0
\end{bmatrix}.$$

Summarizing, we have the following result.

Lemma 1.3.5 *Let* **G** *be a transfer function for a state space system* $\{A, B, C, D\}$ *and* d *a scalar valued polynomial defined by (1.20). Then* $\mathbf{G} = N/d$ *where* N *is an operator valued polynomial if and only if (1.33) holds. Moreover, in this case, N is given by (1.21) and (1.34). When $D = 0$, the coefficients of N are also given by $N_n = 0$ and (1.35) or (1.36).*

Remark 1.3.1 (Cayley-Hamilton Theorem) We have already seen that if **G** is the transfer function for $\{A, B, C, D\}$, then it can be expressed as $\mathbf{G} = N/d$ where d is the characteristic polynomial of A. Hence, the characteristic polynomial of A must satisfy (1.33). In particular, if we consider $B = C = I$ and $i = j = 0$, we obtain that $d(A) = 0$. This yields the Cayley-Hamilton Theorem, namely if A is an operator on a finite dimensional space $\mathcal{X}$, then $d(A) = 0$ where d is the characteristic polynomial of A.

1.4 State space realizations of transfer functions

We say that $\{A \text{ on } \mathcal{X}, B, C, D\}$ is a *state space realization* of **G** if **G** is the transfer function for $\{A, B, C, D\}$, that is, (1.14) holds. The dimension of the state space $\mathcal{X}$ is the dimension of the realization $\{A, B, C, D\}$. The analysis in the previous section shows that all transfer functions are proper rational functions. This is part of the following basic result in system theory.

Theorem 1.4.1 *Let* **G** *be an operator valued function of a complex variable. Then* **G** *admits a finite-dimensional state space realization if and only if* **G** *is a proper rational function.*

PROOF. We only need to prove that a proper rational function **G** has a finite-dimensional state space realization. Our proof is motivated by the form of the matrix A in equation (1.6) of Example 1.2.2. Let **G** be a proper rational function with values in $\mathcal{L}(\mathcal{U}, \mathcal{Y})$. Then **G** admits a decomposition of the form

$$\mathbf{G} = N/d + D \tag{1.37}$$

where N is a polynomial with values in $\mathcal{L}(\mathcal{U}, \mathcal{Y})$ and d is a scalar valued polynomial of the form

$$N(s) = \sum_{k=0}^{n-1} N_k s^k \quad \text{and} \quad d(s) = s^n + \sum_{k=0}^{n-1} a_k s^k. \tag{1.38}$$

Here $D, N_0, \cdots, N_{n-1}$ are operators in $\mathcal{L}(\mathcal{U}, \mathcal{Y})$ while $a_0, \cdots, a_{n-1}$ are scalars. Now let $\mathcal{X} = \oplus_1^n \mathcal{U}$ be the linear space formed by the set of all n-tuples of vectors from $\mathcal{U}$, that is, let $\mathcal{X}$ be the linear space consisting of all vectors of the form $[u_1, u_2, \cdots, u_n]^{tr}$ where u_j is a vector in $\mathcal{U}$ for all integers $j = 1, 2, \cdots, n$. (Recall that tr denotes transpose.) Let A be the block matrix on $\mathcal{X}$ and B the block matrix from $\mathcal{U}$ into $\mathcal{X}$ and C the block matrix from $\mathcal{X}$

into $\mathcal{Y}$ defined by

$$A = \begin{bmatrix} 0 & I & 0 & \cdots & 0 & 0 \\ 0 & 0 & I & \cdots & 0 & 0 \\ \vdots & \vdots & & \ddots & & \vdots \\ \vdots & \vdots & & & \ddots & \vdots \\ 0 & 0 & 0 & \cdots & 0 & I \\ -a_0 I & -a_1 I & -a_2 I & \cdots & -a_{n-2} I & -a_{n-1} I \end{bmatrix}, \quad B = \begin{bmatrix} 0 \\ 0 \\ \vdots \\ \vdots \\ 0 \\ I \end{bmatrix} \tag{1.39}$$

$$C = \begin{bmatrix} N_0 & N_1 & N_2 & \cdots & N_{n-2} & N_{n-1} \end{bmatrix}.$$

Define V as the polynomial with values in $\mathcal{L}(\mathcal{U}, \mathcal{X})$ given by

$$V(s) = \begin{bmatrix} I & sI & s^2 I & \cdots & s^{n-1} I \end{bmatrix}^{tr}.$$

Here I is the identity on $\mathcal{U}$. Then,

$$(sI - A)V(s) = \begin{bmatrix} sI & -I & 0 & \cdots & 0 & 0 \\ 0 & sI & -I & \cdots & 0 & 0 \\ \vdots & \vdots & & \ddots & & \vdots \\ \vdots & \vdots & & & \ddots & \vdots \\ 0 & 0 & 0 & \cdots & sI & -I \\ a_0 I & a_1 I & a_2 I & \cdots & a_{n-2} I & (s+a_{n-1})I \end{bmatrix} \begin{bmatrix} I \\ sI \\ \vdots \\ \vdots \\ s^{n-2} I \\ s^{n-1} I \end{bmatrix} = \begin{bmatrix} 0 \\ 0 \\ \vdots \\ \vdots \\ 0 \\ d(s)I \end{bmatrix}$$

$$= Bd(s).$$

Hence, whenever s is not an eigenvalue of A,

$$(sI - A)^{-1}B = V(s)/d(s).$$

This along with the definition of C, implies that

$$C(sI - A)^{-1}B = \begin{bmatrix} N_0 & N_1 & \cdots & N_{n-1} \end{bmatrix} V(s)/d(s) = N(s)/d(s).$$

The above equation and (1.37) yields $\mathbf{G}(s) = C(sI - A)^{-1}B + D$. Therefore, $\{A, B, C, D\}$ is a finite dimensional realization of $\mathbf{G}$. ∎

Remark 1.4.1 Let $\mathbf{G}$ be a proper rational function with values in $\mathcal{L}(\mathcal{U}, \mathcal{Y})$ of the form (1.37) where N and d are polynomials as defined in (1.38). Moreover, let A, B, C be the block matrices defined in (1.39). Then the proof of the previous theorem shows that $\{A, B, C, D\}$ is a realization of $\mathbf{G}$. In particular, any proper rational function of the form (1.37), (1.38) admits a state space realization whose state dimension is at most $n \cdot \dim \mathcal{U}$.

We now present another finite dimensional state space realization of a proper rational function $\mathbf{G}$ with values in $\mathcal{L}(\mathcal{U}, \mathcal{Y})$. Since $\mathbf{G}$ is proper and rational, it has a power series expansion of the form $\mathbf{G}(s) = \sum_{i=0}^{\infty} s^{-i} G_i$ where each G_i is in $\mathcal{L}(\mathcal{U}, \mathcal{Y})$. Also $\mathbf{G}$ can be expressed as in (1.37) where N and d are polynomials as defined in (1.38). Now let $\mathcal{X} = \oplus_1^n \mathcal{Y}$

be the linear space formed by the set of all n-tuples of vectors from $\mathcal{Y}$. Let A be the block matrix on $\mathcal{X}$ and B the block matrix from $\mathcal{U}$ into $\mathcal{X}$ and C the block matrix from $\mathcal{X}$ into $\mathcal{Y}$ defined by

$$
A = \begin{bmatrix}
0 & I & 0 & \cdots & 0 & 0 \\
0 & 0 & I & \cdots & 0 & 0 \\
\vdots & \vdots & & \ddots & & \vdots \\
\vdots & \vdots & & & \ddots & \vdots \\
0 & 0 & 0 & \cdots & 0 & I \\
-a_0 I & -a_1 I & -a_2 I & \cdots & -a_{n-2} I & -a_{n-1} I
\end{bmatrix}, \quad
B = \begin{bmatrix}
G_1 \\
G_2 \\
\vdots \\
\vdots \\
G_{n-1} \\
G_n
\end{bmatrix}
\tag{1.40}
$$

$$
C = \begin{bmatrix} I & 0 & 0 & \cdots & 0 & 0 \end{bmatrix}.
$$

It follows from (1.27) that $-a_0 G_1 - a_1 G_2 - \cdots - a_{n-1} G_n = G_{n+1}$. This along with the shift structure of A implies that $AB = [G_2 \; G_3 \; \cdots \; G_{n+1}]^{tr}$. By induction with (1.27), one can readily show that, for any integer $k \geq 0$, we have $A^k B = [G_{k+1} \; G_{k+2} \; \cdots \; G_{k+n}]^{tr}$. Using the structure of C yields $C A^k B = G_{k+1}$ for all $k \geq 0$. Recalling Proposition 1.3.3, we see that $\mathbf{G}(s) = C(sI - A)^{-1} B + D$, that is, $\{A, B, C, D\}$ is a realization of $\mathbf{G}$. Thus, any proper rational function of the form (1.37), (1.38) admits a state space realization whose state dimension is at most $n \cdot \dim \mathcal{Y}$.

Remark 1.4.2 By combining Remark 1.4.1 with our previous analysis, we obtain the following result. Any proper rational function $\mathbf{G}$ of the form (1.37), (1.38) with values in $\mathcal{L}(\mathcal{U}, \mathcal{Y})$ admits a state space realization whose state dimension is at most $n \cdot m$ where m is the minimum dimension of the spaces $\mathcal{U}$ and $\mathcal{Y}$.

In many applications such as mechanical systems one encounters higher order vector differential equations of the form

$$
y^{(n)} + A_{n-1} y^{(n-1)} + \cdots + A_1 \dot{y} + A_0 y = u.
\tag{1.41}
$$

Here the input u and output y are functions with values in $\mathcal{U}$ and A_j are operators on $\mathcal{U}$ for $j = 0, 1, \cdots, n-1$. By following Example 1.2.2, set $x_j = y^{(j-1)}$ for $j = 1, 2, \cdots, n$. Let $\mathcal{X}$ be the space defined by $\mathcal{X} = \oplus_1^n \mathcal{U}$. Then a state space realization for this system is given by

$$
\dot{x} = Ax + Bu \quad \text{and} \quad y = Cx
\tag{1.42}
$$

where A on $\mathcal{X}$ and B from $\mathcal{U}$ into $\mathcal{X}$ and C from $\mathcal{X}$ into $\mathcal{U}$ are given by

$$
A = \begin{bmatrix}
0 & I & 0 & \cdots & 0 & 0 \\
0 & 0 & I & \cdots & 0 & 0 \\
\vdots & \vdots & & \ddots & & \vdots \\
\vdots & \vdots & & & \ddots & \vdots \\
0 & 0 & 0 & \cdots & 0 & I \\
-A_0 & -A_1 & -A_2 & \cdots & -A_{n-2} & -A_{n-1}
\end{bmatrix}, \quad
B = \begin{bmatrix}
0 \\
0 \\
\vdots \\
\vdots \\
0 \\
I
\end{bmatrix}
\tag{1.43}
$$

$$
C = \begin{bmatrix} I & 0 & 0 & \cdots & 0 & 0 \end{bmatrix}.
$$

By taking the Laplace transform of (1.41) with all the initial conditions set equal to zero, we obtain $\Omega(s)\mathbf{y}(s) = \mathbf{u}(s)$ where Ω is the operator valued polynomial defined by

$$\Omega(s) = A_0 + A_1 s + A_2 s^2 + \cdots + A_{n-1} s^{n-1} + s^n I. \qquad (1.44)$$

It follows from Exercise 1 below that $\Omega(s)^{-1} = C(sI - A)^{-1}B$. In other words, Ω^{-1} is the transfer function for the system $\{A, B, C, 0\}$ defined in (1.43).

A matrix of the form of A in (1.43) is called a companion matrix or a block companion matrix when $\dim \mathcal{U} > 1$. Companion matrices play a fundamental role in systems and control. If M is any operator on a finite dimensional space, then $\det[M]$ is the determinant of any matrix representation of M. The following result provides a complete characterization of the eigenvalues and eigenvectors for companion matrices.

Lemma 1.4.2 *Let A be the (block) companion matrix on $\oplus_1^n \mathcal{U}$ given in (1.43) and Ω the corresponding operator valued polynomial in (1.44). Then λ is an eigenvalue of A if and only if $\det[\Omega(\lambda)] = 0$. Furthermore, all eigenvectors v corresponding to a specified eigenvalue λ are given by*

$$v = \begin{bmatrix} w \\ \lambda w \\ \vdots \\ \lambda^{n-1} w \end{bmatrix} \qquad (1.45)$$

where w is a nonzero vector in the kernel of $\Omega(\lambda)$.

PROOF. Recall that λ is an eigenvalue of A if and only if $Av = \lambda v$ where v is a nonzero vector in $\mathcal{X} = \oplus_1^n \mathcal{U}$. Clearly, any such vector admits a decomposition of the form

$$v = \begin{bmatrix} w_1 & w_2 & \cdots & w_n \end{bmatrix}^{tr}$$

where w_j is in $\mathcal{U}$ for all j. By using the structure of A, it follows that $Av = \lambda v$ if and only if

$$w_2 = \lambda w_1, w_3 = \lambda w_2, \cdots, w_n = \lambda w_{n-1} \text{ and}$$
$$-A_0 w_1 - A_1 w_2 - \cdots - A_{n-1} w_n = \lambda w_n.$$

These relationships are equivalent to

$$w_2 = \lambda w_1, w_3 = \lambda^2 w_1, \cdots, w_n = \lambda^{n-1} w_1 \text{ and}$$
$$(\lambda^n I + \lambda^{n-1} A_{n-1} + \cdots + \lambda A_1 + A_0) w_1 = 0. \qquad (1.46)$$

By setting $w = w_1$, the first $n-1$ equations show that v admits a representation of the form (1.45). In particular, $v \neq 0$ is equivalent to $w \neq 0$. Furthermore, the last expression in (1.46) shows that λ is an eigenvalue of A if and only if $\Omega(\lambda)w_1 = 0$. Therefore, λ is an eigenvalue of A if and only if $\det[\Omega(\lambda)] = 0$. From the proceeding analysis it should be clear that all the eigenvectors corresponding to a specified eigenvalue λ are given by v in (1.45) where $\Omega(\lambda)w = 0$ and $w \neq 0$. ∎

Scalar transfer functions. The scalar valued case plays an important role in many applications. Let **g** be any scalar valued proper rational function. Then **g** admits a decomposition of the form

$$\mathbf{g}(s) = \frac{c_1 + c_2 s + \cdots + c_n s^{n-1}}{a_0 + a_1 s + \cdots + a_{n-1} s^{n-1} + s^n} + \gamma. \tag{1.47}$$

Let A, B and C be the matrices defined by

$$A = \begin{bmatrix} 0 & 1 & 0 & \cdots & 0 & 0 \\ 0 & 0 & 1 & \cdots & 0 & 0 \\ \vdots & \vdots & & \ddots & & \vdots \\ \vdots & \vdots & & & \ddots & \vdots \\ 0 & 0 & 0 & \cdots & 0 & 1 \\ -a_0 & -a_1 & -a_2 & \cdots & -a_{n-2} & -a_{n-1} \end{bmatrix}, \quad B = \begin{bmatrix} 0 \\ 0 \\ \vdots \\ \vdots \\ 0 \\ 1 \end{bmatrix} \tag{1.48}$$

$$C = \begin{bmatrix} c_1 & c_2 & c_3 & \cdots & c_{n-1} & c_n \end{bmatrix}.$$

Then our previous analysis shows that $\{A, B, C, \gamma\}$ is a realization of **g**. Moreover, λ is an eigenvalue for the companion matrix A in (1.48) if and only if λ is a zero of the polynomial $a_0 + a_1 s + \cdots + a_{n-1} s^{n-1} + s^n$. In this case, all corresponding eigenvectors for A are scalar multiples of $[1, \lambda, \lambda^2, \cdots, \lambda^{n-1}]^{tr}$. Finally, it is noted that the poles of **g** are contained in the eigenvalues of A.

Exercise 1 Suppose $\mathbf{G} = N\Omega^{-1}$ where N and Ω are polynomials of the form

$$N(s) = \sum_{k=0}^{n-1} N_k s^k \quad \text{and} \quad \Omega(s) = s^n I + \sum_{k=0}^{n-1} A_k s^k. \tag{1.49}$$

Here $N_0, \cdots, N_{n-1}$ are operators in $\mathcal{L}(\mathcal{U}, \mathcal{Y})$ while $A_0, \cdots, A_{n-1}$ are operators in $\mathcal{L}(\mathcal{U}, \mathcal{U})$. Let $\mathcal{X}$ be the space defined by $\mathcal{X} = \oplus_1^n \mathcal{U}$. Then show that a state space realization for this system is given by

$$\dot{x} = Ax + Bu \quad \text{and} \quad y = Cx \tag{1.50}$$

where A on $\mathcal{X}$ and B from $\mathcal{U}$ into $\mathcal{X}$ and C from $\mathcal{X}$ into $\mathcal{Y}$ are given by

$$A = \begin{bmatrix} 0 & I & 0 & \cdots & 0 & 0 \\ 0 & 0 & I & \cdots & 0 & 0 \\ \vdots & \vdots & & \ddots & & \vdots \\ \vdots & \vdots & & & \ddots & \vdots \\ 0 & 0 & 0 & \cdots & 0 & I \\ -A_0 & -A_1 & -A_2 & \cdots & -A_{n-2} & -A_{n-1} \end{bmatrix}, \quad B = \begin{bmatrix} 0 \\ 0 \\ \vdots \\ \vdots \\ 0 \\ I \end{bmatrix} \tag{1.51}$$

$$C = \begin{bmatrix} N_0 & N_1 & N_2 & \cdots & N_{n-2} & N_{n-1} \end{bmatrix}.$$

1.5 Notes

All the results in this section are classical. For further results on linear systems see Brockett [21], Chen [26], DeCarlo [33], Delchamps [34], Fuhrmann [47], Kailath [68], Polderman-Willems [98], Rugh [110], Skelton [114] and Sontag [116]. For some references on Hilbert spaces see Akhiezer-Glazman [2], Balakrishnan [8], Conway [30], Gohberg-Goldberg [53], Halmos [59] and Taylor-Lay [117]. For an in-depth study of operator theory and its applications see Gohberg-Goldberg-Kaashoek [54, 55].

Chapter 2

Stability

In this chapter we will present some classical results on state space stability and input-output stability. We will also present some elementary stability results for mechanical systems.

Let us establish some terminology used in this chapter and throughout the monograph. Let P be an operator on a Hilbert space $\mathcal{H}$. Then we say that P is a *positive operator* if $(Ph, h) \geq 0$ for all h in $\mathcal{H}$. The notation $P \geq 0$ means that P is positive. If P is a positive operator, then P is a self-adjoint operator. In general when we write $P \geq Q$ we mean that P and Q are two self-adjoint operators and $P - Q$ is positive. (It is easy to construct examples where $P - Q$ is positive and P and Q are not self-adjoint.) If $\mathcal{H}$ is finite dimensional, then P is positive if and only if P is a self-adjoint operator and all the eigenvalues of P are greater than or equal to zero. Following Halmos [58], we say that P is a *strictly positive* operator if there exists a constant $\delta > 0$ such that $(Ph, h) \geq \delta \|h\|^2$ for all h in $\mathcal{H}$. So, P is strictly positive if and only if P is positive and invertible. If $\mathcal{H}$ is finite dimensional, then P is strictly positive if and only if P is a self-adjoint operator and all the eigenvalues of P are strictly greater than zero. Moreover, in the finite dimensional case, P is strictly positive if and only if $(Ph, h) > 0$ for all nonzero h in $\mathcal{H}$. So, if $\mathcal{H}$ is finite dimensional, then the notation $P > 0$ means that P is strictly positive. Finally, it is noted that our terminology is slightly different from the standard terminology used in linear algebra. If $\mathcal{H} = \mathbb{C}^n$, then our strictly positive operator is usually referred to as a positive definite matrix, and our positive operator is usually defined as a positive semi-definite matrix.

2.1 State space stability

This section is devoted to the stability of systems described by

$$\dot{x} = Ax \tag{2.1}$$

where A is an operator on a finite dimensional Hilbert space $\mathcal{X}$ and $t \geq 0$. We say that the system $\dot{x} = Ax$ is *stable* if

$$\lim_{t \to \infty} x(t) = 0$$

for every solution x to (2.1). Recall that every solution x to (2.1) is given by $x(t) = e^{At}x_0$ where x_0 is a constant vector in $\mathcal{X}$. Therefore, the system $\dot{x} = Ax$ is stable if and only if e^{At}

converges to zero as t tends to infinity. The following basic result completely characterizes the stability of $\dot{x} = Ax$ in terms of the eigenvalues of A.

Theorem 2.1.1 *The system $\dot{x} = Ax$ in a finite dimensional space is stable if and only if all the eigenvalues of A have nonzero negative real part, that is, all the eigenvalues of A are in $\{s \in \mathbb{C} : \Re(s) < 0\}$. In this case, there exists positive constants m and $\alpha > 0$ such that $\|e^{At}\| \leq me^{-\alpha t}$ for all $t \geq 0$.*

PROOF. Recall the Laplace transform result: $\mathcal{L}(e^{At}) = (sI - A)^{-1}$. If L is any operator on the finite dimensional space $\mathcal{X}$, then $\det[L]$ denotes the determinant of any matrix representation of L. Because A is on a finite dimensional space, $(sI - A)^{-1} = N(s)/d(s)$ where N is the algebraic adjoint of A and $d(s) = \det[sI - A]$ is the characteristic polynomial of A. The roots of d are the eigenvalues of A. Moreover, N is a polynomial with values in $\mathcal{L}(\mathcal{X}, \mathcal{X})$ satisfying $\deg N < \deg d$. By computing the partial fraction expansion of N/d, we obtain that

$$(sI - A)^{-1} = \sum_{j=1}^{m} \sum_{k=0}^{r_j} \frac{R_{jk}}{(s - \lambda_j)^{k+1}} \tag{2.2}$$

where $\lambda_1, \lambda_2 \ldots, \lambda_m$ are the distinct roots of d, or equivalently, the eigenvalues of A and R_{jk} are operators on $\mathcal{X}$ for all j, k. Moreover, R_{jr_j} is nonzero for all j. Hence,

$$e^{At} = \sum_{j=1}^{m} \sum_{k=0}^{r_j} \frac{t^k}{k!} e^{\lambda_j t} R_{jk}. \tag{2.3}$$

Therefore, e^{At} approaches zeros as t approaches infinity if and only if $\Re(\lambda_j) < 0$ for all j.

Equation (2.3) implies that

$$\|e^{At}\| \leq \sum_{j=1}^{m} \sum_{k=0}^{r_j} \frac{t^k}{k!} e^{\Re(\lambda_j) t} \|R_{jk}\|$$

for all $t \geq 0$. Now assume that A is stable. From this it readily follows that there exists positive constants m and $\alpha > 0$ such that $\|e^{At}\| \leq me^{-\alpha t}$ for all $t \geq 0$. ∎

Notice that the previous proof provides a method to compute e^{At}. To be more specific, compute the operators R_{jk} in the partial fraction expansion of $(sI - A)^{-1}$. Then e^{At} is given by (2.3).

By a slight abuse of terminology we say that an operator A on a finite dimensional space is stable if all its eigenvalues have nonzero negative real part. Finally, e^{At} is stable if A is stable.

2.2   Mechanical systems

In this section we present a general class of mechanical systems and an elementary stability result. We have already seen a one degree-of-freedom mechanical system in Chapter 1, that is, $m\ddot{q} + c\dot{q} + kq = 0$. Obviously, m, c and k are positive operators on $\mathbb{C}^1$. The following example illustrates how positive operators naturally occur in a two degree-of-freedom mechanical system.

Example 2.2.1 Consider the simple structure illustrated in Figure 2.1. It consists of two

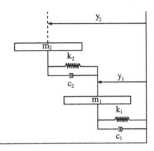

Figure 2.1: A simple structure

"floors" of masses $m_1 > 0$ and $m_2 > 0$ connected by linear springs of spring constants $k_1 > 0$, $k_2 > 0$ and dashpots with damping coefficients $c_1 > 0$, $c_2 > 0$. The scalars y_1, y_2 are the horizontal displacements of the floors from their equilibrium positions. An application of Newton's Second Law to each floor yields:

$$m_1\ddot{y}_1 + (c_1 + c_2)\dot{y}_1 - c_2\dot{y}_2 + (k_1 + k_2)y_1 - k_2 y_2 = 0$$
$$m_2\ddot{y}_2 - c_2\dot{y}_1 + c_2\dot{y}_2 - k_2 y_1 + k_2 y_2 = 0.$$

Let q be the vector in $\mathbb{C}^2$ defined by $q = [y_1, y_2]^{tr}$ where tr denotes the transpose. Then this system can be described by the following second order vector differential equation:

$$M\ddot{q} + C\dot{q} + Kq = 0$$

where the M, C, K are the operators given by

$$M = \begin{bmatrix} m_1 & 0 \\ 0 & m_2 \end{bmatrix}, \qquad C = \begin{bmatrix} c_1 + c_2 & -c_2 \\ -c_2 & c_2 \end{bmatrix}, \qquad K = \begin{bmatrix} k_1 + k_2 & -k_2 \\ -k_2 & k_2 \end{bmatrix}.$$

Since $m_1 > 0$ and $m_2 > 0$, the operator M is strictly positive. Finally, in this example, C and K are also strictly positive.

A general mechanical system with n degrees-of-freedom is described by

$$M\ddot{q} + C\dot{q} + Kq = 0 \qquad (2.4)$$

where q lives in an n-dimensional vector space $\mathcal{Q}$ and M is a strictly positive operator on $\mathcal{Q}$, while C and K are operators on $\mathcal{Q}$. The initial conditions for this differential equation are given by $q(0)$ and $\dot{q}(0)$. Here M is called the *inertia operator*, K is the *stiffness operator*, and C is the *damping operator*. This differential equation can be used to model a large class of linear mechanical systems; see Meirovitch [89]. We say that system (2.4) is stable if $q(t)$ approaches zero as t approaches infinity for every solution q.

To convert the mechanical system in (2.4) to state space form, let $\mathcal{X} = \mathcal{Q} \oplus \mathcal{Q}$ be the $2n$-dimensional linear space consisting of all vectors of the form $[q_1, q_2]^{tr}$ where q_1 and q_2

are vectors in $\mathcal{Q}$. If we set $x = [q, \dot{q}]^{tr}$, then the mechanical system (2.4) admits a state description of the form $\dot{x} = Ax$ where A is the block matrix on $\mathcal{X}$ defined by

$$A = \begin{bmatrix} 0 & I \\ -M^{-1}K & -M^{-1}C \end{bmatrix}. \tag{2.5}$$

We claim that the mechanical system (2.4) is stable if and only its state space representation $\dot{x} = Ax$ is stable. Clearly, if $\dot{x} = Ax$ is stable, then $q(t)$ approaches zero as t tends to infinity, and thus, the corresponding mechanical system is stable. Now assume that the mechanical system is stable. Let λ be any nonzero eigenvalue of A. Because A is a block companion matrix, the eigenvector corresponding to λ is a vector of the form $v = [w, \lambda w]^{tr}$ where w is a nonzero vector in $\mathcal{Q}$; see Lemma 1.4.2. For the initial condition $x(0) = v$, the solution to the differential equation $\dot{x} = Ax$ is given by $x(t) = e^{At}v = [e^{\lambda t}w, \lambda e^{\lambda t}w]^{tr}$. Since q is the first component of x, and the mechanical system is stable, we must have $\Re(\lambda) < 0$. So, all the eigenvalues of A have nonzero negative real part. Therefore, $\dot{x} = Ax$ is stable, which proves our claim.

Theorem 2.2.1 *Let Ω be the second order operator valued polynomial defined by*

$$\Omega(s) = s^2 M + sC + K \tag{2.6}$$

where M, C and K are all operators on $\mathcal{Q}$ with M invertible. Then the corresponding mechanical system in (2.4) is stable if and only if all the roots of the polynomial $\det[\Omega(s)]$ have nonzero negative real part.

PROOF. We have just seen that the mechanical system in (2.4) is stable if and only if the system $\dot{x} = Ax$ is stable where A is given by (2.5). According to Lemma 1.4.2, the eigenvalues of A in (2.5) are given by the set of all complex numbers λ such that $\det[\Omega(\lambda)] = 0$. Hence, $\dot{x} = Ax$ is stable if and only if all the roots of $\det[\Omega]$ have nonzero negative real part. This completes the proof.   ∎

We are now ready to state a basic stability result for mechanical systems.

Theorem 2.2.2 *If M, C and K are all strictly positive operators on $\mathcal{Q}$, then the corresponding mechanical system in (2.4) is stable.*

PROOF. Let Ω be the operator valued polynomial defined in (2.6). From Theorem 2.2.1, the mechanical system in (2.4) is stable if and only if all the roots of the polynomial $\det[\Omega]$ have nonzero negative real part. If λ is a root of $\det[\Omega]$, then $\Omega(\lambda)v = 0$ for some nonzero vector v in $\mathcal{Q}$. Thus,

$$0 = (\Omega(\lambda)v, v) = \lambda^2 (Mv, v) + \lambda (Cv, v) + (Kv, v). \tag{2.7}$$

It is well known (use the quadratic formula) that the roots of a second order polynomial with nonzero positive real coefficients have nonzero negative real parts. Since M, C, and K are positive operators and $v \neq 0$, it follows from (2.7) that $\Re(\lambda) < 0$. Therefore, all roots of $\det[\Omega]$ have nonzero negative real parts and the mechanical system is stable.   ∎

If K is positive and singular, then the mechanical system in (2.4) is not stable. To see this, notice that $\Omega(0) = K$ and thus $\det[\Omega(0)] = 0$. Hence, it follows from Theorem 2.2.1 that the mechanical is not stable. However, the following result demonstrates that stability is possible when C is positive and singular.

Theorem 2.2.3 *Assume that M and K are strictly positive operators on $\mathcal{Q}$ and C is a positive operator on $\mathcal{Q}$. Then the corresponding mechanical system in (2.4) is stable if and only if the kernel of*

$$\begin{bmatrix} K - \sigma M \\ C \end{bmatrix} \tag{2.8}$$

is zero for all real numbers $\sigma > 0$.

PROOF. If v is a nonzero vector in the kernel of $\begin{bmatrix} K - \sigma M & C \end{bmatrix}^{tr}$ for some $\sigma > 0$, then $\Omega(\lambda)v = 0$ for $\lambda = \imath\sqrt{\sigma}$ where $\Omega(\lambda)$ is defined in (2.6). According to Theorem 2.2.1, this readily implies that the mechanical system is not stable.

On the other hand, if the mechanical system in (2.4) is unstable, then there exists a scalar λ and a nonzero vector v in $\mathcal{Q}$ such that $\Re(\lambda) \geq 0$ and $\Omega(\lambda)v = 0$. Thus (2.7) holds. We claim that $(Cv, v) = 0$. Since C is positive, $(Cv, v) \geq 0$. If $(Cv, v) > 0$, then $(\Omega(\lambda)v, v)$ is a quadratic polynomial with nonzero positive coefficients. Thus, $\Re(\lambda) < 0$ which contradicts that fact that $\Re(\lambda) \geq 0$; hence $(Cv, v) = 0$. So, $\lambda^2(Mv, v) + (Kv, v) = 0$ and hence, $\lambda = \pm\imath\sqrt{\sigma}$ where $\sigma = (Kv, v)/(Mv, v)$. Since $(Cv, v) = 0$, we have $Cv = 0$. (To see this, simply notice that $0 = (Cv, v) = ||C^{1/2}v||^2$ where $C^{1/2}$ is the positive square root of C. Therefore, $C^{1/2}v = 0$ and $Cv = C^{1/2}C^{1/2}v = 0$.) We now obtain $(-\sigma M + K)v = \Omega(\lambda)v = 0$. Thus v is in the kernel of $[K - \sigma M, C]^{tr}$. This completes the proof. ∎

Example 2.2.2 Recall the two story structure of Example 2.2.1. There we assumed that both of the damping coefficients c_1 and c_2 were nonzero and positive. However, an application of the above theorem shows that for stability, one only needs one of the damping coefficients to be nonzero and positive, and the other to be greater than or equal to zero.

2.3   Input-output stability

This section is concerned with input-output stability of linear systems. To begin, consider a linear map T defined by

$$y(t) = (Tu)(t) = \int_0^t G(t-\tau)u(\tau)\,d\tau \tag{2.9}$$

where G is a Lebesgue measurable function with values in $\mathcal{L}(\mathcal{U}, \mathcal{Y})$ and the input $u(t)$ has values in $\mathcal{U}$ while the output $y(t)$ has values in $\mathcal{Y}$. Throughout we always assume that both $\mathcal{U}$ and $\mathcal{Y}$ are finite dimensional Hilbert spaces. We say that the input-output map T defined in (2.9) is *stable*, if T defines a bounded operator from $L^2([0,\infty),\mathcal{U})$ into $L^2([0,\infty),\mathcal{Y})$. We say that G is in $L^1(\mathcal{U},\mathcal{Y})$ if G is a Lebesgue measurable function on $[0,\infty)$ with values in $\mathcal{L}(\mathcal{U},\mathcal{Y})$ and its L^1 norm

$$||G||_1 := \int_0^\infty ||G(t)||\,dt \tag{2.10}$$

is finite. This sets the stage for the following useful result.

Proposition 2.3.1 *If G is in $L^1(\mathcal{U},\mathcal{Y})$, then the linear map T in (2.9) is a bounded operator from $L^2([0,\infty),\mathcal{U})$ into $L^2([0,\infty),\mathcal{Y})$. Moreover, $||T|| \leq ||G||_1$.*

PROOF. If $y = Tu$, then an application of the Cauchy-Schwartz inequality yields

$$\|y(t)\|^2 \leq \left(\int_0^t \|G(t-\tau)\| \, \|u(\tau)\| \, d\tau \right)^2$$

$$= \left(\int_0^t \|G(t-\tau)\|^{1/2} \|G(t-\tau)\|^{1/2} \|u(\tau)\| \, d\tau \right)^2$$

$$\leq \int_0^t \|G(t-\tau)\| \, d\tau \cdot \int_0^t \|G(t-\tau)\| \, \|u(\tau)\|^2 \, d\tau$$

$$\leq \|G\|_1 \int_0^t \|G(t-\tau)\| \, \|u(\tau)\|^2 \, d\tau .$$

The Cauchy-Schwartz inequality was used to obtain the second inequality. By integrating the previous inequality and changing the order of integration, we have

$$\int_0^\infty \|y(t)\|^2 \, dt \leq \|G\|_1 \int_0^\infty \int_0^t \|G(t-\tau)\| \, \|u(\tau)\|^2 \, d\tau dt$$

$$= \|G\|_1 \int_0^\infty \int_\tau^\infty \|G(t-\tau)\| \, dt \, \|u(\tau)\|^2 \, d\tau$$

$$= \|G\|_1^2 \, \|u\|^2 .$$

Therefore, $\|y\| \leq \|G\|_1 \|u\|$. This completes the proof. ∎

The following result yields another useful bound for the operator T.

Corollary 2.3.2 *Let G be a Lebesgue measurable function with values in $\mathcal{L}(\mathbb{C}, \mathcal{Y})$ satisfying*

$$\|G(t)\| \leq me^{-2\alpha t} \qquad (t \in [0, \infty)) \tag{2.11}$$

for some $m \geq 0$ and $\alpha > 0$. Then G is in $L^1(\mathbb{C}, \mathcal{Y})$. Furthermore, the linear map T in (2.9) defines a bounded operator from $L^2[0, \infty)$ into $L^2([0, \infty), \mathcal{Y})$ and

$$\|T\| \leq \|G\|_1 \leq \|e^{\alpha t} G\|_2 / \sqrt{2\alpha} \leq m/2\alpha \tag{2.12}$$

where $\| \cdot \|_2$ denotes the $L^2([0, \infty), \mathcal{Y})$ norm.

PROOF. Using the Cauchy-Schwartz inequality, we have

$$\|G\|_1^2 = \left(\int_0^\infty \|G(t)\| \, dt \right)^2$$

$$= \left(\int_0^\infty \|e^{\alpha t} G(t)\| \, e^{-\alpha t} \, dt \right)^2$$

$$\leq \int_0^\infty \|e^{\alpha t} G(t)\|^2 \, dt \cdot \int_0^\infty e^{-2\alpha t} \, dt$$

$$= \|e^{\alpha t} G\|_2^2 / 2\alpha \leq \|me^{-\alpha t}\|_2^2 / 2\alpha = m^2 / 4\alpha^2 .$$

This along with Proposition 2.3.1 readily yields the inequality in (2.12). ∎

Remark 2.3.1 Recall that we have equality in the Cauchy-Schwartz inequality if and only if the corresponding vectors are linearly dependent. From the proof of the above inequality, it should be clear that we have the equality $||G||_1 = ||e^{\alpha t}G||_2/\sqrt{2\alpha}$ in (2.12) if and only if there is a scalar k such that $||G(t)e^{\alpha t}|| = ke^{-\alpha t}$, or equivalently, $||G(t)|| = ke^{-2\alpha t}$.

Now consider the linear system $\{A, B, C, D\}$ defined by

$$\dot{x} = Ax + Bu \quad \text{and} \quad y = Cx + Du \tag{2.13}$$

where A is on a finite dimensional state space $\mathcal{X}$. As before, the input $u(t)$ lives in $\mathcal{U}$ while the output $y(t)$ lives in $\mathcal{Y}$. If the initial condition is zero, then the input-output map R corresponding to (2.13) is given by

$$y(t) = (Ru)(t) = \int_0^t Ce^{A(t-\tau)}Bu(\tau)\,d\tau + Du(t). \tag{2.14}$$

Motivated by this we say that $\{A, B, C, D\}$ is *input-output stable* if R in (2.14) defines a bounded operator from $L^2([0,\infty),\mathcal{U})$ into $L^2([0,\infty),\mathcal{Y})$. Notice that if A is stable, that is, if all the eigenvalues of A have nonzero negative real part, then $G(t) = Ce^{At}B$ is in $L^1(\mathcal{U}, \mathcal{Y})$. Hence, the operator T in (2.9) corresponding to this G is bounded. Thus, $R = T + D$ is bounded. Therefore, if A is stable, then $\{A, B, C, D\}$ is input-output stable. The converse of this fact is not necessarily true. For example, choose $A = 1$ and $B = C = D = 0$. Then clearly A is unstable while $R = 0$ obviously bounded. Summing up our previous analysis yields the following result.

Proposition 2.3.3 *If all the eigenvalues of A have nonzero negative real part, then the system $\{A, B, C, D\}$ is input-output stable.*

2.4 The H^∞ norm

To compute the norm of the input-output operator T in (2.9), we need the Hardy space $H^\infty(\mathcal{U}, \mathcal{Y})$. Throughout $H^\infty(\mathcal{U}, \mathcal{Y})$ is the Hardy space of all functions $\mathbf{G}$ which are analytic for $\Re(s) > 0$, take values in $\mathcal{L}(\mathcal{U}, \mathcal{Y})$ and whose H^∞ norm

$$||\mathbf{G}||_\infty := \sup\{||\mathbf{G}(s)|| : \Re(s) > 0\} \tag{2.15}$$

is finite. Notice that a rational function $\mathbf{G}$ is in $H^\infty(\mathcal{U}, \mathcal{Y})$ if and only if $\mathbf{G}$ is a proper rational function whose poles are in the open left half plane of $\mathbb{C}$, that is, $\{s \in \mathbb{C} : \Re(s) < 0\}$. Now we are ready to state the following classical result; see Chapter IX, Section I of Foias-Frazho [39].

Theorem 2.4.1 *Suppose G is a Lebesgue measurable function with values in $\mathcal{L}(\mathcal{U}, \mathcal{Y})$ and D is in $\mathcal{L}(\mathcal{U}, \mathcal{Y})$. Let R be the linear map defined by*

$$y(t) = (Ru)(t) = \int_0^t G(t-\tau)u(\tau)\,d\tau + Du(t) \tag{2.16}$$

where u is in $L^2([0,\infty),\mathcal{U})$. Then R defines a bounded operator from $L^2([0,\infty),\mathcal{U})$ into $L^2([0,\infty),\mathcal{Y})$ if and only if the Laplace transform $\mathbf{G}$ of G is a function in $H^\infty(\mathcal{U}, \mathcal{Y})$. In this case $||R|| = ||\mathbf{G} + D||_\infty$.

The following result can be used to obtain a proof of Proposition 2.3.1.

Lemma 2.4.2 *If G is in $L^1(\mathcal{U}, \mathcal{Y})$, then its Laplace transform $\mathbf{G}$ is in $H^\infty(\mathcal{U}, \mathcal{Y})$. Moreover,*
$||\mathbf{G}||_\infty \leq ||G||_1$.

PROOF. Recall that the Laplace transform $\mathbf{G}$ of G is defined by

$$\mathbf{G}(s) = \int_0^\infty e^{-st} G(t)\, dt\,.$$

Now let us show that $\mathbf{G}$ analytic for $\Re(s) > 0$. Consider any complex number s with $\Re(s) > 0$. Since $|e^{-st}| \leq 1$, it follows that $||e^{-st}G(t)|| = |e^{-st}|\,||G(t)|| \leq ||G(t)||$. Moreover, there is a finite number β such that $|te^{-st}| \leq \beta$ for all $t \geq 0$; hence $||te^{-st}G(t)|| \leq \beta||G(t)||$. Since G is in $L^1(\mathcal{U}, \mathcal{Y})$, it now follows that the functions on $[0, \infty)$ given by $e^{-st}G(t)$ and its derivative with respect to s are both in $L^1(\mathcal{U}, \mathcal{Y})$. So we can interchange the integral and the derivative to show that the derivative of $\mathbf{G}$ exists for all $\Re(s) > 0$. Hence, $\mathbf{G}$ is analytic in $\Re(s) > 0$. Moreover, we have

$$||\mathbf{G}(s)|| \leq \int_0^\infty |e^{-st}|\, ||G(t)||\, dt \leq ||G||_1\,.$$

Therefore, $||\mathbf{G}||_\infty \leq ||G||_1$ and $\mathbf{G}$ is a function in $H^\infty(\mathcal{U}, \mathcal{Y})$ which proves our claim. ∎

Consider any G in $L^1(\mathcal{U}, \mathcal{Y})$. By the previous lemma, $\mathbf{G}$ is in $H^\infty(\mathcal{U}, \mathcal{Y})$ and $||\mathbf{G}||_\infty \leq ||G||_1$. Theorem 2.4.1 now implies that the operator T in (2.9) defines a bounded operator with $||T|| = ||\mathbf{G}||_\infty$. Hence, $||T|| \leq ||G||_1$. This readily proves Proposition 2.3.1.

The following result is an immediate consequence of Theorem 2.4.1

Theorem 2.4.3 *Let $\mathbf{G}$ be the transfer function for a finite dimensional system $\{A, B, C, D\}$ and R be the input-output map defined by*

$$(Ru)(t) = \int_0^t Ce^{A(t-\tau)}Bu(\tau)\, d\tau + Du(t)\,. \qquad (2.17)$$

Then $\{A, B, C, D\}$ is input-output stable if and only if $\mathbf{G}$ is in $H^\infty(\mathcal{U}, \mathcal{Y})$. Moreover, in this case R is a bounded operator from $L^2([0, \infty), \mathcal{U})$ into $L^2([0, \infty), \mathcal{Y})$ and $||R|| = ||\mathbf{G}||_\infty$. In particular, if all the eigenvalues of A have nonzero negative real part, then $\mathbf{G}$ is in $H^\infty(\mathcal{U}, \mathcal{Y})$ and $\{A, B, C, D\}$ is input-output stable.

If $\mathbf{G}$ is the transfer function for $\{A, B, C, D\}$, then we say that $\mathbf{G}$ is *stable* if all the poles of $\mathbf{G}$ are in the open left half plane, or equivalently, $\mathbf{G}$ is in $H^\infty(\mathcal{U}, \mathcal{Y})$.

If P is any self-adjoint operator on $\mathcal{X}$, then $\lambda_{\max}(P)$ denotes the largest eigenvalue of P. Now let G be a Lebesgue measurable function with values in $\mathcal{L}(\mathcal{U}, \mathcal{Y})$. Then $d_1(G)$ is the scalar defined by

$$d_1(G) = \sup\{||Gu||_1 : u \in \mathcal{U} \text{ and } ||u|| \leq 1\}\,.$$

Note that $d_1(G)$ can be infinite. However, if G is in $L^1(\mathcal{U}, \mathcal{Y})$, then $d_1(G)$ is finite and satisfies $d_1(G) \leq ||G||_1$. If $\mathcal{U} = \mathbb{C}^1$, then $d_1(G) = ||G||_1$. Finally, let N be an operator from $\mathcal{U}$ into $\mathcal{Y}$, then the adjoint of N is denoted by N^*. Recall N^* is the unique operator from $\mathcal{Y}$ into $\mathcal{U}$ defined by $(Nu, y) = (u, N^*y)$ for all u in $\mathcal{U}$ and y in $\mathcal{Y}$. This sets the stage for the following result which is a generalization of Corollary 2.3.2.

Theorem 2.4.4 *Let G be a Lebesgue measurable function with values in $\mathcal{L}(\mathcal{U}, \mathcal{Y})$ where $\mathcal{U}$ is finite dimensional. Suppose that there are positive scalars m and $\alpha > 0$ satisfying $||G(t)|| \leq me^{-2\alpha t}$ for all $t \geq 0$. Then the operator*

$$\Pi = \int_0^\infty e^{2\alpha t} G(t)^* G(t) \, dt \tag{2.18}$$

is a well defined positive operator on $\mathcal{U}$. Moreover, the Laplace transform $\mathbf{G}$ of G is in $H^\infty(\mathcal{U}, \mathcal{Y})$ and satisfies the following bounds:

$$||\mathbf{G}||_\infty^2 \leq d_1(G)^2 \leq \lambda_{\max}(\Pi)/2\alpha. \tag{2.19}$$

PROOF. Consider any unit vector u in $\mathcal{U}$. It follows from the hypotheses that Gu is in $L^1(\mathbb{C}, \mathcal{Y})$. Applying Lemma 2.4.2, we have $||\mathbf{G}u||_\infty \leq ||Gu||_1 \leq d_1(G)$. Thus,

$$||\mathbf{G}||_\infty = \sup\{||\mathbf{G}u||_\infty : u \in \mathcal{U} \text{ and } ||u|| \leq 1\} \leq d_1(G).$$

This readily establishes the first inequality in (2.19). To show that Π is a well defined operator, let u and v be two arbitrary vectors in $\mathcal{U}$. Then, using the exponential bound on $G(t)$, we obtain

$$
\begin{aligned}
|(\Pi u, v)| &= \left| \int_0^\infty e^{2\alpha t} \left(G(t)^* G(t) u, v \right) dt \right| = \left| (e^{\alpha t} Gu, e^{\alpha t} Gv) \right| \\
&\leq ||e^{\alpha t} Gu||_2 \, ||e^{\alpha t} Gv||_2 \leq ||me^{-\alpha t}||_2^2 \, ||u|| \, ||v|| \leq m^2 ||u|| ||v||/2\alpha. \tag{2.20}
\end{aligned}
$$

Hence, $(\Pi u, v)$ is finite for all u and v. Thus, Π is a well defined operator on $\mathcal{U}$. In fact, by choosing $v = \Pi u$ in (2.20), yields $||\Pi|| \leq m^2/2\alpha$. Since $(\Pi u, u) = ||e^{\alpha t} Gu||_2^2 \geq 0$, it follows that Π is a positive operator. To obtain the second inequality in (2.19), let u be a unit vector in $\mathcal{U}$. Then using the Cauchy-Schwartz inequality, we have

$$
\begin{aligned}
||Gu||_1^2 &= \left(\int_0^\infty ||e^{\alpha t} G(t)u|| e^{-\alpha t} \, dt \right)^2 \\
&\leq \int_0^\infty ||e^{\alpha t} G(t)u||^2 \, dt \cdot \int_0^\infty e^{-2\alpha t} \, dt \\
&= (\Pi u, u)/2\alpha \leq \lambda_{\max}(\Pi)/2\alpha.
\end{aligned}
$$

Therefore, $d_1(G)^2 \leq \lambda_{\max}(\Pi)/2\alpha$.   ∎

2.5 Notes

The results in this chapter are classical. For stability results on nonlinear differential equations see Khalil [74] and Vidyasagar [124]. A detailed study of mechanical systems is presented in Meirovitch [89]. For some further results on mechanical systems and feedback control see Meirovitch [90] and Skelton [114]. For some classical results on H^∞ functions see Garnett [50] and Hoffman [64].

Chapter 3

Lyapunov Theory

In this chapter we develop and use some special Lyapunov equations to study the stability of linear systems. These Lyapunov equations naturally lead to some stability bounds for linear systems. Finally, we use Lyapunov techniques to derive some fundamental stability results for linear mechanical systems.

3.1 Basic Lyapunov theory

This section is devoted to basic Lyapunov stability results for state space systems described by

$$\dot{x} = Ax \tag{3.1}$$

where A acts on a finite dimensional vector space $\mathcal{X}$ and $t \geq 0$. Recall that the operator A, or equivalently, the system $\dot{x} = Ax$ is stable if and only if all the eigenvalues of A have nonzero negative real part. Notice that the scalar system $\dot{x} = ax$ is stable if and only if $2\Re(a) = a + \bar{a} < 0$. The generalization of $a + \bar{a}$ to system (3.1) is $A + A^* < 0$. Motivated by this, we say that A or system (3.1) is *dissipative* if $A + A^* < 0$. Clearly, A is dissipative if and only if $2\Re(Ax, x) < 0$ for all nonzero x. If A is dissipative, then A is stable. To see this, consider any eigenvalue λ of A and a corresponding unit eigenvector v, that is, $Av = \lambda v$ and $\|v\| = 1$. Then $0 > 2\Re(Av, v) = 2\Re(\lambda)$. Since this holds for every eigenvalue, A is stable. However, if A is stable, then A it is not necessarily dissipative. For example, consider the operator

$$A = \begin{bmatrix} -1 & 3 \\ 0 & -1 \end{bmatrix}.$$

Clearly, A is stable. The eigenvalues for the operator $A + A^*$ are $\{-5, 1\}$. So, A is not dissipative.

Recall that an operator A is similar to another operator F on $\mathcal{F}$ if there exists an invertible transformation T from $\mathcal{X}$ into $\mathcal{F}$ such that $F = TAT^{-1}$. Since similarity transformations preserve eigenvalues, they also preserve stability. Later we will see that A is stable if and only if A is similar to a dissipative operator. Assume that A is similar to a dissipative operator F, that is, there exists a similarity transform T satisfying

$$TAT^{-1} + T^{-*}A^*T^* = F + F^* < 0.$$

29

The adjoint of the inverse of T is denoted by T^{-*}. Let P be the strictly positive operator on $\mathcal{X}$ defined by $P := T^*T$. (Because T is invertible, it follows that $(Px, x) = \|Tx\|^2 > 0$ for all nonzero x in $\mathcal{X}$, and thus P is strictly positive.) Pre-multiplying and post-multiplying the previous inequality by T^* and T, respectively, yields

$$PA + A^*P < 0. \tag{3.2}$$

So, if A is similar to a dissipative operator, then A is stable and there exists a strictly positive operator P satisfying (3.2). We now show that the existence of a strictly positive operator P satisfying (3.2) guarantees stability.

Lemma 3.1.1 *Let A be an operator on a finite-dimensional vector space and suppose that there exists a strictly positive operator P satisfying (3.2). Then A is stable.*

PROOF. Consider any eigenvalue λ of A. Let v be an eigenvector corresponding to λ, that is, $Av = \lambda v$ where $v \neq 0$. Then

$$0 > ((PA + A^*P)v, v) = (PAv, v) + (Pv, Av) = (\lambda + \bar{\lambda})(Pv, v) \; .$$

Hence, $2\Re(\lambda)(Pv, v) < 0$. Because $P > 0$, we must have $\Re(\lambda) < 0$. Since this holds for every eigenvalue, A is stable. ∎

A strictly positive operator P which satisfies the inequality in (3.2) will be referred to as a *Lyapunov operator* for (3.1) or A. If A is dissipative, then the Lyapunov operator for A is simply the identity operator. Clearly, I is strictly positive. So, the above lemma also shows that all dissipative systems are stable.

Remark 3.1.1 Let A be an operator on a finite-dimensional vector space satisfying

$$AQ + QA^* < 0 \tag{3.3}$$

where Q is strictly positive. Then from the above lemma, A^* is stable, and hence, A is stable.

So far we have shown that if (3.2) holds for some P, then A is stable. Is the converse true? That is, if A is stable, does there exist an operator P such that (3.2) holds. Moreover, if this is true how does one find such an operator P. To answer this question notice that inequality (3.2) is equivalent to

$$PA + A^*P + \Omega = 0 \tag{3.4}$$

where Ω is a strictly positive operator. This linear operator equation is known as a *Lyapunov equation*. So, one approach to looking for Lyapunov operators could be to choose a strictly positive operator Ω and determine whether the Lyapunov equation has a strictly positive solution for P. The following result shows that if A is stable, then (3.4) has a solution P for every Ω.

Lemma 3.1.2 *Let A be stable operator on a finite dimensional vector space. Then for every operator* Ω, *the Lyapunov equation (3.4) has a unique solution P and this solution is given by*

$$P = \int_0^\infty e^{A^*t}\Omega e^{At}\,dt\,. \tag{3.5}$$

Moreover, if Ω *is strictly positive (respectively positive), then P is strictly positive (respectively positive).*

PROOF. First, let us show that P given by (3.5) is a solution to the Lyapunov equation in (3.4). Because A is stable, $\|e^{At}\| \le m e^{-\alpha t}$ for some positive m and $\alpha > 0$; see Theorem 2.1.1. Therefore, the integral in (3.5) exists, and this P is a well defined operator. Recall that

$$e^{At}A = \frac{de^{At}}{dt} \quad \text{and} \quad A^*e^{A^*t} = \frac{de^{A^*t}}{dt}\,.$$

Using this and (3.5), we obtain

$$
\begin{aligned}
PA + A^*P &= \int_0^\infty \left[e^{A^*t}\Omega e^{At}A + A^*e^{A^*t}\Omega e^{At} \right]\,dt \\
&= \int_0^\infty \left[e^{A^*t}\Omega \frac{de^{At}}{dt} + \frac{de^{A^*t}}{dt}\Omega e^{At} \right]\,dt \\
&= \int_0^\infty \frac{d\left(e^{A^*t}\Omega e^{At} \right)}{dt}\,dt \\
&= \lim_{t_1 \to \infty} \int_0^{t_1} \frac{de^{A^*t}\Omega e^{At}}{dt}\,dt \\
&= \lim_{t_1 \to \infty} e^{A^*t_1}\Omega e^{At_1} - \Omega = -\Omega\,.
\end{aligned}
$$

Therefore, the operator P defined in (3.5) is a solution to Lyapunov equation (3.4).

To demonstrate that the solution to this Lyapunov equation is unique, consider any P satisfying (3.4). Then for any $t_1 > 0$, we have

$$
\begin{aligned}
\int_0^{t_1} e^{A^*t}\Omega e^{At}\,dt &= -\int_0^{t_1} e^{A^*t}(PA + A^*P)e^{At}\,dt \\
&= -\int_0^{t_1} \frac{d\left(e^{A^*t}Pe^{At} \right)}{dt}\,dt \\
&= P - e^{A^*t_1}Pe^{At_1}\,.
\end{aligned}
$$

Thus,

$$\int_0^{t_1} e^{A^*t}\Omega e^{At}\,dt = P - e^{A^*t_1}Pe^{At_1}\,. \tag{3.6}$$

By letting t_1 approach infinity and using the fact that A is stable, we see that P is given by (3.5). Hence P is unique.

To complete the proof, assume that Ω is positive. Then for any x in $\mathcal{X}$, we have

$$(Px, x) = \left(\int_0^\infty e^{A^* t} \Omega e^{At} dt \, x, x \right) = \int_0^\infty \left(\Omega e^{At} x, e^{At} x \right) \, dt \geq 0 \,. \tag{3.7}$$

So, if Ω is positive, then P is positive. If Ω is strictly positive, then there exists an $\epsilon > 0$ such that $\Omega \geq \epsilon I$. Using this in (3.7) along with $x \neq 0$, yields

$$(Px, x) = \int_0^\infty \left(\Omega e^{At} x, e^{At} x \right) \, dt \geq \epsilon \int_0^\infty \| e^{At} x \|^2 \, dt > 0 \,.$$

In this case, P is strictly positive. This completes the proof.   ∎

Using the above two lemmas, we can now state the main result of this section.

Theorem 3.1.3 *Let A be an operator on a finite dimensional vector space. Then the following statements are equivalent.*

(a) The system $\dot{x} = Ax$ is stable.

(b) There exist strictly positive operators P and Ω satisfying the Lyapunov equation (3.4).

(c) For any strictly positive operator Ω, the Lyapunov equation (3.4) has a strictly positive solution P. In this case P is the only solution.

(d) The operator A is similar to a dissipative operator.

PROOF.   The first lemma shows that (b) implies (a). The second lemma states that (a) implies (c). Hence, (b) implies (c). To see that (c) implies (b), pick any strictly positive Ω. So, (b) holds and thus, (a), (b), (c) are equivalent.

To complete the proof, we have to prove the equivalence of (d). Recall that all dissipative operators are stable. So, if A is similar to a dissipative operator, then A is stable. In other words, (d) implies (a). On the other hand, if A is stable, then there is a strictly positive operator P such that

$$PA + A^* P + I = 0 \,.$$

Since P is strictly positive, it admits a strictly positive square root denoted by $P^{1/2}$. Multiplying both sides of the previous Lyapunov equation by $P^{-1/2}$ yields

$$P^{1/2} A P^{-1/2} + P^{-1/2} A^* P^{1/2} = -P^{-1} < 0 \,.$$

This shows that A is similar to the dissipative operator $P^{1/2} A P^{-1/2}$.   ∎

Since the stability of A is equivalent to the stability of A^*, one can state the above theorem by replacing P with Q and replacing (3.4) with

$$AQ + QA^* + \Omega = 0 \,. \tag{3.8}$$

This readily yields the following result.

Corollary 3.1.4 *Let A be an operator on a finite dimensional vector space. Then the following statements are equivalent.*

(a) *The system $\dot{x} = Ax$ is stable.*

(b) *There exist strictly positive operators Q and Ω satisfying the Lyapunov equation (3.8).*

(c) *For any strictly positive operator Ω, the Lyapunov equation (3.8) has a strictly positive solution Q. In this case Q is the only solution.*

(d) *The operator A is similar to a dissipative operator.*

Exercise 2 As before, let A be an operator on a finite dimensional vector space. Consider the Lyapunov equation

$$PA + A^*P + 2\alpha P + \Omega = 0 \qquad (3.9)$$

where Ω is some strictly positive operator and α is a positive scalar. Then show that this equation has a strictly positive solution if and only if all the eigenvalues for A are in $\{s \in \mathbb{C} : \Re(s) < -\alpha\}$. In this case, show that the solution to (3.9) is unique.

3.1.1 Lyapunov functions

Let us present a connection between Lyapunov functions and the Lyapunov equation given in (3.4). To this end, consider the nonlinear system given by $\dot{x} = f(x)$ where f is a continuous function mapping $\mathbb{R}^n$ into $\mathbb{R}^n$. For simplicity of presentation in this section only, we assume that $\mathcal{X} = \mathbb{R}^n$ is the real vector space consisting of all n tuples of the form $[x_1, x_2, \cdots, x_n]^{tr}$ where x_j is in $\mathbb{R}$ for $j = 1, 2, \cdots, n$. Let V be a function from $\mathbb{R}^n$ into $\mathbb{R}$. Recall that the derivative of V at x is the linear operator from $\mathbb{R}^n$ into $\mathbb{R}$ defined by

$$DV(x)h := \frac{d}{d\epsilon}V(x + \epsilon h)|_{\epsilon=0} \qquad (h \in \mathbb{R}^n)$$

where ϵ is a scalar. We say that V is a *positive definite function* if V is a continuously differentiable function from $\mathbb{R}^n$ into $\mathbb{R}$ satisfying the following three conditions: $V(x) > 0$ for all nonzero x in $\mathbb{R}^n$ while $V(0) = 0$ and $V(x)$ approaches infinity as $\|x\|$ tends to infinity. For example, if P is a strictly positive matrix on $\mathbb{R}^n$, then the quadratic function given by $V(x) = (Px, x)$ is a positive definite function. (By strictly positive on $\mathbb{R}^n$ we mean that P is a real valued self-adjoint matrix on $\mathbb{R}^n$ and $(Px, x) > 0$ for all nonzero x in $\mathbb{R}^n$.) A function V is a *Lyapunov function* for $\dot{x} = f(x)$ if V is a positive definite function and

$$DV(x)f(x) < 0 \qquad \text{(for all } x \neq 0).$$

Notice that $\dot{V}(x(t)) = DV(x(t))f(x(t))$. So, if V is a Lyapunov function, then $\dot{V}(x(t)) < 0$ for all nonzero $x(t)$, that is, $V(x(t))$ is a decreasing function of t. Using this along with the properties of Lyapunov functions, one can prove the following well known result; see Khalil [74] and Vidyasagar [124].

Theorem 3.1.5 *Consider the system $\dot{x} = f(x)$ where f is a continuous function on $\mathbb{R}^n$. Assume that there exists a Lyapunov function V for $\dot{x} = f(x)$. Then for every initial condition the differential equation $\dot{x} = f(x)$ has a solution. Moreover, every solution x can be extended to the interval $[0, \infty)$ and satisfies*

$$\lim_{t \to \infty} x(t) = 0.$$

Lyapunov functions play an important role in the stability analysis and control design of nonlinear systems; see Khalil [74] and Vidyasagar [124]. In general it can be difficult to construct a Lyapunov function for an arbitrary nonlinear system.

Constructing Lyapunov functions for a stable linear system $\dot{x} = Ax$ is quite simple. To see this, assume that A is a stable operator on $\mathcal{X}$. Let Ω be any strictly positive matrix on $\mathbb{R}^n$ and P the solution to the following Lyapunov equation

$$PA + A^*P + \Omega = 0. \tag{3.10}$$

Theorem 3.1.3 shows that P is strictly positive. Thus, $V(x) = (Px, x)$ is a positive definite function. We claim that this $V(x)$ is a Lyapunov function for the differential equation $\dot{x} = Ax$. To verify this simply notice that $DV(x)h = (Ph, x) + (Px, h)$ where h is in $\mathbb{R}^n$. So, using $f(x) = Ax$ along with (3.10), we obtain

$$DV(x)Ax = (PAx, x) + (Px, Ax) = -(\Omega x, x).$$

Clearly, $-(\Omega x, x) < 0$ for all nonzero x. Therefore, any stable linear system $\dot{x} = Ax$ has a Lyapunov function of the form $V(x) = (Px, x)$. Furthermore, the above analysis shows that each strictly positive operator Ω uniquely determines a quadratic Lyapunov function $V(x) = (Px, x)$ where P is this unique solution to the Lyapunov equation in (3.10).

Exercise 3 Show that $V(x) = x^2$ is a Lyapunov function for $\dot{x} = -x^3$. Compute the general solution to this nonlinear differential equation to verify that $x(t)$ approaches 0 as t approaches infinity for all initial conditions.

3.2   Lyapunov functions and related bounds

In this section we use the Lyapunov equation to establish some bounds on a linear system.

3.2.1   Bounds on e^{At}

In this section we will use the Lyapunov equation in (3.10) to compute some bounds on $\|e^{At}\|$ in terms of the operators P and Ω. This analysis begins with the following fundamental result.

Lemma 3.2.1 *Let A be an operator on a finite dimensional space $\mathcal{X}$ satisfying*

$$A + A^* \leq -2\alpha I \tag{3.11}$$

where $\alpha \geq 0$. Then $\|e^{At}\| \leq e^{-\alpha t}$ for all $t \geq 0$.

PROOF. First assume that A is marginally dissipative, that is, $A + A^* \leq 0$. Now let g be the function defined by $g(t) = \|e^{At}x\|^2$ where x is in $\mathcal{X}$. Then

$$\dot{g}(t) = \frac{d}{dt}(e^{At}x, e^{At}x) = (Ae^{At}x, e^{At}x) + (e^{At}x, Ae^{At}x) = 2\Re(Ae^{At}x, e^{At}x) \leq 0\,.$$

Because $\dot{g}(t) \leq 0$, it follows that g is nonincreasing and its maximum is obtained at zero. Thus,

$$\|e^{At}x\|^2 = g(t) \leq g(0) = \|x\|^2\,.$$

Hence, e^{At} is contractive for all $t \geq 0$, that is, $\|e^{At}\| \leq 1$ for all $t \geq 0$. So, if A is marginally dissipative, then e^{At} is contractive for all $t \geq 0$. However, if $A + A^* \leq -2\alpha I$, then $A + \alpha I$ is marginally dissipative. Therefore, $\|e^{(A+\alpha I)t}\| \leq 1$, or equivalently, $\|e^{At}\| \leq e^{-\alpha t}$ for all $t \geq 0$. This completes the proof. ∎

If A is dissipative, then $A + A^* \leq -2\alpha I$ for some $\alpha > 0$. In this case, the above lemma shows that the solution to the differential equation $\dot{x} = Ax$ with $x(0) = x_0$ satisfies the bound $\|x(t)\| \leq e^{-\alpha t}\|x_0\|$ for all $t \geq 0$. This bound motivated the definition of dissipative. If R is an operator on $\mathcal{X}$, then $\lambda_{max}(R)$ is the largest real eigenvalue of R while $\lambda_{min}(R)$ is the smallest real eigenvalue of R. This sets the stage for the following result which is a generalization of the previous lemma.

Theorem 3.2.2 *Let A be a stable operator on a finite dimensional space $\mathcal{X}$ and Ω a strictly positive operator on $\mathcal{X}$. Let P be the unique solution for the Lyapunov equation (3.10). Then all the eigenvalues of $P^{-1}\Omega$ are nonzero positive real numbers. Moreover, if $\alpha = \lambda_{min}(P^{-1}\Omega)/2$, then*

$$\|e^{At}\| \leq \left(\frac{\lambda_{max}(P)}{\lambda_{min}(P)}\right)^{1/2} e^{-\alpha t}\,. \tag{3.12}$$

PROOF. By multiplying both sides of the Lyapunov equation (3.10) by $P^{-1/2}$, we obtain

$$P^{-1/2}A^*P^{1/2} + P^{1/2}AP^{-1/2} = -P^{-1/2}\Omega P^{-1/2} \leq -2\alpha I \tag{3.13}$$

where 2α is the smallest eigenvalue of $P^{-1/2}\Omega P^{-1/2}$. So, if $F = P^{1/2}AP^{-1/2}$, then F is similar to A and $F + F^* \leq -2\alpha I$. According to the previous lemma, we have $\|e^{Ft}\| \leq e^{-\alpha t}$. Thus,

$$\begin{aligned}\|e^{At}\| &= \|P^{-1/2}e^{Ft}P^{1/2}\| \leq \|P^{-1/2}\|\,\|e^{Ft}\|\,\|P^{1/2}\| \\ &\leq \|P^{-1/2}\|\,\|P^{1/2}\|e^{-\alpha t}\,.\end{aligned}$$

Since $\|P^{1/2}\|^2 = \lambda_{max}(P)$ and $\|P^{-1/2}\|^2 = 1/\lambda_{min}(P)$, we readily obtain the bound in (3.12).

To complete the proof it remains to show that $2\alpha = \lambda_{min}(P^{-1/2}\Omega P^{-1/2})$ also equals $\lambda_{min}(P^{-1}\Omega)$. First notice that $P^{-1/2}\Omega P^{-1/2}$ is similar to $P^{-1}\Omega$. This follows from

$$P^{1/2}(P^{-1}\Omega)P^{-1/2} = P^{-1/2}\Omega P^{-1/2}\,.$$

So, $P^{-1}\Omega$ and $P^{-1/2}\Omega P^{-1/2}$ have the same eigenvalues. In particular, all the eigenvalues of $P^{-1}\Omega$ are nonzero positive real numbers. Furthermore, $\lambda_{min}(P^{-1}\Omega)$ is the smallest eigenvalue of $P^{-1/2}\Omega P^{-1/2}$. Hence, $2\alpha = \lambda_{min}(P^{-1}\Omega)$. This completes the proof. ∎

To obtain some additional insight into Lyapunov functions, let A be a stable operator on $\mathcal{X}$ and Ω a strictly positive operator on $\mathcal{X}$. Let P be the solution to the Lyapunov equation in (3.10). Recall that P is strictly positive. Now let $\mathcal{X}_P$ be the inner product space consisting of all vectors in $\mathcal{X}$ and determined by the inner product $(x, y)_P = (Px, y)$ where x and y are in $\mathcal{X}$. Because P is strictly positive, it follows that $\mathcal{X}_P$ is a well defined Hilbert space. Let U be the "identity" operator mapping $\mathcal{X}$ into $\mathcal{X}_P$ defined by $Ux = x$. Obviously, U is invertible. We claim that

$$\|U\|^2 = \lambda_{\max}(P) \quad \text{and} \quad \|U^{-1}\|^2 = 1/\lambda_{\min}(P) \ . \tag{3.14}$$

The first equality follows from

$$
\begin{aligned}
\|U\|^2 &= \sup\{\|Ux\|_P^2 : \|x\| \leq 1\} \\
&= \sup\{(Px, \ x) : \|x\| \leq 1\} = \lambda_{\max}(P) \ .
\end{aligned}
$$

To verify the second equality notice that

$$
\begin{aligned}
\|U^{-1}\|^2 &= \sup\{\|U^{-1}x\|^2 : \|x\|_P \leq 1\} = \sup\{(x, x) : (Px, x) \leq 1\} \\
&= \sup\{(P^{-1}y, y) : \|y\|^2 \leq 1\} = \lambda_{\max}(P^{-1}) = 1/\lambda_{\min}(P) \ .
\end{aligned}
$$

The third equality was obtained by replacing x by $P^{-1/2}y$. This completes the proof of (3.14).

Now let A_0 be the operator on $\mathcal{X}_P$ defined by $A_0 x = Ax$. Clearly, $A_0 U = UA$ and thus A_0 is similar to A. In particular, $e^{A_0 t} = U e^{At} U^{-1}$. Using the Lyapunov equation in (3.10), we obtain

$$2\Re(A_0 x, x)_P = (PAx, x) + (Px, Ax) = -(\Omega x, x) = -(P^{-1}\Omega x, x)_P \ . \tag{3.15}$$

Since $(P^{-1}\Omega x, x)_P = (\Omega x, x) > 0$ for all nonzero x, it follows that $P^{-1}\Omega$ is a strictly positive operator on the space $\mathcal{X}_P$. In particular, all the eigenvalues of $P^{-1}\Omega$ are nonzero and positive. Using $\alpha = \lambda_{\min}(P^{-1}\Omega)/2$ in (3.15), we obtain

$$A_0 + A_0^* \leq -2\alpha I \ . \tag{3.16}$$

Therefore, A_0 is dissipative. According to Lemma 3.2.1, we have $\|e^{A_0 t}\| \leq e^{-\alpha t}$. By consulting (3.14), we obtain

$$\|e^{At}\| = \|U^{-1} e^{A_0 t} U\| \leq \|U^{-1}\| \ \|U\| \ \|e^{A_0 t}\| = \left(\frac{\lambda_{\max}(P)}{\lambda_{\min}(P)}\right)^{1/2} e^{-\alpha t} \ .$$

This yields another proof of Theorem 3.2.2. Finally, this analysis also shows that any stable operator is similar to a dissipative operator.

3.2.2   Some system bounds

Let A be an operator on a finite dimensional space $\mathcal{X}$ and let $\Gamma_{\max}(A)$ be the real number defined by

$$\Gamma_{\max}(A) = -\max\{\Re(\lambda) : \lambda \text{ is an eigenvalue of } A\} \ .$$

Now consider $\alpha < \Gamma_{\max}(A)$. Then all the eigenvalues of $A + \alpha I$ have nonzero negative real part and thus $A + \alpha I$ is stable. So, if Ω is any positive operator on $\mathcal{X}$, then there exists a unique positive solution P to the Lyapunov equation $P(A + \alpha I) + (A + \alpha I)^* P + \Omega = 0$, or equivalently,

$$PA + A^*P + 2\alpha P + \Omega = 0. \tag{3.17}$$

In fact, according to Lemma 3.1.2, the solution P is given by

$$P = \int_0^\infty e^{(A^* + \alpha I)t} \Omega e^{(A + \alpha I)t} \, dt. \tag{3.18}$$

In particular, if A is stable and α is chosen such that $0 < \alpha < \Gamma_{\max}(A)$, then there is a positive solution to the Lyapunov equation (3.17). This Lyapunov equation is used in the following theorem. To present this theorem recall that if G is any function in $L^1(\mathcal{U}, \mathcal{Y})$, then $d_1(G)$ is defined by

$$d_1(G) := \sup \{ \|Gu\|_1 \; : \; u \in \mathcal{U} \text{ and } \|u\| \leq 1 \}.$$

Throughout it is assumed that both $\mathcal{U}$ and $\mathcal{Y}$ are finite dimensional spaces. (If $\mathcal{U} = \mathbb{C}^1$, then $d_1(G) = \|G\|_1$.) Finally, recall that if G is the impulse response for a finite dimensional stable linear system, then G is in $L^1(\mathcal{U}, \mathcal{Y})$ and thus $d_1(G)$ is finite.

Theorem 3.2.3 *Let G be the impulse response for a stable linear system $\{A, B, C, 0\}$ and $\mathbf{G}$ be its transfer function. Consider $0 < \alpha < \Gamma_{\max}(A)$ and let P be the unique solution to the Lyapunov equation*

$$PA + A^*P + 2\alpha P + (2\alpha)^{-1} C^* C = 0. \tag{3.19}$$

Then we have the following bounds:

$$\|\mathbf{G}\|_\infty \leq d_1(G) \leq \lambda_{\max}(B^*PB)^{1/2}. \tag{3.20}$$

In particular, if $\mathcal{U}$ is one dimensional, then

$$\|\mathbf{G}\|_\infty \leq \|G\|_1 \leq (B^*PB)^{1/2}.$$

PROOF. According to Theorem 2.4.4, we have $\|\mathbf{G}\|_\infty \leq d_1(G)$. This readily proves the first inequality in (3.20). To obtain the second inequality, first notice that because $A + \alpha I$ is stable, P is uniquely determined by (3.18) where $\Omega = (2\alpha)^{-1} C^* C$, that is,

$$P = \frac{1}{2\alpha} \int_0^\infty e^{(A^* + \alpha I)t} C^* C e^{(A + \alpha I)t} \, dt.$$

Let u be any unit vector in $\mathcal{U}$. Then using the previous expression for P along with $G(t) = Ce^{At}B$ and the Cauchy-Schwartz inequality, we have

$$
\begin{aligned}
\|Gu\|_1^2 &= \left(\int_0^\infty \|Ce^{(A+\alpha I)t}Bu\| e^{-\alpha t} \, dt \right)^2 \\
&\leq \int_0^\infty \|Ce^{(A+\alpha I)t}Bu\|^2 \, dt \cdot \int_0^\infty e^{-2\alpha t} \, dt \\
&= \frac{1}{2\alpha} \int_0^\infty \left(B^* e^{(A^*+\alpha I)t} C^* C e^{(A+\alpha I)t} Bu, u \right) dt \\
&= (B^*PBu, u) \;\; \leq \;\; \lambda_{\max}(B^*PB).
\end{aligned}
$$

Therefore, $d_1(G)^2 \leq \lambda_{\max}(B^*PB)$. This completes the proof.   ■

Notice that the above result can also be proven by applying Theorem 2.4.4 with $\Pi = 2\alpha B^*PB$.

Exercise 4 Let A be an operator on a finite dimensional vector space $\mathcal{X}$ and Ω a strictly positive operator on $\mathcal{X}$. Consider the Lyapunov equation

$$PA + A^*P + 2\alpha P + \Omega = 0. \tag{3.21}$$

Then show that

$$\Gamma_{\max}(A) = \sup\{\alpha : \text{there exists a } P > 0 \text{ solving (3.21)}\}.$$

Exercise 5 Let G be the impulse response for a stable system $\{A \text{ on } \mathcal{X}, B, C, 0\}$ and let **G** be its transfer function. Suppose that Ω is a strictly positive operator on $\mathcal{X}$ and P is the unique solution to the Lyapunov equation (3.10). For $\alpha = \lambda_{\min}(P^{-1}\Omega)/2$, show that

$$\|\mathbf{G}\|_\infty \leq \|G\|_1 \leq \alpha^{-1}\|B\|\|C\|\sqrt{\lambda_{\max}(P)/\lambda_{\min}(P)}.$$

3.3   Lyapunov functions for mechanical systems

In this section we use Lyapunov techniques to obtain another proof of the stability result for mechanical systems presented in Theorem 2.2.2. As before, we illustrate the results with the simple model of a two story structure considered in Section 2.2. Recall that this system is described by the following set of differential equations

$$
\begin{array}{rcrcrcrcrcl}
m_1\ddot{y}_1 & + & (c_1+c_2)\dot{y}_1 & - & c_2\dot{y}_2 & + & (k_1+k_2)y_1 & - & k_2y_2 & = & 0 \\
m_2\ddot{y}_2 & - & c_2\dot{y}_1 & + & c_2\dot{y}_2 & - & k_2y_1 & + & k_2y_2 & = & 0.
\end{array}
$$

As before, m_1, m_2, c_1, c_2, k_1 and k_2 are all nonzero positive scalars. Recall that if $q = [y_1, y_2]^{tr}$, then this system can be described by the following second order vector differential equation:

$$M\ddot{q} + C\dot{q} + Kq = 0 \tag{3.22}$$

where the strictly positive operators M, C and K are given by

$$M = \begin{bmatrix} m_1 & 0 \\ 0 & m_2 \end{bmatrix}, \quad C = \begin{bmatrix} c_1+c_2 & -c_2 \\ -c_2 & c_2 \end{bmatrix} \text{ and } K = \begin{bmatrix} k_1+k_2 & -k_2 \\ -k_2 & k_2 \end{bmatrix}.$$

The kinetic energy for this system is given by

$$\frac{m_1\dot{y}_1^2 + m_2\dot{y}_2^2}{2} = (M\dot{q}, \dot{q})/2$$

and the potential energy is given by

$$\frac{k_1y_1^2 + k_2(y_2-y_1)^2}{2} = (Kq, q)/2.$$

Consider now a general mechanical system described by (3.22) where $q(t)$ is a vector which describes the configuration of the system. As before, assume that M, K, and C are strictly positive. The *kinetic energy* of the system is given by $(M\dot{q}, \dot{q})/2$, and the *potential energy* of the system is given by $(Kq, q)/2$. The total system energy is

$$(Kq, q)/2 + (M\dot{q}, \dot{q})/2.$$

Recall that, if we define the state vector $x = [q, \dot{q}]^{tr}$, then the mechanical system in (3.22) has a state space description of the form $\dot{x} = Ax$ where

$$A = \begin{bmatrix} 0 & I \\ -M^{-1}K & -M^{-1}C \end{bmatrix}. \tag{3.23}$$

We now look for a Lyapunov operator P which guarantees the stability of A. For a candidate consider the operator associated with the total energy, that is,

$$P = \frac{1}{2} \begin{bmatrix} K & 0 \\ 0 & M \end{bmatrix}. \tag{3.24}$$

Clearly, P is strictly positive. Moreover, a simple calculation shows that

$$PA + A^*P + \Omega = 0 \tag{3.25}$$

where

$$\Omega = \begin{bmatrix} 0 & 0 \\ 0 & C \end{bmatrix}.$$

Since Ω is positive and singular, we cannot infer that $\dot{x} = Ax$ is stable with our current Lyapunov results. However, later Lyapunov results will show that this P is sufficient to establish stability.

To obtain a stability result with our existing Lyapunov theory, consider the following candidate Lyapunov operator

$$P = \frac{1}{2} \begin{bmatrix} K + \epsilon C & \epsilon M \\ \epsilon M & M \end{bmatrix} = \frac{1}{2} \begin{bmatrix} K & 0 \\ 0 & M \end{bmatrix} + \frac{\epsilon}{2} \begin{bmatrix} C & M \\ M & 0 \end{bmatrix}.$$

For sufficiently small $\epsilon > 0$, this P is strictly positive. A simple calculation verifies that the Lyapunov equation in (3.25) holds where Ω is now given by

$$\Omega = \begin{bmatrix} \epsilon K & 0 \\ 0 & C - \epsilon M \end{bmatrix}.$$

For sufficiently small $\epsilon > 0$, the operator $C - \epsilon M$ is strictly positive, and thus, Ω is also strictly positive. Hence, $\dot{x} = Ax$ is stable. So, if the operators M, C, and K are all strictly positive, then the corresponding mechanical system is stable. This yields a Lyapunov based proof of Theorem 2.2.2.

3.4 Notes

The results in this section are classical. For further results on Lyapunov functions and their applications in control systems see Khalil [74] and Vidyasagar [124]. For some further results on mechanical systems see Meirovitch [89, 90] and Skelton [114].

Chapter 4

Observability

In this chapter we study the concept of observability for linear systems. The classical tests for observability are given. The connections between observability, stability and Lyapunov equations are presented. Finally, some of the proofs in this chapter use the Projection Theorem and classical least squares results from Hilbert space. These results are reviewed in the appendix on least squares.

4.1 Observability

This section is devoted to observability of linear systems. Consider the linear system

$$\dot{x} = Ax + Bu \qquad \text{and} \qquad y = Cx + Du \qquad (4.1)$$

where A is an operator on a finite dimensional space $\mathcal{X}$ and B maps $\mathcal{U}$ into $\mathcal{X}$, while the operator C maps $\mathcal{X}$ into $\mathcal{Y}$ and D maps $\mathcal{U}$ into $\mathcal{Y}$. This system is said to be *observable* over an interval $[0, t_1]$ (with $t_1 > 0$), if given the input $u(t)$ and output $y(t)$ over this interval, one can uniquely determine the state trajectory $x(t)$ on this interval. Clearly, the state $x(t)$ over $[0, t_1]$ and the initial state $x(0)$ uniquely determine each other. Therefore, the system $\{A, B, C, D\}$ is observable over $[0, t_1]$ if and only if given the input $u(t)$ and the output $y(t)$ over $[0, t_1]$ one can uniquely determine the initial state $x(0)$.

Recall that the solutions to the differential equation in (4.1) satisfy

$$Ce^{At}x(0) = g(t) := y(t) - \int_0^t Ce^{A(t-\tau)}Bu(\tau)\, d\tau - Du(t). \qquad (4.2)$$

Since both $u(t)$ and $y(t)$ are known, equation (4.2) shows that the system $\{A, B, C, D\}$ is observable over $[0, t_1]$ if and only if given the function $g(t)$ over $[0, t_1]$ one can uniquely determine the initial state $x(0)$ from $g(t) = Ce^{At}x(0)$. Therefore, the observability of the system $\{A, B, C, D\}$ depends only on the pair $\{C, A\}$ and is independent of the operators B, D and the input u. Motivated by this, we say that the pair $\{C, A\}$ is observable over $[0, t_1]$ if the system $\dot{x} = Ax$ and $y = Cx$ is observable over $[0, t_1]$, that is, given the output $y(t) = Cx(t)$ over the interval $[0, t_1]$ one can uniquely determine the initial state $x(0)$. Obviously, the system $\{A, B, C, D\}$ is observable if and only if the pair $\{C, A\}$ is observable.

41

Let T be the operator from $\mathcal{X}$ into $L^2([0, t_1], \mathcal{Y})$ defined by

$$(Tx)(t) = Ce^{At}x \qquad (x \in \mathcal{X}). \tag{4.3}$$

Notice that the pair $\{C, A\}$ is observable over $[0, t_1]$ if and only if the operator T is one to one, or equivalently, $\ker T = \{0\}$. (The kernel or null space of an operator is denoted by ker.) Throughout, we call $\mathcal{X}_{\bar{o}} := \ker T$ the *unobservable subspace* for the pair $\{C, A\}$ or system (4.1). We claim that $\mathcal{X}_{\bar{o}}$ is an invariant subspace for A, that is, $A\mathcal{X}_{\bar{o}} \subset \mathcal{X}_{\bar{o}}$. To see this, assume that $x \in \mathcal{X}_{\bar{o}}$, that is, $Tx = 0$. Then differentiating $Ce^{At}x = 0$ yields $0 = Ce^{At}Ax = (TAx)(t)$. Hence, Ax is also in $\mathcal{X}_{\bar{o}}$ and thus $\mathcal{X}_{\bar{o}}$ is invariant for A.

The *observability matrix* associated with $\{C, A\}$ is the block matrix defined by

$$W_o = \begin{bmatrix} C \\ CA \\ \vdots \\ CA^{n-1} \end{bmatrix} \tag{4.4}$$

where n is the dimension of $\mathcal{X}$. The following result shows that W_o and T have the same kernel, and hence, the observability of $\{C, A\}$ is independent of the interval.

Lemma 4.1.1 *Let T be the operator from a n-dimensional space $\mathcal{X}$ into $L^2([0, t_1], \mathcal{Y})$ defined by (4.3) where A is an operator on $\mathcal{X}$ and C maps $\mathcal{X}$ into $\mathcal{Y}$. Then*

$$\mathcal{X}_{\bar{o}} := \ker T = \bigcap \left\{ \ker CA^k : k = 0, 1, 2, \cdots \right\} = \ker W_o. \tag{4.5}$$

Moreover, $\mathcal{X}_{\bar{o}}$ is an invariant subspace for A.

PROOF. Notice that x is in the kernel of T if and only if $Ce^{At}x = 0$ for $0 \leq t \leq t_1$. By using the power series expansion for e^{At}, we see that x is in the kernel of T if and only if

$$\sum_{k=0}^{\infty} \frac{CA^k x}{k!} t^k = 0 \qquad (\text{for } 0 \leq t \leq t_1). \tag{4.6}$$

Because a power series is zero if and only if the coefficients of t^k are all zero, it follows that x is in the kernel of T if and only if $CA^k x = 0$ for all $k \geq 0$. This proves the second equality in (4.5). By the Cayley-Hamilton Theorem, A^k can be expressed as a linear combination of $\{I, A, \cdots, A^{(n-1)}\}$ for each integer $k \geq 0$. In particular,

$$\ker W_o = \left\{ x \in \mathcal{X} : CA^k x = 0 \text{ for all } k \geq 0 \right\}.$$

Hence, x is in the kernel of T if and only if x is in the kernel of the observability matrix W_o. Therefore, T and W_o have the same kernel. We have already shown that $\mathcal{X}_{\bar{o}}$ is an invariant subspace for A.   ∎

Clearly, the kernel of W_o is independent of the interval $[0, t_1]$. According to the previous lemma $\ker T = \ker W_o$, and hence, $\ker T$ is also independent of the interval $[0, t_1]$. Since the pair $\{C, A\}$ is observable over $[0, t_1]$ if and only if $\ker T = \{0\}$, we see that observability is independent of the time interval $[0, t_1]$. So, from now on we drop the interval $[0, t_1]$ when referring to observability of the pair $\{C, A\}$ or the system $\{A, B, C, D\}$.

Throughout $\mathcal{X}_o$ denotes the orthogonal complement of the unobservable subspace $\mathcal{X}_{\bar{o}}$, that is, $\mathcal{X}_o = \mathcal{X} \ominus \mathcal{X}_{\bar{o}}$. The subspace $\mathcal{X}_o$ is called the *observable subspace* associated with the pair $\{C, A\}$. Since $\mathcal{X}_{\bar{o}} = \ker T = \ker W_o$, we have

$$\mathcal{X}_o = (\ker T)^\perp = (\ker W_o)^\perp = \operatorname{ran} W_o^* ; \tag{4.7}$$

see Lemma 16.2.1 in the Appendix. It follows that the pair $\{C, A\}$ is observable if and only if ran $W_o^* = \mathcal{X}$. Using the Cayley-Hamilton Theorem, we can also obtain the following characterization of the observable subspace:

$$\mathcal{X}_o = \operatorname{ran} W_o^* = \bigvee_{k=0}^{n-1} A^{*k} C^* \mathcal{Y} = \bigvee_{k=0}^{\infty} A^{*k} C^* \mathcal{Y}. \tag{4.8}$$

Recall that $\bigvee$ denotes the linear span. Summing up this analysis yields the following classical result.

Theorem 4.1.2 *Suppose that A is an operator on a n-dimensional space $\mathcal{X}$ and C is an operator mapping $\mathcal{X}$ into $\mathcal{Y}$. Let W_o be the observability matrix defined in (4.4). Then the following statements are equivalent.*

(i) The pair $\{C, A\}$ is observable.

(ii) The operator W_o is one to one, that is, $\ker W_o = \{0\}$.

(iii) The operator W_o^* is onto, that is, $\operatorname{ran} W_o^* = \mathcal{X}$.

(iv) The rank of W_o is n.

Corollary 4.1.3 *Let A be an operator on a n-dimensional space $\mathcal{X}$ and C an operator mapping $\mathcal{X}$ into $\mathbb{C}^1$. Then the pair $\{C, A\}$ is observable if and only if its observability matrix W_o is nonsingular.*

Example 4.1.1 Consider a mechanical system consisting of two small blocks of equal mass m constrained to move without friction along a horizontal line and connected together by a linear spring of coefficient $k > 0$; see Figure 4.1.1. Letting q_1 and q_2 denote the displacements

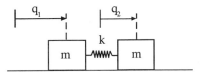

Figure 4.1: A mechanical system

of the blocks from an equilibrium configuration, an application of Newton's second law yields

$$\begin{aligned} m\ddot{q}_1 &= k(q_2 - q_1) \\ m\ddot{q}_2 &= -k(q_2 - q_1). \end{aligned}$$

Consider the state variables $x_1 = q_1, x_2 = q_2, x_3 = \dot{q}_1$ and $x_4 = \dot{q}_2$. Then this system has the state space description $\dot{x} = Ax$ where

$$A = \begin{bmatrix} 0 & 0 & 1 & 0 \\ 0 & 0 & 0 & 1 \\ -k/m & k/m & 0 & 0 \\ k/m & -k/m & 0 & 0 \end{bmatrix}.$$

We will consider here two options for the measured output.

First consider $y = \frac{1}{2}(q_1 + q_2)$ which represents the displacement of the mass center from its reference equilibrium position. Then $y = Cx$ where

$$C = \begin{bmatrix} \frac{1}{2} & \frac{1}{2} & 0 & 0 \end{bmatrix}.$$

A simple calculation reveals that rank $W_o = 2$. So, the pair $\{C, A\}$ is not observable. Physically this makes sense for the following reason. Since the sum of external forces in the horizontal direction is zero, the acceleration of the mass center is always zero. Hence, if the displacement and velocity of the mass center are initially zero, then the displacement of the mass center is zero for all time. However, this does not mean that q_1 and q_2 are zero. The two masses can oscillate about the mass center.

For a second choice of measured output consider $y = q_1$. In this case

$$C = \begin{bmatrix} 1 & 0 & 0 & 0 \end{bmatrix}.$$

A simple calculation reveals that rank $W_o = 4$. Hence, the pair $\{C, A\}$ is observable.

Exercise 6 Recall that the equation of motion for the damped linear oscillator is given by

$$m\ddot{q} + c\dot{q} + kq = u$$

where the mass $m > 0$ while the damping c and spring constant k are non-negative. If $x_1 = q$ and $\dot{x}_1 = \dot{q}$, then a state space description for this system is given by

$$\begin{aligned} \dot{x}_1 &= x_2 \\ \dot{x}_2 &= -(k/m)\, x_1 - (c/m)\, x_2 + u/m. \end{aligned}$$

Show that the following holds.

(a) Position measurement: If $y = x_1$, then we have observability.

(b) Velocity measurement: If $y = x_2$, then we have observability if and only if $k \neq 0$.

(c) Acceleration measurement: If $y = \dot{x}_2$, then we have observability if and only if $k \neq 0$.

4.2 Unobservable eigenvalues and the PBH test

As before, let A be an operator on a n-dimensional space $\mathcal{X}$ and C an operator from $\mathcal{X}$ to $\mathcal{Y}$ and W_o the observability matrix defined in (4.4). Recall that the unobservable subspace $\mathcal{X}_{\bar{o}} = \ker W_o$ is an invariant subspace for A, that is, $A\mathcal{X}_{\bar{o}} \subset \mathcal{X}_{\bar{o}}$. We say that λ is an *unobservable eigenvalue* and v is an *unobservable eigenvector* for the pair $\{C, A\}$ if v is a nonzero vector in $\mathcal{X}_{\bar{o}}$ and $Av = \lambda v$. So, the set of all unobservable eigenvalues and eigenvectors is precisely the set of all eigenvalues and eigenvectors for the operator $A_{\bar{o}}$ on $\mathcal{X}_{\bar{o}}$ defined by $A_{\bar{o}}f = Af$ for $f \in \mathcal{X}_{\bar{o}}$. Since $\{C, A\}$ is observable if and only if $\mathcal{X}_{\bar{o}} = \{0\}$, it follows that $\{C, A\}$ is observable if and only if $\{C, A\}$ has no unobservable eigenvalues. Notice that if $\{v, \lambda\}$ is an unobservable eigenvector eigenvalue pair of $\{C, A\}$, then $Ce^{At}v = Ce^{\lambda t}v = 0$. In particular, $x(t) = e^{\lambda t}v$ is a solution of

$$\dot{x} = Ax \text{ and } y = Cx$$

where $y(t) = 0$ for all t. Clearly, one cannot distinguish this output y from the trivial solution for y where the initial condition is $x(0) = 0$. The following result known as a Popov-Belevitch-Hautus (PBH) Lemma provides a useful characterization of unobservable eigenvalues.

Lemma 4.2.1 (PBH Observability Lemma.) *A complex number λ is an unobservable eigenvalue of the pair $\{C, A \text{ on } \mathcal{X}\}$ with eigenvector v if and only if v is a nonzero vector in the kernel of*

$$\Gamma_\lambda = \begin{bmatrix} A - \lambda I \\ C \end{bmatrix}.$$

In particular, the pair $\{C, A\}$ is observable if and only if $\ker \Gamma_\lambda = \{0\}$ for all complex numbers λ.

PROOF. Assume that v is a nonzero vector in the kernel of Γ_λ. Then

$$Av = \lambda v \quad \text{and} \quad Cv = 0.$$

Clearly, λ is an eigenvalue of A with eigenvector v. It now follows that $CA^kv = \lambda^kCv = 0$ for all positive integers k; hence, v is in $\ker W_o$. Therefore, λ is an unobservable eigenvalue of the pair $\{C, A\}$ with eigenvector v.

Now assume that λ is an unobservable eigenvalue of the pair $\{C, A\}$ with eigenvector v. Hence, $Av = \lambda v$ and $Cv = 0$. From this it readily follows that v is a nonzero vector in the kernel of Γ_λ. This completes the proof. ∎

Recall that $\{C, A \text{ on } \mathcal{X}\}$ is similar to another pair $\{\tilde{C}, \tilde{A} \text{ on } \tilde{\mathcal{X}}\}$ if there exists an invertible operator R from $\tilde{\mathcal{X}}$ onto $\mathcal{X}$ satisfying $R\tilde{A} = AR$ and $\tilde{C} = CR$. Clearly, observability is preserved under a similarity transformation. In many applications, the output y is scalar and A is similar to a diagonal matrix $\tilde{A}$. In this case, the PBH lemma can be used to obtain the following observability result.

Proposition 4.2.2 *Consider the pair of matrices*

$$
A = \begin{bmatrix} \lambda_1 & 0 & \cdots & 0 \\ 0 & \lambda_2 & & 0 \\ \vdots & & \ddots & \vdots \\ 0 & 0 & \cdots & \lambda_n \end{bmatrix} \qquad and \qquad C = \begin{bmatrix} c_1 & c_2 & \cdots & c_n \end{bmatrix} \tag{4.9}
$$

where $\{\lambda_1, \lambda_2, \cdots, \lambda_n\}$ and $\{c_1, c_2, \cdots, c_n\}$ are scalars. Then $\{C, A\}$ is observable if and only if $\{\lambda_1, \lambda_2, \cdots, \lambda_n\}$ are distinct and $c_i \neq 0$ for $i = 1, 2, \cdots, n$.

PROOF. For any scalar λ the matrix Γ_λ in the PBH Lemma is given by

$$
\Gamma_\lambda = \begin{bmatrix} \lambda_1 - \lambda & 0 & \cdots & 0 \\ 0 & \lambda_2 - \lambda & & 0 \\ \vdots & & \ddots & \vdots \\ 0 & 0 & \cdots & \lambda_n - \lambda \\ c_1 & c_2 & \cdots & c_n \end{bmatrix}.
$$

If $c_i = 0$, then with $\lambda = \lambda_i$, the i-th column of Γ_λ is zero. Hence, Γ_λ has a nontrivial kernel and by the PBH Lemma, $\{C, A\}$ is unobservable. If $\lambda_i = \lambda_j$, then with $\lambda = \lambda_i$, the i-th and j-th columns are linearly dependent. By the PBH Lemma, $\{C, A\}$ is unobservable.

On the other hand, assume that $\{\lambda_j\}_1^n$ are distinct and $c_i \neq 0$ for $i = 1, 2, \cdots, n$. Clearly, the columns of Γ_λ are linearly independent when λ is not an eigenvalue of A. Now consider $\lambda = \lambda_i$. Because $c_i \neq 0$, the i-th column of Γ_λ is linearly independent of the other $n - 1$ linearly independent columns. So, the kernel of Γ_λ is zero for all λ. According the PBH Lemma, $\{C, A\}$ is observable.

One can also prove the above result using the operator T from $\mathcal{X}$ into $L^2[0, t_1]$ defined in (4.3). Recall that the pair $\{C, A\}$ is observable if and only if the kernel of T is zero. Because $\mathcal{X} = \mathbb{C}^n$ and A is a diagonal matrix, $Ce^{At}x = \sum_{i=1}^n c_i e^{\lambda_i t} x_i$, where $x_1, x_2, \cdots, x_n$ are the components of x. Therefore, the kernel of T is zero if and only if the set $\{c_i e^{\lambda_i t} : i = 1, 2, \cdots, n\}$ is linearly independent in $L^2[0, t_1]$. Clearly, this set is linearly independent if and only if $\{\lambda_j\}_1^n$ are distinct and $c_i \neq 0$ for $i = 1, 2, \cdots, n$. ∎

Exercise 7 Let A and C be the matrices given in (4.9). Then show that the corresponding observability matrix W_o admits a factorization of the form $W_o = V^{tr}\Lambda$ where V is the Vandermonde matrix on $\mathbb{C}^n$ generated by $\{\lambda_j\}_1^n$; see (16.49), and $\Lambda = \operatorname{diag}(c_1, c_2, \cdots, c_n)$ is a diagonal matrix on $\mathbb{C}^n$. Using this result give another proof of Proposition 4.2.2.

4.3   An observability least squares problem

In this section we will use some classical results on least squares optimization to solve an observability optimization problem. These classical least squares optimization results are reviewed in the Appendix. Recall that the basic observability problem associated with the system $\{A, B, C, D\}$ in (4.1) is given a function g over an interval $[0, t_1]$, find an initial state x_0 to satisfy $g(t) = Ce^{At}x_0$. Clearly, this equation does not have a solution for an arbitrary function g. If g is given by (4.2), there is at least one solution for x_0. Furthermore, the solution is unique if and only if $\{C, A\}$ is observable. If the solution is not unique one may search for a solution of minimum norm. Moreover, in many systems and control problems the measured output g is corrupted by noise. Hence, one searches for an initial state x_0 of minimum norm such that $Ce^{At}x_0$ approximates $g(t)$ as close as possible. One mathematical formulation of this problem is the following *observability least squares optimization problem:* Given an output g in $L^2([0, t_1], \mathcal{Y})$, find an initial state $\hat{x}_0$ to solve the following optimization problem

$$\|\hat{x}_0\| = \inf \|x_0\| \text{ subject to } \|g - Ce^{At}x_0\|_{L^2}^2 = \inf \left\{ \int_0^{t_1} \|g(t) - Ce^{At}x\|^2 \, dt \, : \, x \in \mathcal{X} \right\}. \quad (4.10)$$

The *finite time observability Gramian* defined by

$$P(t) = \int_0^t e^{A^*\sigma} C^* C e^{A\sigma} \, d\sigma \quad (4.11)$$

plays a fundamental role in solving this problem. The following lemma permits one to compute $P(t)$ by solving an ordinary differential equation.

Lemma 4.3.1 *Let $P(t)$ be the finite time observability Gramian given in (4.11). Then $P(t)$ is positive for all $t \geq 0$. Moreover, $P(t)$ is the solution to the differential equation*

$$\dot{P} = A^*P + PA + C^*C \quad (4.12)$$

subject to the initial condition $P(0) = 0$.

PROOF. To verify that $P(t)$ is positive, simply observe that for all x in $\mathcal{X}$, we have

$$(P(t)x, x) = \int_0^t (e^{A^*\sigma} C^* C e^{A\sigma} x, x) \, d\sigma = \int_0^t \|Ce^{A\sigma}x\|^2 \, d\sigma \geq 0.$$

So, $P(t)$ is positive. Notice that the definition of $P(t)$ in (4.11) gives $\dot{P} = e^{A^*t} C^* C e^{At}$. Moreover,

$$\begin{aligned}
A^*P + PA &= \int_0^t \left(A^* e^{A^*\sigma} C^* C e^{A\sigma} + e^{A^*\sigma} C^* C e^{A\sigma} A \right) d\sigma = \int_0^t \frac{d}{d\sigma} e^{A^*\sigma} C^* C e^{A\sigma} \, d\sigma \\
&= e^{A^*\sigma} C^* C e^{A\sigma} \Big|_0^t = e^{A^*t} C^* C e^{At} - C^*C \\
&= \dot{P} - C^*C.
\end{aligned}$$

This yields the differential equation in (4.12). Obviously, from (4.11) the initial condition is $P(0) = 0$. ∎

Let T be a finite rank operator from $\mathcal{H}$ into $\mathcal{K}$. Then the restricted inverse T^{-r} of T is the unique operator from $\mathcal{K}$ into $\mathcal{H}$ defined by $T^{-r}z = \hat{h}$ where $\hat{h}$ is the unique element in $(\ker T)^{\perp}$ such that $T\hat{h} = P_{\mathcal{R}}z$ and $P_{\mathcal{R}}$ is the orthogonal projection onto the range of T; see Section 16.2 in the Appendix. We are now ready to present the main result of this section.

Theorem 4.3.2 *Let A be an operator on a finite dimensional vector space $\mathcal{X}$ and C an operator mapping $\mathcal{X}$ into $\mathcal{Y}$ and g a specified vector in $L^2([0, t_1], \mathcal{Y})$. Let $P(t_1)$ be the finite time observability Gramian defined in (4.11) or (4.12). Then the following holds.*

(i) *The solution to the observability optimization problem in (4.10) is unique and is given by*

$$\hat{x}_0 = P(t_1)^{-r} \int_0^{t_1} e^{A^* t} C^* g(t)\, dt\,. \tag{4.13}$$

(ii) *The pair $\{C, A\}$ is observable if and only if $P(t_1)$ is strictly positive.*

(iii) *If the pair $\{C, A\}$ is observable, then the observability optimization problem in (4.10) reduces to the optimization problem*

$$\|g - Ce^{At}\hat{x}_0\|_{L^2}^2 = \inf\left\{ \int_0^{t_1} \|g(t) - Ce^{At}x\|^2\, dt : x \in \mathcal{X} \right\}, \tag{4.14}$$

and the corresponding optimal initial state is given by

$$\hat{x}_0 = P(t_1)^{-1} \int_0^{t_1} e^{A^* t} C^* g(t)\, dt\,. \tag{4.15}$$

PROOF. As before, let T be the operator from $\mathcal{X}$ into $L^2([0, t_1], \mathcal{Y})$ defined by $Tx_0 = Ce^{At}x_0$ for $x_0 \in \mathcal{X}$. Let $P_\mathcal{R}$ be the orthogonal projection onto the range $\mathcal{R}$ of T; see Section 16.1 in the Appendix. By employing the Projection Theorem, the observability optimization problem in (4.10) is equivalent to the following minimum norm optimization problem:

$$\|\hat{x}_0\| = \inf \{\|x_0\| : x_0 \in \mathcal{X} \text{ and } Tx_0 = P_\mathcal{R}g\}\,. \tag{4.16}$$

According to Corollary 16.5.2 in Section 16.5, the solution to this optimization problem is unique and is given by

$$\hat{x}_0 = (T^*T)^{-r}T^*g\,. \tag{4.17}$$

We claim that T^* is the operator mapping $L^2([0, t_1], \mathcal{Y})$ into $\mathcal{X}$ given by

$$T^* f = \int_0^{t_1} e^{A^* t} C^* f(t)\, dt \qquad (f \in L^2([0, t_1], \mathcal{Y}))\,. \tag{4.18}$$

To verify (4.18), simply notice that for any x in $\mathcal{X}$ and f in $L^2([0, t_1], \mathcal{Y})$, we have

$$(x, T^* f)_\mathcal{X} = (Tx, f)_{L^2} = \int_0^{t_1} (Ce^{At}x, f(t))_\mathcal{Y}\, dt = (x, \int_0^{t_1} e^{A^* t} C^* f(t)\, dt)_\mathcal{X}\,. \tag{4.19}$$

(The inner product on a Hilbert space $\mathcal{H}$ is denoted by $(g, h)_\mathcal{H}$.) Therefore, the adjoint T^* of T is given by (4.18). Finally, by combining the definition of T with its adjoint in (4.18), we see that $T^*T = P(t_1)$ where $P(t_1)$ is the finite time observability Gramian given in (4.11). Using $T^*T = P(t_1)$ along with (4.17) and (4.18), we see that the solution to the observability optimization problem in (4.10) is given by (4.13). This proves Part (i).

Recall that the pair $\{C, A\}$ is observable if and only if the operator T is one to one, or equivalently, T^*T is strictly positive. Since $T^*T = P(t_1)$ it follows that the pair $\{C, A\}$ is observable if and only if $P(t_1)$ is strictly positive. Hence, Part (ii) holds.

Since the restricted inverse of $P(t_1)$ becomes the actual inverse when $P(t_1)$ is strictly positive, formula (4.15) follows when the pair $\{C, A\}$ is observable. If $\{C, A\}$ is observable, T is one to one, and thus, the optimization problem in (4.16) reduces to $\|g - T\hat{x}_0\| = \inf \|g - T\mathcal{X}\|$. So, when $\{C, A\}$ is observable, the optimization problem in (4.10) reduces to the optimization problem in (4.14). This completes the proof. ∎

Since $T^*T = P(t_1)$, it follows that $\|T\|^2$ equals the maximum eigenvalue of $P(t_1)$. The singular values of T are the square root of the eigenvalues of $P(t_1)$.

Finally, it is noted that Theorem 4.3.2 can be used to solve the original observability problem posed at the beginning of this chapter. To be specific, let $\{A, B, C, D\}$ be the linear system described in (4.1) where the input $u(t)$ and the output $y(t)$ are known over $[0, t_1]$. Then, the initial state $\hat{x}_0$ of smallest norm which corresponds to the above data is given by (4.13) where g is computed according (4.2).

Remark 4.3.1 To see why the operator $P(t)$ defined in (4.11) is called a Gramian, consider the case when $\mathcal{X} = \mathbb{C}^n$. Let $\{e_1, e_2, \cdots, e_n\}$ be the standard orthonormal basis for $\mathbb{C}^n$, that is, the i-th element of e_i is one and all the other components of e_i are zero. Let ψ_i be the vectors in $L^2([0, t_1], \mathcal{Y})$ given by $\psi_i(t) = Ce^{At}e_i$ for $i = 1, 2, \cdots, n$. Recall from equation (16.35) in Section 16.3, that by definition the ij-th element of the Gram matrix associated with the vectors $\{\psi_1, \psi_2, \cdots, \psi_n\}$ is given by $G_{ij} = (\psi_j, \psi_i)$. Thus,

$$
\begin{aligned}
G_{ij} &= (\psi_j, \psi_i) = \int_0^{t_1} (Ce^{At}e_j, Ce^{At}e_i)\, dt = \left(\int_0^{t_1} e^{A^*t}C^*Ce^{At}\, dt\, e_j, e_i \right) \\
&= (P(t_1)e_j, e_i) = P_{ij}(t_1).
\end{aligned}
$$

Hence, $P(t_1) = G$. Let T be the operator defined in (4.3) and set $x = [x_1, x_2, \cdots, x_n]^{tr}$. Then $Tx = \sum_1^n x_i\psi_i$. Clearly, T is one to one if and only if the vectors $\{\psi_i\}_1^n$ are linearly independent. Therefore, the pair $\{C, A\}$ is observable if and only if the set $\{\psi_i\}_1^n$ is linearly independent. Recall that the Gram matrix G is strictly positive if and only if $\{\psi_i\}_1^n$ is linearly independent. So, the Gram interpretation provides another way of showing that the observability of $\{C, A\}$ is equivalent to $P(t_1)$ being strictly positive.

Exercise 8 Consider the *Sobolev* space $\mathcal{H}_1$ consisting of the set of all differentiable functions with values in $\mathcal{Y}$ under the inner product

$$
(f, h) = \int_0^{t_1} (f(t), h(t))\, dt + \int_0^{t_1} (\dot{f}(t), \dot{h}(t))\, dt. \tag{4.20}
$$

As before, let A be an operator on $\mathcal{X}$, while C maps $\mathcal{X}$ into $\mathcal{Y}$. Consider the following observability optimization problem: Given a specified function g in $\mathcal{H}_1$, find the optimal initial state $\hat{x}_0$ in $\mathcal{X}$ satisfying

$$
\|\hat{x}_0\| = \inf \|x_0\| \quad \text{subject to} \quad \|g - Ce^{At}x_0\|_{\mathcal{H}_1} = \inf \{\|g - Ce^{At}x\|_{\mathcal{H}_1} : x \in \mathcal{X}\}.
$$

Show that the optimal solution $\hat{x}_0$ to this optimization problem is

$$\hat{x}_0 = (P(t_1) + A^*P(t_1)A)^{-r}\left[\int_0^{t_1} e^{A^*t}C^*g(t)\,dt + \int_0^{t_1} A^*e^{A^*t}C^*\dot{g}(t)\,dt\right]$$

where $P(t_1)$ is the finite time observability Gramian given in (4.11). Moreover, let T be the operator from $\mathcal{X}$ into $\mathcal{H}_1$ defined by $Tx_0 = Ce^{At}x_0$. Then show that T^*T equals $P(t_1) + A^*P(t_1)A$. So, the pair $\{C, A\}$ is observable if and only if $P(t_1) + A^*P(t_1)A$ is strictly positive for any $t_1 > 0$.

4.4   Stability and observability

This section is devoted to the observability of stable systems. Consider the pair $\{C, A\}$ where A is a stable operator on a finite dimensional vector space $\mathcal{X}$ and C maps $\mathcal{X}$ into $\mathcal{Y}$. Let T be the operator from $\mathcal{X}$ into $L^2([0, \infty), \mathcal{Y})$ defined by

$$(Tx)(t) = Ce^{At}x \qquad (x \in \mathcal{X}). \tag{4.21}$$

Because A is stable, Tx is in $L^2([0, \infty), \mathcal{Y})$ for all x in $\mathcal{X}$. So, T is a finite rank linear map from $\mathcal{X}$ into $L^2([0, \infty), \mathcal{Y})$. Since any finite rank linear map acting between two Hilbert spaces is bounded, T is bounded. Hence, T is a well defined operator. To directly verify that T is bounded, recall that because A is stable, $\|e^{At}\| \leq me^{-\alpha t}$ for some positive m and $\alpha > 0$. Thus, $\|Tx\| \leq m\|C\|\|e^{-\alpha t}\|_{L^2}\|x\|$. This implies that $\|T\| \leq m\|C\|/\sqrt{2\alpha}$. Clearly, the pair $\{C, A\}$ is observable if and only if the operator T is one to one, or equivalently, $\ker T = \{0\}$. Obviously, Lemma 4.1.1 holds with $[0, t_1]$ replaced with $[0, \infty)$, and thus, the unobservable space $\mathcal{X}_{\bar{o}} = \ker T = \ker W_o$.

Consider the linear system $\{A, B, C, D\}$ described by (4.1) where A is a stable operator on $\mathcal{X}$. In this case, $Ce^{At}x(0)$ is in $L^2([0, \infty), \mathcal{Y})$. So, the vector $g = Tx(0)$ defined in (4.2) is also in $L^2([0, \infty), \mathcal{Y})$. Recall that the observability problem is to determine an initial state x_0 given g, that is, solve the equation $g = Tx_0$. Moreover, this equation has a unique solution if and only if T is one to one, or equivalently, $\{C, A\}$ is observable. If the measurement of the output y is corrupted by noise, then g may not be in the range of T. In this case, it makes sense to find a vector $\hat{x}_0$ with the smallest possible norm to minimize $\|g - T\mathcal{X}\|$. Therefore, when A is stable, this naturally leads the following infinite horizon observability least squares optimization problem: Given a vector g in $L^2([0, \infty), \mathcal{Y})$, find an initial state $\hat{x}_0$ to solve the optimization problem

$$\|\hat{x}_0\| = \inf \|x_0\| \text{ subject to } \|g - Ce^{At}x_0\|_{L^2}^2 = \inf\left\{\int_0^\infty \|g(t) - Ce^{At}x\|^2\,dt : x \in \mathcal{X}\right\}. \tag{4.22}$$

The *observability Gramian* for the pair $\{C, A\}$ is the operator on $\mathcal{X}$ defined by

$$P = \int_0^\infty e^{A^*\sigma}C^*Ce^{A\sigma}\,d\sigma. \tag{4.23}$$

Notice that if $P(t)$ is the finite time observability Gramian defined in (4.11), then $P(t) \to P$ as t approaches infinity. Moreover, because A is stable, P is a positive (bounded) operator on $\mathcal{X}$. According to Lemma 3.1.2 this P is the unique solution to the following Lyapunov equation:

$$A^*P + PA + C^*C = 0.\tag{4.24}$$

The observability Gramian P plays a fundamental role in the following result.

Theorem 4.4.1 *Let A be a stable operator on a finite dimensional vector space $\mathcal{X}$ and C an operator mapping $\mathcal{X}$ into $\mathcal{Y}$ and g a specified vector in $L^2([0,\infty),\mathcal{Y})$. Let P be the observability Gramian for the pair $\{C,A\}$. Then the following holds.*

(i) The solution to the observability optimization problem in (4.22) is unique and is given by

$$\hat{x}_0 = P^{-r}\int_0^\infty e^{A^*t}C^*g(t)\,dt.\tag{4.25}$$

(ii) The pair $\{C,A\}$ is observable if and only if P is strictly positive.

(iii) If the pair $\{C,A\}$ is observable, then the observability optimization problem reduces to

$$\|g - Ce^{At}\hat{x}_0\|_{L^2}^2 = \inf\Big\{\int_0^\infty \|g(t) - Ce^{At}x\|^2\,dt : x \in \mathcal{X}\Big\},\tag{4.26}$$

and the corresponding optimal initial state is given by

$$\hat{x}_0 = P^{-1}\int_0^\infty e^{A^*t}C^*g(t)\,dt.\tag{4.27}$$

PROOF. Let $P_\mathcal{R}$ be the orthogonal projection onto the range $\mathcal{R}$ of T. By employing the Projection Theorem, the infinite horizon observability optimization problem equivalent to the following minimum norm optimization problem

$$\|\hat{x}_0\| = \inf\{\|x_0\| : x_0 \in \mathcal{X} \text{ and } Tx_0 = P_\mathcal{R}g\}.\tag{4.28}$$

According to Corollary 16.5.2 in Section 16.5, the solution to this problem is unique and is given by

$$\hat{x}_0 = (T^*T)^{-r}T^*g.\tag{4.29}$$

By replacing t_1 with ∞ in (4.19), it follows that T^* is the operator mapping $L^2([0,\infty),\mathcal{Y})$ into $\mathcal{X}$ given by

$$T^*g = \int_0^\infty e^{A^*t}C^*g(t)\,dt \qquad (g \in L^2([0,\infty),\mathcal{Y})).\tag{4.30}$$

Finally, by combining the definition of T with (4.30), we see that $T^*T = P$. Using this along with (4.29) and (4.30), we obtain that the solution to the observability optimization problem (4.22) is given by (4.25). This proves Part (i).

Recall that the pair $\{C,A\}$ is observable if and only if the operator T is one to one, or equivalently, T^*T is strictly positive. Since $T^*T = P$ it follows that the pair $\{C,A\}$ is observable if and only if P is strictly positive. This verifies Part (ii).

Because the restricted inverse of P becomes the actual inverse when P is strictly positive, formula (4.27) follows when the pair $\{C,A\}$ is observable. If $\{C,A\}$ is observable, then T is one to one, and thus, the optimization problem in (4.28) reduces to $\|g - T\hat{x}_0\| = \inf \|g - T\mathcal{X}\|$. So, when $\{C,A\}$ is observable, the optimization problem in (4.22) reduces to the optimization problem in (4.26). ∎

Remark 4.4.1 Let P be the observability Gramian for the pair $\{C, A\}$ where A is stable. Let T be the operator from $\mathcal{X}$ into $L^2([0, \infty), \mathcal{Y})$ defined in (4.21). Then $\|T\|^2$ is the largest eigenvalue of P. Moreover, the singular values of T are the square root of the eigenvalues of P. This follows from the fact that $T^*T = P$.

The following result uses the observability Lyapunov equation in (4.24) to determine the stability of an observable system.

Theorem 4.4.2 *The following statements are equivalent for an observable pair $\{C, A\}$.*

(a) The system $\dot{x} = Ax$ is stable.

(b) There exists a strictly positive operator P satisfying the Lyapunov equation

$$A^*P + PA + C^*C = 0. \tag{4.31}$$

PROOF. Part (ii) of the previous theorem shows that (a) implies (b). Now assume that (b) holds, that is, there exists a positive operator P satisfying the Lyapunov equation (4.31). Consider any eigenvalue λ of A. Let $v \neq 0$ be an eigenvector corresponding to λ, that is, $Av = \lambda v$. Then

$$-(C^*Cv, v) = ((A^*P + PA)v, v) = (Pv, Av) + (PAv, v) = (\bar{\lambda} + \lambda)(Pv, v).$$

Hence, $-\|Cv\|^2 = 2\Re(\lambda)(Pv, v)$. Since $P > 0$, we must have $(Pv, v) > 0$. This implies that $\Re(\lambda) \leq 0$. To prove stability, it remains to show that $\Re(\lambda) < 0$. If $\Re(\lambda) = 0$, then $\|Cv\|^2 = 0$, or equivalently, $Cv = 0$. It now follows that $CA^k v = \lambda^k Cv = 0$ for all integers $k \geq 0$. Since $\{C, A\}$ is observable, $v = 0$. This contradicts the fact that v is nonzero. So, we have $\Re(\lambda) \neq 0$. Therefore, A is stable. ∎

Exercise 9 Consider the mechanical system

$$M\ddot{q} + C\dot{q} + Kq = 0 \tag{4.32}$$

where M, C and K are all strictly positive operators on a finite dimensional space $\mathcal{Q}$; see Section 3.3. By using $x = [q, \dot{q}]^{tr}$, this mechanical system has a state space description of the form $\dot{x} = Ax$ where

$$A = \begin{bmatrix} 0 & I \\ -M^{-1}K & -M^{-1}C \end{bmatrix}.$$

Let P and Ω be the positive matrices defined by

$$P = \frac{1}{2}\begin{bmatrix} K & 0 \\ 0 & M \end{bmatrix} \quad \text{and} \quad \Omega = \begin{bmatrix} 0 & 0 \\ 0 & C \end{bmatrix}.$$

Recall that P corresponds to the total energy of the system. Clearly, P is strictly positive. Moreover, a simple calculation shows that

$$PA + A^*P + \Omega = 0.$$

Using this Lyapunov equation and Theorem 4.4.2 show that the mechanical system in (4.32) is stable.

4.5 Notes

All the results in this section are classical results in linear systems theory. The history of observability and linear systems is well established in Kailath [68], Rugh [110] and elsewhere. So, we will not develop a historical account of observability and linear control systems. For further results on linear systems see Brockett [21], Chen [26], DeCarlo [33], Delchamps [34], Kailath [68], Polderman-Willems [98], Rugh [110], Skelton [114] and Sontag [116]. Using operator techniques to solve observability and controllability problems is also classical; see Balakrishnan [8], Brockett [21], Fuhrmann [47], Naylor-Sell [93], Luenberger [85] and Porter [100]. The PBH test was independently developed by Popov [99], Belevitch [14] and Hautus [61].

For simplicity of presentation only we concentrated on time invariant systems. Many of the results can be easily extended to the time varying case. To see this consider the time varying system $\dot{x} = A(t)x$ and $y(t) = C(t)x(t)$ where A is a continuous function with values in $\mathcal{L}(\mathcal{X}, \mathcal{X})$ and C is a continuous function with values in $\mathcal{L}(\mathcal{X}, \mathcal{Y})$. Let T be the operator from $\mathcal{X}$ into $L^2([t_0, t_1], \mathcal{Y})$ defined by $(Tx_0)(t) = C(t)\Phi(t, t_0)x_0$ where x_0 is in $\mathcal{X}$ and $\Phi(t, \tau)$ is the state transition operator for A. Then the time varying pair $\{C, A\}$ is *observable over the interval* $[t_0, t_1]$ if the operator T is one to one. The time varying pair $\{C, A\}$ can be observable over one interval and not observable over another interval. Obviously, the time varying pair $\{C, A\}$ is observable over the interval $[t_0, t_1]$ if and only if T^* is onto, or equivalently, T^*T is strictly positive. The adjoint of T is the operator mapping $L^2([t_0, t_1], \mathcal{Y})$ into $\mathcal{X}$ given by

$$T^*f = \int_{t_0}^{t_1} \Phi(t, t_0)^*C(t)^*f(t)\,dt \qquad (f \in L^2([t_0, t_1], \mathcal{Y})). \tag{4.33}$$

So, the time varying pair $\{C, A\}$ is observable over the interval $[t_0, t_1]$ if and only if

$$P(t_0) := T^*T = \int_{t_0}^{t_1} \Phi(t, t_0)^*C(t)^*C(t)\Phi(t, t_0)\,dt$$

is a strictly positive operator on $\mathcal{X}$. Recall that $\partial\Phi(t, \tau)/\partial\tau = -\Phi(t, \tau)A(\tau)$. Using this along with Leibnitz's Rule, it follows that $P(t_0)$ can be obtained by solving the following differential equation backwards in time

$$\dot{P} + A(t)^*P + PA(t) + C(t)^*C(t) = 0 \qquad (P(t_1) = 0). \tag{4.34}$$

Therefore, the time varying pair $\{C, A\}$ is observable over the interval $[t_0, t_1]$ if and only if $P(t_0)$ is strictly positive where $P(t)$ is the solution to the differential equation in (4.34). The observability optimization problem and its solution can also be extended to the time varying setting. In this case, the observability optimization problem becomes: Let g be a specified vector in $L^2([t_0, t_1], \mathcal{Y})$ and $P_\mathcal{R}$ be the orthogonal projection onto the range of T, then find an optimal initial state $\hat{x}_0$ in $\mathcal{X}$ such that

$$\|\hat{x}_0\| = \inf\{\|x_0\| : P_\mathcal{R}g = Tx_0\}.$$

The solution to this optimization problem is unique and is given by

$$\hat{x}_0 = (T^*T)^{-r}T^*g = P(t_0)^{-r}\int_{t_0}^{t_1} \Phi(t, t_0)^*C(t)^*g(t)\,dt.$$

Chapter 5

Controllability

This chapter is devoted to the controllability of linear systems. Operator techniques are used to solve a controllability least squares optimization problem. Finally, some of the proofs in this chapter use the Projection Theorem and classical least squares results from Hilbert space. These results are reviewed in the Appendix.

5.1   Controllability

This section presents some basic controllability results. Consider the linear system

$$\dot{x} = Ax + Bu \tag{5.1}$$

where A is an operator on a finite dimensional space $\mathcal{X}$ and B maps $\mathcal{U}$ into $\mathcal{X}$. We denote this system by $\{A, B\}$. Roughly speaking, we say that this system is controllable over an interval $[t_0, t_1]$ (with $t_1 > t_0$) if its state can be "driven" from any initial state to any terminal state over the interval by the appropriate choice of control input. To be more precise, $\{A, B\}$ is *controllable* over $[t_0, t_1]$ if for every pair of states x_0, x_1 in $\mathcal{X}$, there is a control u in $L^2([t_0, t_1], \mathcal{U})$ such that the solution x of

$$\dot{x}(t) = Ax(t) + Bu(t) \qquad \text{with} \qquad x(t_0) = x_0 \tag{5.2}$$

satisfies $x(t_1) = x_1$. Since system (5.1) is time-invariant, it follows that an input u drives the state from x_0 to x_1 over the interval $[t_0, t_1]$ if and only if the corresponding "shifted" input $\tilde{u}$, given by $\tilde{u}(t) = u(t + t_0)$, drives the state from x_0 to x_1 over the interval $[0, t_1 - t_0]$. Hence, system (5.1) is controllable over $[t_0, t_1]$ if and only if it is controllable over $[0, t_1 - t_0]$. Therefore, without loss of generality, we consider $t_0 = 0$.

Recall that, with $t_0 = 0$, the solutions to the differential equation in (5.2) satisfy

$$x(t_1) - e^{At_1} x_0 = \int_0^{t_1} e^{A(t_1 - \tau)} Bu(\tau) \, d\tau . \tag{5.3}$$

Let T be the operator from $L^2([0, t_1], \mathcal{U})$ into $\mathcal{X}$ defined by

$$Tu = \int_0^{t_1} e^{A(t_1 - \tau)} Bu(\tau) \, d\tau . \tag{5.4}$$

Notice that the pair $\{A, B\}$ is controllable over $[0, t_1]$ if and only if for every pair of states x_0, x_1 in $\mathcal{X}$, there is a control u in $L^2([0, t_1], \mathcal{U})$ such that $x_1 - e^{At_1}x_0 = Tu$. Since x_1 is an arbitrary vector in $\mathcal{X}$, it follows that $\{A, B\}$ is controllable if and only if the operator T is onto, or equivalently, $\operatorname{ran} T = \mathcal{X}$. (The range of an operator is denoted by ran.)

Throughout we call $\mathcal{X}_c := \operatorname{ran} T$ the *controllable subspace* for the pair $\{A, B\}$ or system (5.1). The *controllability matrix* associated with $\{A, B\}$ is the block matrix defined by

$$W_c = \begin{bmatrix} B & AB & \cdots & A^{n-1}B \end{bmatrix} \tag{5.5}$$

where n is the dimension of $\mathcal{X}$. The following result shows that W_c and T have the same range, and hence, controllability of $\{A, B\}$ is independent of the interval.

Lemma 5.1.1 *Let A be an operator on a n-dimensional space $\mathcal{X}$ and B an operator from $\mathcal{U}$ into $\mathcal{X}$. Let T be the operator from $L^2([0, t_1], \mathcal{U})$ into $\mathcal{X}$ defined in (5.4) and W_c the controllability matrix for $\{A, B\}$ in (5.5). Then*

$$\mathcal{X}_c := \operatorname{ran} T = \operatorname{span}\{A^k B \mathcal{U} : k = 0, 1, 2, \cdots\} = \operatorname{ran} W_c. \tag{5.6}$$

Moreover, $\mathcal{X}_c$ is an invariant subspace for A.

PROOF. We claim that the adjoint T^* of T is the operator from $\mathcal{X}$ into $L^2([0, t_1], \mathcal{U})$ defined by

$$(T^*x)(t) = B^* e^{A^*(t_1 - t)} x \qquad (x \in \mathcal{X}). \tag{5.7}$$

To verify this notice that for u in $L^2([0, t_1], \mathcal{U})$ and x in $\mathcal{X}$, we have

$$\begin{aligned}
(u, T^*x)_{L^2} &= (Tu, x)_{\mathcal{X}} = \left(\int_0^{t_1} e^{A(t_1 - \tau)} Bu(\tau) \, d\tau, x \right)_{\mathcal{X}} = \int_0^{t_1} (e^{A(t_1 - t)} Bu(t), x)_{\mathcal{X}} \, dt \\
&= \int_0^{t_1} (u(t), B^* e^{A^*(t_1 - t)} x)_{\mathcal{U}} \, dt.
\end{aligned} \tag{5.8}$$

Hence, T^* is given by (5.7).

By using the formula for T^* in (5.7) and $(\operatorname{ran} T)^\perp = \ker T^*$, we see that x is in $(\operatorname{ran} T)^\perp$ if and only if $B^* e^{A^*(t_1 - t)} x = 0$ for all $0 \leq t \leq t_1$; see Lemma 16.2.1. By setting $\sigma = t_1 - t$ and using the power series expansion for $e^{A\sigma}$, the vector x is in $(\operatorname{ran} T)^\perp$ if and only if

$$0 = B^* e^{A^* \sigma} x = \sum_{k=0}^{\infty} \frac{B^* A^{*k} \sigma^k x}{k!} \qquad (\text{for all } 0 \leq \sigma \leq t_1).$$

Recall that a power series is zero if and only if all of its coefficients of σ^k are zero for all $k \geq 0$. Thus, x is in $(\operatorname{ran} T)^\perp$ if and only if $B^* A^{*k} x = 0$ for all $k \geq 0$. Since the kernel of $B^* A^{*k}$ is the orthogonal complement of the range of $A^k B$, it now follows that x is in $(\operatorname{ran} T)^\perp$ if and only if x is orthogonal to $A^k B \mathcal{U}$ for all $k \geq 0$. So, $(\operatorname{ran} T)^\perp$ is the orthogonal complement of $\bigvee_0^\infty A^k B \mathcal{U}$. In other words, we obtain

$$\operatorname{ran} T = \operatorname{span}\{A^k B \mathcal{U} : k = 0, 1, 2, \cdots\}, \tag{5.9}$$

that is, the second equality in (5.6) holds. This equation readily shows that $\mathcal{X}_c = \operatorname{ran} T$ is an invariant subspace for A. By the Cayley-Hamilton Theorem, A^k can be expressed as a linear combination of $\{I, A, \cdots, A^{(n-1)}\}$ for each integer $k \geq 0$. Therefore, the range of T equals $\operatorname{span}\{A^k B \mathcal{U} : k = 0, 1, 2, \cdots, n - 1\} = \operatorname{ran} W_c$. This yields (5.6). ∎

Remark 5.1.1 If x_0 and x_1 are any two vectors in the controllable subspace $\mathcal{X}_c$, then there always exists an input u which drives the state from x_0 to x_1 over the interval $[0, t_1]$. To see this, first notice that since $\mathcal{X}_c$ is an invariant subspace for A, the vector $e^{At}x_0$ is in $\mathcal{X}_c$ and thus $x_1 - e^{At_1}x_0$ is also in $\mathcal{X}_c$. Since $\mathcal{X}_c = \operatorname{ran} T$, there exists an input u in $L^2([0, t_1], \mathcal{U})$ such that $x_1 - e^{At_1}x_0 = Tu$. According to (5.3) and (5.4), this input drives the state from x_0 to x_1 over the interval $[0, t_1]$.

Clearly, the range of W_c is independent of the interval $[0, t_1]$. According to the previous lemma, $\operatorname{ran} T = \operatorname{ran} W_c$, and hence, the range of T is also independent of the interval $[0, t_1]$. Since the pair $\{A, B\}$ is controllable over $[0, t_1]$ if and only if $\operatorname{ran} T = \mathcal{X}$, we see that controllability is independent of the time interval $[0, t_1]$. So, from now on we drop the interval $[0, t_1]$ when referring to controllability of the pair $\{A, B\}$. Since W_c and T have the same range, it also follows that the pair $\{A, B\}$ is controllable if and only if $\operatorname{ran} W_c = \mathcal{X}$.

We let $\mathcal{X}_{\bar{c}}$ denote the orthogonal complement of the controllable subspace $\mathcal{X}_c$ and we call it the *uncontrollable subspace* associated with $\{A, B\}$. Obviously, the pair $\{A, B\}$ is controllable if and only if $\mathcal{X}_{\bar{c}}$ is zero. Since $\mathcal{X}_c = \operatorname{ran} T = \operatorname{ran} W_c$, we have

$$\mathcal{X}_{\bar{c}} = (\operatorname{ran} T)^{\perp} = (\operatorname{ran} W_c)^{\perp} = \ker W_c^* . \tag{5.10}$$

Hence, the pair $\{A, B\}$ is controllable if and only if $\ker W_c^* = \{0\}$. Using Cayley-Hamilton, we can also obtain the following characterization of the uncontrollable subspace:

$$\mathcal{X}_{\bar{c}} = \ker W_c^* = \bigcap_{k=0}^{n-1} \ker B^* A^{*k} = \bigcap_{k=0}^{\infty} \ker B^* A^{*k} . \tag{5.11}$$

Summing up this analysis yields the following classical result.

Theorem 5.1.2 *Let A be an operator on a n-dimensional space $\mathcal{X}$, while B is an operator mapping $\mathcal{U}$ into $\mathcal{X}$ and W_c is the controllability matrix in (5.5) associated with $\{A, B\}$. Then the following statements are equivalent.*

(i) The pair $\{A, B\}$ is controllable.

(ii) The operator W_c is onto, that is, $\operatorname{ran} W_c = \mathcal{X}$.

(iii) The rank of W_c is n.

Corollary 5.1.3 *Suppose that A is an operator on a finite dimensional space $\mathcal{X}$ and B is an operator mapping $\mathbb{C}^1$ into $\mathcal{X}$. Then the pair $\{A, B\}$ is controllable if and only if its controllability matrix W_c is nonsingular.*

Duality. As before, let A be an operator on a n-dimensional space $\mathcal{X}$, and C an operator from $\mathcal{X}$ into $\mathcal{Y}$. Recall that the observability matrix for the pair $\{C, A\}$ is the block matrix defined by

$$W_o = \begin{bmatrix} C \\ CA \\ \vdots \\ CA^{n-1} \end{bmatrix} . \tag{5.12}$$

Moreover, the pair $\{C, A\}$ is observable if and only if W_o is one to one. The unobservable subspace $\mathcal{X}_{\bar{o}}$ for the pair $\{C, A\}$ is the kernel of W_o. The observable subspace $\mathcal{X}_o$ for $\{C, A\}$ is the orthogonal complement of $\mathcal{X}_{\bar{o}}$. So, $\mathcal{X}_o$ equals the range of W_o^*. By replacing $\{C, A\}$ with $\{B^*, A^*\}$, it follows that the controllability matrix for $\{A, B\}$ is the adjoint of the observability matrix for $\{B^*, A^*\}$. In particular, the controllable subspace for $\{A, B\}$ is the observable subspace for $\{B^*, A^*\}$, and the uncontrollable subspace for $\{A, B\}$ is the unobservable subspace for $\{B^*, A^*\}$. Hence, the pair $\{A, B\}$ is controllable if and only if $\{B^*, A^*\}$ is observable. Obviously, the pair $\{C, A\}$ is observable if and only if $\{A^*, C^*\}$ is controllable. Because of this, we say that controllability is the dual of observability.

5.2   Uncontrollable eigenvalues and the PBH test

In this section we present the PBH test for controllability. To this end, notice that if $\mathcal{H}$ is an invariant subspace for an operator M on $\mathcal{K}$, then the orthogonal complement $\mathcal{H}^\perp$ of $\mathcal{H}$ is an invariant subspace for M^*. To see this, let h be in $\mathcal{H}$ and g be in $\mathcal{H}^\perp$. Then $0 = (Mh, g) = (h, M^*g)$. Hence, $M^*\mathcal{H}^\perp$ is orthogonal to $\mathcal{H}$. So, $\mathcal{H}^\perp$ is an invariant subspace for M^*.

As before, let A be an operator on a n dimensional space $\mathcal{X}$ and B an operator from $\mathcal{U}$ to $\mathcal{X}$ and W_c the controllability matrix defined in (5.5). Recall that the controllable subspace $\mathcal{X}_c = \operatorname{ran} W_c$ is an invariant subspace for A, that is, $A\mathcal{X}_c \subset \mathcal{X}_c$. So, the uncontrollable subspace $\mathcal{X}_{\bar{c}}$ is an invariant subspace for A^*. We say that λ is an *uncontrollable eigenvalue* for the pair $\{A, B\}$ if there is a nonzero vector v in $\mathcal{X}_{\bar{c}}$ such that $A^*v = \bar{\lambda}v$. In this case v is called the *uncontrollable eigenvector* for $\{A, B\}$. Thus, λ is an uncontrollable eigenvalue for the pair $\{A, B\}$ if and only if λ is an eigenvalue for the operator $A_{\bar{c}}$ on $\mathcal{X}_{\bar{c}}$ defined by $A_{\bar{c}}^*x = A^*x$ for $x \in \mathcal{X}_{\bar{c}}$. Obviously, $\{A, B\}$ is controllable if and only if its uncontrollable space $\mathcal{X}_{\bar{c}}$ is zero. Therefore, $\{A, B\}$ is controllable if and only if $\{A, B\}$ has no uncontrollable eigenvalues.

Now assume that λ is an uncontrollable eigenvalue for the pair $\{A, B\}$ and let v in $\mathcal{X}_{\bar{c}}$ be an eigenvector for A^* corresponding to $\bar{\lambda}$. By computing the inner product of $\dot{x} = Ax + Bu$ with v and using the fact that $\operatorname{ran} B \subset \operatorname{ran} W_c = \mathcal{X}_c$, we obtain

$$\frac{d(x, v)}{dt} = (\dot{x}, v) = (Ax, v) = (x, A^*v) = \lambda(x, v).$$

The solution to this differential equation is given by

$$(x(t), v) = e^{\lambda t}(x(0), v). \tag{5.13}$$

So, if the initial state has a nonzero component in the direction of v, then (5.13) shows that regardless of the input u, the resulting state trajectory has a component proportional to $e^{\lambda t}$ in the direction of v. Since v is orthogonal to the range of T for any t_1, one cannot drive an initial state with a nonzero v component to an arbitrary vector x_1 at time t_1.

We now show that λ is an uncontrollable eigenvalue for the pair $\{A, B\}$ if and only if $\bar{\lambda}$ is an unobservable eigenvalue for the pair $\{B^*, A^*\}$. Recall that $\mathcal{X}_{\bar{o}} = \ker W_c^*$ is precisely the unobservable subspace for $\{B^*, A^*\}$. Since $\ker W_c^* = (\operatorname{ran} W_c)^\perp = \mathcal{X}_{\bar{c}}$, it follows that $\bar{\lambda}$ is an unobservable eigenvalue for $\{B^*, A^*\}$ if and only if λ is an uncontrollable eigenvalue for the pair $\{A, B\}$. By combining these observations with the PBH observability Lemma 4.2.1, we readily obtain the following PBH controllability lemma.

Lemma 5.2.1 (PBH controllability lemma.) *A complex number λ is an uncontrollable eigenvalue of the pair $\{A$ on $\mathcal{X}, B\}$ if and only if*

$$\text{rank}\begin{bmatrix} A - \lambda I & B \end{bmatrix} < \dim \mathcal{X}.$$

In particular, the pair $\{A, B\}$ is controllable if and only if $\text{rank}\begin{bmatrix} A - \lambda I & B \end{bmatrix} = \dim \mathcal{X}$ for all complex numbers λ.

The following result readily follows from Proposition 4.2.2 along with the fact that $\{A, B\}$ is controllable if and only if $\{B^*, A^*\}$ is observable.

Proposition 5.2.2 *Consider the pair of matrices*

$$A = \begin{bmatrix} \lambda_1 & 0 & \cdots & 0 \\ 0 & \lambda_2 & & 0 \\ \vdots & & \ddots & \vdots \\ 0 & 0 & \cdots & \lambda_n \end{bmatrix}, \qquad B = \begin{bmatrix} b_1 \\ b_2 \\ \vdots \\ b_n \end{bmatrix}$$

where $\{\lambda_1, \cdots, \lambda_n\}$ and $\{b_1, \cdots, b_n\}$ are scalars. Then the pair $\{A, B\}$ is controllable if and only if $\{\lambda_1, \cdots, \lambda_n\}$ are distinct and $b_i \neq 0$ for $i = 1, \cdots, n$.

For some further insight into this proposition, we present a proof which is based on the Vandermonde matrix. For A and B with the above structure, one can readily show that the corresponding controllability matrix W_c is given by

$$W_c = \begin{bmatrix} b_1 & 0 & 0 & \cdots & 0 \\ 0 & b_2 & 0 & \cdots & 0 \\ 0 & 0 & b_3 & & 0 \\ \vdots & \vdots & & \ddots & \vdots \\ 0 & 0 & 0 & \cdots & b_n \end{bmatrix} \begin{bmatrix} 1 & \lambda_1 & \lambda_1^2 & \cdots & \lambda_1^{n-1} \\ 1 & \lambda_2 & \lambda_2^2 & \cdots & \lambda_2^{n-1} \\ 1 & \lambda_3 & \lambda_3^2 & \cdots & \lambda_3^{n-1} \\ \vdots & \vdots & \vdots & & \vdots \\ 1 & \lambda_n & \lambda_n^2 & \cdots & \lambda_n^{n-1} \end{bmatrix}.$$

Notice that the last matrix is precisely the square Vandermonde matrix generated by the scalars $\{\lambda_j\}_1^n$. According to Remark 16.4.1, this Vandermonde matrix is nonsingular if and only if $\{\lambda_j\}_1^n$ are distinct. It now follows that W_c is nonsingular if and only if $\{\lambda_j\}_1^n$ are distinct and $b_i \neq 0$ for $i = 1, \cdots, n$. This proves Proposition 5.2.2.

5.3 A controllability least squares problem

Recall that the basic controllability problem associated with the system $\{A, B\}$ in (5.1) is given two vectors x_0 and x_1 in $\mathcal{X}$ find an input u in $L^2([0, t_1], \mathcal{U})$ such that for $x(0) = x_0$ we have $x(t_1) = x_1$, or equivalently,

$$x_1 - e^{At_1}x_0 = \int_0^{t_1} e^{A(t_1 - \tau)} Bu(\tau)\, d\tau. \tag{5.14}$$

If a solution for u exists, then it is not unique. To see this, let T be the operator from $L^2([0, t_1], \mathcal{U})$ into $\mathcal{X}$ defined in (5.4). Since T maps an infinite dimensional space into a finite dimensional space, its kernel is infinite dimensional. In particular, this shows that if there is an input u satisfying (5.14), that is, $x_1 - e^{At_1}x_0 = Tu$, then (5.14) admits infinitely many solutions. Furthermore, a solution exists for every x_0 and x_1 if and only if $\{A, B\}$ is controllable.

If we cannot reach x_1 exactly, then we would like to choose a input u such that $x(t_1)$ comes as close as possible to the specified terminal state x_1. A more interesting problem is to find an input u with the smallest energy over the class of all inputs u which drive the state $x(t_1)$ as close as possible to x_1. One mathematical formulation of this problem is the following *controllability least squares optimization problem:* Given the vectors x_0 and x_1 in $\mathcal{X}$, find an input $\hat{u}$ in $L^2([0, t_1], \mathcal{U})$, to solve the following optimization problem:

$$||\hat{u}||^2 = \inf\left\{\int_0^{t_1} ||u(t)||^2 \, dt : \hat{x}(t_1) = e^{At_1}x_0 + \int_0^{t_1} e^{A(t_1-\tau)}Bu(\tau)\, d\tau\right\}$$

where $\hat{x}(t_1)$ is the unique vector in $\mathcal{X}$ satisfying   (5.15)

$$||x_1 - \hat{x}(t_1)|| = \inf\{||x_1 - x(t_1)|| : \dot{x} = Ax + Bu \text{ and } x(0) = x_0\}.$$

If the pair $\{A, B\}$ is controllable, then there exists an input u such that $x(t_1) = x_1$. In this case, $x_1 = \hat{x}(t_1)$ and the previous optimization problem reduces to finding an input $\hat{u}$ such that

$$||\hat{u}||^2 = \inf\left\{\int_0^{t_1} ||u(t)||^2 \, dt : x_1 = e^{At_1}x_0 + \int_0^{t_1} e^{A(t_1-\tau)}Bu(\tau)\, d\tau\right\}.$$ (5.16)

The *finite time controllability Gramian* defined by

$$Q(t) = \int_0^t e^{A\sigma}BB^* e^{A^*\sigma}\, d\sigma$$ (5.17)

plays a fundamental role in solving this problem. By replacing $\{C, A\}$ in Lemma 4.3.1 with $\{B^*, A^*\}$, we obtain the following result which permits one to compute $Q(t)$ by solving an ordinary differential equation.

Lemma 5.3.1 *Let $Q(t)$ be the finite time controllability Gramian given in (5.17). Then $Q(t)$ is positive. Moreover, $Q(t)$ is the solution to the differential equation*

$$\dot{Q} = AQ + QA^* + BB^*$$ (5.18)

subject to the initial condition $Q(0) = 0$.

We are now ready to present the main result of this section.

Theorem 5.3.2 *Let A be an operator on a finite dimensional vector space $\mathcal{X}$ and B an operator mapping $\mathcal{U}$ into $\mathcal{X}$ and x_0, x_1 be specified vectors in $\mathcal{X}$. Finally, let $Q(t_1)$ be the finite time controllability Gramian defined in (5.17). Then the following holds.*

   (i) The solution to the controllability optimization problem in (5.15) is unique and given by

$$\hat{u}(t) = B^* e^{A^*(t_1-t)}Q(t_1)^{-r}(x_1 - e^{At_1}x_0).$$ (5.19)

(ii) The pair $\{A, B\}$ is controllable if and only if $Q(t_1)$ is strictly positive.

(iii) If $\{A, B\}$ is controllable, then the controllability optimization problem in (5.15) reduces to the optimization problem in (5.16) and the corresponding optimal input is given by

$$\hat{u}(t) = B^* e^{A^*(t_1 - t)} Q(t_1)^{-1}(x_1 - e^{At_1} x_0) . \qquad (5.20)$$

PROOF. As before, let T be the operator from $L^2([0, t_1], \mathcal{U})$ into $\mathcal{X}$ defined in (5.4). Let $P_{\mathcal{R}}$ be the orthogonal projection onto the range $\mathcal{R}$ of T. Notice that $\mathcal{R} = \mathcal{X}_c$. By employing the Projection Theorem, the controllability optimization problem in (5.15) is equivalent to the following minimum norm optimization problem:

$$||\hat{u}|| = \inf \left\{ ||u|| : u \in L^2([0, t_1], \mathcal{U}) \text{ and } Tu = P_{\mathcal{R}}(x_1 - e^{At_1} x_0) \right\} . \qquad (5.21)$$

According to Corollary 16.5.2 in Section 16.5, the solution to this problem is unique and given by

$$\hat{u} = T^*(TT^*)^{-r}(x_1 - e^{At_1} x_0) . \qquad (5.22)$$

Recall that T^* is the operator from $\mathcal{X}$ into $L^2([0, t_1], \mathcal{U})$ given by $(T^* x)(t) = B^* e^{A^*(t_1 - t)} x$; see (5.7). By combining this with the definition of T, we see that

$$TT^* = \int_0^{t_1} e^{A(t_1 - \tau)} BB^* e^{A^*(t_1 - \tau)} \, d\tau = \int_0^{t_1} e^{A\sigma} BB^* e^{A^* \sigma} \, d\sigma = Q(t_1) \qquad (5.23)$$

where $Q(t_1)$ is the finite time controllability Gramian given in (5.17). Using $TT^* = Q(t_1)$ along with (5.22) and the expression for T^*, we obtain that the solution to the controllability optimization problem (5.15) is given by (5.19). This proves Part (i).

To verify Part (ii), recall that the pair $\{A, B\}$ is controllable if and only if the operator T is onto, or equivalently, TT^* is strictly positive. Since $TT^* = Q(t_1)$, it follows that the pair $\{A, B\}$ is controllable if and only if $Q(t_1)$ is strictly positive. To prove Part (iii) simply recall that the restricted inverse of $Q(t_1)$ becomes the actual inverse when $Q(t_1)$ is strictly positive. So, formula (5.20) follows (5.19) when the pair $\{A, B\}$ is controllable. ∎

Recall from Remark 5.1.1 that if x_0 and x_1 are any two vectors in the controllable subspace $\mathcal{X}_c$, then there always exists an input u which drives the state from x_0 to x_1 over the interval $[0, t_1]$. In this case, $x_1 - e^{At_1} x_0$ is in ran T. In fact, the input $\hat{u}$ of smallest possible norm which drives the state from x_0 to x_1 is given by (5.19) in Theorem 5.3.2.

Since $TT^* = Q(t_1)$, it follows that $||T||^2$ equals the largest eigenvalue of $Q(t_1)$. Moreover, the singular values of T are the square root of the nonzero eigenvalues of $Q(t_1)$. Finally, it is noted that by using the formula for $\hat{u}$ in (5.22), we obtain

$$||\hat{u}||^2 = (Q(t_1)^{-r}(x_1 - e^{At_1} x_0), x_1 - e^{At_1} x_0) . \qquad (5.24)$$

5.4 Stability and controllability

This section is devoted to the controllability of stable systems. Consider the linear system $\{A, B\}$ described by (5.1) where A is a stable operator on a finite dimensional vector space

$\mathcal{X}$. Consider any terminal time t_1 and let T be the operator from $L^2((-\infty, t_1], \mathcal{U})$ into $\mathcal{X}$ defined by

$$Tu = \int_{-\infty}^{t_1} e^{A(t_1-\tau)} Bu(\tau) \, d\tau \qquad (u \in L^2((-\infty, t_1], \mathcal{U})). \tag{5.25}$$

Because A is stable, T is bounded, and thus, T is a well defined linear operator. Since A is stable, $\|e^{At}\| \le me^{-\alpha t}$ for some positive m and $\alpha > 0$. An application of the Cauchy-Schwartz inequality, yields

$$\|Tu\| \le \int_{-\infty}^{t_1} \|e^{A(t_1-\tau)}\| \, \|Bu(\tau)\| \, d\tau \le m\|B\| \int_{-\infty}^{t_1} e^{-\alpha(t_1-\tau)} \|u(\tau)\| \, d\tau$$

$$\le m\|B\| \left[\int_{-\infty}^{t_1} e^{-2\alpha(t_1-\tau)} \, d\tau \right]^{1/2} \|u\|_{L^2} = m\|B\| \|u\|/\sqrt{2\alpha}.$$

Therefore, $\|T\| \le m\|B\|/\sqrt{2\alpha}$, and T is a well defined linear operator.

We say that the pair $\{A, B\}$ is controllable over the interval $(-\infty, t_1]$ if T is onto. By mimicking the calculation in (5.8), it follows that the adjoint T^* of T is the operator from $\mathcal{X}$ to $L^2((-\infty, t_1], \mathcal{U})$ given by

$$(T^*x)(t) = B^* e^{A^*(t_1-t)} x \qquad (x \in \mathcal{X}). \tag{5.26}$$

Using this one may readily show that Lemma 5.1.1 holds where $[0, t_1]$ is replaced with $(-\infty, t_1]$, and thus, the controllable subspace $\mathcal{X}_c = \operatorname{ran} T = \operatorname{ran} W_c$. Hence, the controllability of $\{A, B\}$ over $(-\infty, t_1]$ is equivalent to our previous notion of controllability for $\{A, B\}$. So, as before, we drop the interval when discussing controllability.

Because A is stable, we can define the following infinite horizon controllability least squares optimization problem: Given a vector $x_1 \in \mathcal{X}$, find an input $\hat{u}$ to solve the following optimization problem

$$\|\hat{u}\|^2 = \inf \left\{ \int_{-\infty}^{t_1} \|u(t)\|^2 \, dt : \hat{x}(t_1) = \int_{-\infty}^{t_1} e^{A(t_1-\tau)} Bu(\tau) \, d\tau \right\}$$

where $\hat{x}(t_1)$ is the unique vector in $\mathcal{X}$ satisfying $\qquad$ (5.27)

$$\|x_1 - \hat{x}(t_1)\| = \inf \left\{ \|x_1 - x(t_1)\| : x(t_1) = \int_{-\infty}^{t_1} e^{A(t_1-\tau)} Bu(\tau) \, d\tau \text{ and } u \in L_2((-\infty, t_1], \mathcal{U}) \right\}.$$

The *controllability Gramian* for the pair $\{A, B\}$ is the operator on $\mathcal{X}$ defined by

$$Q = \int_0^\infty e^{A\sigma} BB^* e^{A^*\sigma} \, d\sigma. \tag{5.28}$$

Notice that if $Q(t)$ is the finite time controllability Gramian defined in (5.17), then $Q(t) \to Q$ as t approaches infinity. According to Lemma 3.1.2 this Q is the unique solution to the following Lyapunov equation:

$$AQ + QA^* + BB^* = 0. \tag{5.29}$$

The controllability Gramian Q plays a fundamental role in the solution to the infinite horizon controllability least squares optimization problem.

Theorem 5.4.1 *Let A be a stable operator on a finite dimensional vector space $\mathcal{X}$ while B is an operator mapping $\mathcal{U}$ into $\mathcal{X}$ and x_1 is a specified vector in $\mathcal{X}$. Let Q be the controllability Gramian for $\{A, B\}$. Then the following holds.*

(i) *The solution to the controllability optimization problem in (5.27) is unique and given by*

$$\hat{u}(t) = B^* e^{A^*(t_1 - t)} Q^{-r} x_1 \,. \tag{5.30}$$

(ii) *The pair $\{A, B\}$ is controllable if and only if Q is strictly positive.*

(iii) *If $\{A, B\}$ is controllable, then the controllability optimization problem in (5.27) reduces to*

$$\|\hat{u}\|^2 = \inf \left\{ \int_{-\infty}^{t_1} \|u(t)\|^2 \, dt : x_1 = \int_{-\infty}^{t_1} e^{A(t_1 - \tau)} B u(\tau) \, d\tau \right\} \tag{5.31}$$

and the corresponding optimal input is given by

$$\hat{u}(t) = B^* e^{A^*(t_1 - t)} Q^{-1} x_1 \,. \tag{5.32}$$

PROOF. Let $P_{\mathcal{R}}$ be the orthogonal projection onto the range $\mathcal{R} = \mathcal{X}_c$ of T. By employing the Projection Theorem, the infinite horizon controllability optimization problem is equivalent to the following minimum norm optimization problem

$$\|\hat{u}\| = \inf \left\{ \|u\| : u \in L_2((-\infty, t_1], \mathcal{U}) \text{ and } Tu = P_{\mathcal{R}} x_1 \right\} \,. \tag{5.33}$$

According to Corollary 16.5.2 in Section 16.5, the solution to this problem is unique and given by

$$\hat{u} = T^* (TT^*)^{-r} x_1 \,. \tag{5.34}$$

By combining the definition of T in (5.25) with its adjoint in (5.26), we obtain $TT^* = Q$; see (5.28). Using this along with (5.34) and (5.26), we see that the solution to the controllability optimization problem in (5.27) is given by (5.30). This proves Part(i).

To verify Part (ii) recall that the pair $\{A, B\}$ is controllable if and only if the operator T is onto, or equivalently, TT^* is strictly positive. Since $TT^* = Q$, it follows that the pair $\{A, B\}$ is controllable if and only if Q is strictly positive. Hence, Part (ii) holds.

Because the restricted inverse of Q becomes the actual inverse when Q is strictly positive, formula (5.32) follows when the pair $\{A, B\}$ is controllable. Finally, if $\{A, B\}$ is controllable, T is onto, and thus, the optimization problem in (5.33) reduces to $\|\hat{u}\| = \inf\{\|u\| : Tu = x_1\}$. So, the optimization problem in (5.27) reduces to the optimization problem in (5.31) when $\{A, B\}$ is controllable. ∎

Remark 5.4.1 Let A be a stable operator on a finite dimensional vector space $\mathcal{X}$ and B an operator mapping $\mathcal{U}$ into $\mathcal{X}$. Let Q be the controllability Gramian for $\{A, B\}$, and T the operator from $L^2((-\infty, t_1], \mathcal{U})$ into $\mathcal{X}$ defined in (5.25). Then using $TT^* = Q$, it follows that $\|T\|^2$ equals the largest eigenvalue of Q. Moreover, the singular values of T are the nonzero eigenvalues of Q. Recall that the uncontrollable subspace $\mathcal{X}_{\bar{c}}$ for the pair $\{A, B\}$ is given by $\mathcal{X}_{\bar{c}} = (\operatorname{ran} T)^{\perp} = \ker T^*$. Since $TT^* = Q$, the kernel of Q equals $\mathcal{X}_{\bar{c}}$. Finally, by using the formula for $\hat{u}$ in (5.34) along with $TT^* = Q$, we obtain $\|\hat{u}\|^2 = (Q^{-r} x_1, x_1)$.

Recall that $\{A, B\}$ is controllable if and only if $\{B^*, A^*\}$ is observable. By using this duality in Theorem 4.4.2, we readily obtain the following result.

Theorem 5.4.2 *Let $\{A, B\}$ be a controllable pair. Then the following statements are equivalent.*

(a) *The system $\dot{x} = Ax$ is stable.*

(b) *There exists a strictly positive operator Q satisfying the Lyapunov equation*

$$AQ + QA^* + BB^* = 0. \tag{5.35}$$

Exercise 10 Let $\{\alpha_j\}_1^n$ be a set complex numbers in the open right half plane. Moreover, let A be the diagonal matrix on $\mathbb{C}^n$ and B the column vector given by

$$A = -\text{diag}\,(\alpha_1, \alpha_2, \cdots, \alpha_n) \qquad \text{and} \qquad B = [b_1,\, b_2,\, \cdots,\, b_n]^{tr}.$$

Let Q be the $n \times n$ matrix defined by

$$Q = \begin{bmatrix} \frac{b_1 \bar{b}_1}{\alpha_1 + \bar{\alpha}_1} & \frac{b_1 \bar{b}_2}{\alpha_1 + \bar{\alpha}_2} & \cdots & \frac{b_1 \bar{b}_n}{\alpha_1 + \bar{\alpha}_n} \\[2ex] \frac{b_2 \bar{b}_1}{\alpha_2 + \bar{\alpha}_1} & \frac{b_2 \bar{b}_2}{\alpha_2 + \bar{\alpha}_2} & \cdots & \frac{b_2 \bar{b}_n}{\alpha_2 + \bar{\alpha}_n} \\[2ex] \vdots & \vdots & & \vdots \\[2ex] \frac{b_n \bar{b}_1}{\alpha_n + \bar{\alpha}_1} & \frac{b_n \bar{b}_2}{\alpha_n + \bar{\alpha}_2} & \cdots & \frac{b_n \bar{b}_n}{\alpha_n + \bar{\alpha}_n} \end{bmatrix}. \tag{5.36}$$

A matrix of the form (5.36) is called a Pick matrix. Show that Q is the unique solution to the Lyapunov equation $AQ + QA^* + BB^* = 0$. In particular, Q is positive. Moreover, show that Q is strictly positive if and only if $\{\alpha_j\}_1^n$ are distinct and $b_i \neq 0$ for all $i = 1, 2, \cdots, n$.

5.5   Notes

All the results in this section are classical results in linear systems theory. The history of controllability and linear systems is well established in Kailath [68], Rugh [110] and elsewhere. So, we will not present a historical account of linear systems. For further results on linear systems see Brockett [21], Chen [26], DeCarlo [33], Delchamps [34], Kailath [68], Polderman-Willems [98], Rugh [110], Skelton [114] and Sontag [116]. Using operator techniques to solve controllability problems is also classical; see Balakrishnan [8], Brockett [21], Fuhrmann [47], Naylor-Sell [93], Luenberger [85] and Porter [100]. The PBH test was independently developed by Popov [99], Belevitch [14] and Hautus [61].

For simplicity of presentation only we concentrated on time invariant systems. Many of these results can be easily extended to the time varying case. To see this consider the time varying system $\dot{x} = A(t)x + B(t)u$ where A is a continuous function with values in

$\mathcal{L}(\mathcal{X}, \mathcal{X})$ and B is a continuous function with values in $\mathcal{L}(\mathcal{U}, \mathcal{X})$. Let T be the operator from $L^2([t_0, t_1], \mathcal{U})$ into $\mathcal{X}$ defined by

$$Tu = \int_{t_0}^{t_1} \Phi(t_1, \tau) B(\tau) u(\tau)\, d\tau \qquad (u \in L^2([t_0, t_1], \mathcal{U}))$$

where $\Phi(t, \tau)$ is the state transition operator for A. Then the time varying pair $\{A, B\}$ is *controllable over the interval* $[t_0, t_1]$ if the operator T is onto. The time varying pair $\{A, B\}$ can be controllable over one interval and not controllable over another interval. Obviously, the pair $\{A, B\}$ is controllable over the interval $[t_0, t_1]$ if and only if T^* is one to one, or equivalently, TT^* is strictly positive. The adjoint of T is the operator mapping $\mathcal{X}$ into $L^2([t_0, t_1], \mathcal{U})$ given by

$$(T^* f)(t) = B(t)^* \Phi(t_1, t)^* f \qquad (f \in \mathcal{X}). \tag{5.37}$$

So, the time varying pair $\{A, B\}$ is controllable over the interval $[t_0, t_1]$ if and only if

$$P(t_1) := TT^* = \int_{t_0}^{t_1} \Phi(t_1, t) B(t) B(t)^* \Phi(t_1, t)^*\, dt$$

is a strictly positive operator on $\mathcal{X}$. Notice that $P(t)$ can be obtained by solving the following differential equation

$$\dot{P} = A(t)P + PA(t)^* + B(t)B(t)^* \qquad (P(t_0) = 0). \tag{5.38}$$

Therefore, the time varying pair $\{A, B\}$ is controllable over the interval $[t_0, t_1]$ if and only if $P(t_1)$ is strictly positive where $P(t)$ is the solution to the differential equation in (5.38). The controllability optimization problem and its solution can also be extended to the time varying setting. In this case, the controllability optimization problem becomes: Let x_1 be a specified vector in $\mathcal{X}$ and $P_{\mathcal{R}}$ be the orthogonal projection onto the range of T, then find an optimal control $\hat{u}$ in $L^2([t_0, t_1], \mathcal{U})$ such that

$$\|\hat{u}\| = \inf\{\|u\| : P_{\mathcal{R}} x_1 = Tu\}.$$

The solution to this optimization problem is unique and is given by

$$\hat{u} = T^* (TT^*)^{-r} x_1 = B(t)^* \Phi(t_1, t)^* P(t_1)^{-r} x_1.$$

Chapter 6

Controllable and Observable Realizations

In this chapter we will use the controllable and observable subspaces, to obtain the controllable and observable decomposition for a linear system. Some connections to semi-invariant subspaces for linear operators will also be given. Finally, it is shown that if $\mathbf{G}$ is the transfer function of a controllable and observable system, then the polynomial formed by the poles of $\mathbf{G}$ is the minimal polynomial for the corresponding state space system operator.

6.1 Invariant subspaces

This section is devoted to invariant subspaces and their matrix representations. Let M be a linear operator mapping $\mathcal{X}$ into $\mathcal{Y}$ and let $\mathcal{H}$ be a subspace of $\mathcal{X}$. (In this section, the vector spaces can be infinite dimensional.) The notation $M|\mathcal{H}$ denotes the restriction of the operator M to $\mathcal{H}$, that is, $M|\mathcal{H}$ is the operator from $\mathcal{H}$ to $\mathcal{Y}$ defined by $(M|\mathcal{H})x = Mx$ where x is in $\mathcal{H}$. Consider any operator N whose range is contained in $\mathcal{H}$. Then $N = P_{\mathcal{H}}N$. (The orthogonal projection onto a subspace $\mathcal{H}$ is denoted by $P_{\mathcal{H}}$.) Moreover, $MN = (M|\mathcal{H})N = (M|\mathcal{H})(P_{\mathcal{H}}N)$. Suppose the kernel of M contains $\mathcal{H}^{\perp}$ and N is any operator whose range is in $\mathcal{X}$. Then, $MN = MP_{\mathcal{H}}N$ and since the range of $P_{\mathcal{H}}N$ is in $\mathcal{H}$, we have $MN = (M|\mathcal{H})(P_{\mathcal{H}}N)$. Finally, let M be an operator on $\mathcal{X}$. Then we say that S is the *compression* of M to $\mathcal{H}$ if S is the operator on $\mathcal{H}$ defined by $S = P_{\mathcal{H}}M|\mathcal{H}$. Clearly, $\|S\| \leq \|M\|$.

Let M be a linear operator mapping $\mathcal{X}$ into $\mathcal{Y}$ and suppose that $\mathcal{X} = \mathcal{X}_1 \oplus \mathcal{X}_2 \oplus \cdots \oplus \mathcal{X}_n$ is an orthogonal decomposition for $\mathcal{X}$ while $\mathcal{Y} = \mathcal{Y}_1 \oplus \mathcal{Y}_2 \oplus \cdots \oplus \mathcal{Y}_m$ is an orthogonal decomposition for $\mathcal{Y}$ where $\{\mathcal{X}_j\}_1^n$ and $\{\mathcal{Y}_i\}_1^m$ are subspaces of $\mathcal{X}$ and $\mathcal{Y}$, respectively. We say that

$$
\begin{bmatrix} M_{11} & M_{12} & \cdots & M_{1n} \\ M_{21} & M_{22} & \cdots & M_{2n} \\ \vdots & \vdots & & \vdots \\ M_{m1} & M_{m2} & \cdots & M_{mn} \end{bmatrix} : \begin{bmatrix} \mathcal{X}_1 \\ \mathcal{X}_2 \\ \vdots \\ \mathcal{X}_n \end{bmatrix} \longrightarrow \begin{bmatrix} \mathcal{Y}_1 \\ \mathcal{Y}_2 \\ \vdots \\ \mathcal{Y}_m \end{bmatrix}
$$

is a *matrix representation* for M if M_{ij} is the operator mapping $\mathcal{X}_j$ into $\mathcal{Y}_i$ defined by $M_{ij} = P_{\mathcal{Y}_i}M|\mathcal{X}_j$. Recall that a unitary operator W is an isometry whose range is onto, or equivalently, $W^* = W^{-1}$. (An isometry V is an operator from $\mathcal{V}$ into $\mathcal{K}$ satisfying $V^*V = I$.)

Obviously, there exists unitary operators U_1 and U_2 such that $U_1 M U_2$ equals its matrix representation. So, we will not distinguish between M and its matrix representation, that is, by a slight abuse of notation we will sometimes use the same symbol for the operator and its matrix representation. Note that the matrix representation for the orthogonal projection $P_{\mathcal{H}}$ onto $\mathcal{H}$ is given by

$$\begin{bmatrix} I & 0 \\ 0 & 0 \end{bmatrix} : \begin{bmatrix} \mathcal{H} \\ \mathcal{H}^{\perp} \end{bmatrix} \longrightarrow \begin{bmatrix} \mathcal{H} \\ \mathcal{H}^{\perp} \end{bmatrix}. \tag{6.1}$$

Now let A be an operator on $\mathcal{X}$. Recall that $\mathcal{H}$ is an *invariant subspace* for A if $\mathcal{H}$ is a subspace of $\mathcal{X}$ satisfying $A\mathcal{H} \subset \mathcal{H}$, or equivalently, $AP_{\mathcal{H}} = P_{\mathcal{H}} A P_{\mathcal{H}}$. Since $A\mathcal{H} \subset \mathcal{H}$ if and only if $(P_{\mathcal{H}^{\perp}})A|\mathcal{H} = 0$, it follows that $\mathcal{H}$ is an invariant subspace for A if and only if A admits a matrix representation of the form

$$\begin{bmatrix} A_{11} & A_{12} \\ 0 & A_{22} \end{bmatrix} : \begin{bmatrix} \mathcal{H} \\ \mathcal{H}^{\perp} \end{bmatrix} \longrightarrow \begin{bmatrix} \mathcal{H} \\ \mathcal{H}^{\perp} \end{bmatrix}. \tag{6.2}$$

Here A_{11} is the operator on $\mathcal{H}$ given by $A_{11} = P_{\mathcal{H}} A | \mathcal{H} = A | \mathcal{H}$, while A_{22} is the compression of A to $\mathcal{H}^{\perp}$. Finally, notice that, for any integer $k \geq 0$, we have $A_{11}^k = P_{\mathcal{H}} A^k | \mathcal{H} = A^k | \mathcal{H}$.

We have already seen that $\mathcal{H}$ is an invariant subspace for A if and only if its orthogonal complement $\mathcal{H}^{\perp}$ is an invariant subspace for A^*. We say that $\mathcal{H}$ is a *co-invariant subspace* for A if $\mathcal{H}$ is an invariant subspace for A^*, or equivalently, $\mathcal{H}^{\perp}$ is an invariant subspace for A. Hence, $\mathcal{H}$ is a co-invariant subspace for A if and only if A admits a matrix representation of the form

$$\begin{bmatrix} A_{11} & 0 \\ A_{21} & A_{22} \end{bmatrix} : \begin{bmatrix} \mathcal{H} \\ \mathcal{H}^{\perp} \end{bmatrix} \longrightarrow \begin{bmatrix} \mathcal{H} \\ \mathcal{H}^{\perp} \end{bmatrix}. \tag{6.3}$$

Here A_{22} is the operator on $\mathcal{H}^{\perp}$ given by $A_{22} = P_{\mathcal{H}^{\perp}} A | \mathcal{H}^{\perp} = A | \mathcal{H}^{\perp}$ while A_{11} is the compression of A to $\mathcal{H}$. Finally, $\mathcal{H}$ is a co-invariant subspace for A if and only if $P_{\mathcal{H}} A = P_{\mathcal{H}} A P_{\mathcal{H}}$. Notice also that, for any integer $k \geq 0$, we have $A_{11}^k = P_{\mathcal{H}} A^k | \mathcal{H}$.

We say that $\mathcal{H}$ is a *reducing subspace* for A if $\mathcal{H}$ is an invariant subspace for both A and A^*. Therefore, $\mathcal{H}$ is a reducing subspace if and only if $\mathcal{H}$ is both an invariant subspace and a co-invariant subspace for A. Moreover, $\mathcal{H}$ is a reducing subspace for A if and only if $\mathcal{H}^{\perp}$ is a reducing subspace for A. Clearly, $\mathcal{H}$ is a reducing subspace for A if and only if A admits a matrix representation of the form

$$\begin{bmatrix} A_{11} & 0 \\ 0 & A_{22} \end{bmatrix} : \begin{bmatrix} \mathcal{H} \\ \mathcal{H}^{\perp} \end{bmatrix} \longrightarrow \begin{bmatrix} \mathcal{H} \\ \mathcal{H}^{\perp} \end{bmatrix}. \tag{6.4}$$

Notice that $\mathcal{H}$ is a reducing subspace for A if and only if A commutes with the orthogonal projection $P_{\mathcal{H}}$, that is, $AP_{\mathcal{H}} = P_{\mathcal{H}} A$.

To complete this section, we introduce the notion of a dilation and a semi-invariant subspace. Let A be an operator on $\mathcal{X}$. Then we say that A is a *dilation* of an operator T on $\mathcal{H}$ if T is the compression of A to $\mathcal{H}$ and

$$P_{\mathcal{H}} A^k | \mathcal{H} = T^k \qquad (\text{for } k = 1, 2, 3, \cdots). \tag{6.5}$$

In other words, A is a dilation of T if and only if T^k is the compression of A^k to $\mathcal{H}$ for all integers $k \geq 1$. In this case, if q is a polynomial, then (6.5) implies that $q(T) = P_{\mathcal{H}} q(A) | \mathcal{H}$.

Moreover, using the expansion $(sI - A)^{-1} = \sum_0^\infty A^k / s^{n+1}$ for $|s| > \|A\|$, we have $(sI - T)^{-1} = P_{\mathcal{H}}(sI - A)^{-1}|\mathcal{H}$. Finally, it is noted that dilation theory plays a fundamental role in operator theory; see Gohberg-Goldberg-Kaashoek [55], Foias-Frazho [39] and Sz.-Nagy-Foias [120].

As before, let A be an operator on $\mathcal{X}$. Then $\mathcal{H}$ is a *semi-invariant subspace* for A if there exists two invariant subspaces $\mathcal{M}$ and $\mathcal{N}$ for A satisfying $\mathcal{H} = \mathcal{N} \ominus \mathcal{M}$ where $\mathcal{M} \subset \mathcal{N}$. In other words, $\mathcal{H}$ is a semi-invariant subspace for A if and only if $\mathcal{N} = \mathcal{M} \oplus \mathcal{H}$ where $\mathcal{M}$ and $\mathcal{N}$ are two invariant subspaces for A. In this case, $\mathcal{M}^\perp$ and $\mathcal{N}^\perp$ are two invariant subspaces for A^* satisfying $\mathcal{N}^\perp \subset \mathcal{M}^\perp$. Using $\mathcal{M} \oplus \mathcal{H} \oplus \mathcal{N}^\perp = \mathcal{X}$, we see that $\mathcal{H} = \mathcal{M}^\perp \ominus \mathcal{N}^\perp$. So, $\mathcal{H}$ is also semi-invariant for A^*. Therefore, $\mathcal{H}$ is semi-invariant for A if and only if $\mathcal{H}$ is semi-invariant for A^*. It is easy to show that $\mathcal{H}$ is semi-invariant for A if and only if A admits a matrix representation of the form

$$A = \begin{bmatrix} * & * & * \\ 0 & T & * \\ 0 & 0 & * \end{bmatrix} : \begin{bmatrix} \mathcal{M} \\ \mathcal{H} \\ \mathcal{N}^\perp \end{bmatrix} \longrightarrow \begin{bmatrix} \mathcal{M} \\ \mathcal{H} \\ \mathcal{N}^\perp \end{bmatrix} \tag{6.6}$$

where T is the compression of A to $\mathcal{H}$. If A is an upper triangular block matrix of the form in (6.6), then T is the compression of A to $\mathcal{H}$, and (6.5) holds, that is, A is a dilation of T. This proves half of the following result due to Sarason [111].

Proposition 6.1.1 *Let A be an operator on $\mathcal{X}$ and T the compression of A to $\mathcal{H}$. Then $\mathcal{H}$ is a semi-invariant subspace for A if and only if A is a dilation of T.*

PROOF. Assume that A is a dilation of T, that is, (6.5) holds. Let $\mathcal{N}$ be the following "controllable " subspace generated by A and $\mathcal{H}$

$$\mathcal{N} = \bigvee_{k=0}^\infty A^k \mathcal{H}.$$

Here $\bigvee_{k=0}^\infty \mathcal{H}_k$ denotes the closed linear span of the subspaces $\mathcal{H}_k$ for all integers $k \geq 0$. Obviously, $\mathcal{N}$ is an invariant subspace for A. To complete the proof it is sufficient to show that $\mathcal{M} = \mathcal{N} \ominus \mathcal{H}$ is invariant for A. To this end, notice that (6.5) yields

$$P_{\mathcal{H}} A A^k h = T T^k h = T P_{\mathcal{H}} A^k h \qquad \text{(for all } k \geq 0 \text{ and } h \in \mathcal{H}).$$

Hence, $P_{\mathcal{H}} A | \mathcal{N} = T P_{\mathcal{H}} | \mathcal{N}$. Consider any m in $\mathcal{M}$. Then m is in $\mathcal{N}$ and $P_{\mathcal{H}} m = 0$. Hence, Am is in $\mathcal{N}$ and

$$P_{\mathcal{H}}(Am) = (P_{\mathcal{H}} A | \mathcal{N}) m = (T P_{\mathcal{H}} | \mathcal{N}) m = T P_{\mathcal{H}} m = 0.$$

Since Am is in $\mathcal{N}$ and $P_{\mathcal{H}}(Am) = 0$, it follows that Am is in $\mathcal{M}$. Thus $\mathcal{M}$ is invariant for A. Therefore, $\mathcal{H}$ is a is semi-invariant subspace for A. ∎

6.2 The controllable and observable decomposition

In this section we present the controllable and observable decomposition of a linear system. Consider the following continuous time system

$$\dot{x} = Ax + Bu \qquad \text{and} \qquad y = Cx + Du \tag{6.7}$$

where A is an operator on a finite dimensional space $\mathcal{X}$ and B maps $\mathcal{U}$ into $\mathcal{X}$ while C maps $\mathcal{X}$ into $\mathcal{Y}$ and D maps $\mathcal{U}$ into $\mathcal{Y}$. Recall that the transfer function $\mathbf{G}$ for $\{A, B, C, D\}$ is the proper rational function defined by

$$\mathbf{G}(s) = C(sI - A)^{-1}B + D. \tag{6.8}$$

Recall also that $\{A, B, C, D\}$ is a *realization* for a proper rational function $\mathbf{G}$ if $\mathbf{G}$ is the transfer function for $\{A, B, C, D\}$, that is, (6.8) holds. In this section, we will use certain invariant subspaces to extract from $\{A, B, C, D\}$ a controllable and observable realization $\{A_{co}, B_{co}, C_{co}, D_{co}\}$ of the same transfer function $\mathbf{G}$, that is,

$$\mathbf{G}(s) = C_{co}(sI - A_{co})^{-1}B_{co} + D_{co}$$

where $\{A_{co}, B_{co}\}$ is controllable and $\{C_{co}, A_{co}\}$ is observable.

A controllable realization.   First we present a specific controllable realization of $\mathbf{G}$. To this end, recall that the controllable subspace $\mathcal{X}_c$ for the pair $\{A, B\}$ is given by

$$\mathcal{X}_c = \bigvee_{k=0}^{\infty} A^k B\mathcal{U} = \bigvee \{A^k B\mathcal{U} : 0 \leq k < \dim \mathcal{X}\}. \tag{6.9}$$

We have seen that $\mathcal{X}_c$ is an invariant subspace for A. So, A admits a matrix representation of the form:

$$A = \begin{bmatrix} A_c & * \\ 0 & A_{\bar{c}} \end{bmatrix} : \begin{bmatrix} \mathcal{X}_c \\ \mathcal{X}_{\bar{c}} \end{bmatrix} \longrightarrow \begin{bmatrix} \mathcal{X}_c \\ \mathcal{X}_{\bar{c}} \end{bmatrix} \tag{6.10}$$

where $\mathcal{X}_{\bar{c}}$ is the uncontrollable subspace defined by $\mathcal{X}_{\bar{c}} = \mathcal{X} \ominus \mathcal{X}_c$. Here A_c is the operator on $\mathcal{X}_c$ defined by $A_c = A|\mathcal{X}_c$ and $A_{\bar{c}}$ is the compression of A to $\mathcal{X}_{\bar{c}}$. Furthermore, the operators B and C admit matrix representations of the form:

$$B = \begin{bmatrix} B_c \\ 0 \end{bmatrix} : \mathcal{U} \longrightarrow \begin{bmatrix} \mathcal{X}_c \\ \mathcal{X}_{\bar{c}} \end{bmatrix} \quad \text{and} \quad C = \begin{bmatrix} C_c & C_{\bar{c}} \end{bmatrix} : \begin{bmatrix} \mathcal{X}_c \\ \mathcal{X}_{\bar{c}} \end{bmatrix} \longrightarrow \mathcal{Y}, \tag{6.11}$$

respectively. Since the range of B is contained in $\mathcal{X}_c$, we obtain that $B_c = P_{\mathcal{X}_c}B = B$. Also, $B_{\bar{c}} = P_{\mathcal{X}_{\bar{c}}}B = 0$, that is, the second entry in the matrix representation of B is zero.

We claim that $\{A_c, B_c, C_c, D\}$ is a controllable realization of $\mathbf{G}$. Because $\mathcal{X}_c$ is invariant for A, we have $A_c^k = A^k|\mathcal{X}_c$ for all integers $k \geq 0$. Since the range of B is in $\mathcal{X}_c$, it follows that $A^k B = (A^k|\mathcal{X}_c)B = A_c^k B_c$ for all integers $k \geq 0$. Hence,

$$\bigvee_{k=0}^{\infty} A_c^k B_c \mathcal{U} = \bigvee_{k=0}^{\infty} A^k B\mathcal{U} = \mathcal{X}_c.$$

So, $\{A_c^k B_c \mathcal{U}\}_0^{\infty}$ spans $\mathcal{X}_c$ and $\{A_c, B_c\}$ is controllable. For any integer $k \geq 0$, the range of $A^k B$ is in $\mathcal{X}_c$ and $A^k B = A_c^k B_c$. Hence, $CA^k B = (C|\mathcal{X}_c)A^k B = C_c A_c^k B_c$. It now follows that $C(sI - A)^{-1}B = C_c(sI - A_c)^{-1}B_c$. Thus,

$$\mathbf{G}(s) = C(sI - A)^{-1}B + D = C_c(sI - A_c)^{-1}B_c + D.$$

Therefore, $\{A_c, B_c, C_c, D\}$ is a controllable realization of $\mathbf{G}$.

We claim that if $\{C, A\}$ is observable, then $\{C_c, A_c\}$ is observable. To prove this it is sufficient to first show that the unobservable subspace $\mathcal{X}_{c\bar{o}}$ for $\{C_c, A_c\}$ is the intersection of the controllable subspace $\mathcal{X}_c$ and the unobservable subspace $\mathcal{X}_{\bar{o}}$ for $\{A, B, C, D\}$, that is,

$$\mathcal{X}_{c\bar{o}} = \mathcal{X}_c \bigcap \mathcal{X}_{\bar{o}}. \tag{6.12}$$

We call $\mathcal{X}_{c\bar{o}}$ the *controllable/unobservable* subspace for $\{A, B, C, D\}$. Indeed, if $\{C, A\}$ is observable, then $\mathcal{X}_{\bar{o}} = \{0\}$, and hence, $\mathcal{X}_c \bigcap \mathcal{X}_{\bar{o}} = \{0\}$, or equivalently, $\{C_c, A_c\}$ is observable. According to Lemma 4.1.1, the subspace

$$\mathcal{X}_{c\bar{o}} = \bigcap_{k=0}^{\infty} \ker C_c A_c^k.$$

Since $\mathcal{X}_c$ is invariant for A, we see that $C_c A_c^k = (C|\mathcal{X}_c) A^k |\mathcal{X}_c = C A^k |\mathcal{X}_c$. This readily implies that $\ker C_c A_c^k = \mathcal{X}_c \bigcap \ker C A^k$ and

$$\mathcal{X}_{c\bar{o}} = \bigcap_{k=0}^{\infty} \mathcal{X}_c \bigcap \ker C A^k = \mathcal{X}_c \bigcap \left(\bigcap_{k=0}^{\infty} \ker C A^k \right) = \mathcal{X}_c \bigcap \mathcal{X}_{\bar{o}}.$$

Obviously, $\mathcal{X}_c$ and $\mathcal{X}_{\bar{o}}$ are invariant subspaces for A. Since the intersection of two invariant subspaces is an invariant subspace, $\mathcal{X}_{c\bar{o}}$ is an invariant subspace for A. The above analysis yields the following result.

Proposition 6.2.1 *Suppose* $\mathbf{G}$ *is the transfer function for* $\{A, B, C, D\}$ *and* $\mathcal{X}_c$ *is the controllable subspace for* $\{A, B\}$. *Let* A_c *on* $\mathcal{X}_c$ *and* B_c *from* $\mathcal{U}$ *into* $\mathcal{X}_c$ *and* C_c *from* $\mathcal{X}_c$ *into* $\mathcal{Y}$ *be the operators defined by*

$$A_c = A|\mathcal{X}_c \quad and \quad B_c = B \quad and \quad C_c = C|\mathcal{X}_c. \tag{6.13}$$

Then $\{A_c, B_c, C_c, D\}$ *is a controllable realization of* $\mathbf{G}$ *whose unobservable subspace* $\mathcal{X}_{c\bar{o}}$ *is the invariant subspace of* A *given by* $\mathcal{X}_c \bigcap \mathcal{X}_{\bar{o}}$. *In particular, if* $\{C, A\}$ *is observable, then* $\{C_c, A_c\}$ *is observable.*

An observable realization. We now obtain a specific observable realization of a transfer function $\mathbf{G}$. Let $\mathbf{G}$ be the transfer function for $\{A, B, C, D\}$, and let $\mathcal{X}_{\bar{o}}$ be the unobservable subspace for the pair $\{C, A\}$, that is,

$$\mathcal{X}_{\bar{o}} = \bigcap_{k=0}^{\infty} \ker C A^k. \tag{6.14}$$

Clearly, $\mathcal{X}_{\bar{o}}$ is a invariant subspace for A. So, A admits a matrix representation of the form:

$$A = \begin{bmatrix} A_{\bar{o}} & * \\ 0 & A_o \end{bmatrix} : \begin{bmatrix} \mathcal{X}_{\bar{o}} \\ \mathcal{X}_o \end{bmatrix} \longrightarrow \begin{bmatrix} \mathcal{X}_{\bar{o}} \\ \mathcal{X}_o \end{bmatrix} \tag{6.15}$$

where $\mathcal{X}_o = \mathcal{X} \ominus \mathcal{X}_{\bar{o}}$ is the observable subspace for $\{C, A\}$. Notice that $\mathcal{X}_o$ is a co-invariant subspace for A. In other words, $\mathcal{X}_o$ is an invariant subspace for A^*. The operator A_o is the compression of A to $\mathcal{X}_o$, that is, A_o is the operator on $\mathcal{X}_o$ defined by $A_o = P_o A | \mathcal{X}_o$ where P_o is the orthogonal projection onto $\mathcal{X}_o$. The operator $A_{\bar{o}}$ on $\mathcal{X}_{\bar{o}}$ is given by the restriction of A to $\mathcal{X}_{\bar{o}}$. Furthermore, the operators B and C admit matrix representations of the form:

$$ B = \begin{bmatrix} B_{\bar{o}} \\ B_o \end{bmatrix} : \mathcal{U} \longrightarrow \begin{bmatrix} \mathcal{X}_{\bar{o}} \\ \mathcal{X}_o \end{bmatrix} \qquad \text{and} \qquad C = \begin{bmatrix} 0 & C_o \end{bmatrix} : \begin{bmatrix} \mathcal{X}_{\bar{o}} \\ \mathcal{X}_o \end{bmatrix} \longrightarrow \mathcal{Y}, \qquad (6.16) $$

respectively, where $B_o = P_o B$ and $C_o = C | \mathcal{X}_o$. Since $\mathcal{X}_{\bar{o}}$ is contained in the kernel of C, the first entry of C is zero, that is, $C_{\bar{o}} = C | \mathcal{X}_{\bar{o}} = 0$. Because $\mathcal{X}_o$ is invariant for A^*, it follows that $A_o^* = A^* | \mathcal{X}_o$.

We claim that $\{A_o, B_o, C_o, D\}$ is an observable realization of $\mathbf{G}$. Since $\mathcal{X}_o$ is a co-invariant subspace for A and $A_o = P_o A | \mathcal{X}_o$, it follows that $A_o^k = P_o A^k | \mathcal{X}_o$ for every integer $k \geq 0$. Because $C | \mathcal{X}_{\bar{o}} = 0$, we obtain that $(CA^k) | \mathcal{X}_o = (C | \mathcal{X}_o)(P_o A^k | \mathcal{X}_o) = C_o A_o^k$. Hence, the unobservable subspace associated with $\{C_o, A_o\}$ is given by

$$ \bigcap_{k=0}^{\infty} \ker C_o A_o^k = \bigcap_{k=0}^{\infty} \ker(CA^k | \mathcal{X}_o) = \left(\bigcap_{k=0}^{\infty} \ker CA^k \right) \bigcap \mathcal{X}_o = \mathcal{X}_{\bar{o}} \bigcap \mathcal{X}_o = \{0\} . $$

Since the unobservable subspace of $\{C_o, A_o\}$ is $\{0\}$, this pair is observable. For every non-negative integer k, we have $(CA^k) | \mathcal{X}_{\bar{o}} = 0$. Using $(CA^k) | \mathcal{X}_o = C_o A_o^k$, we obtain $CA^k B = (CA^k) | \mathcal{X}_o P_o B = C_o A_o^k B_o$. It now follows that

$$ \mathbf{G}(s) = C(sI - A)^{-1} B + D = C_o(sI - A_o)^{-1} B_o + D . $$

Hence, $\{A_o, B_o, C_o, D\}$ is a realization of $\mathbf{G}$.

We now demonstrate that the uncontrollable subspace $\mathcal{X}_{o\bar{c}}$ for $\{A_o, B_o\}$ is the intersection of the observable subspace $\mathcal{X}_o$ and uncontrollable subspace $\mathcal{X}_{\bar{c}}$ for $\{A, B, C, D\}$. Recall that

$$ \mathcal{X}_{o\bar{c}} = \bigcap_{k=0}^{\infty} \ker B_o^* A_o^{*k} . $$

Since $\mathcal{X}_o$ is invariant for A^* and $B_o^* = B^* | \mathcal{X}_o$, we see that $B_o^* A_o^{*k} = (B^* | \mathcal{X}_o) A^{*k} | \mathcal{X}_o = B^* A^{*k} | \mathcal{X}_o$. Hence, $\ker B_o^* A_o^{*k} = \mathcal{X}_o \bigcap \ker B^* A^{*k}$ and

$$ \mathcal{X}_{o\bar{c}} = \bigcap_{k=0}^{\infty} \ker B_o^* A_o^{*k} = \bigcap_{k=0}^{\infty} \left(\mathcal{X}_o \bigcap \ker B^* A^{*k} \right) = \mathcal{X}_o \bigcap \left(\bigcap_{k=0}^{\infty} \ker B^* A^{*k} \right) = \mathcal{X}_o \bigcap \mathcal{X}_{\bar{c}} , $$

that is, the uncontrollable subspace $\mathcal{X}_{o\bar{c}}$ for $\{A_o, B_o\}$ is the intersection of the observable subspace $\mathcal{X}_o$ and uncontrollable subspace $\mathcal{X}_{\bar{c}}$ for $\{A, B, C, D\}$. Since $\mathcal{X}_o$ and $\mathcal{X}_{\bar{c}}$ are co-invariant subspaces for A, it follows that their intersection $\mathcal{X}_{o\bar{c}}$ is also co-invariant for A. We have just demonstrated the following result which is the dual of Proposition 6.2.1.

Proposition 6.2.2 *Suppose* **G** *is the transfer function for* $\{A, B, C, D\}$ *and* P_o *is the orthogonal projection onto* $\mathcal{X}_o$ *the observable subspace for* $\{C, A\}$. *Let* A_o *on* $\mathcal{X}_o$ *and* B_o *from* $\mathcal{U}$ *into* $\mathcal{X}_o$ *and* C_o *from* $\mathcal{X}_o$ *into* $\mathcal{Y}$ *be the operators defined by*

$$A_o = P_o A | \mathcal{X}_o \quad and \quad B_o = P_o B \quad and \quad C_o = C | \mathcal{X}_o. \tag{6.17}$$

Then $\{A_o, B_o, C_o, D\}$ *is an observable realization of* **G** *whose uncontrollable subspace* $\mathcal{X}_{o\bar{c}}$ *is the co-invariant subspace for* A *given by* $\mathcal{X}_o \cap \mathcal{X}_{\bar{c}}$. *In particular, if* $\{A, B\}$ *is controllable, then* $\{A_o, B_o\}$ *is controllable.*

PROOF. One can also obtain a proof using duality. Notice that the transfer function for the system $\{A^*, C^*, B^*, D^*\}$ is given by $\mathbf{G}(\bar{s})^*$ and by duality $\mathcal{X}_o$ is the controllable subspace for $\{A^*, C^*\}$. Since $\mathcal{X}_o$ is an invariant subspace for A^*, we have $A_o^* = A^* | \mathcal{X}_o$. Moreover, $C_o^* = C^*$ and $B_o^* = B^* | \mathcal{X}_o$. According to Proposition 6.2.1, the system $\{A_o^*, C_o^*, B_o^*, D^*\}$ is a controllable realization of $\mathbf{G}(\bar{s})^*$. By taking adjoints and employing duality, we see that $\{A_o, B_o, C_o, D\}$ is an observable realization of **G**. By duality $\mathcal{X}_{o\bar{c}}$ is the unobservable subspace for $\{B_o^*, A_o^*\}$ and $\mathcal{X}_{\bar{c}}$ is the unobservable subspace for $\{B^*, A^*\}$. By applying the previous proposition to the system $\{A^*, C^*, B^*, D^*\}$, yields $\mathcal{X}_{o\bar{c}} = \mathcal{X}_o \cap \mathcal{X}_{\bar{c}}$. ∎

A controllable and observable realization. Using Proposition 6.2.1, one can extract a controllable realization of **G** and using Proposition 6.2.2, one can extract a observable realization of **G**. Now let us combine these results to obtain a realization of **G** which is both controllable and observable. To this end, let $\{A_c \text{ on } \mathcal{X}_c, B_c, C_c, D\}$ be the controllable realization of **G** given by Proposition 6.2.1. Recall $\mathcal{X}_{c\bar{o}}$, the unobservable subspace of $\{C_c, A_c\}$, which we call the controllable/unobservable subspace of $\{A, B, C, D\}$. Let $\mathcal{X}_{co}$ be the observable subspace for $\{C_c, A_c\}$, that is, $\mathcal{X}_{co} = \mathcal{X}_c \ominus \mathcal{X}_{c\bar{o}}$. We call $\mathcal{X}_{co}$ the *controllable/observable subspace* for $\{A, B, C, D\}$. Note that $\mathcal{X}_{co}$ is given by

$$\mathcal{X}_{co} = \bigvee_{k=0}^{\infty} A_c^{*k} C_c^* \mathcal{Y} = \bigvee \{A_c^{*k} C_c^* \mathcal{Y} : 0 \leq k < \dim \mathcal{X}_c\}. \tag{6.18}$$

Obviously, $\mathcal{X}_{co}$ is an invariant subspace for A_c^*. Also, $\mathcal{X}_c = \mathcal{X}_{co} \oplus \mathcal{X}_{c\bar{o}}$. This readily implies that the operator A_c admits a matrix representation of the form:

$$A_c = \begin{bmatrix} A_{c\bar{o}} & * \\ 0 & A_{co} \end{bmatrix} : \begin{bmatrix} \mathcal{X}_{c\bar{o}} \\ \mathcal{X}_{co} \end{bmatrix} \longrightarrow \begin{bmatrix} \mathcal{X}_{c\bar{o}} \\ \mathcal{X}_{co} \end{bmatrix}. \tag{6.19}$$

The operator A_{co} on $\mathcal{X}_{co}$ is the compression of A_c to the controllable/observable subspace $\mathcal{X}_{co}$. In this setting the matrix representations for B_c and C_c are given by

$$B_c = \begin{bmatrix} B_{c\bar{o}} \\ B_{co} \end{bmatrix} : \mathcal{U} \longrightarrow \begin{bmatrix} \mathcal{X}_{c\bar{o}} \\ \mathcal{X}_{co} \end{bmatrix} \quad and \quad C_c = \begin{bmatrix} 0 & C_{co} \end{bmatrix} : \begin{bmatrix} \mathcal{X}_{c\bar{o}} \\ \mathcal{X}_{co} \end{bmatrix} \longrightarrow \mathcal{Y}. \tag{6.20}$$

It follows from Proposition 6.2.2 that $\{A_{co}, B_{co}, C_{co}, D\}$ is a controllable and observable realization of **G**.

By embedding the matrix representations for A_c, B_c and C_c in (6.19) and (6.20) into the matrix representations for A, B and C in (6.10) and (6.11), we see that the operators A, B and C admit matrix representations of the form

$$
A = \begin{bmatrix} A_{c\bar{o}} & * & * \\ 0 & A_{co} & * \\ 0 & 0 & A_{\bar{c}} \end{bmatrix} : \begin{bmatrix} \mathcal{X}_{c\bar{o}} \\ \mathcal{X}_{co} \\ \mathcal{X}_{\bar{c}} \end{bmatrix} \longrightarrow \begin{bmatrix} \mathcal{X}_{c\bar{o}} \\ \mathcal{X}_{co} \\ \mathcal{X}_{\bar{c}} \end{bmatrix}
$$

(6.21)

$$
B = \begin{bmatrix} B_{c\bar{o}} \\ B_{co} \\ 0 \end{bmatrix} : \mathcal{U} \longrightarrow \begin{bmatrix} \mathcal{X}_{c\bar{o}} \\ \mathcal{X}_{co} \\ \mathcal{X}_{\bar{c}} \end{bmatrix} \quad \text{and} \quad C = \begin{bmatrix} 0 & C_{co} & C_{\bar{c}} \end{bmatrix} : \begin{bmatrix} \mathcal{X}_{c\bar{o}} \\ \mathcal{X}_{co} \\ \mathcal{X}_{\bar{c}} \end{bmatrix} \longrightarrow \mathcal{Y}.
$$

By using the matrix representations in (6.21), we see that the state space system in (6.7) admits a decomposition of the form

$$
\begin{bmatrix} \dot{x}_{c\bar{o}} \\ \dot{x}_{co} \\ \dot{x}_{\bar{c}} \end{bmatrix} = \begin{bmatrix} A_{c\bar{o}} & * & * \\ 0 & A_{co} & * \\ 0 & 0 & A_{\bar{c}} \end{bmatrix} \begin{bmatrix} x_{c\bar{o}} \\ x_{co} \\ x_{\bar{c}} \end{bmatrix} + \begin{bmatrix} B_{c\bar{o}} \\ B_{co} \\ 0 \end{bmatrix} u
$$

(6.22a)

$$
y = \begin{bmatrix} 0 & C_{co} & C_{\bar{c}} \end{bmatrix} \begin{bmatrix} x_{c\bar{o}} \\ x_{co} \\ x_{\bar{c}o} \end{bmatrix} + Du.
$$

(6.22b)

The state space representation in (6.22) or (6.25) below is usually referred to as the controllable and observable decomposition of the state space. This decomposition decomposes the state space $\mathcal{X}$ into its controllable and unobservable part $\mathcal{X}_{c\bar{o}}$, the controllable and observable part $\mathcal{X}_{co}$ and finally its uncontrollable part $\mathcal{X}_{\bar{c}}$. Summing up the previous analysis we obtain the following result which allows us to extract a controllable and observable realization from any system $\{A, B, C, D\}$.

Theorem 6.2.3 *Let* **G** *be the transfer function for* $\{A, B, C, D\}$ *and* P_{co} *the orthogonal projection onto* $\mathcal{X}_{co}$ *the controllable/observable subspace of* $\{A, B, C, D\}$. *Let* A_{co} *on* $\mathcal{X}_{co}$ *and* B_{co} *mapping* $\mathcal{U}$ *into* $\mathcal{X}_{co}$ *and* C_{co} *mapping* $\mathcal{X}_{co}$ *into* $\mathcal{Y}$ *be the operators defined by*

$$
A_{co} = P_{co}A|\mathcal{X}_{co} \quad \text{and} \quad B_{co} = P_{co}B \quad \text{and} \quad C_{co} = C|\mathcal{X}_{co}.
$$

(6.23)

Then $\{A_{co}, B_{co}, C_{co}, D\}$ *is a controllable and observable realization of* **G**.

Since $\mathcal{X}_c$ and $\mathcal{X}_{\bar{o}}$ are both invariant subspaces for A, it follows that $\mathcal{X}_{c\bar{o}} = \mathcal{X}_c \cap \mathcal{X}_{\bar{o}}$ is also an invariant subspace for A. Recall that $\mathcal{X}_{co} = \mathcal{X}_c \ominus \mathcal{X}_{c\bar{o}}$. Therefore, $\mathcal{X}_{co}$ is the semi-invariant subspace for A given by

$$
\mathcal{X}_{co} = \mathcal{X}_c \ominus (\mathcal{X}_c \cap \mathcal{X}_{\bar{o}}).
$$

(6.24)

One can even decompose the uncontrollable subspace $\mathcal{X}_{\bar{c}}$ further into the unobservable and observable subspaces associated with $\{C_{\bar{c}}, A_{\bar{c}}\}$. We call these subspaces the *uncontrollable/unobservable subspace* and the *uncontrollable/observable subspace* associated with $\{A, B, C, D\}$ and denote them by $\mathcal{X}_{\bar{c}\bar{o}}$ and $\mathcal{X}_{\bar{c}o}$ respectively. To be more specific, let

$$
\mathcal{X}_{\bar{c}o} = \bigvee_{k=0}^{\infty} A_{\bar{c}}^{*k} C_{\bar{c}}^{*} \mathcal{Y}
$$

and $\mathcal{X}_{\bar{c}} = \mathcal{X}_{\bar{c}\bar{o}} \oplus \mathcal{X}_{\bar{c}o}$. The operators $A_{\bar{c}}$, $B_{\bar{c}}$ and $C_{\bar{c}}$ admit matrix representations of the form

$$
A_{\bar{c}} = \begin{bmatrix} A_{\bar{c}\bar{o}} & * \\ 0 & A_{\bar{c}o} \end{bmatrix} : \begin{bmatrix} \mathcal{X}_{\bar{c}\bar{o}} \\ \mathcal{X}_{\bar{c}o} \end{bmatrix} \longrightarrow \begin{bmatrix} \mathcal{X}_{\bar{c}\bar{o}} \\ \mathcal{X}_{\bar{c}o} \end{bmatrix}
$$

$$
B_{\bar{c}} = \begin{bmatrix} 0 \\ 0 \end{bmatrix} : \mathcal{U} \longrightarrow \begin{bmatrix} \mathcal{X}_{\bar{c}\bar{o}} \\ \mathcal{X}_{\bar{c}o} \end{bmatrix} \quad \text{and} \quad C_{\bar{c}} = \begin{bmatrix} 0 & C_{\bar{c}o} \end{bmatrix} : \begin{bmatrix} \mathcal{X}_{\bar{c}\bar{o}} \\ \mathcal{X}_{\bar{c}o} \end{bmatrix} \longrightarrow \mathcal{Y}.
$$

Notice that the first entry in $C_{\bar{c}}$ is zero because the range of $C_{\bar{c}}^*$ is contained in $\mathcal{X}_{\bar{c}o}$. By inserting these matrix representations into (6.22) we obtain the following controllable and observable decomposition of the state space representation (6.7)

$$
\begin{bmatrix} \dot{x}_{c\bar{o}} \\ \dot{x}_{co} \\ \dot{x}_{\bar{c}\bar{o}} \\ \dot{x}_{\bar{c}o} \end{bmatrix} = \begin{bmatrix} A_{c\bar{o}} & * & * & * \\ 0 & A_{co} & * & * \\ 0 & 0 & A_{\bar{c}\bar{o}} & * \\ 0 & 0 & 0 & A_{\bar{c}o} \end{bmatrix} \begin{bmatrix} x_{c\bar{o}} \\ x_{co} \\ x_{\bar{c}\bar{o}} \\ x_{\bar{c}o} \end{bmatrix} + \begin{bmatrix} B_{c\bar{o}} \\ B_{co} \\ 0 \\ 0 \end{bmatrix} u \qquad (6.25\text{a})
$$

$$
y = \begin{bmatrix} 0 & C_{co} & 0 & C_{\bar{c}o} \end{bmatrix} \begin{bmatrix} x_{c\bar{o}} \\ x_{co} \\ x_{\bar{c}\bar{o}} \\ x_{\bar{c}o} \end{bmatrix} + Du. \qquad (6.25\text{b})
$$

Notice that in this representation any one or even all of the spaces $\mathcal{X}_{c\bar{o}}$, $\mathcal{X}_{\bar{c}\bar{o}}$ or $\mathcal{X}_{\bar{c}o}$ can be zero. However, the controllable and observable space $\mathcal{X}_{co}$ is zero if and only if the transfer function $\mathbf{G} = D$ is constant. Finally, it is noted that for A, the subspace $\mathcal{X}_{c\bar{o}}$ is invariant, $\mathcal{X}_{\bar{c}o}$ is co-invariant, while $\mathcal{X}_{co}$ and $\mathcal{X}_{\bar{c}\bar{o}}$ are both semi-invariant.

6.2.1 A minimal realization procedure

To complete this section, we demonstrate how one can use the singular value decomposition to extract a controllable and observable realization from a finite dimensional system. Recall that a system $\{A \text{ on } \mathcal{X}, B, C, D\}$ is unitarily equivalent to $\{\tilde{A} \text{ on } \tilde{\mathcal{X}}, \tilde{B}, \tilde{C}, \tilde{D}\}$ if there exists a unitary operator W mapping $\tilde{\mathcal{X}}$ onto $\mathcal{X}$ such that $AW = W\tilde{A}$, $B = W\tilde{B}$, $CW = \tilde{C}$, and $D = \tilde{D}$. Clearly, unitary equivalence preserves stability, controllability, and observability.

Procedure. Let $\{A \text{ on } \mathcal{X}, B, C, D\}$ be an n-dimensional realization of $\mathbf{G}$. Let $U_c \Lambda V_c^*$ be a singular value decomposition of the corresponding controllability matrix W_c, that is,

$$
W_c = \begin{bmatrix} B & AB & \cdots & A^{n-1}B \end{bmatrix} = U_c \Lambda V_c^*. \qquad (6.26)
$$

(A review of the singular value decomposition is given in Section 16.6 in the Appendix. In our singular value decomposition $U \Lambda V^*$ the operators U and V are isometries and Λ is an invertible diagonal matrix consisting of the nonzero singular values.) Let n_c be the rank of W_c, or equivalently, assume that W_c has n_c (nonzero) singular values. Then U_c is an isometry from $\mathbb{C}^{n_c}$ into $\mathcal{X}$. Let $\{\tilde{A}_c \text{ on } \mathbb{C}^{n_c}, \tilde{B}_c, \tilde{C}_c, D\}$ be the system defined by

$$
\tilde{A}_c = U_c^* A U_c, \qquad \tilde{B}_c = U_c^* B \qquad \text{and} \qquad \tilde{C}_c = C U_c. \qquad (6.27)
$$

Let $U_{co}\Lambda_{co}V_{co}^*$ be a singular value decomposition of the observability matrix associated with the pair $\{\tilde{C}_c, \tilde{A}_c\}$, that is,

$$W_{co} = \begin{bmatrix} \tilde{C}_c \\ \tilde{C}_c\tilde{A}_c \\ \vdots \\ \tilde{C}_c\tilde{A}_c^{n_c-1} \end{bmatrix} = U_{co}\Lambda_{co}V_{co}^*. \tag{6.28}$$

Let n_{co} be the number of (nonzero) singular values of W_{co}. Then V_{co} is an isometry from $\mathbb{C}^{n_{co}}$ into $\mathbb{C}^{n_c}$. Let $\{\tilde{A}_{co}$ on $\mathbb{C}^{n_{co}}, \tilde{B}_{co}, \tilde{C}_{co}, D\}$ be the system defined by

$$\tilde{A}_{co} = V_{co}^*\tilde{A}_cV_{co}, \qquad \tilde{B}_{co} = V_{co}^*\tilde{B}_c \qquad \text{and} \qquad \tilde{C}_{co} = \tilde{C}_cV_{co}. \tag{6.29}$$

Using (6.27) it follows that

$$\tilde{A}_{co} = V_{co}^*U_c^*AU_cV_{co}, \qquad \tilde{B}_{co} = V_{co}^*U_c^*B \qquad \text{and} \qquad \tilde{C}_{co} = CU_cV_{co}. \tag{6.30}$$

Then $\{\tilde{A}_{co}, \tilde{B}_{co}, \tilde{C}_{co}, D\}$ is a controllable and observable realization of $\mathbf{G}$ and is unitarily equivalent to the system $\{A_{co}, B_{co}, C_{co}, D\}$ defined in Theorem 6.2.3.

We now verify that $\{\tilde{A}_{co}, \tilde{B}_{co}, \tilde{C}_{co}, D\}$ is a controllable and observable realization of $\mathbf{G}$. Since the controllable subspace $\mathcal{X}_c$ equals the range of W_c, the orthogonal projection P_c onto $\mathcal{X}_c$ is given by $P_c = U_cU_c^*$. (If U is any isometry, then UU^* is the orthogonal projection onto the range of U; see Lemma 16.2.3.) Let $\{A_c$ on $\mathcal{X}_c, B_c, C_c, D\}$ be the controllable realization of $\mathbf{G}$ defined in Proposition 6.2.1. Let W be the unitary operator from $\mathbb{C}^{n_c}$ onto $\mathcal{X}_c$ defined by $Wx = U_cx$. Then,

$$\begin{aligned} A_cW &= P_cAU_c = WU_c^*AU_c = W\tilde{A}_c \\ B_c &= P_cB = WU_c^*B = W\tilde{B}_c \\ C_cW &= CU_c = \tilde{C}_c. \end{aligned} \tag{6.31}$$

Therefore, $\{\tilde{A}_c$ on $\mathbb{C}^{n_c}, \tilde{B}_c, \tilde{C}_c, D\}$ is unitarily equivalent to $\{A_c, B_c, C_c, D\}$ and must be a controllable realization of $\mathbf{G}$.

Let $\tilde{\mathcal{X}}_{co}$ be the observable subspace associated with $\{\tilde{C}_c, \tilde{A}_c\}$, that is, $\tilde{\mathcal{X}}_{co} = \bigvee_0^\infty \tilde{A}_c^{*k}\tilde{C}_c^*\mathcal{Y}$. Using $W\tilde{A}_c^* = A_c^*W$ and $W^*C_c^* = \tilde{C}_c^*$, we see that the controllable/observable subspace $\mathcal{X}_{co}$ of the system $\{A_c, B_c, C_c, D\}$ is given by

$$\mathcal{X}_{co} = \bigvee_{k=0}^\infty A_c^{*k}C_c^*\mathcal{Y} = \bigvee_{k=0}^\infty W\tilde{A}_c^{*k}W^*C_c^*\mathcal{Y} = U_c\tilde{\mathcal{X}}_{co}.$$

So, the isometry U_c maps $\tilde{\mathcal{X}}_{co}$ onto the subspace $\mathcal{X}_{co}$. Since $\tilde{\mathcal{X}}_{co}$ equals the range of W_{co}^* and $W_{co}^* = V_{co}\Lambda_{co}U_{co}^*$ is a singular value decomposition of W_{co}^*, it follows that V_{co} is an isometry from $\mathbb{C}^{n_{co}}$ into $\mathbb{C}^{n_c}$ whose range is $\tilde{\mathcal{X}}_{co}$. Thus, U_cV_{co} is an isometry from $\mathbb{C}^{n_{co}}$ into $\mathcal{X}$ whose range is $\mathcal{X}_{co}$. In particular, $P_{co} = U_cV_{co}V_{co}^*U_c^*$ is the orthogonal projection onto $\mathcal{X}_{co}$. Let Z

be the unitary operator from $C^{n_{co}}$ onto $\mathcal{X}_{co}$ defined by $Z = U_c V_{co}$. Then using (6.30), we obtain

$$
\begin{aligned}
A_{co} Z &= P_{co} A Z = Z V_{co}^* U_c^* A Z = Z \tilde{A}_{co} \\
B_{co} &= P_{co} B = Z V_{co}^* U_c^* B = Z \tilde{B}_{co} \\
C_{co} Z &= C U_c V_{co} = \tilde{C}_{co} .
\end{aligned}
$$

Hence, $\{\tilde{A}_{co}, \tilde{B}_{co}, \tilde{C}_{co}, D\}$ is unitarily equivalent to $\{A_{co}, B_{co}, C_{co}, D\}$. Therefore, the system $\{\tilde{A}_{co} \text{ on } \mathbb{C}^{n_{co}}, \tilde{B}_{co}, \tilde{C}_{co}, D\}$ is a controllable and observable realization of $\mathbf{G}$.

6.3 The minimal polynomial and realizations

In this section we will show that if $\{A, B, C, D\}$ is a controllable and observable realization of $\mathbf{G}$, then λ is a pole of $\mathbf{G}$ if and only if λ is an eigenvalue of A. Moreover, the roots (multiplicities included) of the minimal polynomial of A are the poles of $\mathbf{G}$.

To this end, let A be an operator on a finite dimensional space $\mathcal{X}$. Recall that a polynomial m is the *minimal polynomial* of A if m is the monic polynomial of the lowest degree such that $m(A) = 0$. The minimal polynomial of A is unique and is denoted by m_A. If p is another minimal polynomial for A, then $q = p - m_A$ satisfies $q(A) = 0$. Since q has lower degree than the minimal polynomial, it follows that $q(s) \equiv 0$, and thus, $p = m_A$. So, the minimal polynomial is unique.

Moreover, if p is any nontrivial polynomial satisfying $p(A) = 0$, then m_A divides p, that is, $p = m_A r$ where r is a polynomial. Since $\deg p \geq \deg m_A$, the Euclidean algorithm shows that $p = q + m_A r$ where q and r are polynomials and $\deg q < \deg m_A$. Using $m_A(A) = 0$, we have

$$ 0 = p(A) = q(A) + m_A(A) r(A) = q(A) . $$

Hence, $q(A) = 0$. Because m_A is the minimal polynomial of A and $\deg q < \deg m_A$, it follows that $q(s) \equiv 0$. Thus, $p = m_A r$ and m_A divides p.

As before, let A be an operator on a finite dimensional space $\mathcal{X}$. We claim that λ is an eigenvalue of A if and only if λ is a root of the minimal polynomial of A, that is, $m_A(\lambda) = 0$. Recall from the Cayley Hamilton Theorem that $p(A) = 0$ where $p(s) = \det[sI - A]$ is the characteristic polynomial of A. Hence, the minimal polynomial of A divides the characteristic polynomial of A. From this it follows that all the roots of the minimal polynomial of A are also roots of the characteristic polynomial for A. Hence, every root of the minimal polynomial of A is an eigenvalue of A. We now show the converse, that is, if λ is an eigenvalue for A, then $m_A(\lambda) = 0$. Consider any eigenvalue λ of A and let v be any eigenvector corresponding. Using $Av = \lambda v$ we have $0 = m_A(A)v = m_A(\lambda)v$. Because v is nonzero, $m_A(\lambda)$ must be zero. Hence, λ is a root of m_A.

Since the roots of the minimal polynomial m_A of A correspond to the eigenvalues of A,

$$ m_A(s) = \prod_{i=1}^{l} (s - \lambda_i)^{k_i} $$

where $\{\lambda_1, \lambda_2, \cdots, \lambda_l\}$ are the distinct eigenvalues of A and k_i is the multiplicity of λ_j as a root of m_A. Finally, we see that if each root the characteristic polynomial is a simple root (that is, it has multiplicity one), then the minimal polynomial equals the characteristic polynomial.

We say that a complex number λ is a *pole of order j* of a rational transfer function $\mathbf{G}$ if λ is a pole of $\mathbf{G}$ and j is the smallest integer such that

$$(s - \lambda)^j \mathbf{G}(s)$$

is analytic at $s = \lambda$. Let m be the polynomial whose roots are the poles (multiplicity included) of $\mathbf{G}$, that is,

$$m(s) = \prod_{i=1}^{l}(s - \lambda_i)^{k_i} \tag{6.32}$$

where $\{\lambda_1, \lambda_2, \cdots, \lambda_l\}$ are the distinct poles of $\mathbf{G}$ and k_i is the order of λ_i. Obviously, $m\mathbf{G}$ is analytic everywhere. Since a rational analytic function must be a polynomial, it follows that $\mathbf{G} = M/m$ where M is an operator valued polynomial. Now assume that $\mathbf{G}$ is a rational function of the form $\mathbf{G} = N/d$ where N is an operator valued polynomial and d is a scalar valued polynomial. Then the poles (multiplicity included) of $\mathbf{G}$ must be contained in the zeros of d. Hence, m divides d, that is, $d = mr$ where r is a polynomial. So, m is the unique monic scalar valued polynomial of the lowest possible degree appearing in the denominator of $\mathbf{G}$. To be precise, m is the only monic scalar valued polynomial such that $\mathbf{G} = M/m$ where M is an operator valued polynomial and $\deg m \leq \deg d$ for any $N/d = \mathbf{G}$ where N is an operator valued polynomial and d is a scalar valued polynomial. We now show that the poles of $\mathbf{G}$ are precisely the eigenvalues of any controllable and observable realization of $\mathbf{G}$.

Theorem 6.3.1 *Let $\{A, B, C, D\}$ be a controllable and observable finite dimensional realization of $\mathbf{G}$. Then the following holds.*

(i) *The transfer function $\mathbf{G}$ admits a decomposition of the form $\mathbf{G} = N/d$ where N is an operator valued polynomial and d is a scalar valued polynomial if and only if $d(A) = 0$.*

(ii) *The minimal polynomial m for A is given by (6.32) where $\{\lambda_1, \lambda_2, \cdots, \lambda_l\}$ are the distinct poles of $\mathbf{G}$ and k_i is the order of λ_i.*

(iii) *A complex number λ is a pole of $\mathbf{G}$ if and only if λ is an eigenvalue of A.*

(iv) *The operator A is stable if and only if all the poles of $\mathbf{G}$ have nonzero negative real parts.*

PROOF. For the moment assume that Part (i) holds. Let m_A be the minimal polynomial of A and m the polynomial in (6.32) formed by the poles of $\mathbf{G}$ including their multiplicity. Since $\mathbf{G} = N/m$ for some operator valued polynomial N, Part (i) shows that $m(A) = 0$. Hence, m_A divides m. Because $m_A(A) = 0$, Part (i) also implies that $\mathbf{G} = M/m_A$ for some operator valued polynomial M. Recall that if $\mathbf{G} = N/d$ where N is an operator valued polynomial and d is a scalar valued polynomial, then m divides d. So, m must divide m_A. Therefore, $m_A = m$ and Part(ii) holds. Parts (iii) and (iv) follow from the fact that λ is a **root** of the minimal polynomial of A if and only if λ is an eigenvalue of A.

To prove Part (i), we first suppose that $\mathbf{G}$ admits a decomposition of the form $\mathbf{G} = N/d$ where N is an operator valued polynomial and d is a scalar valued polynomial. Without loss of generality we can assume that d is a monic polynomial given by $d(s) = \Sigma_{i=0}^n d_i s^i$. Recall from (1.26) in Chapter 1 that

$$d_0 G_k + d_1 G_{k+1} + \cdots + d_{n-1} G_{k+n-1} + G_{k+n} = 0 \qquad (k \geq 1) \qquad (6.33)$$

where G_i is the coefficient of $1/s^i$ in the power series expansion (1.19) of $\mathbf{G}$. Since $G_i = CA^{i-1}B$ for $i \geq 1$ condition (6.33) is equivalent to

$$CA^{k-1} d(A) B = 0 \qquad (k \geq 1). \qquad (6.34)$$

Since A commutes with $d(A)$, this implies that

$$CA^i d(A) A^j B = 0 \qquad (i, j \geq 0). \qquad (6.35)$$

Consider any integer $i \geq 0$. Since $CA^i d(A) A^j B = 0$ for all integers $j \geq 0$, it follows from the controllability of $\{A, B\}$ that $CA^i d(A) = 0$. Since $CA^i d(A) = 0$ for all integers $i \geq 0$, it now follows from the observability of $\{C, A\}$ that $d(A) = 0$.

Now consider any monic polynomial d for which $d(A) = 0$. Clearly, (6.34) and hence (6.33) or (1.26) holds. Now let $N(s) = \Sigma_{i=0}^n N_i s^i$ where the operators $\{N_i\}_0^n$ are computed from (1.24). Since relationships (1.24) and (1.26) are equivalent to (1.22), we obtain $N = d\mathbf{G}$. Therefore, $\mathbf{G} = N/d$ which proves Part (i). ∎

If $\{A, B, C, D\}$ is a controllable and observable realization of $\mathbf{G}$, then Theorem 6.3.1 shows that λ is a pole of G if and only if λ is an eigenvalue of A. In particular, $\mathbf{G}$ is analytic in $\Re(s) \geq 0$ if and only if A is stable.

Let $\{A, B, C, D\}$ be a controllable and observable realization of a transfer $\mathbf{G}$ with values in $\mathcal{L}(\mathbb{C}^j, \mathbb{C}^k)$. As before, assume that $\mathbf{G} = N/d$ where N is an operator valued polynomial and d is a scalar valued polynomial. Let d_r be the greatest common divisor of all the entries of N and d. Then according to Theorem 6.3.1, the minimal polynomial for A is given by $m_A = d/d_r$.

Remark 6.3.1 (Minimal polynomial) Consider any operator A on $\mathbb{C}^n$. Let d_r be the greatest common divisor of all the minors of order $n - 1$ of $sI - A$. In other words, d_r is the greatest common divisor of all the entries of adj $(sI - A)$. Clearly, the polynomial d_r is also a divisor of the characteristic polynomial d of A. Obviously, $\{A, I, I, 0\}$ is a controllable and observable realization of $(sI - A)^{-1} = \text{adj} \, (sI - A)/d(s)$. It now follows from Theorem 6.3.1 that the minimal polynomial m_A of A is given by $m_A = d/d_r$. As a consequence of this result, one can readily show that the minimal polynomial of a companion matrix of the form

$$A = \begin{bmatrix} 0 & 1 & 0 & \cdots & 0 \\ 0 & 0 & 1 & \cdots & 0 \\ \vdots & \vdots & & \ddots & \vdots \\ 0 & 0 & 0 & \cdots & 1 \\ -a_0 & -a_1 & -a_2 & \cdots & -a_{n-1} \end{bmatrix} \qquad (6.36)$$

is the same as the characteristic polynomial of the matrix. To see this, note that the minor of

$$sI - A = \begin{bmatrix} s & -1 & 0 & \cdots & 0 \\ 0 & s & -1 & \cdots & 0 \\ \vdots & \vdots & & \ddots & \vdots \\ 0 & 0 & 0 & \cdots & -1 \\ a_0 & a_1 & a_2 & \cdots & s + a_{n-1} \end{bmatrix}$$

corresponding to a_0 is simply the determinant of a lower triangular matrix whose diagonal elements are all equal to -1. Since this minor is ± 1, it follows that the greatest common divisor of all the minors of order $n-1$ is a constant. Hence, the minimal polynomial of A is the same as the characteristic polynomial of A which is given by

$$d(s) = a_0 + a_1 s + \cdots + a_{n-1} s^{n-1} + s^n. \tag{6.37}$$

For another proof of this fact, let d be the previous polynomial in (6.37) and A the corresponding companion matrix in (6.36). Let B be the column vector in $\mathbb{C}^n$ and C be the $1 \times n$ row vector in defined by

$$B = \begin{bmatrix} 0 & 0 & \cdots & 0 & 1 \end{bmatrix}^{tr} \quad \text{and} \quad C = \begin{bmatrix} 1 & 0 & \cdots & 0 & 0 \end{bmatrix}.$$

Recall that $\{A, B, C, D\}$ is a realization of $1/d$; see equations (1.47) and (1.48) in Section 1.4. It is easy to verify that this system is controllable and observable. Therefore, $\{A, B, C, D\}$ is a controllable and observable realization of $1/d$. Obviously, the poles of $1/d$ are precisely the zeros of d. According to Theorem 6.3.1, the polynomial d must be the minimal polynomial of A. Finally, because A is an $n \times n$ matrix and the degree of the minimal polynomial of A is n, it follows that d is also the characteristic polynomial for A.

Exercise 11 Let A be the block companion matrix defined by

$$A = \begin{bmatrix} 0 & I & 0 & \cdots & 0 \\ 0 & 0 & I & \cdots & 0 \\ \vdots & \vdots & & \ddots & \vdots \\ 0 & 0 & 0 & \cdots & I \\ -a_0 I & -a_1 I & -a_2 I & \cdots & -a_{n-1} I \end{bmatrix} \quad \text{on} \quad \begin{bmatrix} \mathcal{Y} \\ \mathcal{Y} \\ \vdots \\ \mathcal{Y} \\ \mathcal{Y} \end{bmatrix} \tag{6.38}$$

where $\mathcal{Y}$ is a k-dimensional space. Show that the polynomial d in (6.37) is the minimal polynomial for A. Moreover, show that d^k is the characteristic polynomial for A.

6.4   The inverse of a transfer function

Now suppose that $\{A, B, C, D\}$ is a realization of a proper rational function $\mathbf{G}$ with D invertible. Then whenever s is not an eigenvalue of $A - BD^{-1}C$, the operator $\mathbf{G}(s)$ is invertible and

$$\mathbf{G}(s)^{-1} = D^{-1} - D^{-1}C(sI - A + BD^{-1}C)^{-1}BD^{-1}. \tag{6.39}$$

In other words,

$$\{A - BD^{-1}C,\ BD^{-1},\ -D^{-1}C,\ D^{-1}\} \tag{6.40}$$

is a realization of $\mathbf{G}^{-1}$.

To see this, let $\Phi(s) = (sI - A)^{-1}$. Using the identity $(I + T)^{-1} = I - (I + T)^{-1}T$ it follows, except possibly for a finite number of values for s, that

$$
\begin{aligned}
\mathbf{G}(s)^{-1} &= (D + C\Phi(s)B)^{-1} = D^{-1}(I + C\Phi(s)BD^{-1})^{-1}\\
&= D^{-1} - D^{-1}(I + C\Phi(s)BD^{-1})^{-1}C\Phi(s)BD^{-1}\\
&= D^{-1} - D^{-1}C(I + \Phi(s)BD^{-1}C)^{-1}\Phi(s)BD^{-1}\\
&= D^{-1} - D^{-1}C(sI - A + BD^{-1}C)^{-1}BD^{-1}\,.
\end{aligned}
$$

In achieving the fourth equality we used $(I + RQ)^{-1}R = R(I + QR)^{-1}$. Therefore, (6.39) holds and $\{A - BD^{-1}C,\ BD^{-1},\ -D^{-1}C,\ D^{-1}\}$ is a realization of $\mathbf{G}^{-1}$. In particular, this shows that $\mathbf{G}^{-1}$ is proper rational function.

For a state space proof of this result recall that $\mathbf{G}$ is the transfer function of the system

$$\dot{x} = Ax + Bu \quad \text{and} \quad y = Cx + Du\,. \tag{6.41}$$

To be precise, when all the initial conditions are set equal to zero, then $\mathbf{y} = \mathbf{G}\mathbf{u}$. (Recall that $\mathbf{f}$ denotes the Laplace transform f.) If D is invertible, then $u = -D^{-1}Cx + D^{-1}y$. Substituting this into (6.41), yields

$$\dot{x} = (A - BD^{-1}C)x + BD^{-1}y \quad \text{and} \quad u = -D^{-1}Cx + D^{-1}y\,. \tag{6.42}$$

By taking the Laplace transform with all the initial conditions set equal to zero, we arrive at

$$\mathbf{u} = \left(D^{-1} - D^{-1}C(sI - A + BD^{-1}C)^{-1}BD^{-1}\right)\mathbf{y}\,.$$

Since $\mathbf{y} = \mathbf{G}\mathbf{u}$, this implies that the inverse of $\mathbf{G}$ exists and is given by (6.39). This analysis proves Part (i) of the following result.

Proposition 6.4.1 *Let* $\Sigma = \{A, B, C, D\}$ *be a realization for a proper rational function* $\mathbf{G}$ *with* D *invertible. Then the following holds.*

(i) *The system* $\Sigma_1 = \{A - BD^{-1}C,\ BD^{-1},\ -D^{-1}C,\ D^{-1}\}$ *is a realization for* $\mathbf{G}^{-1}$.

(ii) *The system* Σ *is controllable, respectively observable, if and only if* Σ_1 *is controllable, respectively observable.*

(iii) *If* $\{A, B, C, D\}$ *is controllable and observable, then* λ *is a pole of* $\mathbf{G}^{-1}$ *if and only if* λ *is an eigenvalue of* $A - BD^{-1}C$. *In this case,* $A - BD^{-1}C$ *is stable if and only if* $\mathbf{G}^{-1}$ *is analytic in the closed right half plane.*

PROOF. Part (ii) follows from Lemma 6.4.2 below. If Σ is a controllable and observable realization, then Σ_1 is a controllable and observable realization of $\mathbf{G}^{-1}$. So, Part (iii) is a consequence of Theorem 6.3.1. ∎

Lemma 6.4.2 *Consider a system* $\Sigma = \{A$ *on* $\mathcal{X}, B, C, D\}$ *where* B *maps* $\mathcal{U}$ *into* $\mathcal{X}$ *and* C *maps* $\mathcal{X}$ *into* $\mathcal{Y}$. *Let* R *be any operator from* $\mathcal{Y}$ *into* $\mathcal{U}$. *Then* Σ *is controllable, respectively observable, if and only if* $\{A - BRC, B, C, D\}$ *is controllable, respectively observable.*

PROOF. Recall that the PBH test shows that a pair $\{A, B\}$ is controllable if and only if the rank of $[A - \lambda I, B]$ equals the dimension of the state space for all complex numbers λ; see Lemma 5.2.1. Notice that

$$\begin{bmatrix} A - \lambda I & B \end{bmatrix} \text{ and } \begin{bmatrix} A - BRC - \lambda I & B \end{bmatrix}$$

have the same rank. Therefore, $\{A, B\}$ is controllable if and only if $\{A - BRC, B\}$ is controllable. One can also prove this fact by noting that $\{A^n B \mathcal{U}\}_0^\infty$ and $\{(A - BRC)^n B \mathcal{U}\}_0^\infty$ span the same space. Because $\{C, A\}$ is observable if and only if the pair $\{A^*, C^*\}$ is controllable, it follows that $\{C, A\}$ is observable if and only if the pair $\{C, A - BRC\}$ is observable. ∎

6.5   Notes

The controllable and observable decomposition is due to Gilbert [51] and Kalman [71]. This decomposition is a classical result in linear systems. Obviously, the controllable and observable subspace $\mathcal{X}_{co}$ is a semi-invariant subspace for A. The concept of a semi-invariant subspace is due to Sarason [111]. For more on minimal polynomials, invariant polynomials and their relationship to companion matrices see Gantmacher [48].

Chapter 7

More Realization Theory

This chapter is devoted to realization theory, that is, finding state space models for proper rational transfer functions. It is noted that only the results in Sections 7.1 and 7.2 are used in the rest of the monograph. The remaining sections of this chapter are of independent interest.

Consider the state space system of the form

$$\dot{x} = Ax + Bu \text{ and } y = Cx + Du \tag{7.1}$$

where A is an operator on a finite dimensional space $\mathcal{X}$ and B is an operator mapping $\mathcal{U}$ into $\mathcal{X}$, while C maps $\mathcal{X}$ into $\mathcal{Y}$ and D maps $\mathcal{U}$ into $\mathcal{Y}$. Throughout it is always assumed that the input space $\mathcal{U}$ and output space $\mathcal{Y}$ are finite dimensional Hilbert spaces. Recall that $\{A, B, C, D\}$ is a realization of $\mathbf{G}$ if $\mathbf{G}$ is the transfer function for (7.1), that is,

$$\mathbf{G}(s) = C(sI - A)^{-1}B + D. \tag{7.2}$$

The transfer function $\mathbf{G}$ for any finite dimensional system is a proper rational function. Moreover, $\mathbf{G}$ is a strictly proper rational function if and only if D equals zero. This naturally leads to the following realization problem which is the subject of this chapter: Given an operator valued proper rational function $\mathbf{G}$, find a finite dimensional realization for $\mathbf{G}$. In particular, we are interested in finding a minimal realization for $\mathbf{G}$, that is, a realization of the lowest possible state dimension.

In this chapter we use the shift operator to demonstrate that any proper rational function admits a finite dimensional realization. Moreover, we show that all minimal realizations of the same transfer function are similar. Furthermore, a realization is minimal if and only if it is controllable and observable. Finally, we will also present the Kalman-Ho partial realization algorithm.

7.1 The restricted backward shift realization

We say that a realization of a transfer function $\mathbf{G}$ is a *minimal realization* if the dimension of its state space is less than or equal to the state space dimension of any other realization of $\mathbf{G}$. Clearly, all minimal realizations of $\mathbf{G}$ have the same state space dimension. The dimension of

83

the state space for any minimal realization of $\mathbf{G}$ is called the *McMillan degree of* $\mathbf{G}$. Using block companion matrices, Section 1.4 demonstrates how to construct finite dimensional realizations for any proper rational transfer function. However, these realizations may not be minimal. This leads to the following basic realization problem. Given any proper rational function $\mathbf{G}$, find all minimal realizations for $\mathbf{G}$.

Two systems $\Sigma_1 = \{A_1 \text{ on } \mathcal{X}_1, B_1, C_1, D_1\}$ and $\Sigma_2 = \{A_2 \text{ on } \mathcal{X}_2, B_2, C_2, D_2\}$ are *similar* if there exists an invertible operator T mapping $\mathcal{X}_1$ onto $\mathcal{X}_2$ satisfying

$$TA_1 = A_2T, \qquad TB_1 = B_2, \qquad C_1 = C_2T, \qquad D_1 = D_2. \tag{7.3}$$

In this case, we say that T *intertwines* Σ_1 with Σ_2. It is easy to show that similar realizations have the same transfer function. However, two realizations of the same transfer function are not necessarily similar. In fact, they may not even have the same state dimension. Clearly, the similarity relationship is transitive, that is, if realization Σ_1 is similar to realization Σ_2 and realization Σ_2 is similar to realization Σ_3, then realization Σ_1 is similar to realization Σ_3.

Let $\mathbf{G}$ be a proper rational function with values in $\mathcal{L}(\mathcal{U}, \mathcal{Y})$. Then $\mathbf{G}$ admits a power series expansion of the form $\mathbf{G}(s) = \sum_0^\infty G_i/s^i$ where G_i is in $\mathcal{L}(\mathcal{U}, \mathcal{Y})$ for all integers $i \geq 0$. Proposition 1.3.3 shows that $\{A, B, C, D\}$ is a realization for $\mathbf{G}$ if and only if

$$G_0 = D \qquad \text{and} \qquad G_i = CA^{i-1}B \qquad \text{(for all } i \geq 1 \text{)}. \tag{7.4}$$

Obviously, the transfer function $\mathbf{G}$ and the sequence of operators $\{G_i\}_0^\infty$ uniquely determine each other. Our theoretical developments will require the notion of a realization whose state space is infinite dimensional. So, without loss of generality we say that $\{A, B, C, D\}$ is a *realization of a sequence of operators* $\{G_i\}_0^\infty$ if (7.4) holds.

We now show how one can readily construct a realization of $\{G_i\}_0^\infty$ using the backward shift operator. To begin, consider any proper rational function $\mathbf{G}$ whose values are linear operators from $\mathcal{U}$ into $\mathcal{Y}$. Let $l_+(\mathcal{Y})$ be the linear space consisting of all infinite sequences with values in $\mathcal{Y}$, that is, $l_+(\mathcal{Y})$ consists of all vectors of the form

$$f = \begin{bmatrix} f_1 \\ f_2 \\ f_3 \\ \vdots \end{bmatrix} \qquad \text{(with } f_i \in \mathcal{Y} \text{ for all } i\text{)}. \tag{7.5}$$

Obviously, $l_+(\mathcal{Y})$ is an infinite dimensional linear space. We do not need a topological structure or an inner product on $l_+(\mathcal{Y})$. Now let S be *the backward shift operator* on $l_+(\mathcal{Y})$ defined by

$$S \begin{bmatrix} f_1 \\ f_2 \\ f_3 \\ \vdots \end{bmatrix} = \begin{bmatrix} f_2 \\ f_3 \\ f_4 \\ \vdots \end{bmatrix} = \begin{bmatrix} 0 & I & 0 & 0 & \cdots \\ 0 & 0 & I & 0 & \cdots \\ 0 & 0 & 0 & I & \cdots \\ \vdots & \vdots & \vdots & & \ddots \end{bmatrix} \begin{bmatrix} f_1 \\ f_2 \\ f_3 \\ \vdots \end{bmatrix}. \tag{7.6}$$

Let E be *the evaluation operator* mapping $l_+(\mathcal{Y})$ into $\mathcal{Y}$, picking out the first component of the vector f in $l_+(\mathcal{Y})$, that is, $Ef = f_1$, or equivalently,

$$E = \begin{bmatrix} I & 0 & 0 & \cdots \end{bmatrix}. \tag{7.7}$$

Finally, let L be the *initialization operator* mapping $\mathcal{U}$ into $l_+(\mathcal{Y})$ defined by

$$L = \begin{bmatrix} G_1 \\ G_2 \\ G_3 \\ \vdots \end{bmatrix}, \tag{7.8}$$

where $\{G_i\}_0^\infty$ are the coefficients obtained from the power series expansion $\mathbf{G}(s) = \sum_0^\infty s^{-i} G_i$. It is easy to verify that

$$G_i = ES^{i-1}L \quad \text{(for all} \quad i \geq 1 \text{)}.$$

Therefore, $\{S \text{ on } l_+(\mathcal{Y}), L, E, G_0\}$ is a realization of $\mathbf{G}$. However, this realization has an infinite dimensional state space $l_+(\mathcal{Y})$ and consequently is not minimal.

To obtain a minimal realization, let $\mathcal{H}_G$ be the controllable subspace of $l_+(\mathcal{Y})$ defined by

$$\mathcal{H}_G = \text{span} \left\{ S^i L \mathcal{U} : i = 0, 1, 2, \cdots \right\}. \tag{7.9}$$

Clearly, $\mathcal{H}_G$ is an invariant subspace for S, that is, $S\mathcal{H}_G \subset \mathcal{H}_G$. Let $\{A_G \text{ on } \mathcal{H}_G, B_G, C_G, D_G\}$ be the system defined by

$$A_G = S|\mathcal{H}_G, \quad B_G = L, \quad C_G = E|\mathcal{H}_G \quad \text{and} \quad D_G = G_0, \tag{7.10}$$

where B_G maps $\mathcal{U}$ into $\mathcal{H}_G$ and C_G maps $\mathcal{H}_G$ into $\mathcal{Y}$. Using the fact that $\mathcal{H}_G$ is an invariant subspace for S and the range of L is contained in $\mathcal{H}_G$, it follows that

$$C_G A_G^{i-1} B_G = ES^{i-1}L = G_i \quad \text{(for all } i \geq 1 \text{)}. \tag{7.11}$$

Therefore, $\{A_G, B_G, C_G, D_G\}$ is a realization for $\mathbf{G}$. We call this system the *restricted backward shift realization* of $\mathbf{G}$, because the operator A_G is the backward shift S restricted to the invariant subspace $\mathcal{H}_G$. This sets the stage for the following result.

Theorem 7.1.1 *Let $\mathbf{G}$ be a proper rational transfer function. Then the restricted backward shift realization of $\mathbf{G}$ is a minimal realization. In particular, the McMillan degree of $\mathbf{G}$ equals the dimension of $\mathcal{H}_G$. Moreover, all minimal realizations of $\mathbf{G}$ are similar.*

PROOF. Let $\{A, B, C, D\}$ be any realization of $\mathbf{G}$. Consider the observability operator W_o mapping $\mathcal{X}$ into $l_+(\mathcal{Y})$ defined by

$$W_o = \begin{bmatrix} C \\ CA \\ CA^2 \\ \vdots \end{bmatrix}. \tag{7.12}$$

It is easy to verify that $SW_o = W_o A$. This implies that $S^i W_o = W_o A^i$ for all integers $i \geq 0$. Since $\{A, B, C, D\}$ is a realization of $\mathbf{G}$, we have $G_i = CA^{i-1}B$ for all $i \geq 1$; see (7.4). It now follows that $L = W_oB$. Hence, $S^iL = S^iW_oB = W_oA^iB$ for all $i \geq 0$. This along with the definition of $\mathcal{H}_G$, yields

$$
\begin{aligned}
\mathcal{H}_G &= \text{span}\left\{S^iL\mathcal{U} : i = 0,1,2,\cdots\right\} = \text{span}\left\{W_oA^iB\mathcal{U} : i = 0,1,2,\cdots\right\} \\
&= W_o\,\text{span}\left\{A^iB\mathcal{U} : i = 0,1,2,\cdots\right\}.
\end{aligned}
\tag{7.13}
$$

This readily implies that $\mathcal{H}_G \subset \text{ran}\,W_o$. Because W_o maps $\mathcal{X}$ into $l_+(\mathcal{Y})$, it follows that $\dim \mathcal{H}_G \leq \dim \mathcal{X}$. Therefore, the restricted backward shift realization $\{A_G, B_G, C_G, D_G\}$ is a minimal realization of $\mathbf{G}$. .

Now assume that $\{A \text{ on } \mathcal{X}, B, C, D\}$ is also a minimal realization of $\mathbf{G}$. Since $\mathbf{G}$ is rational, $\mathcal{X}$ must be finite dimensional. In this case, $\dim \mathcal{H}_G = \dim \mathcal{X}$. Since $\mathcal{H}_G \subset \text{ran}\,W_o$, the operator W_o must map $\mathcal{X}$ one to one and onto $\mathcal{H}_G$. Let T be the invertible operator mapping $\mathcal{X}$ onto $\mathcal{H}_G$ given by $T = W_o$. By employing the definitions of A_G, B_G and C_G, we obtain

$$
TA = W_oA = SW_o = A_GT, \quad C = EW_o = C_GT \text{ and } B_G = L = W_oB = TB.
\tag{7.14}
$$

Therefore, any minimal realization $\{A, B, C, D\}$ of $\mathbf{G}$ is similar to the restricted backward shift realization of $\mathbf{G}$. Since the similarity relationship is transitive, all minimal realizations of $\mathbf{G}$ are similar.   ∎

Let A be an operator on a finite dimensional space $\mathcal{X}$, while B is an operator mapping $\mathcal{U}$ into $\mathcal{X}$ and C an operator mapping $\mathcal{X}$ into $\mathcal{Y}$. Recall that the pair $\{A, B\}$ is *controllable* if and only if $\mathcal{X} = \text{span}\{A^nB\mathcal{U} : n = 0,1,2,\cdots\}$, or equivalently, $\mathcal{X} = \text{ran}\,([\begin{array}{cccc} B & AB & A^2B & \cdots \end{array}])$. Notice that the pair $\{C, A\}$ is *observable* if and only if the operator W_o defined in (7.12) is one to one. We now show that the restricted backward shift realization is controllable and observable. Controllability follows from the definition of $\mathcal{H}_G$, that is,

$$
\text{span}\left\{A_G^iB_G\mathcal{U} : i = 0,1,2,\cdots\right\} = \text{span}\left\{S^iL\mathcal{U} : i = 0,1,2,\cdots\right\} = \mathcal{H}_G.
\tag{7.15}
$$

Observability follows from the fact that

$$
\begin{bmatrix} C_G \\ C_GA_G \\ C_GA_G^2 \\ \vdots \end{bmatrix} = \begin{bmatrix} E \\ ES \\ ES^2 \\ \vdots \end{bmatrix} |\mathcal{H}_G = I|\mathcal{H}_G.
\tag{7.16}
$$

Obviously, $I|\mathcal{H}_G$ is one to one. Therefore, the restricted backward shift realization is controllable and observable. In fact, the restricted backward shift realization is the realization one obtains by extracting the controllable part from the observable realization $\{S, L, E, G_0\}$; see Proposition 6.2.1. So, the controllability and observability of the restricted backward shift realization also follows from Proposition 6.2.1. We now are ready to show the equivalence between minimality, controllability and observability.

Theorem 7.1.2 *Let* $\Sigma = \{A, B, C, D\}$ *be a realization of a proper rational transfer function* $\mathbf{G}$. *Then* Σ *is a minimal realization of* $\mathbf{G}$ *if and only if* Σ *is both controllable and observable. Moreover, all controllable and observable realizations of* $\mathbf{G}$ *are similar.*

PROOF. Suppose that $\Sigma = \{A, B, C, D\}$ is a minimal realization of **G**. Then it follows from Theorem 7.1.1 that Σ is similar to the restricted backward shift realization. Because the restricted backward shift realization is controllable and observable, the system Σ is also controllable and observable. So, minimality implies controllability and observability.

Now suppose that Σ is controllable and observable. By definition, the observability condition implies that the operator W_o is one to one. Furthermore, using controllability along with equation (7.13), we see that

$$\mathcal{H}_G = W_o \operatorname{span}\{A^i B \mathcal{U} : i = 0, 1, 2, \cdots\} = W_o \mathcal{X} .$$

In other words, $\mathcal{H}_G = \operatorname{ran} W_o$. Therefore W_o maps $\mathcal{X}$ one to one and onto $\mathcal{H}_G$. In particular, $\mathcal{X}$ and $\mathcal{H}_G$ have the same finite dimension. Because the restricted backward shift realization is minimal, Σ is minimal. Hence, a controllability and observability realization is minimal. This completes the proof. ∎

Remark 7.1.1 Let **G** be a proper rational function with values in $\mathcal{L}(\mathcal{U}, \mathcal{Y})$. Then **G** $= N/d + D$ where D is an operator in $\mathcal{L}(\mathcal{U}, \mathcal{Y})$ and N is a polynomial with values in $\mathcal{L}(\mathcal{U}, \mathcal{Y})$ while d is a scalar valued polynomial of the form

$$\begin{aligned} N(s) &= N_0 + sN_1 + \cdots + s^{n-1}N_{n-1} \\ d(s) &= a_0 + sa_1 + \cdots + s^{n-1}a_{n-1} + s^n . \end{aligned} \tag{7.17}$$

Let $\mathcal{X} = \oplus_1^n \mathcal{U}$ be the Hilbert space formed by the set of all vectors of the form $[u_1, u_2, \cdots, u_n]^{tr}$ where u_i is in $\mathcal{U}$ for all $i = 1, 2, \cdots, n$. Let $\{A \text{ on } \mathcal{X}, B, C, D\}$ be the system consisting of the block matrices defined by

$$A = \begin{bmatrix} 0 & I & 0 & \cdots & 0 \\ 0 & 0 & I & \cdots & 0 \\ \vdots & \vdots & & \ddots & \vdots \\ 0 & 0 & 0 & \cdots & I \\ -a_0 I & -a_1 I & -a_2 I & \cdots & -a_{n-1}I \end{bmatrix} \text{ and } B = \begin{bmatrix} 0 \\ 0 \\ \vdots \\ 0 \\ I \end{bmatrix}$$

$$C = \begin{bmatrix} N_0 & N_1 & N_2 & \cdots & N_{n-1} \end{bmatrix} . \tag{7.18}$$

Obviously, the pair $\{A, B\}$ is controllable. So, by consulting Section 1.4 it follows that $\{A, B, C, D\}$ is a controllable realization of **G** $= N/d + D$. So, one can construct a minimal realization for **G** by simply extracting the observable part $\{A_o \text{ on } \mathcal{X}_o, B_o, C_o, D\}$ from the system $\{A, B, C, D\}$ in (7.18); see Proposition 6.2.2.

The scalar valued rational case plays an important role in many applications. We say that two polynomials p and d are *co-prime* if p and d have no common zeros. If f is a rational function of the form $f = p/d$ where p and d are two co-prime polynomials, then the poles of f are precisely the zeros of d including their multiplicity.

Proposition 7.1.3 *Let* **g** $= p/d$ *be a scalar valued proper rational function where p and d are two co-prime polynomials. Then the McMillan degree of* **g** *equals the degree of d. Moreover, if $\{A, B, C, D\}$ is a minimal realization of* **g**, *then d equals both the characteristic and minimal polynomial for A up to a constant, that is, $\gamma d(s) = m_A(s) = \det[sI - A]$ where γ is a constant.*

PROOF. Remark 7.1.1 shows that there exists a realization of $\mathbf{g}$ whose state dimension equals the degree of d. Let $\{A \text{ on } \mathcal{X}, B, C, D\}$ be a minimal realization of $\mathbf{g}$. Then $\dim \mathcal{X} \leq \deg d$. Using (1.16), it follows that $\mathbf{g}(s) = C(sI - A)^{-1}B + D = a(s)/b(s)$ where a is a polynomial and $b(s) = \det[sI - A]$ is the characteristic polynomial for A. Furthermore, $a/b = p/d$. Now assume that $\dim \mathcal{X} < \deg d$. Since $\deg b = \dim \mathcal{X}$ and $a/b = p/d$, it follows that p and d must have at least one common zero. This contradicts the hypothesis that p and d have no common zeros. Hence, $\dim \mathcal{X} = \deg d$. Therefore, the McMillan degree of $\mathbf{g}$ equals the degree of d.

Because p and d are co-prime, the poles of $\mathbf{g}$ are the zeros of d including their multiplicity. According to Theorem 6.3.1, the polynomial d is the minimal polynomial for A up to a constant. This completes the proof.   ■

Remark 7.1.2 Let $\mathbf{G} = p/d + \delta$ be a scalar valued proper rational function where p and d are two co-prime polynomials of the form

$$
\begin{aligned}
p(s) &= c_0 + sc_1 + \cdots + s^{n-1}c_{n-1} \\
d(s) &= a_0 + sa_1 + \cdots + s^{n-1}a_{n-1} + s^n
\end{aligned}
\tag{7.19}
$$

and δ is a scalar. Let $\{A \text{ on } \mathbb{C}^n, B, C, \delta\}$ be the system defined by

$$
A = \begin{bmatrix}
0 & 1 & 0 & \cdots & 0 \\
0 & 0 & 1 & \cdots & 0 \\
\vdots & \vdots & & \ddots & \vdots \\
0 & 0 & 0 & \cdots & 1 \\
-a_0 & -a_1 & -a_2 & \cdots & -a_{n-1}
\end{bmatrix} \quad \text{and } B = \begin{bmatrix} 0 \\ 0 \\ \vdots \\ 0 \\ 1 \end{bmatrix}
$$

$$
C = \begin{bmatrix} c_0 & c_1 & c_2 & \cdots & c_{n-1} \end{bmatrix}.
\tag{7.20}
$$

By combining Remark 7.1.1 with Proposition 7.1.3, it follows that $\{A, B, C, \delta\}$ in (7.20) is a minimal realization of $\mathbf{G}$.

Exercise 12 As before, let A be an operator on a finite dimensional vector space $\mathcal{X}$. A vector b in $\mathcal{X}$ is *cyclic* for A if the span of $\{A^j b\}_0^\infty$ equals $\mathcal{X}$. If A admits a cyclic vector, then show that the minimal polynomial for A equals the characteristic polynomial for A.

Clearly, $b = [0, 0, \cdots, 0, 1]^{tr}$ is a cyclic vector for the companion matrix A in (6.36). So, this result also shows that the minimal polynomial for a companion matrix equals its characteristic polynomial.

Exercise 13 Consider any matrix A on $\mathbb{C}^n$. Recall that a complex number λ is an eigenvalue for A if and only if $\bar{\lambda}$ is an eigenvalue for A^*, the conjugate transpose of A. In general, this is not true for an operator on an infinite dimensional space. To see this consider the backward shift operator S on $l_+(\mathcal{Y})$ where $\mathcal{Y} = \mathbb{C}^1$. Show that every complex number λ is an eigenvalue for S. What are the corresponding eigenvectors? Show also that the conjugate transpose S^* of the shift has no eigenvalues.

7.2 System Hankel operators

In this section we show that the McMillan degree of a transfer function $\mathbf{G}$ equals the rank of a certain Hankel matrix associated with $\mathbf{G}$. Recall that a Hankel matrix is a matrix whose entries $\{a_{i,j}\}$ satisfy $a_{i,j} = a_{i+j}$ for all i and j. To this end, let $\mathbf{G} = \sum_0^\infty s^{-i} G_i$ be the power series expansion for $\mathbf{G}$. Then the Hankel matrix H associated with $\mathbf{G}$ is the block Hankel matrix defined by

$$H = \begin{bmatrix} G_1 & G_2 & G_3 & \cdots \\ G_2 & G_3 & G_4 & \cdots \\ G_3 & G_4 & G_5 & \cdots \\ \vdots & \vdots & \vdots & \end{bmatrix}. \tag{7.21}$$

Now let $l_{c+}(\mathcal{U})$ be the subspace of $l_+(\mathcal{U})$ consisting of the vectors in $l_+(\mathcal{U})$ with compact support, that is, the set of all vectors $f = [f_1, f_2, f_3, \cdots]^{tr}$ in $l_+(\mathcal{U})$ where f_j is nonzero for only a finite number of indices j. Because every vector in $l_{c+}(\mathcal{U})$ has compact support, H is a well defined linear operator from $l_{c+}(\mathcal{U})$ into $l_+(\mathcal{Y})$. Notice that H can also be expressed as

$$H = \begin{bmatrix} L & SL & S^2L & \cdots \end{bmatrix} \tag{7.22}$$

where S is the backward shift operator on $l_+(\mathcal{Y})$ and L is the initialization operator from $\mathcal{U}$ into $l_+(\mathcal{Y})$ defined in (7.8). This along with the definition of the state space $\mathcal{H}_G$ for the restricted backward shift realization, yields

$$\mathcal{H}_G = \mathrm{span}\left\{ S^i LU : i = 0, 1, 2, \cdots \right\} = \mathrm{ran}\left(\begin{bmatrix} L & SL & S^2L & \cdots \end{bmatrix} \right) = \mathrm{ran}\, H. \tag{7.23}$$

Therefore, the range of the Hankel operator H is precisely the state space for the restricted backward shift realization for $\mathbf{G}$. Because the restricted backward shift realization is minimal, the state dimension of any minimal realization of $\mathbf{G}$ equals the rank of the Hankel operator H. This readily yields the following result.

Theorem 7.2.1 *Let H be the Hankel matrix in (7.21) generated by the proper rational function $\mathbf{G}$. Then the McMillan degree of $\mathbf{G}$ equals the rank of H.*

Consider any realization $\{A \text{ on } \mathcal{X}, B, C, D\}$ for $\mathbf{G}$ and let W_c be the controllability operator from $l_{c+}(\mathcal{U})$ into $\mathcal{X}$ defined by

$$W_c = \begin{bmatrix} B & AB & A^2B & \cdots \end{bmatrix}. \tag{7.24}$$

Using $G_i = CA^{i-1}B$ for all integers $i \geq 1$, it follows that the Hankel matrix H in (7.21) admits a factorization of the form

$$H = \begin{bmatrix} CB & CAB & CA^2B & \cdots \\ CAB & CA^2B & CA^3B & \cdots \\ CA^2B & CA^3B & CA^4B & \cdots \\ \vdots & \vdots & \vdots & \end{bmatrix} = W_o W_c, \tag{7.25}$$

where W_o is the operator from $\mathcal{X}$ into $l_+(\mathcal{Y})$ into defined by (7.12). Letting n be the dimension of $\mathcal{X}$, it follows from the Cayley-Hamilton Theorem that every column of H is a linear combination of the first n columns of H. Similarly, every row of H is a linear combination of the first n rows of H. Hence, the rank of H equals the rank of the following $n \times n$ block matrix:

$$H_n = \begin{bmatrix} CB & CAB & CA^2B & \cdots & CA^{n-1}B \\ CAB & CA^2B & CA^3B & \cdots & CA^nB \\ CA^2B & CA^3B & CA^4B & \cdots & CA^{n+1}B \\ \vdots & \vdots & \vdots & & \vdots \\ CA^{n-1}B & CA^nB & CA^{n+1}B & \cdots & CA^{2n-2}B \end{bmatrix}.$$

In other words, the rank of the Hankel operator H equals the rank of the $n \times n$ block matrix H_n contained in the upper left hand corner of H.

Remark 7.2.1 Recall that a transfer function $\mathbf{G}$ admits a finite dimensional realization if and only if $\mathbf{G}$ is proper and rational. Since the rank of the Hankel operator H in (7.21) corresponding to $\mathbf{G}$ equals the McMillan degree of $\mathbf{G}$, it follows that H *has finite rank if and only if* $\mathbf{G}$ *is rational*. Let us establish this result directly. Clearly, if H has finite rank, then the restricted backward shift realization of $\mathbf{G}$ is finite dimensional, and thus, $\mathbf{G}$ is rational.

If $\mathbf{G}$ is a proper rational function, then $\mathbf{G} = N/d$ where N is an operator valued polynomial and d is a scalar valued polynomial of the form (1.21). According to the equations in (1.27) any column of H, after the first $n = \deg d$ columns, is a linear combination of the preceding n columns, and hence, every column of H is a linear combination of the first n columns of H. Therefore, the rank of H is less that or equal to $n \dim \mathcal{U}$ where $\mathbf{G}$ has values in $\mathcal{L}(\mathcal{U}, \mathcal{Y})$. Hence, H has finite rank. Equation (1.27) also shows that any row of H, after the first n rows, is a linear combination of the preceding n rows, and thus, every row of H is a linear combination of the first n rows of H. Therefore,

$$\operatorname{rank} H \leq \deg d \min\{\dim \mathcal{U}, \dim \mathcal{Y}\}. \tag{7.26}$$

The above analysis also shows that the rank of H equals the rank of H_n where

$$H_n = \begin{bmatrix} G_1 & G_2 & \cdots & G_n \\ G_2 & G_3 & \cdots & G_{n+1} \\ \vdots & \vdots & & \vdots \\ G_n & G_{n+1} & \cdots & G_{2n-1} \end{bmatrix}. \tag{7.27}$$

Thus, the McMillan degree of $\mathbf{G}$ equals the rank of H_n which is bounded above by the right hand side of (7.26). Finally, it is noted that if $\mathbf{G}$ is a scalar valued rational function, then (7.26) also shows that McMillan degree of $\mathbf{G}$ is less than or equal to the degree of d.

Bounded Hankel operator. Let H be the Hankel operator determined by the transfer function $(s + 2)^{-1}$. Notice that H is an unbounded operator from $l_+^2(\mathcal{U})$ into $l_+^2(\mathcal{Y})$, even though the minimal realization $\{-2, 1, 1, 0\}$ for $(s + 2)^{-1}$ is a stable. Recall that $l_+^2(\mathcal{F})$ is the Hilbert space formed by the set of all square summable unilateral infinite tuples of the form $f = [f_1, f_2, f_3, \cdots]^{tr}$ with values in $\mathcal{F}$, that is, f_j is in $\mathcal{F}$ for all integers $j \geq 1$ and $\|f\|^2 = \sum_1^\infty \|f_j\|^2$ is finite. The following result provides necessary and sufficient conditions for a Hankel operator to be a bounded operator.

Proposition 7.2.2 *Let H be the Hankel operator determined by a proper rational function* $\mathbf{G}$ *with values in $\mathcal{L}(\mathcal{U}, \mathcal{Y})$. Then H is a bounded operator from $l_+^2(\mathcal{U})$ into $l_+^2(\mathcal{Y})$ if and only if all the poles of $\mathbf{G}$ are inside the open unit disc $\{s : |s| < 1\}$.*

PROOF. Assume that H is a bounded operator. Then $\mathcal{H}_G$ is a subspace of $l_+^2(\mathcal{Y})$. If f is in $l_+^2(\mathcal{Y})$, then $S^n f$ approaches zero in the $l_+^2(\mathcal{Y})$ topology as n tends to infinity. To see this, let $f = [f_1, f_2, f_3, \cdots]^{tr}$. Then using the fact that $\|f\|^2 = \sum_1^\infty \|f_j\|^2$ is finite, we have

$$\|S^n f\|^2 = \sum_{j=n+1}^\infty \|f_j\|^2 \to 0$$

as n tends to infinity. Hence, the sequence $\{S^n f\}_0^\infty$ approaches zero. Because H is bounded, $\mathcal{H}_G$ is an invariant subspace for S contained in $l_+^2(\mathcal{Y})$. So, for all h in $\mathcal{H}_G$, we see that $A_G^n h = S^n h$ approaches zero in the $l_+^2(\mathcal{Y})$ topology as n tends to infinity. This implies that all the eigenvalues of A_G are in the open unit disc. Since the restricted backward shift realization $\{A_G, B_G, C_G, G_0\}$ is a minimal realization of $\mathbf{G}$, all the poles of $\mathbf{G}$ must also be in the open unit disc; see Theorem 6.3.1.

Assume that all the poles of $\mathbf{G}$ are in the open unit disc. Let $\{A \text{ on } \mathcal{X}, B, C, G_0\}$ be a minimal realization of $\mathbf{G}$. Then H admits a factorization of the form $H = W_o W_c$ where W_o is defined in (7.12) while W_c is defined in (7.24). Because all the poles of $\mathbf{G}$ are in the open unit disc, all the eigenvalues of A are in the open unit disc. This implies that $\|A^k\| \le mr^k$ for all integers $k \ge 0$ where m is a positive scalar and $0 \le r < 1$. By using the geometric series, with x in $\mathcal{X}$, we obtain

$$\|W_o x\|^2 = \sum_{k=0}^\infty \|CA^k x\|^2 \le m^2 \|C\|^2 \sum_{k=0}^\infty r^{2k} \|x\|^2 = m^2 \|C\|^2 \|x\|^2 / (1 - r^2).$$

Hence, $\|W_o\|^2 \le m^2 \|C\|^2 / (1-r^2)$, and thus, W_o is a bounded operator. A similar calculation shows that $\|W_c^*\|^2 \le m^2 \|B\|^2 / (1 - r^2)$. So, W_c is also a bounded operator. Therefore, $H = W_o W_c$ is a bounded operator. ∎

7.3 Realizations and factoring Hankel matrices

In this section we present a method to compute the minimal realization for a transfer function by factoring a certain Hankel matrix. To this end, consider any pair r, m of positive integers and any sequence $\{G_i : 1 \le i \le r+m-1\}$ of operators in $\mathcal{L}(\mathcal{U}, \mathcal{Y})$. Let $H_{r,m}$ be the Hankel operator generated by this sequence, that is,

$$H_{r,m} = \begin{bmatrix} G_1 & G_2 & \cdots & G_m \\ G_2 & G_3 & \cdots & G_{m+1} \\ \vdots & \vdots & & \vdots \\ G_r & G_{r+1} & \cdots & G_{r+m-1} \end{bmatrix}. \tag{7.28}$$

Notice that if the sequence $\{G_i : 1 \le i \le r+m-1\}$ is a subsequence of an infinite sequence $\{G_i\}_0^\infty$, then $H_{r,m}$ is the $r \times m$ block Hankel operator contained in the upper left hand corner of the infinite block Hankel operator H generated by the infinite sequence and defined in (7.21). We will also use the notation $H_r = H_{r,r}$ which is consistent with our previous notation. Our first observation is that if H is the infinite block Hankel operator generated by a sequence $\{G_i\}_0^\infty$, and r and m are any integers for which rank $H_{r,m} = $ rank H, then rank $H_{j,k} = $ rank $H_{r,m}$ for all integers $j \geq r$ and $k \geq m$. This follows from the fact that the rank of any submatrix of a matrix is less than or equal to the rank of the original matrix. Finally, recall that if $\{G_i\}_0^\infty$ is the coefficient sequence for the power series expansion of a proper rational function N/d where N and d are polynomials with d scalar, then rank $H_{n,n} = $ rank H where $n = \deg d$; see Section 7.2.

If $\{A, B, C, D\}$ is any realization of $\{G_i\}_0^\infty$, then using $G_0 = D$ and $G_i = CA^{i-1}B$ for all integers $i \geq 0$, we obtain

$$H_{r,m} = \begin{bmatrix} C \\ CA \\ \vdots \\ CA^{r-1} \end{bmatrix} \begin{bmatrix} B & AB & \cdots & A^{m-1}B \end{bmatrix} \tag{7.29}$$

for all positive integers r and m. This shows that the operator $H_{r,m}$ admits a factorization of the form $H_{r,m} = W_{or}W_{cm}$ where

$$W_{or} = \begin{bmatrix} C \\ CA \\ \vdots \\ CA^{r-1} \end{bmatrix} \quad \text{and} \quad W_{cm} = \begin{bmatrix} B & AB & \cdots & A^{m-1}B \end{bmatrix}.$$

Consider now a minimal realization $\{A \text{ on } \mathbb{C}^n, B, C, D\}$ of $\{G_i\}_0^\infty$. Then rank $H = n$ and the observability operator W_{on} is one to one with domain $\mathbb{C}^n$ while the controllability operator W_{cn} is onto $\mathbb{C}^n$. Let r be the smallest integer for which W_{or} is one to one and let m be the smallest integer for which W_{cm} is onto. Then for $j \geq r$ and $k \geq m$, (7.29) shows that the operator $H_{j,k}$ admits a factorization of the form $H_{j,k} = W_{oj}W_{ck}$ where the W_{oj} is one to one with domain $\mathbb{C}^n$ while W_{ck} is onto $\mathbb{C}^n$. Hence, rank $H_{j,k} = n = $ rank H whenever $j \geq r$ and $k \geq m$.

Note that rank $H_{k+1} = $ rank H_k for some integer k, does not necessarily imply that rank H_k is the dimension of the minimal realization for **G**. To see this consider the transfer function

$$\mathbf{G}(s) = \frac{1}{s} + \frac{1}{s^2} + \frac{1}{s^3}.$$

Then $H_1 = 1$ and

$$H_2 = \begin{bmatrix} 1 & 1 \\ 1 & 1 \end{bmatrix}.$$

Hence, H_1 and H_2 both have rank one. However, every minimal realization of **G** is of order three. Finally, it is noted that the rank of H_3 is three.

As before, let $\mathbf{G}(s) = \sum_0^\infty s^{-i}G_i$ be the power series expansion for a proper rational function **G** with values in $\mathcal{L}(\mathcal{U}, \mathcal{Y})$. To obtain a minimal realization for the sequence $\{G_i\}_0^\infty$, let H be the Hankel operator generated by $\{G_i\}_0^\infty$. Assume that the rank of H is n, or equivalently, n is the McMillan degree of $\mathbf{G}$. Let $WV = H$ be any factorization of H where the operator W from $\mathbb{C}^n$ into $l_+(\mathcal{Y})$ is one to one, while the linear map V is onto $\mathbb{C}^n$. Since $\mathbf{G}$ has a controllable and observable realization with state space $\mathbb{C}^n$, it follows from equations (7.12),(7.24), (7.25) and $G_i = CA^{i-1}B$ for all integers $i \geq 0$ that such a factorization always exists. In fact, $H = W_o W_c$. Now let $\{A_G, B_G, C_G, G_0\}$ be the restricted backward shift realization of $\mathbf{G}$ introduced in Section 7.1. Here $A_G = S|\mathcal{H}_G$ where S is the shift operator on the sequence space $l_+(\mathcal{Y})$ defined in (7.6). The state space $\mathcal{H}_G$ is precisely the range of the Hankel operator H. Since W is one to one and V is onto, the range of W is $\mathcal{H}_G$. Hence, W defines a similarity transformation from $\mathbb{C}^n$ onto $\mathcal{H}_G$. So, the system $\{A \text{ on } \mathbb{C}^n, B, C, G_0\}$, defined by

$$SW = WA \text{ and } B_G = L = WB \text{ and } C = EW = C_G W \,, \tag{7.30}$$

is similar to the restricted backward realization, and thus, it is also a minimal realization of the transfer function $\mathbf{G}$. The operators E and L are defined in (7.7) and (7.8), respectively. Finally, it is noted that W and V admit matrix representations of the form

$$W = \begin{bmatrix} W_1 \\ W_2 \\ W_3 \\ \vdots \end{bmatrix} \text{ and } V = \begin{bmatrix} V_1 & V_2 & V_3 & \cdots \end{bmatrix} \tag{7.31}$$

where W_i maps $\mathbb{C}^n$ into $\mathcal{Y}$ and V_i maps $\mathcal{U}$ into $\mathbb{C}^n$ for all integers $i \geq 0$.

Recall that for any positive integer k and any vector space $\mathcal{F}$, the notation $\oplus_1^k \mathcal{F}$ refers to the linear space formed by the set of k-tuples with components in $\mathcal{F}$, that is, the set of all vectors of the form $[f_1, f_2, \cdots, f_k]^{tr}$ where f_j is in $\mathcal{F}$ for all $j = 1, 2, \cdots, k$. Let P_k be the operator mapping $l_+(\mathcal{Y})$ onto $\oplus_1^k \mathcal{Y}$ which selects the first k components of a vector f in $l_+(\mathcal{Y})$, that is,

$$P_k \begin{bmatrix} f_1 \\ f_2 \\ f_3 \\ \vdots \end{bmatrix} = \begin{bmatrix} f_1 \\ f_2 \\ \vdots \\ f_k \end{bmatrix}.$$

Let Φ_k be the operator embedding $\oplus_1^k \mathcal{U}$ into the linear space $l_{c+}(\mathcal{U})$, that is,

$$\Phi_k \begin{bmatrix} u_1 \\ u_2 \\ \vdots \\ u_k \end{bmatrix} = \begin{bmatrix} u_1 \\ u_2 \\ \vdots \\ u_k \\ 0 \\ 0 \\ \vdots \end{bmatrix}.$$

For any positive integers $j \geq 1$ and $k \geq 1$, we have

$$P_j W = \begin{bmatrix} W_1 \\ W_2 \\ \vdots \\ W_j \end{bmatrix} \text{ and } V\Phi_k = \begin{bmatrix} V_1 & V_2 & \cdots & V_k \end{bmatrix} \text{ and } H_{j,k} = P_j W V \Phi_k. \tag{7.32}$$

Since $L = H\Phi_1$, we have $WB = L = WV\Phi_1$. Because W is one to one, $B = V\Phi_1 = V_1$, that is, B is the first block of V. Also, $C = EW = P_1W = W_1$, that is, C is the first block of W.

Consider now any positive integers j and k such that rank $H_{j,k} = n = $ rank H. Since $H_{j,k} = P_jWV\Phi_k$ and P_jW has domain $\mathbb{C}^n$ it follows that the operator P_jW is one to one. By applying P_j to the first equation in (7.30), we obtain $P_jSW = P_jWA$. Hence,

$$A = (P_jW)^{-r}P_jSW = ((P_jW)^*P_jW)^{-1}(P_jW)^*P_jSW.$$

Here $(P_jW)^{-r}$ is the restricted inverse of P_jW. The last equality follows because P_jW is one to one. Since

$$P_jSW = \begin{bmatrix} W_2 \\ W_3 \\ \vdots \\ W_{j+1} \end{bmatrix},$$

a minimal realization $\{A, B, C, G_0\}$ for $\mathbf{G}$ is given by

$$A = \left(\sum_{i=1}^{j} W_i^*W_i\right)^{-1}\left(\sum_{i=1}^{j} W_i^*W_{i+1}\right) \quad \text{and} \quad B = V_1 \quad \text{and} \quad C = W_1. \qquad (7.33)$$

Finally, it is noted that this minimal realization can be computed from the operators $P_{j+1}W$ and $V\Phi_k$ appearing in the factorization $H_{j+1,k} = P_{j+1}WV\Phi_k$.

Consider any positive integers j and k for which rank $H_{j,k} = $ rank $H = n$. Let $YX = H_{j+1,k}$ be any factorization of $H_{j+1,k}$ where Y is a one to one operator whose domain is $\mathbb{C}^n$ and X is onto $\mathbb{C}^n$. Then Y and X admit a block matrix representation of the form

$$Y = \begin{bmatrix} Y_1 \\ Y_2 \\ \vdots \\ Y_{j+1} \end{bmatrix} \quad \text{and} \quad X = \begin{bmatrix} X_1 & X_2 & \cdots & X_k \end{bmatrix} \qquad (7.34)$$

where each Y_i maps $\mathbb{C}^n$ into $\mathcal{Y}$ and each X_i maps $\mathcal{U}$ into $\mathbb{C}^n$. We now claim that the system $\{A, B, C, G_0\}$ defined by

$$A = \left(\sum_{i=1}^{j} Y_i^*Y_i\right)^{-1}\left(\sum_{i=1}^{j} Y_i^*Y_{i+1}\right) \quad \text{and} \quad B = X_1 \quad \text{and} \quad C = Y_1 \qquad (7.35)$$

is a minimal realization of $\{G_i\}_0^\infty$.

To verify this it is sufficient to show that there exists a factorization $H = WV$ of H where W is one to one with domain $\mathbb{C}^n$ and V is onto $\mathbb{C}^n$ with $P_{j+1}W = Y$ and $V\Phi_k = X$. In this case, the formulas for A, B and C in (7.33) and (7.35) are equivalent. Hence, $\{A, B, C, G_0\}$ in (7.35) is a minimal realization of $\mathbf{G}$. To establish these facts, consider any factorization $H = W_oW_c$ where W_o maps $\mathbb{C}^n$ one to one into $l_+(\mathcal{Y})$ and W_c is onto $\mathbb{C}^n$. The operators W_c and W_o can be constructed from any minimal realization of $\mathbf{G}$ whose state space is $\mathbb{C}^n$; see (7.12), (7.24) and (7.25). Then $YX = H_{j+1,k} = P_{j+1}W_oW_c\Phi_k$. Because $H_{j+1,k}$ has rank

n it follows that $P_{j+1}W_o$ is one to one and $W_c\Phi_k$ is onto. Notice that if $RT = MN$ where R and M are one to one, and T and N are onto, then $R = M\Omega$ and $T = \Omega^{-1}N$ where Ω is an invertible transformation. Hence, there exists is an invertible transformation Ω such that $Y = P_{j+1}W_o\Omega$ and $X = \Omega^{-1}W_c\Phi_k$. Letting $W = W_o\Omega$ and $V = \Omega^{-1}W_c$, we see that W is one to one with domain $\mathbb{C}^n$ and V is onto $\mathbb{C}^n$. Also, $H = WV$ while $Y = P_{j+1}W$ and $X = V\Phi_k$. This verifies our claim, that is, the system $\{A, B, C, G_0\}$ defined in (7.35) is a minimal realization of $\mathbf{G}$.

Finally, it is noted that

$$
H_{j+1,k} = \begin{bmatrix} G_1 & G_2 & \cdots & G_k \\ G_2 & G_3 & \cdots & G_{k+1} \\ \vdots & \vdots & & \vdots \\ G_{j+1} & G_{j+2} & \cdots & G_{j+k} \end{bmatrix}.
\tag{7.36}
$$

Thus, the formation of $H_{j+1,k}$ requires only the first $j + k$ elements of the sequence $\{G_i\}_1^\infty$. Summing up the previous analysis yields the following result.

Theorem 7.3.1 Let $\mathbf{G} = \sum_0^\infty s^{-i}G_i$ be the power series expansion for any proper rational function, and assume that the McMillan degree of $\mathbf{G}$ is n. Let j and k any positive integers such that $\operatorname{rank} H_{j,k} = n$. Let $YX = H_{j+1,k}$ be any factorization of $H_{j+1,k}$ where Y is one to one with domain $\mathbb{C}^n$ and X is onto $\mathbb{C}^n$. Then $\{A, B, C, G_0\}$ given by (7.35) is a minimal realization of $\mathbf{G}$. In particular, if $\mathbf{G} = N/d$ where N is an operator valued polynomial and d is a scalar valued polynomial, then one can choose j and k to be the degree of d.

Remark 7.3.1 Assume that $\mathbf{G}$ is a proper rational function of the form $\mathbf{G} = N/d$ where N is an operator valued polynomial and d is a scalar valued polynomial. Then one can use Lemma 1.3.2 to compute the coefficients $\{G_i\}_0^{j+k}$ in the power series expansion $\mathbf{G}(s) = \sum_0^\infty s^{-i}G_i$. Using $\{G_i\}_0^{j+k}$, the above theorem provides an algorithm to compute a minimal realization for $\mathbf{G}$.

Remark 7.3.2 The singular value decomposition provides an efficient method to compute a factorization $H_{j+1,k} = YX$ where Y is one to one and X is onto. To see this, let $U\Lambda V^*$ be a singular value decomposition of $H_{j+1,k}$ where U and V are isometries and Λ is a strictly positive diagonal matrix; see Section 16.6 in the Appendix. Then setting $Y = U$ and $X = \Lambda V^*$ yields the desired factorization of $H_{j+1,k}$. Finally, it is noted that in applications of Theorem 7.3.1 to experimentally obtained data, one does not have to choose the shortest sequence $\{G_i\}_1^m$ for which $\operatorname{rank} H_{j,k} = n$ and $j + k = m$. In practice one may want to choose the integers j and k so that $j + k \geq m$ and $\{G_i\}_1^{j+k}$ includes all the data available from the experiments. This takes advantage of the singular value decomposition of $H_{j+1,k}$ and helps reduce the effects of any noise associated with the experimental data; see Damen-Van den Hof-Hajdasinski [32].

As before, let H be the Hankel matrix determined by a proper rational function $\mathbf{G}$ with values in $\mathcal{L}(\mathcal{U}, \mathcal{Y})$. The application of Theorem 7.3.1 appears to works well when the rank of H is small and $j + k$ is not too "large", or when H is a bounded operator from $l_+^2(\mathcal{U})$

into $l_+^2(\mathcal{Y})$. If H is not a bounded operator, then the norm of $H_{j+1,k}$ approaches infinity as j and k become large. In this case, $H_{j+1,k}$ can become ill conditioned for large j and k, which leads to numerical problems when implementing Theorem 7.3.1. However, if H is a bounded operator, then the norm of $H_{j+1,k}$ is bounded for all j and k, and one can use Theorem 7.3.1 to compute a minimal realization for $\mathbf{G}$ even when the rank of H is "large". Proposition 7.2.2 shows that H is a bounded operator if and only if all the poles of $\mathbf{G}$ are inside the open unit disc $\{s : |s| < 1\}$. In other words, if $\{A, B, C, D\}$ is a minimal realization of $\mathbf{G}$, then H is a bounded operator if and only if all the eigenvalues of A are in the open unit disc; see Theorem 6.3.1. If all the eigenvalues of A are in the open unit disc, then A is discrete time stable. So, Theorem 7.3.1 is useful for computing minimal realizations for discrete time stable systems. Finally, it is noted that when all the poles of $\mathbf{G}$ are inside the open unit disc, then one can use the fast Fourier transform to calculate the coefficients $\{G_i\}$ in the power series expansion of $\mathbf{G}$.

7.4   Partial realizations and the Kalman-Ho Algorithm

In this section we present the Kalman-Ho Algorithm to construct a realization from a finite set of data. As before, let $\mathbf{G} = \sum_0^\infty s^{-i} G_i$ be the power series expansion for a proper rational function $\mathbf{G}$ with values in $\mathcal{L}(\mathcal{U}, \mathcal{Y})$. A state space system $\{A, B, C, D\}$ is called a *partial realization of order m* for a transfer function $\mathbf{G}$ or a sequence $\{G_i\}_0^\infty$ if

$$G_0 = D \quad \text{and} \quad G_i = CA^{i-1}B \quad (\text{for } 1 \leq i \leq m). \tag{7.37}$$

When (7.37) holds, we also say that $\{A, B, C, D\}$ is a partial realization of $\{G_i\}_0^m$. The system $\{A, B, C, D\}$ is a *minimal partial realization* of $\{G_i\}_0^m$ if it is a partial realization of $\{G_i\}_0^m$ of the lowest state dimension. Obviously, a minimal partial realization is controllable and observable. However, not all minimal partial realizations of the same sequence $\{G_i\}_0^m$ are similar. For example, $\{0, 1, 1, 0\}$ and $\{1, 1, 1, 0\}$ are both minimal partial realizations of the sequence $\{0, 1\}$ and they are not similar. For another example, the minimal partial realizations

$$\left\{ \begin{bmatrix} 0 & 1 \\ 0 & 0 \end{bmatrix}, \begin{bmatrix} 0 \\ 1 \end{bmatrix}, [\, 1 \ \ 0 \,], 0 \right\} \quad \text{and} \quad \left\{ \begin{bmatrix} 0 & 1 \\ 1 & 0 \end{bmatrix}, \begin{bmatrix} 0 \\ 1 \end{bmatrix}, [\, 1 \ \ 0 \,], 0 \right\}$$

of the sequence $\{0, 0, 1\}$ are not similar.

Remark 7.4.1 Let $\Sigma_j = \{A_j \text{ on } \mathcal{X}_j, B_j, C_j, D_j\}$, for $j = 1, 2$, be two partial realizations of the sequence $\{G_i : 0 \leq i \leq 2r\}$. Moreover, assume that $r \geq \dim \mathcal{X}_j$ for $j = 1, 2$. Then Σ_1 and Σ_2 have the same transfer function $\mathbf{G}$, that is, $\mathbf{G}(s) = C_j(sI - A_j)^{-1}B_j + D_j$ for $j = 1, 2$.
To see this, let $\{A \text{ on } \mathcal{X}_1 \oplus \mathcal{X}_2, B, C, 0\}$ be the state space system defined by

$$A = \begin{bmatrix} A_1 & 0 \\ 0 & A_2 \end{bmatrix} \quad \text{and} \quad B = \begin{bmatrix} B_1 \\ B_2 \end{bmatrix} \quad \text{and} \quad C = [\, C_1 \ \ -C_2 \,]. \tag{7.38}$$

Because both Σ_1 and Σ_2 are partial realizations of $\{G_i : 0 \leq i \leq 2r\}$, it follows that $CA^iB = 0$ for $0 \leq i \leq 2r - 1$. However, the degree of the characteristic polynomial for A

must be less than or equal to $2r$. By the Cayley-Hamilton Theorem, this readily implies that $CA^iB = 0$ for all integers $i \geq 0$. Hence, $C_1A_1^iB_1 = C_2A_2^iB_2$ for all integers $i \geq 0$. Obviously, $D_1 = G_0 = D_2$. Therefore, Σ_1 and Σ_2 have the same transfer function.

As before, let $\{G_i\}_0^m$ be a sequence of operators in $\mathcal{L}(\mathcal{U}, \mathcal{Y})$. Then we say that $\{F_i\}_0^\infty$ is *an extension* of $\{G_i\}_0^m$ if $\{F_i\}_0^\infty$ is a sequence of operators in $\mathcal{L}(\mathcal{U}, \mathcal{Y})$ satisfying $F_i = G_i$ for $i = 0, 1, \cdots, m$. Suppose that $\{G_i\}_0^m$ has an extension $\{G_i\}_0^\infty$ whose McMillan degree equals the rank of $H_{j,k}$ where $m = j + k$. (The Hankel matrix $H_{j,k}$ is defined in (7.28).) In other words, assume that the rank of the Hankel matrix H associated with $\{G_i\}_0^\infty$ equals the rank of $H_{j,k}$. Then this is the only extension of $\{G_i\}_0^{j+k}$ whose McMillan degree equals the rank of $H_{j,k}$. Using only the finite sequence $\{G_i\}_0^{j+k}$, Theorem 7.3.1 provides an explicit realization for any infinite sequence extending $\{G_i\}_0^{j+k}$ with McMillan degree equal to the rank of $H_{j,k}$. Hence, this infinite sequence extending $\{G_i\}_0^{j+k}$ is unique. Moreover, if $\{A, B, C, G_0\}$ is a minimal realization of this unique extension $\{G_i\}_0^\infty$, then $\{A, B, C, G_0\}$ is a minimal partial realization of $\{G_i\}_0^{j+k}$. To prove that this partial realization is minimal, let $\tilde{H}$ be the block Hankel matrix generated by any partial realization of $\{G_i\}_0^{j+k}$. Then $\tilde{H}$ contains $H_{j,k}$ in its upper left hand corner. Hence, the rank of $\tilde{H}$ is greater than or equal to that of $H_{j,k}$ which is precisely the state space dimension of $\{A, B, C, G_0\}$. Since the rank of the Hankel operator $\tilde{H}$ equals the state dimension of its minimal realization, the partial realization $\{A, B, C, G_0\}$ is minimal. Furthermore, since all minimal realizations of the same transfer function are similar and $\{G_i\}_0^{j+k}$ has a unique extension, it follows that all minimal partial realizations of $\{G_i\}_0^{j+k}$ are similar. Finally, because H and $H_{j,k}$ have the same rank, the block Hankel matrices $H_{j,k}$, $H_{j+1,k}$ and $H_{j,k+1}$ all have the same rank. The following lemma uses this rank condition to establish the existence of a unique extension for a partial sequence.

Lemma 7.4.1 *Let $\{G_i : 0 \leq i \leq j + k\}$ be a sequence of operators mapping $\mathcal{U}$ into $\mathcal{Y}$. Assume that the block Hankel matrices $H_{j,k}$, $H_{j+1,k}$ and $H_{j,k+1}$ formed by this sequence have the same rank. Then there exists a unique extension $\{G_i\}_0^\infty$ of the sequence $\{G_i\}_0^{j+k}$ whose McMillan degree equals the rank of $H_{j,k}$. In particular, if $\{A, B, C, G_0\}$ is a minimal realization of this unique extension $\{G_i\}_0^\infty$, then $\{A, B, C, G_0\}$ is a minimal partial realization of $\{G_i\}_0^{j+k}$. Finally, all minimal partial realizations of $\{G_i\}_0^{j+k}$ are similar.*

PROOF. To complete the proof it remains to show that there exists an extension $\{G_i\}_0^\infty$ of the sequence $\{G_i\}_0^{j+k}$ whose McMillan degree equals the rank of $H_{j,k}$. Since $H_{j,k}$ and $H_{j,k+1}$ have the same rank, it follows that there exists operators $\{\beta_i\}_1^k$ on $\mathcal{U}$ such that

$$
\begin{bmatrix} G_{k+1} \\ G_{k+1} \\ \vdots \\ G_{k+j} \end{bmatrix} = \begin{bmatrix} G_1 & G_2 & \cdots & G_k \\ G_2 & G_3 & \cdots & G_{k+1} \\ \vdots & \vdots & & \vdots \\ G_j & G_{j+1} & \cdots & G_{j+k-1} \end{bmatrix} \begin{bmatrix} \beta_1 \\ \beta_2 \\ \vdots \\ \beta_k \end{bmatrix}.
$$

Now let G_{j+k+1} be the operator in $\mathcal{L}(\mathcal{U}, \mathcal{Y})$ defined by

$$G_{j+k+1} = G_{j+1}\beta_1 + G_{j+2}\beta_2 + \cdots + G_{j+k}\beta_k.$$

Then the range of the last column of $H_{j+1,k+1}$ is contained in the range of the preceding columns of $H_{j+1,k+1}$ which is precisely the range of $H_{j+1,k}$. Hence, $H_{j+1,k+1}$ and $H_{j+1,k}$ have the same rank, which equals the rank of $H_{j,k+1}$. Since $H_{j,k+1}$ corresponds to the first j rows of $H_{j+1,k+1}$ and these two matrices have the same rank, it follows that there exist operators $\{\alpha_i\}_1^j$ on $\mathcal{Y}$ satisfying

$$
\begin{bmatrix} G_{j+1} & G_{j+2} & \cdots & G_{j+k+1} \end{bmatrix} = \begin{bmatrix} \alpha_1 & \alpha_2 & \cdots & \alpha_j \end{bmatrix} \begin{bmatrix} G_1 & G_2 & \cdots & G_{k+1} \\ G_2 & G_3 & \cdots & G_{k+2} \\ \vdots & \vdots & & \vdots \\ G_j & G_{j+1} & \cdots & G_{j+k} \end{bmatrix}. \tag{7.39}
$$

This yields

$$
\begin{aligned}
G_{j+1} &= \alpha_1 G_1 + \alpha_2 G_2 + \cdots + \alpha_j G_j \\
G_{j+2} &= \alpha_1 G_2 + \alpha_2 G_3 + \cdots + \alpha_j G_{j+1} \\
&\vdots \\
G_{j+k+1} &= \alpha_1 G_{k+1} + \alpha_2 G_{k+2} + \cdots + \alpha_j G_{k+j}.
\end{aligned} \tag{7.40}
$$

Now we simply recursively define the operators $\{G_i : i > j + k + 1\}$ by

$$
G_i = \alpha_1 G_{i-j} + \alpha_2 G_{i-j+1} + \cdots + \alpha_j G_{i-1} \qquad (\text{for } i > j + k + 1). \tag{7.41}
$$

Let H be the Hankel operator formed by the infinite sequence $\{G_i\}_{i=0}^\infty$. Then it follows from (7.40) and (7.41) that $H_{j,k+1}$ and $H_{\infty,k+1}$ have the same rank which equals the rank of $H_{j,k}$. Since $H_{j,k}$ is contained in upper left corner of both $H_{\infty,k}$ and $H_{\infty,k+1}$, it follows that $H_{\infty,k}$ and $H_{\infty,k+1}$ have the same rank. This implies that the range of the $k + 1$-th column of H is contained in the range of the $H_{\infty,k}$. Notice that the i-th column of H is simply $S^{i-1}L$ where S is the backward shift on $l_+(\mathcal{Y})$ and L is $H_{\infty,1}$, that is, the first block column of H. Hence,

$$
S^k L \mathcal{U} \subset \operatorname{ran} H_{\infty,k} = \operatorname{span}\{S^i L \mathcal{U} : i = 0, 1, \cdots, k - 1\}.
$$

According to Lemma 7.6.1, it now follows that $S^i L \mathcal{U} \subset \operatorname{ran} H_{\infty,k}$ for all integers $i \geq 0$. Therefore, H and $H_{\infty,k}$ have the same rank which equals the rank of $H_{j,k}$. ∎

Remark 7.4.2 Notice that for scalar sequences, $H_{j+1,j}$ is the transpose of $H_{j,j+1}$, and hence, these two matrices have the same rank. So, when implementing Lemma 7.4.1 in the scalar case with $j = k$, one only has to check that the rank of $H_{j,j}$ equals the rank of $H_{j+1,j}$ or $H_{j,j+1}$. The following example shows that this may not be true for non-scalar sequences. Consider the transfer function

$$
\mathbf{G}(s) = \frac{1}{s} \begin{bmatrix} 1 \\ 0 \end{bmatrix} + \frac{1}{s^2} \begin{bmatrix} 0 \\ 1 \end{bmatrix}.
$$

Then the matrices $H_{1,1} = G_1 = \begin{bmatrix} 1 & 0 \end{bmatrix}^{tr}$ and

$$
H_{2,1} = \begin{bmatrix} G_1 \\ G_2 \end{bmatrix} = \begin{bmatrix} 1 \\ 0 \\ 0 \\ 1 \end{bmatrix}
$$

both have rank one. However,

$$H_{1,2} \; = \; [\, G_1 \;\; G_2 \,] = \begin{bmatrix} 1 & 0 \\ 0 & 1 \end{bmatrix}$$

has rank two. If $j \neq k$, then even for scalar sequences one has to check the rank of all three matrices $H_{j,k}$, $H_{j+1,k}$ and $H_{j,k+1}$. For example, consider the transfer function $\mathbf{G}(s) = 1/s^2$. Then $G_1 = 0$, $G_2 = 1$ and $G_3 = 0$. Considering $j = 2$ and $k = 1$,

$$H_{2,1} = \begin{bmatrix} 0 \\ 1 \end{bmatrix} \;\; \text{and} \;\; H_{3,1} = \begin{bmatrix} 0 \\ 1 \\ 0 \end{bmatrix} \;\; \text{while} \;\; H_{2,2} = \begin{bmatrix} 0 & 1 \\ 1 & 0 \end{bmatrix} .$$

Although $H_{2,1}$ and $H_{3,1}$ both have rank one, the rank of $H_{2,2}$ equals two.

Let us now present a state space proof of Lemma 7.4.1. Since $H_{j,k}$ corresponds to the first j rows of $H_{j+1,k}$ and these two matrices have the same rank, it follows that there exist operators $\{\alpha_i\}_1^j$ on $\mathcal{Y}$ such that

$$[\, G_{j+1} \;\; G_{j+2} \;\; \cdots \;\; G_{j+k} \,] = [\, \alpha_1 \;\; \alpha_2 \;\; \cdots \;\; \alpha_j \,] \begin{bmatrix} G_1 & G_2 & \cdots & G_k \\ G_2 & G_3 & \cdots & G_{k+1} \\ \vdots & \vdots & & \vdots \\ G_j & G_{j+1} & \cdots & G_{j+k-1} \end{bmatrix} . \tag{7.42}$$

(Let us note that, in contrast to the previous proof, we do not have to first generate G_{j+k+1}.) Let A, B and C be the block matrices defined by

$$A \; = \; \begin{bmatrix} 0 & I & 0 & \cdots & 0 \\ 0 & 0 & I & \cdots & 0 \\ \vdots & \vdots & \vdots & & \vdots \\ 0 & 0 & 0 & \cdots & I \\ \alpha_1 & \alpha_2 & \alpha_3 & \cdots & \alpha_j \end{bmatrix} \;\; \text{and} \;\; B = \begin{bmatrix} G_1 \\ G_2 \\ \vdots \\ G_j \end{bmatrix} . \tag{7.43}$$

$$C \; = \; [\, I \;\; 0 \;\; 0 \;\; \cdots \;\; 0 \;\; 0 \,] .$$

Notice that A is a block companion matrix on $\oplus_1^j \mathcal{Y}$. Using the shift structure of A along with (7.42), we obtain

$$A \begin{bmatrix} G_1 & G_2 & \cdots & G_k \\ G_2 & G_3 & \cdots & G_{k+1} \\ \vdots & \vdots & & \vdots \\ G_j & G_{j+1} & \cdots & G_{j+k-1} \end{bmatrix} = \begin{bmatrix} G_2 & G_3 & \cdots & G_{k+1} \\ G_3 & G_4 & \cdots & G_{k+2} \\ \vdots & \vdots & & \vdots \\ G_{j+1} & G_{j+2} & \cdots & G_{j+k} \end{bmatrix} . \tag{7.44}$$

Notice that B is the first column of the Hankel matrix on the left of the equal sign, and thus, AB is the first column of the matrix on the right of the equal sign. The operator AB also appears as the second column of the Hankel matrix on the left of the equal sign, and hence,

A^2B is the second column of the matrix on the right of the equal sign. Continuing in this fashion shows that

$$
\begin{bmatrix} B & AB & \cdots & A^k B \end{bmatrix} = \begin{bmatrix} G_1 & G_2 & \cdots & G_{k+1} \\ G_2 & G_3 & \cdots & G_{k+2} \\ \vdots & \vdots & & \vdots \\ G_j & G_{j+1} & \cdots & G_{j+k} \end{bmatrix} = H_{j,k+1} .
\tag{7.45}
$$

It immediately follows from (7.45) that $CA^{i-1}B = G_i$ for $i = 1, 2, \cdots, k+1$. Consider $k + 1 < i \leq j + k$ and notice that $A^k B$ is the last column of $H_{j,k+1}$. Using the structure of C and the shift structure of A, we have that $CA^{i-1}B = CA^{i-k-1}A^k B = G_i$. It now follows that $\{A, B, C, G_0\}$ is a partial realization of $\{G_i : 0 \leq i \leq j + k\}$. So, if $G_i = CA^{i-1}B$ for all integers $i \geq 0$, then $\{A, B, C, G_0\}$ is a realization of the sequence $\{G_i\}_0^\infty$ which extends $\{G_i\}_0^{j+k}$.

We now claim that $\operatorname{rank} H_{j,k} = \operatorname{rank} H$ where H is the infinite Hankel matrix generated by $\{G_i\}_0^\infty$. According to equation (7.45),

$$
\begin{bmatrix} B, & AB & \cdots & A^{k-1}B \end{bmatrix} = H_{j,k} .
\tag{7.46}
$$

Since $H_{j,k}$ and $H_{j,k+1}$ have the same rank they also have the same range. So, if $\mathcal{X}_c$ is the range of $H_{j,k}$, then equation (7.45) also shows that $A^k B \mathcal{U} \subset \mathcal{X}_c$. By consulting Lemma 7.6.1, we obtain

$$
\mathcal{X}_c = \operatorname{span}\{A^i B \mathcal{U} : i = 0, 1, 2, \cdots\} .
$$

In other words, $\mathcal{X}_c$ is the controllable subspace for the pair $\{A, B\}$. Let $\{A_c \text{ on } \mathcal{X}_c, B_c, C_c, G_0\}$ be the system defined by

$$
A_c = A | \mathcal{X}_c, \quad B_c = B \quad \text{and} \quad C_c = C | \mathcal{X}_c .
\tag{7.47}
$$

Here A_c is on $\mathcal{X}_c$ and B_c map $\mathcal{U}$ into $\mathcal{X}_c$ while C_c maps $\mathcal{X}_c$ into $\mathcal{Y}$. Proposition 6.2.1 shows that $\{A_c, B_c, C_c, G_0\}$ is a controllable realization of $\{G_i\}_0^\infty$. Because the McMillan degree of $\{G_i\}_0^\infty$ equals the rank of H, we obtain $\operatorname{rank} H_{j,k} \leq \operatorname{rank} H \leq \dim \mathcal{X}_c = \operatorname{rank} H_{j,k}$. So, we have equality. In particular, the block matrices $H_{j,k}$ and H have the same rank. Since $\dim \mathcal{X}_c = \operatorname{rank} H$, the system $\{A_c, B_c, C_c, G_0\}$ is a minimal realization. For another proof of this fact, simply notice that pair $\{C, A\}$ is observable. By consulting Proposition 6.2.1, it also follows that $\{A_c, B_c, C_c, G_0\}$ is a minimal realization for $\{G_i\}_0^\infty$. Moreover, the McMillan degree of $\{G_i\}_0^\infty$ equals the rank of $H_{j,k}$. Therefore, $\{A_c, B_c, C_c, G_0\}$ is a minimal partial realization of $\{G_i\}_0^{j+k}$.

The above analysis shows that the system $\{A_c, B_c, C_c, G_0\}$ in (7.47) is a minimal partial realization of $\{G_i\}_0^{j+k}$. Clearly this system is controllable and observable. Moreover, one can easily compute $\{A_c, B_c, C_c, G_0\}$ by extracting the controllable part from the companion partial realization $\{A, B, C, G_0\}$ of $\{G_i\}_0^{j+k}$ in (7.43). Furthermore, this realization procedure provides another method of obtaining a minimal realization of an infinite sequence $\{G_i\}_0^\infty$ whose associated Hankel matrix H has finite rank. To see this, simple choose any finite subsequence $\{G_i\}_0^{j+k}$ such that $H_{j,k}$ and H have the same rank. Then the realization $\{A_c, B_c, C_c, G_0\}$ constructed in (7.47) is a minimal realization of $\{G_i\}_0^\infty$.

Now assume that $j = k$ and the sequence $\{G_i\}$ is scalar valued. Then equations (7.43) and (7.44) along with the above analysis yield the following result.

Proposition 7.4.2 *Let $\{G_i : 0 \leq i \leq 2n\}$ be a sequence of scalars satisfying $\operatorname{rank} H_n = \operatorname{rank} H_{n+1} = n$. Let M be the matrix on $\mathbb{C}^n$ defined by*

$$
M = \begin{bmatrix} G_2 & G_3 & \cdots & G_{n+1} \\ G_3 & G_4 & \cdots & G_{n+2} \\ \vdots & \vdots & & \vdots \\ G_{n+1} & G_{n+2} & \cdots & G_{2n} \end{bmatrix}. \tag{7.48}
$$

Then a minimal partial realization $\{A, B, C, G_0\}$ of $\{G_i : 0 \leq i \leq 2n\}$ is given by

$$
A = M H_n^{-1} \quad \text{and} \quad B = \begin{bmatrix} G_1 & G_2 & \cdots & G_n \end{bmatrix}^{tr} \quad \text{and} \quad C = \begin{bmatrix} 1 & 0 & \cdots & 0 \end{bmatrix}. \tag{7.49}
$$

In this case, A is a companion matrix of the form

$$
A = \begin{bmatrix} 0 & 1 & 0 & \cdots & 0 \\ 0 & 0 & 1 & \cdots & 0 \\ \vdots & \vdots & \vdots & & \vdots \\ 0 & 0 & 0 & \cdots & 1 \\ \alpha_1 & \alpha_2 & \alpha_3 & \cdots & \alpha_n \end{bmatrix}. \tag{7.50}
$$

Let $\mathbf{G} = \sum_0^\infty s^{-i} G_i$ be the power series expansion for a scalar valued proper rational function whose McMillan degree equals n. Then the system $\{A, B, C, G_0\}$ in (7.49) is a minimal realization of $\mathbf{G}$. In particular, one can obtain a minimal realization of $\mathbf{G}$ from the first $2n + 1$ terms $\{G_i\}_0^{2n}$ in its power series expansion. Let us present an elementary proof of this fact. Consider any minimal realization $\{A_1 \text{ on } \mathcal{X}_1, B_1, C_1, D\}$ of $\mathbf{G}$. Let W_{on} and W_{cn} be the observability and controllability matrices associated with this system, that is, W_{on} and W_{cn} are the matrices defined by

$$
W_{on} = \begin{bmatrix} C_1 \\ C_1 A_1 \\ \vdots \\ C_1 A_1^{n-1} \end{bmatrix} \quad \text{and} \quad W_{cn} = \begin{bmatrix} B_1 & A_1 B_1 & \cdots & A_1^{n-1} B_1 \end{bmatrix}.
$$

Because the pair $\{C_1, A_1\}$ is observable, W_{on} is an invertible operator from $\mathcal{X}_1$ onto $\mathbb{C}^n$. Let $\{A \text{ on } \mathbb{C}^n, B, C, D\}$ be the minimal realization of $\mathbf{G}$ which is similar to $\{A_1, B_1, C_1, D\}$ through W_{on}, that is,

$$
A W_{on} = W_{on} A_1 \quad \text{and} \quad B = W_{on} B_1 \quad \text{and} \quad C W_{on} = C_1. \tag{7.51}
$$

Using $C_1 A_1^{i-1} B_1 = G_i$, it follows that $H_n = W_{on} W_{cn}$ and $M = W_{on} A_1 W_{cn}$. Since $A W_{on} = W_{on} A_1$, we have

$$
A H_n = A W_{on} W_{cn} = W_{on} A_1 W_{cn} = M.
$$

Thus, $A H_n = M$. Because the pair $\{A_1, B_1\}$ is controllable, W_{cn} is invertible. So, $H_n = W_{on} W_{cn}$ is invertible, and we obtain $A = M H_n^{-1}$. Using $C_1 A_1^{i-1} B_1 = G_i$ once again, we have

$$
B = W_{on} B_1 = \begin{bmatrix} G_1 & G_2 & \cdots & G_n \end{bmatrix}^{tr}.
$$

Finally, since $CW_{on} = C_1$ we see that $C = [\ 1\ \ 0\ \ \cdots\ \ 0\ ]$.

To complete the proof it remains to show that A is a companion matrix, that is, a matrix of the form (7.50). To this end, let E_{n-1} be the operator from $\mathbb{C}^n$ onto $\mathbb{C}^{n-1}$ which picks out the first $n-1$ components of each element of $\mathbb{C}^n$, that is, $E_{n-1}[\ x_1\ \ \cdots\ \ x_{n-1}\ \ x_n\ ]^{tr} = [\ x_1\ \ \cdots\ \ x_{n-1}\ ]^{tr}$. Then using $A = MH_n^{-1}$, we obtain $E_{n-1}AH_n = E_{n-1}M = E_{n-1}ZH_n$ where Z is a companion matrix on $\mathbb{C}^n$. Since H_n is invertible, $E_{n-1}A = E_{n-1}Z$. So, the first $n-1$ rows of A and Z are identical. Therefore, A is a companion matrix.

7.4.1   The Kalman-Ho Algorithm

Consider any positive integers j, k and any finite sequence of operators $\{G_i\}_0^{j+k}$ mapping $\mathcal{U}$ into $\mathcal{Y}$. Assume that the block Hankel matrices $H_{j,k}$, $H_{j+1,k}$ and $H_{j,k+1}$ formed by this sequence have the same rank. Then we can compute a minimal partial realization $\{A, B, C, D\}$ of $\{G_i\}_0^{j+k}$ by the following procedure.

Kalman-Ho Algorithm.   Letting n be the rank of $H_{j,k}$, compute a factorization for $H_{j+1,k}$ of the form $H_{j+1,k} = YX$ where Y maps $\mathbb{C}^n$ one to one into $\oplus_1^{j+1}\mathcal{Y}$ and X maps $\oplus_1^k\mathcal{U}$ onto $\mathbb{C}^n$. Decompose Y and X as follows

$$Y = \begin{bmatrix} Y_1 \\ Y_2 \\ \vdots \\ Y_{j+1} \end{bmatrix} \quad \text{and} \quad X = [\ X_1\ \ X_2\ \ \cdots\ \ X_k\ ] \qquad (7.52)$$

where each Y_i maps $\mathbb{C}^n$ into $\mathcal{Y}$ and X_i maps $\mathcal{U}$ into $\mathbb{C}^n$. Then a minimal partial realization $\{A, B, C, G_0\}$ of $\{G_i\}_0^{j+k}$ is given by

$$A = \left(\sum_{i=1}^j Y_i^*Y_i\right)^{-1}\left(\sum_{i=1}^j Y_i^*Y_{i+1}\right) \quad \text{and} \quad B = X_1 \quad \text{and} \quad C = Y_1. \qquad (7.53)$$

PROOF. According to Lemma 7.4.1, there exists a unique extension $\{G_i\}_0^\infty$ of $\{G_i\}_0^{j+k}$ whose McMillan degree equals the rank of $H_{j,k}$. By Theorem 7.3.1, the system $\{A, B, C, G_0\}$ in (7.53) is a minimal realization of the infinite sequence. Therefore, this system is a minimal partial realization of $\{G_i\}_0^{j+k}$.   ∎.

7.5   Matrix representation of operators

In the next section we use matrix representations to extract a minimal realization in matrix form directly from the restricted backward shift realization. To this end, let us review how to obtain matrix representations for finite dimensional linear operators. Let A be a linear operator on a finite dimensional vector space $\mathcal{X}$. We say that M is a *matrix representation* of A if there exists an invertible operator P mapping $\mathbb{C}^n$ onto $\mathcal{X}$ satisfying

$$AP = PM.$$

In other words, a matrix representation of A is an operator on $\mathbb{C}^n$ which is similar to A. Hence, all matrix representations of A are similar.

We now show that every finite dimensional operator admits a matrix representation. To this end, let $\{\phi_1, \phi_2, \cdots, \phi_n\}$ be any basis for $\mathcal{X}$ and P be the operator mapping $\mathbb{C}^n$ onto $\mathcal{X}$ defined by

$$P = \begin{bmatrix} \phi_1 & \phi_2 & \cdots & \phi_n \end{bmatrix}. \tag{7.54}$$

Clearly, P is one to one and onto. It should be clear that

$$M = P^{-1}AP \tag{7.55}$$

is a matrix representation for A. In this case we say that M is a matrix representation for A with respect to the basis $\{\phi_i\}_1^n$. Let $\{e_i\}_1^n$ be the standard orthonormal basis for $\mathbb{C}^n$, that is, e_i is the vector whose i-th component is one and all other components are zero. Then we can identify M with the matrix

$$\begin{bmatrix} m_{11} & m_{12} & \cdots & m_{1n} \\ m_{21} & m_{22} & \cdots & m_{2n} \\ \vdots & \vdots & & \vdots \\ m_{n1} & m_{n2} & \cdots & m_{nn} \end{bmatrix},$$

where the entries m_{ij} are uniquely given by

$$Me_j = \sum_{i=1}^n m_{ij}\, e_i \qquad (\text{for } j = 1, 2, \cdots, n). \tag{7.56}$$

Using $AP = PM$, we obtain

$$A\phi_j = APe_j = PMe_j = P\sum_{i=1}^n m_{ij}e_i = \sum_{i=1}^n m_{ij}Pe_i = \sum_{i=1}^n m_{ij}\phi_i.$$

It now follows that the entries m_{ij} for the matrix M are given by

$$A\phi_j = \sum_{i=1}^n m_{ij}\phi_i \qquad (\text{for } j = 1, 2, \cdots, n). \tag{7.57}$$

Since $\{\phi_i\}_1^n$ is a basis, the entries m_{ij} of are uniquely determined by (7.57). In this case, we say that we have identified the basis $\{\phi_i\}_1^n$ for $\mathcal{X}$ with the standard basis $\{e_i\}_1^n$ for $\mathbb{C}^n$. Comparing (7.56) and (7.57), we see that the action of M on the standard basis is exactly the same as the action of A on the basis $\{\phi_i\}_1^n$. Finally, it is noted that the matrix representation of a linear operator is not unique. It depends upon the chosen basis. Since there is an infinite number of bases, there is an infinite number of matrix representations for A.

Example 7.5.1 Let A be a linear operator on a vector space $\mathcal{X}$ of finite dimension n. Suppose that A has n distinct eigenvalues, $\{\lambda_1, \lambda_2, \cdots, \lambda_n\}$ with corresponding eigenvectors,

$\{\phi_1, \phi_2, \cdots, \phi_n\}$. It is well known that the eigenvectors corresponding to distinct eigenvalues are linearly independent; see for example Halmos [58] and Horn-Johnson [65]. Hence, $\{\phi_1, \phi_2, \cdots, \phi_n\}$ is a basis for $\mathcal{X}$. Since

$$A\phi_j = \lambda_j \phi_j \qquad (\text{for } j = 1, 2, \cdots, n),$$

it follows from (7.57) that the matrix representation M of A with respect to the above basis of eigenvectors is given by the following diagonal matrix:

$$M = \begin{bmatrix} \lambda_1 & 0 & \cdots & 0 \\ 0 & \lambda_2 & \cdots & 0 \\ \vdots & \vdots & \ddots & \vdots \\ 0 & 0 & \cdots & \lambda_n \end{bmatrix}.$$

Example 7.5.2 Let $\mathcal{X}$ be the vector space consisting of the set of all polynomials $p(x)$ of degree less than or equal to 4. Let A be the differentiation operator on $\mathcal{X}$, that is, $Ap = \frac{d}{dx}p$. Clearly, A is a linear operator. Let us obtain a matrix representation of A with respect to the basis $\{\phi_1, \phi_2, \cdots, \phi_5\}$ given by $\phi_1(x) = 1$, $\phi_2(x) = x$, $\phi_3(x) = x^2$, $\phi_4(x) = x^3$ and $\phi_5(x) = x^4$. Since

$$\begin{aligned} A\phi_1 &= 0 \\ A\phi_2 &= 1 = 1\phi_1 \\ A\phi_3 &= 2x = 2\phi_2 \\ A\phi_4 &= 3x^2 = 3\phi_3 \\ A\phi_5 &= 4x^3 = 4\phi_4, \end{aligned}$$

it follows from (7.57) that the matrix representation M of the differential operator A with respect to the above basis is given by

$$M = \begin{bmatrix} 0 & 1 & 0 & 0 & 0 \\ 0 & 0 & 2 & 0 & 0 \\ 0 & 0 & 0 & 3 & 0 \\ 0 & 0 & 0 & 0 & 4 \\ 0 & 0 & 0 & 0 & 0 \end{bmatrix}.$$

The operator P mapping $\mathbb{C}^5$ into $\mathcal{X}$ is given by $P = \begin{bmatrix} 1 & x & x^2 & x^3 & x^4 \end{bmatrix}$, and thus,

$$\begin{aligned} AP = \frac{d}{dx}P &= \begin{bmatrix} 0 & 1 & 2x & 3x^2 & 4x^3 \end{bmatrix} \\ &= \begin{bmatrix} 1 & x & x^2 & x^3 & x^4 \end{bmatrix} M = PM. \end{aligned}$$

Therefore, as expected, $AP = PM$.

Example 7.5.3 Let A be an operator on a n dimensional linear space $\mathcal{X}$ and b an operator mapping $\mathbb{C}$ into $\mathcal{X}$. Suppose further that the pair $\{A, b\}$ is controllable and define the vectors $\{\phi_i\}_1^n$ by $\phi_1 = b$, $\phi_2 = Ab$, $\phi_3 = A^2b$, $\cdots$, $\phi_n = A^{n-1}b$. Since $\{A, b\}$ is controllable, it follows that $\{\phi_i\}_1^n$ is a basis for $\mathcal{X}$. Let us find a matrix representation M for the operator A with respect to this basis. To this end, notice that

$$
\begin{aligned}
A\phi_1 &= Ab = \phi_2 \\
A\phi_2 &= A^2b = \phi_3 \\
&\vdots \\
A\phi_{n-1} &= A^{n-1}b = \phi_n \\
A\phi_n &= A^nb.
\end{aligned}
$$

Hence, for $j = 1, 2, \cdots, n-1$, we obtain $A\phi_j = \phi_{j+1}$. This yields the identification $Me_j = e_{j+1}$ for $j = 1, 2, \cdots, n-1$. Now, by the Cayley-Hamilton Theorem,

$$
A^n + a_{n-1}A^{n-1} + \cdots + a_1A + a_0I = 0,
$$

where $\lambda^n + a_{n-1}\lambda^{n-1} + \cdots + a_1\lambda + a_0 = \det[\lambda I - A]$ is the characteristic polynomial for A. Therefore,

$$
A\phi_n = A^nb = -\sum_{i=0}^{n-1} a_iA^ib = \sum_{i=0}^{n-1} -a_i\phi_{i+1}.
$$

This shows, by identification, that $Me_n = -\sum_{i=0}^{n-1} a_ie_{i+1}$. Therefore, the matrix representation M for the operator A with respect to the basis $\{\phi_i\}_1^n$ is given by

$$
M = \begin{bmatrix} Me_1 & Me_2 & \cdots & Me_n \end{bmatrix} = \begin{bmatrix} 0 & 0 & 0 & \cdots & -a_0 \\ 1 & 0 & 0 & \cdots & -a_1 \\ 0 & 1 & 0 & \cdots & -a_2 \\ \vdots & \vdots & & \ddots & \vdots \\ 0 & 0 & 0 & \cdots & -a_{n-1} \end{bmatrix}.
$$

Finally, if P is the invertible operator mapping $\mathbb{C}^n$ into $\mathcal{X}$ defined by $P = \begin{bmatrix} \phi_1 & \phi_2 & \cdots & \phi_n \end{bmatrix}$, then $AP = PM$.

Exercise 14 Let $\mathcal{X}$ be the vector space consisting of the set of all polynomials $p(x)$ of degree less than or equal to four. Let $A = \frac{d}{dx}$ be the differentiation operator on $\mathcal{X}$. Consider the basis $\{\phi_i\}_1^5$ for $\mathcal{X}$ given by $\phi_1(x) = x^4$, $\phi_2(x) = x^3$, $\phi_3(x) = x^2$, $\phi_4(x) = x$ and $\phi_5(x) = 1$. Find the matrix representation for A with respect to this basis.

Exercise 15 Let $\mathcal{X}$ be the vector space consisting of the set of all polynomials $p(x)$ of degree less than or equal to k. Let $A = \frac{d}{dx}$ be the differentiation operator on $\mathcal{X}$. If ϕ is any polynomial in $\mathcal{X}$, then the Taylor series expansion shows that

$$
\phi(x + t) = \sum_{n=0}^{\infty} \frac{t^n}{n!} \frac{d^n\phi}{dx^n}(x).
$$

Recall that $e^{At} = \sum_0^\infty A^nt^n/n!$. Since $A\phi = \frac{d\phi}{dx}$, it follows that $(e^{At}\phi)(x) = \phi(x+t)$. In particular, the solution to the differential equation $\dot{f} = Af$ subject to the initial condition $f(0) = \phi$ is given by $f(t)(x) = e^{At}f(0) = \phi(x+t)$. In fact, $\dot{f} = Af$ can be viewed as a partial differential equation $\frac{\partial f}{\partial t} = \frac{\partial f}{\partial x}$. Find the matrix representation M for A with respect to the basis $\{1, x, x^2, \cdots, x^k\}$.

7.6   Shift realizations for proper rational functions

In this section, we show how one can obtain a minimal state space realization for a proper rational transfer function $\mathbf{G}$, by simply finding a matrix representation for the restricted backward shift realization. As before, $\mathbf{G}$ takes values in $\mathcal{L}(\mathcal{U}, \mathcal{Y})$ where $\mathcal{U}$ and $\mathcal{Y}$ are finite dimensional vector spaces. To begin, let $\mathcal{R}_+(\mathcal{Y})$ be the linear space consisting of the set of all strictly proper rational functions with values in $\mathcal{Y}$. If f is in $\mathcal{R}_+(\mathcal{Y})$, then f admits a power series expansion of the form

$$f(s) = \sum_{i=1}^{\infty} s^{-i} f_i \qquad (f_i \in \mathcal{Y}). \tag{7.58}$$

In this setting, the backward shift operator S on $\mathcal{R}_+(\mathcal{Y})$ is defined by

$$(Sf)(s) = sf(s) - (sf(s))_{\infty} \qquad (\text{for } f \in \mathcal{R}_+(\mathcal{Y})), \tag{7.59}$$

where $(sf(s))_{\infty} = \lim_{s \to \infty} sf(s) = f_1$. To compute $(sf(s))_{\infty}$, let $f = p/d$ where p is a polynomial with values in $\mathcal{Y}$ and d is a scalar valued polynomial satisfying $\deg p < \deg d = n$. Then $(sf(s))_{\infty} = f_1 = p_{n-1}/d_n$ where d_n is the coefficient of s^n for d and p_{n-1} is the coefficient of s^{n-1} for p. Notice that if $f(s) = \sum_1^{\infty} s^{-i} f_i$, then $Sf = \sum_1^{\infty} s^{-i} f_{i+1}$. So, the backward shift S on $\mathcal{R}_+(\mathcal{Y})$ acts on the coefficients $\{f_i\}_1^{\infty}$ in exactly the same way as the backward shift operator defined in equation (7.6). By a slight abuse of notation, we will use the symbol S to represent both the backward shift operator on the sequence space $l_+(\mathcal{Y})$ and the backward shift operator defined on the space of strictly proper rational functions $\mathcal{R}_+(\mathcal{Y})$. It will be clear from the context which backward shift operator we are using.

We can move the definition of the evaluation operator E (see equation (7.7)) and the initialization operator L (recall equation (7.8)) to the $\mathcal{R}_+(\mathcal{Y})$ space. In this setting, the evaluation operator E is now the operator mapping $\mathcal{R}_+(\mathcal{Y})$ into $\mathcal{Y}$ defined by

$$Ef = (sf(s))_{\infty} = f_1 \qquad (\text{for } f \in \mathcal{R}_+(\mathcal{Y})). \tag{7.60}$$

Let $\mathbf{G} = \sum_0^{\infty} s^{-j} G_j$ be a proper rational function with values in $\mathcal{L}(\mathcal{U}, \mathcal{Y})$. Then the initialization operator L mapping $\mathcal{U}$ into $\mathcal{R}_+(\mathcal{Y})$ is given by

$$(Lu)(s) = \mathbf{G}(s)u - (\mathbf{G}(s)u)_{\infty} = \sum_{j=1}^{\infty} \frac{G_j u}{s^j} \qquad (\text{for } u \in \mathcal{U}). \tag{7.61}$$

Finally, the direct transmission term $D(G)$ is given by $D(G) = (\mathbf{G}(s))_{\infty} = G_0$. Obviously, $G_i = ES^{i-1}L$ for all integers $i \geq 1$. Therefore, $\{S \text{ on } \mathcal{R}_+(\mathcal{Y}), L, E, G_0\}$ is a realization of $\mathbf{G}$.

To obtain the restricted backward shift realization in this context, consider the controllable subspace $\mathcal{H}(G)$ of $\mathcal{R}_+(\mathcal{Y})$ defined by

$$\mathcal{H}(G) = \text{span}\{S^k L\mathcal{U} : k = 0, 1, 2, \cdots\} = \text{span}\{S^k(\mathbf{G} - G_0)\mathcal{U} : k = 0, 1, 2, \cdots\}.$$

Clearly, $\mathcal{H}(G)$ is an invariant subspace for S. In this setting, the restricted backward shift realization of $\mathbf{G}$ is now given by $\{A(G) \text{ on } \mathcal{H}(G), B(G), C(G), D(G)\}$ where

$$A(G) = S|\mathcal{H}(G), \quad B(G) = L, \quad C(G) = E|\mathcal{H}(G) \text{ and } D(G) = G_0. \tag{7.62}$$

Obviously, $\{A(G), B(G), C(G), G_0\}$ is similar to the restricted backward shift realization $\{A_G$ on $\mathcal{H}_G, B_G, C_G, G_0\}$ of $\mathbf{G}$ defined in (7.10). In fact, if T is the operator from $\mathcal{H}(G)$ into $\mathcal{H}_G$ defined by $Tf = [f_1, f_2, f_3, \cdots]^{tr}$ where $f = \sum_1^{\infty} s^{-i} f_i$, then T is a similarity transformation which intertwines these two systems. Finally, it is noted that the notation A_G is reserved for the restricted shift on the subspace $\mathcal{H}_G$ of infinite tuples, while $A(G)$ is the restricted shift on the subspace $\mathcal{H}(G)$ of strictly proper rational functions. By obtaining a matrix representation for the restricted backward shift realization, one can readily obtain a minimal realization of $\mathbf{G}$ whose state space is $\mathbb{C}^n$. Perhaps the best way to explain this is through an illustrative example.

Example 7.6.1 Let us find a minimal realization for the following proper rational function

$$\mathbf{G}(s) = \frac{s+1}{3 + 5s + 6s^2 + 5s^3 + s^4} = \frac{s+1}{d(s)}, \tag{7.63}$$

where d is the denominator of $\mathbf{G}$. Since $\mathbf{G}$ is a strictly proper rational function $(\mathbf{G}(s))_\infty = 0$. Because the input space $\mathcal{U} = \mathbb{C}^1$, the invariant subspace $\mathcal{H}(G)$ for S is given by $\mathcal{H}(G) = \text{span}\{S^k\mathbf{G} : k = 0, 1, \cdots\}$. Our first step is to find the dimension n of $\mathcal{H}(G)$ and a basis $\{\phi\}_{i=1}^n$ for $\mathcal{H}(G)$. To this end, let $\phi_i = S^{i-1}\mathbf{G}$ for $i = 1, 2, \cdots, n$. Thus, $\phi_1 = \mathbf{G}$. To obtain ϕ_2, we apply the backward shift operator to $\mathbf{G}$

$$\phi_2(s) = (S\mathbf{G})(s) = s\mathbf{G}(s) - (s\mathbf{G}(s))_\infty = \frac{s^2 + s}{d(s)} - 0 = \frac{s^2 + s}{d(s)}.$$

Since ϕ_2 is not a linear combination of ϕ_1, we have $n \geq 2$. In a similar manner, ϕ_3 is computed by

$$\phi_3(s) = (S^2\mathbf{G})(s) = (S\phi_2)(s) = s\phi_2(s) - (s\phi_2(s))_\infty = \frac{s^3 + s^2}{d(s)}.$$

Since ϕ_3 is not a linear combination of ϕ_1 and ϕ_2, we have $n \geq 3$. We now notice that

$$\begin{aligned}
\phi_4(s) &= (S^3\mathbf{G})(s) = S\phi_3(s) = s\phi_3(s) - (s\phi_3(s))_\infty = \frac{s^4 + s^3}{d(s)} - 1 \\
&= \frac{-3 - 5s - 6s^2 - 4s^3}{d(s)} \\
&= -4\phi_3(s) - 2\phi_2(s) - 3\phi_1(s),
\end{aligned}$$

that is, ϕ_4 is a linear combination of ϕ_1, ϕ_2 and ϕ_3. In particular, $S^3\mathbf{G}$ is contained in the space spanned by $\{S^j\mathbf{G}\}_0^2$. Therefore, $\{\phi_1, \phi_2, \phi_3\}$ is a basis for $\mathcal{H}(G)$; see Lemma 7.6.1 below. Hence, the dimension of $\mathcal{H}(G)$ is three. In other words, the McMillan degree of $\mathbf{G}$ is three.

To find a matrix representation for the restricted backward shift realization, the basis $\{\phi_1, \phi_2, \phi_3\}$ is identified with the standard orthonormal basis $\{e_1, e_2, e_3\}$ for $\mathbb{C}^3$. Let P be the operator mapping $\mathbb{C}^3$ onto $\mathcal{H}(G)$ defined by

$$P = [\; \phi_1 \quad \phi_2 \quad \phi_3 \;].$$

Since $\{\phi_1, \phi_2, \phi_3\}$ is a basis, P is invertible. Moreover, the matrix representation $\{A, B, C, D\}$ for the restricted backward shift realization is given by

$$PA = A(G)P, \quad B(G) = PB \quad \text{and} \quad C(G)P = C.$$

To compute a matrix representation A on $\mathbb{C}^3$ for $A(G) = S|\mathcal{H}(G)$, notice that the previous derivation yields

$$A(G)\phi_1 = \phi_2, \quad A(G)\phi_2 = \phi_3 \quad \text{and} \quad A(G)\phi_3 = -3\phi_1 - 2\phi_2 - 4\phi_3.$$

Therefore, $Ae_1 = e_2$ and $Ae_2 = e_3$ while $Ae_3 = -3e_1 - 2e_2 - 4e_3$. Consequently, the matrix representation for the operator $A(G) = S|\mathcal{H}(G)$ with respect to P is given by

$$A = \begin{bmatrix} 0 & 0 & -3 \\ 1 & 0 & -2 \\ 0 & 1 & -4 \end{bmatrix}.$$

Recall that $PBu = B(G)u = Lu = \phi_1 u = Pe_1 u$ for all u in $\mathcal{U} = \mathbb{C}^1$. Therefore, $PB = Pe_1$, and thus, $B = e_1 = [\ 1 \ \ 0 \ \ 0\ ]^*$. The matrix C is computed as follows:

$$C = C(G)P = EP = E\begin{bmatrix} \phi_1 & \phi_2 & \phi_3 \end{bmatrix} = \begin{bmatrix} 0 & 0 & 1 \end{bmatrix}.$$

One may check that $\{A, B, C\}$ is in fact a realization for $\mathbf{G}$, that is,

$$\mathbf{G}(s) = C(sI - A)^{-1}B.$$

Notice that the dimension of the minimal realization is three, even though the original description of the function $\mathbf{G}$ has a fourth order polynomial in the denominator. One can easily check that there is a "pole-zero cancellation" in $\mathbf{G}$. So, the backward shift realization automatically obtained the "pole-zero cancellation".

Lemma 7.6.1 *Let R be an operator on a vector space $\mathcal{R}$ and $\mathcal{V}$ a subset of $\mathcal{R}$. If $R^{k+1}\mathcal{V}$ is contained in the linear span of $\{R^j\mathcal{V}\}_0^k$, then*

$$span\{R^j\mathcal{V} : j = 0, 1, 2, \cdots\} = span\{R^j\mathcal{V} : j = 0, 1, 2, \cdots, k\}. \tag{7.64}$$

PROOF. Since $R^{k+1}\mathcal{V}$ is contained in the linear span of $\{R^j\mathcal{V}\}_0^k$, we have

$$R^{k+2}\mathcal{V} = RR^{k+1}\mathcal{V} \subset R\,span\{R^j\mathcal{V}\}_{j=0}^k \subset span\{R^j\mathcal{V}\}_{j=0}^k.$$

Hence, $R^{k+2}\mathcal{V}$ is contained in the linear span of $\{R^j\mathcal{V}\}_0^k$. By continuing in this fashion, it follows that $R^{k+m}\mathcal{V}$ is contained in the linear span of $\{R^j\mathcal{V}\}_0^k$ for any integer $m \geq 0$. Therefore, equation (7.64) holds. ∎

Consider any strictly proper rational function f in $\mathcal{R}_+(\mathcal{Y})$ of the form $f(s) = p(s)/d(s)$ where p is a polynomial with values in $\mathcal{Y}$ and d is a scalar valued polynomial of the form

$$\begin{aligned} p(s) &= p_0 + p_1 s + \cdots + p_{n-1}s^{n-1} \\ d(s) &= d_0 + d_1 s + \ldots + d_{n-1}s^{n-1} + s^n. \end{aligned} \tag{7.65}$$

Notice that Sf is also a rational function and is given by

$$
\begin{aligned}
(Sf)(s) &= sf(s) - (sf(s))_\infty = \frac{sp(s) - d(s)p_{n-1}}{d(s)} \\
&= \frac{-d_0 p_{n-1} + (p_0 - d_1 p_{n-1})s + \cdots + (p_{n-2} - d_{n-1}p_{n-1})s^{n-1}}{d(s)} \\
&= \frac{q_0 + q_1 s + \cdots + q_{n-1}s^{n-1}}{d(s)}
\end{aligned}
\tag{7.66}
$$

where $\{q_j\}_0^{n-1}$ are vectors in $\mathcal{Y}$. So, by applying S^n to a strictly proper rational function, we obtain a strictly proper rational function with the same denominator d. Recall that a transfer function $\mathbf{G}$ is a proper rational function of the form N/d, where d is a scalar valued polynomial of degree n and N is an operator valued polynomial of degree at most n. Therefore, $L = \mathbf{G} - (\mathbf{G}(s))_\infty$ is a strictly proper rational function of the form M/d, where M is an operator valued polynomial of degree at most $n - 1$. Thus,

$$
\mathcal{H}(G) = \text{span}\{S^k L\mathcal{U} : k = 0, 1, 2, \cdots\} \subset \text{span}\{s^j \mathcal{Y}/d(s) : j = 0, 1, \cdots, n - 1\}. \tag{7.67}
$$

In other words, $\mathcal{H}(G)$ is contained in the subspace of $\mathcal{R}_+(\mathcal{Y})$ consisting of all rational functions of the form q/d where q is a polynomial with values in $\mathcal{Y}$ of degree at most $n-1$. This is another demonstration of the fact that the McMillan degree of $\mathbf{G}$ is always less than or equal to $n \times \dim(\mathcal{Y})$, where n is the degree of d. In particular, this also shows any rational function $\mathbf{G}$ admits a finite dimensional realization. On the other hand, since the transfer function of any finite dimensional system $\{A, B, C, D\}$ is a proper rational function, we obtain another proof of the fact that a transfer function $\mathbf{G}$ admits a finite dimensional realization if and only if $\mathbf{G}$ is a proper rational function.

Example 7.6.2 Let us use the restricted backward shift realization to find a minimal realization of the form $\{A \text{ on } \mathbb{C}^k, B, C, D\}$ for

$$
\mathbf{G}(s) = \frac{1}{d(s)} \begin{bmatrix} s & s(s+1)^2 \\ -s(s+1)^2 & -s(s+1)^2 \end{bmatrix} \tag{7.68}
$$

where $d(s) = (s + 1)^2(s + 2)^2 = s^4 + 6s^3 + 13s^2 + 12s + 4$. This transfer function $\mathbf{G}(s)$ was taken from page 444 in Kailath [68]. First notice that $\mathbf{G}$ is strictly proper, and, thus, $D = (\mathbf{G})_\infty = 0$. Recall that if p/d is a strictly proper rational function, then $S(p/d)$ is also a strictly proper rational function of the form q/d; see (7.66). Therefore, $\mathcal{H}(G) \subseteq \mathcal{H}_d$ where $\mathcal{H}_d$ is the eight dimensional subspace of $\mathcal{R}_+(\mathbb{C}^2)$ consisting of all strictly proper rational functions ϕ of the form

$$
\phi(s) = \frac{1}{d(s)} \begin{bmatrix} a_3 s^3 + a_2 s^2 + a_1 s + a_0 \\ b_3 s^3 + b_2 s^2 + b_1 s + b_0 \end{bmatrix}
$$

where $z = [a_3, \ a_2, \ a_1, \ a_0, \ b_3, \ b_2, \ b_1, \ b_0]^{tr}$ is an arbitrary vector in $\mathbb{C}^8$. Clearly ϕ and z uniquely determine each other. So, we say that this z is the vector in $\mathbb{C}^8$ associated with ϕ in $\mathcal{H}_d$. In fact, the mapping $\Phi\phi = z$ defines a similarity transform from $\mathcal{H}_d$ onto $\mathbb{C}^8$. (Moreover, one can turn $\mathcal{H}_d$ into a Hilbert space by using the inner product $(\phi, \ \psi) = (z, \ g)$ where g is the vector in $\mathbb{C}^8$ associated with ψ in $\mathcal{H}_d$.) Hence, $\mathcal{H}(G) \subset \mathcal{H}_d$ corresponds to a subspace of $\mathbb{C}^8$.

To obtain a matrix representation for the restricted backward shift realization, first let $e_1 = [1, 0]^{tr}$ and $e_2 = [0, 1]^{tr}$. To find a basis $\{\phi_i\}_1^n$ for $\mathcal{H}(G)$, set

$$\phi_1 = \mathbf{G}e_1 = \frac{1}{d(s)} \begin{bmatrix} s \\ -s^3 - 2s^2 - s \end{bmatrix}$$

and let z_1 be the vector in $\mathbb{C}^8$ associated with ϕ_1, that is,

$$z_1 = [0, \ 0, \ 1, \ 0, \ -1, \ -2, \ -1, \ 0]^{tr} \ .$$

A simple calculation shows that (one can use deconv in Matlab); see [60].

$$\begin{aligned} S\phi_1 &= \frac{1}{d(s)} \begin{bmatrix} s^2 \\ 4s^3 + 12s^2 + 12s + 4 \end{bmatrix} := \phi_2 \\ z_2 &= [0, \ 1, \ 0, \ 0, \ 4, \ 12, \ 12, \ 4]^{tr} \end{aligned}$$

where z_2 is the vector $\mathbb{C}^8$ associated with ϕ_2. Now we use the fact that Φ is a similarity transform. Since z_1 and z_2 are linearly independent in $\mathbb{C}^8$, the vectors ϕ_1 and ϕ_2 are linearly independent in $\mathcal{H}(G)$. Applying S to ϕ_2 yields

$$\begin{aligned} S\phi_2 &= \frac{1}{d(s)} \begin{bmatrix} s^3 \\ -12s^3 - 40s^2 - 44s - 16 \end{bmatrix} := \phi_3 \\ z_3 &= [1, \ 0, \ 0, \ 0, \ -12, \ -40, \ -44, \ -16]^{tr} \end{aligned}$$

where z_3 is the vector in $\mathbb{C}^8$ associated with ϕ_3. Because $\{z_1, \ z_2, \ z_3\}$ are linearly independent in $\mathbb{C}^8$, the vectors $\{\phi_1, \ \phi_2, \ \phi_3\}$ are linearly independent in $\mathcal{H}(G)$. Applying S to ϕ_3 gives

$$\begin{aligned} S\phi_3 &= \frac{1}{d(s)} \begin{bmatrix} -6s^3 - 13s^2 - 12s - 4 \\ 32s^3 + 112s^2 + 128s + 48 \end{bmatrix} := \phi_4 \\ z_4 &= [-6, \ -13, \ -12, \ -4, \ 32, \ 112, \ 128, \ 48]^{tr} \end{aligned}$$

where z_4 is the vector in $\mathbb{C}^8$ associated with ϕ_4. Since $\{z_1, \ z_2, \ z_3, \ z_4\}$ are linearly independent, $\{\phi_1, \ \phi_2, \ \phi_3, \ \phi_4\}$ are linearly independent in $\mathcal{H}(G)$. Finally,

$$\begin{aligned} S\phi_4 &= \frac{1}{d(s)} \begin{bmatrix} 23s^3 + 66s^2 + 68s + 24 \\ -80s^3 - 288s^2 - 336s - 128 \end{bmatrix} \\ z &= [23, \ 66, \ 68, \ 24, \ -80, \ -288, \ -336, \ -128]^{tr} \end{aligned}$$

where z is the vector in $\mathbb{C}^8$ associated with $S\phi_4$. In this case, z is in the span of $\{z_1, z_2, z_3, z_4\}$. In fact, $z = -4z_1 - 12z_2 - 13z_3 - 6z_4$. Hence,

$$S\phi_4 = -4\phi_1 - 12\phi_2 - 13\phi_3 - 6\phi_4 \ .$$

Therefore, $\{\phi_1, \ \phi_2, \ \phi_3, \ \phi_4\}$ is a basis for $\mathrm{span}\{S^n\mathbf{G}e_1 : n = 0, \ 1, \ 2, \ \ldots\}$; see Lemma 7.6.1.

Now let us continue on to the $\mathbf{G}e_2$ component and set

$$\phi_5 = \mathbf{G}e_2 = \frac{1}{d(s)} \begin{bmatrix} s^3 + 2s^2 + s \\ -s^3 - 2s^2 - s \end{bmatrix}$$

$$z_5 = [1,\ 2,\ 1,\ 0,\ -1,\ -2,\ -1,\ 0]^{tr}$$

where z_5 is the vector in $\mathbb{C}^8$ associated with ϕ_5. Since $\{z_1,\ z_2,\ z_3,\ z_4,\ z_5\}$ are linearly independent in $\mathbb{C}^8$, the vectors $\{\phi_1,\ \phi_2,\ \phi_3,\ \phi_4,\ \phi_5\}$ are linearly independent in $\mathcal{H}(G)$. Applying S to ϕ_5 yields

$$S\phi_5 = \frac{1}{d(s)} \begin{bmatrix} -4s^3 - 12s^2 - 12s - 4 \\ 4s^3 + 12s^2 + 12s + 4 \end{bmatrix}$$

$$g = [-4,\ -12,\ -12,\ -4,\ 4,\ 12,\ 12,\ 4]^{tr}$$

where g is the vector in $\mathbb{C}^8$ associated with $S\phi_5$. Finally, g is in the span of $\{z_j\}_1^5$. In fact, $g = 2z_1 + 5z_2 + 4z_3 + z_4 - 2z_5$. Thus,

$$S\phi_5 = 2\phi_1 + 5\phi_2 + 4\phi_3 + \phi_4 - 2\phi_5 \ .$$

According to Lemma 7.6.1, the vectors $\{\phi_j\}_1^5$ form a basis for $\mathcal{H}(G) = \text{span}\{S^k \mathbf{G}\mathbb{C}^2 : k \geq 0\}$. Therefore, the state dimension of the minimal realization for $\mathbf{G}$ is five.

To find a matrix representation $\{A \text{ on } \mathbb{C}^5,\ B,\ C,\ 0\}$ for the restricted backward shift realization $\{A(G), B(G), C(G), 0\}$ of $\mathbf{G}$, let P be the similarity transformation from $\mathbb{C}^5$ onto $\mathcal{H}(G)$ defined by

$$P = [\phi_1,\ \phi_2,\ \phi_3,\ \phi_4,\ \phi_5] \ .$$

Let $A = P^{-1}A(G)P$ and $B(G) = PB$ and $C = EP$. Then $\{A, B, C, 0\}$ is a minimal realization of $\mathbf{G}$. Because $A(G) = S|\mathcal{H}(G)$, we have

$$A(G)\phi_1 = \phi_2,\ A(G)\phi_2 = \phi_3,\ A(G)\phi_3 = \phi_4$$
$$A(G)\phi_4 = -4\phi_1 - 12\phi_2 - 13\phi_3 - 6\phi_4$$
$$A(G)\phi_5 = 2\phi_1 + 5\phi_2 + 4\phi_3 + \phi_4 - 2\phi_5 \ .$$

Using $PA = A(G)P$ it follows that

$$A = \begin{bmatrix} 0 & 0 & 0 & -4 & 2 \\ 1 & 0 & 0 & -12 & 5 \\ 0 & 1 & 0 & -13 & 4 \\ 0 & 0 & 1 & -6 & 1 \\ 0 & 0 & 0 & 0 & -2 \end{bmatrix} \quad \text{and} \quad B = \begin{bmatrix} 1 & 0 \\ 0 & 0 \\ 0 & 0 \\ 0 & 0 \\ 0 & 1 \end{bmatrix} \ .$$

The matrix for B is computed from $\mathbf{G} = PB$ along with $\mathbf{G}e_1 = \phi_1$ and $\mathbf{G}e_2 = \phi_5$, that is, $PB = \mathbf{G} = [\phi_1, \phi_5]$. So, B must be given by the above matrix. Finally, using the equation $C = EP = [E\phi_1,\ E\phi_2,\ E\phi_3,\ E\phi_4,\ E\phi_5]$, we obtain

$$C = \begin{bmatrix} 0 & 0 & 1 & -6 & 1 \\ -1 & 4 & -12 & 32 & -1 \end{bmatrix} \ .$$

Therefore, $\{A, B, C, 0\}$ is a minimal realization of $\mathbf{G}$.

By mimicking the procedure in the previous example, one can readily compute a minimal realization $\{A, B, C, D\}$ for any proper rational function $\mathbf{G} = N/d$ with values in $\mathcal{L}(\mathbb{C}^m, \mathbb{C}^k)$ where A and B are upper triangular matrices of the form

$$
A = \begin{bmatrix}
A_1 & * & * & \cdots & * & * \\
0 & A_2 & * & \cdots & * & * \\
\vdots & \vdots & \vdots & & \vdots \\
0 & 0 & 0 & \cdots & A_{n-1} & * \\
0 & 0 & 0 & \cdots & 0 & A_n
\end{bmatrix},
$$

$$
B = \begin{bmatrix}
* & b_1 & * & * & * & * & \cdots & * & * & * \\
0 & 0 & 0 & b_2 & * & * & \cdots & * & * & * \\
0 & 0 & 0 & 0 & 0 & b_3 & \cdots & * & * & * \\
\vdots & \vdots & \vdots & \vdots & \vdots & & & \vdots & \vdots \\
0 & 0 & 0 & 0 & 0 & 0 & \cdots & * & * & * \\
0 & 0 & 0 & 0 & 0 & 0 & \cdots & 0 & b_n & *
\end{bmatrix}.
\tag{7.69}
$$

Here A_j^* for $j = 1, 2, \cdots, n$ are all companion matrices and the minimal polynomial m_j for A_j divides d, that is, $m_j q_j = d$ for some polynomial q_j. (This follows from the fact that the minimal polynomial for A divides d.) Moreover, the pair $\{A_j, b_j\}$ is controllable for all j. Later we will show how one can use the upper triangular structure of A and B, to compute a gain K to arbitrarily place the eigenvalues of $A - BK$ at r locations in the complex place, where r is the dimension of the state space for A.

Exercise 16 Find a minimal realization for the transfer function

$$
\mathbf{G}(s) = \frac{s^2 + 3s + 2}{s^5 + 5s^4 + 10s^3 + 10s^2 + 5s + 2}.
\tag{7.70}
$$

Exercise 17 Find a minimal realization for the 2×3 transfer function

$$
\mathbf{G}(s) = \frac{1}{s^2 + 3s + 2} \begin{bmatrix} 1 & 2(s+1) & 1 \\ 1 & 2(s+1) & 2 \end{bmatrix}.
\tag{7.71}
$$

Exercise 18 Using the properties of the backward shift operator given in equation (7.66), write a computer program to compute a minimal realization for any proper rational function with values in $\mathcal{L}(\mathbb{C}^k, \mathbb{C}^m)$.

Exercise 19 Let S be the backward shift operator on $\mathcal{R}_+(\mathcal{Y})$. Then show that any complex number λ is an eigenvalue for S. Moreover, show that λ is an eigenvalue with eigenvector f for S if and only if

$$
f(s) = (s - \lambda)^{-1} y
$$

where y is a nonzero vector in $\mathcal{Y}$.

7.7 Jordan form realizations

In this section we will obtain a minimal realization in Jordan form for a scalar valued transfer function. To this end, let $\mathbf{G}$ be a proper rational scalar function given by the following partial fraction expansion

$$\mathbf{G}(s) = D + \frac{\gamma_1}{s - \lambda_1} + \frac{\gamma_2}{s - \lambda_2} + \cdots + \frac{\gamma_n}{s - \lambda_n}, \qquad (7.72)$$

where $\{\lambda_j\}_1^n$ are distinct scalars. Furthermore, we assume that $\gamma_i \neq 0$ for all i. Notice that this is equivalent to $\mathbf{G}(s) = p(s)/d(s)$, where p and d are two polynomials with no common roots and $d = \prod_1^n (s - \lambda_j)$. According to Proposition 7.1.3, the McMillan degree of $\mathbf{G}$ is n. To obtain a minimal realization for $\mathbf{G}$, let A be the diagonal matrix on $\mathbb{C}^n$ and B the $n \times 1$ column matrix and C the $1 \times n$ row matrix defined by

$$A = \begin{bmatrix} \lambda_1 & 0 & \cdots & 0 \\ 0 & \lambda_2 & \cdots & 0 \\ \vdots & & \ddots & \vdots \\ 0 & 0 & \cdots & \lambda_n \end{bmatrix} \text{ and } B = \begin{bmatrix} \gamma_1 \\ \gamma_2 \\ \vdots \\ \gamma_n \end{bmatrix} \qquad (7.73)$$

$$C = \begin{bmatrix} 1 & 1 & \cdots & 1 \end{bmatrix}.$$

Then it is easy to verify that $\{A, B, C, D\}$ is a minimal realization of $\mathbf{G}$. Clearly, $\{A, B, C, D\}$ is a realization of $\mathbf{G}$, that is, $\mathbf{G} = C(sI - A)^{-1}B + D$. Because the McMillan degree of $\mathbf{G}$ is n, this realization is minimal. The controllability and observability of $\{A, B, C, D\}$ also follows from the PBH test; see Proposition 5.2.2 and Proposition 4.2.2.

Jordan Models. In this section, we will use the partial fraction expansion to obtain a minimal realization in Jordan form for a scalar valued transfer function with repeated roots. Because computing repeated roots is numerically sensitive, the results in this section are mainly of academic interest. To begin, let $\mathbf{G}$ be the scalar valued transfer function defined by

$$\mathbf{G}(s) = \frac{\gamma_1}{s - \lambda} + \frac{\gamma_2}{(s - \lambda)^2} + \cdots + \frac{\gamma_r}{(s - \lambda)^r}, \qquad (7.74)$$

with $\gamma_r \neq 0$. To obtain a minimal realization for this transfer function, let J be the Jordan matrix on $\mathbb{C}^r$ and B the $r \times 1$ column matrix and C the $1 \times r$ row matrix defined by

$$J = \begin{bmatrix} \lambda & 1 & 0 & \cdots & 0 & 0 \\ 0 & \lambda & 1 & \cdots & 0 & 0 \\ \vdots & \vdots & \ddots & & \vdots & \vdots \\ \vdots & \vdots & & \ddots & \vdots & \vdots \\ 0 & 0 & 0 & \cdots & \lambda & 1 \\ 0 & 0 & 0 & \cdots & 0 & \lambda \end{bmatrix} \text{ and } B = \begin{bmatrix} \gamma_1 \\ \gamma_2 \\ \vdots \\ \vdots \\ \gamma_{r-1} \\ \gamma_r \end{bmatrix} \qquad (7.75)$$

$$C = \begin{bmatrix} 1 & 0 & 0 & \cdots & 0 & 0 \end{bmatrix}.$$

We claim that $\{J, B, C, 0\}$ is a minimal realization of $\mathbf{G}$. To see this, let $\phi_j = (s - \lambda)^{-j}$ for $j = 1, 2, \cdots, r$. Then the inverse of $sI - J$ is given by

$$(sI - J)^{-1} = \begin{bmatrix} \phi_1 & \phi_2 & \phi_3 & \cdots & \phi_{r-1} & \phi_r \\ 0 & \phi_1 & \phi_2 & \cdots & \phi_{r-2} & \phi_{r-1} \\ \vdots & \vdots & \ddots & & \vdots & \vdots \\ \vdots & \vdots & & \ddots & \vdots & \vdots \\ 0 & 0 & 0 & \cdots & \phi_1 & \phi_2 \\ 0 & 0 & 0 & \cdots & 0 & \phi_1 \end{bmatrix}. \tag{7.76}$$

To verify this simply notice that $(sI - J)M(s) = I$ where $M(s)$ is the matrix on the right hand side of (7.76). Using (7.76), it readily follows that $\{J, B, C, 0\}$ is a realization of $\mathbf{G}$, that is, $\mathbf{G} = C(sI - J)^{-1}B$. Since γ_r is nonzero, Proposition 7.1.3 shows that r is the McMillan degree of $\mathbf{G}$, and thus, $\{J, B, C, 0\}$ is a minimal realization of $\mathbf{G}$. Now let us directly verify that this realization is controllable and observable. Because γ_r is nonzero the rank of $[J - \mu I, B]$ equals r for all complex numbers μ. By the PBH controllability test the pair $\{J, B\}$ is controllable. Notice that the kernel of

$$\begin{bmatrix} J - \mu I \\ C \end{bmatrix}$$

is zero for all complex numbers μ. By the PBH observability test the pair $\{C, J\}$ is observable. Therefore, $\{J, B, C, 0\}$ is a minimal realization of $\mathbf{G}$.

Let $\mathbf{G}$ be a scalar valued proper rational function of the form

$$\mathbf{G}(s) = D + \sum_{i=1}^{m} \sum_{j=1}^{r_i} \frac{\gamma_{i,j}}{(s - \lambda_i)^j} \tag{7.77}$$

where $\{\lambda_i\}_1^m$ are distinct complex numbers and $\gamma_{i,r_i} \neq 0$ for $i = 1, 2, \cdots, m$. This implies that $\mathbf{G} = p/d$ where p and d are co-prime polynomials and

$$d(s) = \prod_{i=1}^{m} (s - \lambda_i)^{r_i}. \tag{7.78}$$

Proposition 7.1.3 shows that the McMillan degree of $\mathbf{G}$ equals $n = \sum_1^m r_i$. We claim that a minimal realization for $\mathbf{G}$ in Jordan form is given by $\{A \text{ on } \mathbb{C}^n, B, C, D\}$ where

$$A = \mathrm{diag}\,(J_1, J_2, \cdots, J_m), \tag{7.79}$$

J_i is the $r_i \times r_i$ block Jordan matrix corresponding to λ_i. The matrix B is the $n \times 1$ column vector given by

$$B = \begin{bmatrix} \gamma_{1,1} & \cdots & \gamma_{1,r_1} & \gamma_{2,1} & \cdots & \gamma_{2,r_2} & \cdots & \cdots & \gamma_{m,1} & \cdots & \gamma_{m,r_m} \end{bmatrix}^{tr}. \tag{7.80}$$

The matrix C is the $1 \times n$ row matrix given by

$$C = \begin{bmatrix} \delta_1 & \delta_2, & \cdots & \delta_{m-1} & \delta_m \end{bmatrix}^{tr} \tag{7.81}$$

where $\delta_j = [1, 0, 0, \cdots, 0]$ is the $1 \times r_j$ row vector defined by placing one in the first position and zeros everywhere else. Using (7.76), one can readily verify that

$$\mathbf{G}(s) = C(sI - A)^{-1}B + D.$$

Therefore, $\{A, B, C, D\}$ is a realization of $\mathbf{G}$ whose state dimension equals n. Because the n is also the McMillan degree of $\mathbf{G}$, we immediately see that $\{A, B, C, D\}$ is a minimal realization of $\mathbf{G}$. In particular, $\{A, B, C, D\}$ is controllable and observable. One can also use the PBH tests to show that this system is controllable and observable. Since $\gamma_{i,r_i} \neq 0$ for $i = 1, 2, \cdots, m$, the rank of $[A - \lambda I, B]$ equals n for all complex numbers λ. By the PBH controllability test the pair $\{A, B\}$ is controllable. Notice that the kernel of

$$\left[\begin{array}{c} A - \lambda I \\ C \end{array} \right]$$

is zero for all complex numbers λ. By the PBH observability test the pair $\{C, A\}$ is observable.

Exercise 20 Let $\mathbf{G}$ be a proper rational function with values in $\mathcal{L}(\mathcal{U}, \mathcal{Y})$ given by the following partial fraction expansion

$$\mathbf{G}(s) = D + \sum_{i=1}^{n} (s - \lambda_i)^{-1} \Gamma_i, \tag{7.82}$$

where $\{\lambda_i\}_1^n$ are distinct scalars and $\{\Gamma_i\}_1^n$ is a sequence of operators with values in $\mathcal{L}(\mathcal{U}, \mathcal{Y})$. Furthermore, assume that $\Gamma_i \neq 0$ for all i. Let $\Gamma_i = C_i B_i$ be any factorization of Γ_i where the operator B_i mapping $\mathcal{U}$ into $\mathcal{X}_i$ is onto and the operator C_i mapping $\mathcal{X}_i$ into $\mathcal{Y}$ is one to one. In fact, one can use the singular value decomposition to compute this factorization. Let A be the block diagonal matrix on $\oplus_1^n \mathcal{X}_i$ and B the block column matrix from $\mathcal{U}$ into $\oplus_1^n \mathcal{X}_i$ and C the block row vector from $\oplus_1^n \mathcal{X}_i$ into $\mathcal{Y}$ defined by

$$A = \left[\begin{array}{cccc} \lambda_1 I_1 & 0 & \cdots & 0 \\ 0 & \lambda_2 I_2 & \cdots & 0 \\ \vdots & & \ddots & \vdots \\ 0 & 0 & \cdots & \lambda_n I_n \end{array} \right] \text{ and } B = \left[\begin{array}{c} B_1 \\ B_2 \\ \vdots \\ B_n \end{array} \right] \tag{7.83}$$

$$C = \left[\begin{array}{cccc} C_1 & C_2 & \cdots & C_n \end{array} \right].$$

(The identity on $\mathcal{X}_i$ is denoted by I_i.) Then show that $\{A, B, C, D\}$ is a minimal realization for $\mathbf{G}$.

Exercise 21 Consider the following transfer function

$$\mathbf{G}(s) = \frac{3}{s - 1} + \frac{5}{(s + 2)^2} + \frac{-2}{s + 2} + \frac{4}{(s + 1)^3}. \tag{7.84}$$

Find a minimal realization for $\mathbf{G}$ in Jordan form.

7.8 Notes

The shift operator plays a fundamental role in operator theory; see Gohberg-Goldberg-Kaashoek [55], Nikolskii [94] and Sz.-Nagy-Foias [120]. The shift operator also plays a basic role in interpolation theory; see Foias-Frazho [39], Foias-Frazho-Gohberg-Kaashoek [41] and Rosenblum-Rovnyak [104]. The restricted backward shift realization presented in Section 7.1 was taken from Fuhrmann [47] and Helton [62]. The shift operator and the backward shift realization is also used to solve infinite dimensional realization problems in Hilbert space; see Fuhrmann [47] for a history of this subject area and further results. Shift operators can also be used to solve a bilinear realization problem; see Frazho [45], Wong [128] and Rugh [110]. For a more classical approach to realization theory see Chen [26], Kailath [68], Rugh [110] and Skelton [114]. Bart-Gohberg-Kaashoek [12] uses realization theory to solve some factorization problems in operator theory. Finally, our approach to the Kalman-Ho algorithm was taken form Damen-Van den Hof-Hajdasinski [32] and Kalman-Falb-Arbib [73]. A special type of realization introduced by Moore [91], and called balanced realizations, are used to solve model reduction problems. For some results on model reduction and balanced realizations see Zhou-Doyle-Glover [131].

Model reduction plays an important role in applications. A stable realization $\{A, B, C, D\}$ is *balanced* if it has the same controllability and observability Gramian Λ where Λ is a diagonal matrix consisting of decreasing entries, that is,

$$A\Lambda + \Lambda A^* + BB^* = 0 \quad \text{and} \quad A^*\Lambda + \Lambda A + C^*C = 0$$

where Λ is a diagonal matrix of the form $\Lambda = \text{diag}\{\sigma_j\}$ and $\sigma_1 \geq \sigma_2 \geq \cdots \geq \sigma_n$. Any stable minimal realization can be converted to a balanced realization by a similarity transformation.

For a simple model reduction procedure based on a balanced realization, consider the transfer function $\mathbf{G}$ determined by the minimal stable state space system $\{A \text{ on } \mathcal{X}, B, C, D\}$ whose state dimension is n. To convert this system to a balanced realization, let P be the observability Gramian for the pair $\{C, A\}$. Recall that P is determined by the Lyapunov equation

$$A^*P + PA + C^*C = 0. \tag{7.85}$$

Because A is stable and the pair $\{C, A\}$ is observable, P is strictly positive. Let $P^{1/2}$ be the positive square root of P. Multiplying both sides of equation (7.85) by $P^{-1/2}$, we obtain

$$(P^{1/2}AP^{-1/2})^* + (P^{1/2}AP^{-1/2}) + P^{-1/2}C^*CP^{-1/2} = 0.$$

Now let $\{A_1, B_1, C_1, D\}$ be the realization determined by

$$A_1 = P^{1/2}AP^{-1/2}, \quad B_1 = P^{1/2}B, \quad \text{and} \quad C_1 = CP^{-1/2}.$$

Since $\{A_1, B_1, C_1, D\}$ is similar to $\{A, B, C, D\}$, the system $\{A_1, B_1, C_1, D\}$ is also a minimal realization for $\mathbf{G}$. Moreover, by construction the identity operator I is the observability Gramian for $\{C_1, A_1\}$, that is, $A_1^* + A_1 + C_1^*C_1 = 0$.

Let Q be the controllability Gramian for the pair $\{A_1, B_1\}$, that is,

$$A_1Q + QA_1^* + B_1B_1^* = 0. \tag{7.86}$$

Because A_1 is stable and the pair $\{A_1, B_1\}$ is controllable, Q is strictly positive. Let $Q = U^*\Lambda^2 U$ be the spectral decomposition for Q where U is a unitary operator from $\mathbb{C}^n$ onto $\mathcal{X}$ and $\Lambda^2 = \text{diag}\{\sigma_j^2\}_1^n$ is a diagonal matrix on $\mathbb{C}^n$ satisfying $\sigma_1 \geq \sigma_2 \geq \cdots \geq \sigma_n$. Notice that $\{\sigma_j^2\}_1^n$ are the eigenvalues of Q. By employing this decomposition in (7.86), we arrive at

$$(UA_1U^*)\Lambda^2 + \Lambda^2(UA_1U^*)^* + UB_1B_1^*U^* = 0.$$

Now let $\{A_2, B_2, C_2, D\}$ be the realization determined by

$$A_2 = UA_1U^*, \quad B_2 = UB_1, \quad \text{and} \quad C_2 = C_1U^*.$$

Since $\{A_2, B_2, C_2, D\}$ is similar to $\{A_1, B_1, C_1, D\}$, the system $\{A_2, B_2, C_2, D\}$ is a minimal realization for $\mathbf{G}$. Moreover, by construction the operator Λ^2 is the controllability Gramian for $\{A_2, B_2\}$, that is, $A_2\Lambda^2 + \Lambda^2 A_2^* + B_2 B_2^* = 0$, and I is the observability Gramian for $\{C_2, A_2\}$, that is, $A_2^* + A_2 + C_2^* C_2 = 0$. To obtain the final similarity transformation observe that

$$(\Lambda^{-1/2}A_2\Lambda^{1/2})\Lambda + \Lambda(\Lambda^{-1/2}A_2\Lambda^{1/2})^* + \Lambda^{-1/2}B_2B_2^*\Lambda^{-1/2} = 0$$
$$(\Lambda^{-1/2}A_2\Lambda^{1/2})^*\Lambda + \Lambda(\Lambda^{-1/2}A_2\Lambda^{1/2}) + \Lambda^{1/2}C_2^*C_2\Lambda^{1/2} = 0.$$

Now let $\{A_3, B_3, C_3, D\}$ be the minimal realization determined by

$$A_3 = \Lambda^{-1/2}A_2\Lambda^{1/2}, \quad B_3 = \Lambda^{-1/2}B_2, \quad \text{and} \quad C_3 = C_2\Lambda^{1/2}.$$

Then $\{A_3, B_3, C_3, D\}$ is a balanced minimal stable realization for $\mathbf{G}$. Moreover, Λ is the controllability and observability Gramian for $\{A_3, B_3, C_3, D\}$, that is,

$$A_3\Lambda + \Lambda A_3^* + B_3 B_3^* = 0 \quad \text{and} \quad A_3^*\Lambda + \Lambda A_3 + C_3^* C_3 = 0.$$

Finally, it is noted that $\{A_3, B_3, C_3, D\}$ is similar to $\{A, B, C, D\}$.

To complete this model reduction procedure, let r be an integer in $[1, n]$. Consider the following matrix decompositions of A_3, B_3 and C_3

$$A_3 = \begin{bmatrix} A_r & A_{12} \\ A_{21} & A_{22} \end{bmatrix} \text{ on } \begin{bmatrix} \mathbb{C}^r \\ \mathbb{C}^{n-r} \end{bmatrix}, \quad B_3 = \begin{bmatrix} B_r \\ B_{21} \end{bmatrix} : \mathcal{U} \to \begin{bmatrix} \mathbb{C}^r \\ \mathbb{C}^{n-r} \end{bmatrix}$$

$$C_3 = \begin{bmatrix} C_r & C_{12} \end{bmatrix} : \begin{bmatrix} \mathbb{C}^r \\ \mathbb{C}^{n-r} \end{bmatrix} \to \mathcal{Y}.$$

Then $\{A_r, B_r, C_r, D\}$ is a reduced order model of dimension r for the system $\{A, B, C, D\}$. Finally, let $\mathbf{G}_r$ be the r dimensional realization corresponding to $\{A_r, B_r, C_r, D\}$. Then $\mathbf{G}_r$ can be used as a r dimensional reduced order model for the transfer function $\mathbf{G}$. This reduced order model is fairly accurate when σ_r is much larger than σ_{r+1}; see Glover [52] and Zhou-Doyle-Glover [131] for further results on model reduction.

Chapter 8

State Feedback and Stabilizability

This chapter is devoted to state feedback and stabilization of linear systems. Lyapunov techniques are used to show that a state space system is stabilizable if and only if all of its uncontrollable eigenvalues are stable. Then state feedback is used, on a controllable system, to place the eigenvalues of the corresponding closed loop system at a set of specified locations in the complex plane.

8.1 State feedback and stabilizability

In this section we introduce the concept of state feedback. To this end, consider the system

$$\dot{x} = Ax + Bu. \tag{8.1}$$

As before, A is an operator on $\mathcal{X}$ while B maps $\mathcal{U}$ into $\mathcal{X}$. The spaces $\mathcal{X}$ and $\mathcal{U}$ are finite dimensional. Suppose that the *open loop system* $\dot{x} = Ax$ is unstable. A natural question is the following: Can one choose a controller generating the control input u such that the resulting system is stable, that is, $x(t)$ approaches zero as t tends towards infinity. As will be seen later in the chapter, we do not have to consider controllers that are nonlinear functions of the state to solve this problem. A first natural choice for a controller is simply given by $u(t) = -Kx(t)$ where K is an operator from $\mathcal{X}$ into $\mathcal{U}$. In this case, $u(t)$ is a linear function of the state $x(t)$ at each instance of time t, and thus, $u(t) = -Kx(t)$ is called a *static* (or, *memoryless*) *state feedback controller*. The operator K is called the gain and the minus sign is a convention. This leads to the following question. Can we find an operator K such that $A - BK$ is stable? Motivated by this we say that the system in (8.1) is *stabilizable* if there exists an operator K from $\mathcal{X}$ into $\mathcal{U}$ such that $A - BK$ is stable. Finally, we say that the pair $\{A, B\}$ is stabilizable if the system in (8.1) is stabilizable.

A related question is, under what conditions is the pair $\{A, B\}$ stabilizable? In Section 8.2 we will use Lyapunov techniques to show that if $\{A, B\}$ is controllable, then it is stabilizable. In subsequent sections, for a controllable pair $\{A, B\}$, we will develop a procedure to compute a gain K which places the eigenvalues of $A - BK$ at any n specified locations in the complex plane, where n is the dimension of the state space. This will provide us with a computational method for stabilization and solve a classical eigenvalue placement problem.

Notice that controllability is not necessary for stabilizability, that is, it is possible for a system to be stabilizable but not controllable. To see this, consider a system with A stable and B equal to zero. For a system which is neither controllable nor stabilizable, consider $A = 1$ and $B = 0$. The following example illustrates how certain eigenvalues of A play a role in stabilizability.

Example 8.1.1 Consider the system $\dot{x} = Ax + Bu$ where

$$A = \begin{bmatrix} \alpha & 0 \\ 0 & 0 \end{bmatrix} \quad \text{and} \quad B = \begin{bmatrix} 0 \\ 1 \end{bmatrix}.$$

Clearly, this system is uncontrollable. For any gain $K = [k_1 \quad k_2]$, the matrix

$$A - BK = \begin{bmatrix} \alpha & 0 \\ -k_1 & -k_2 \end{bmatrix}$$

has eigenvalues α and $-k_2$. Notice that regardless of the gain K, the eigenvalue α of A is also an eigenvalue of $A - BK$. So, if $\alpha \geq 0$, then the pair $\{A, B\}$ is not stabilizable. However, if $\alpha < 0$, then $A - BK$ is stable for $k_2 > 0$. In this case, the pair $\{A, B\}$ is stabilizable.

8.2 Simple stabilizing controllers

In this section we use Lyapunov techniques to obtain simple stabilizing controllers for controllable systems. Throughout this section $\{A, B\}$ is a controllable pair where A is an operator on $\mathcal{X}$ and B is an operator from $\mathcal{U}$ into $\mathcal{X}$. To begin, consider any positive $t_1 > 0$ and let

$$Q = \int_{-t_1}^{0} e^{At} BB^* e^{A^*t} \, dt = \int_{0}^{t_1} e^{-At} BB^* e^{-A^*t} \, dt. \qquad (8.2)$$

The above operator can be regarded as the finite time controllability Gramian over the interval $[0, t_1]$ associated with $\{-A, B\}$. Since $\{A, B\}$ is controllable, Q is strictly positive; see Theorem 5.3.2. We now show that the state feedback gain $K = B^*Q^{-1}$ yields a stable closed loop operator $A - BK$, that is, this gain results in a stabilizing controller. Using the fact that the derivative of e^{At} is Ae^{At}, we obtain

$$
\begin{aligned}
AQ + QA^* &= \int_{-t_1}^{0} \left[Ae^{At} BB^* e^{A^*t} + e^{At} BB^* e^{A^*t} A^* \right] dt \\
&= \int_{-t_1}^{0} \frac{d\left(e^{At} BB^* e^{A^*t} \right)}{dt} \, dt = BB^* - e^{-At_1} BB^* e^{-A^*t_1}.
\end{aligned}
$$

This yields the following Lyapunov equation

$$AQ + QA^* - BB^* = -e^{-At_1} BB^* e^{-A^*t_1}.$$

By combing this equation with $A - BK = A - BB^*Q^{-1}$, we have

$$(A - BB^*Q^{-1})Q + Q(A - BB^*Q^{-1})^* = AQ + QA^* - 2BB^* =$$
$$-BB^* - e^{-At_1} BB^* e^{-A^*t_1} = -\tilde{B}\tilde{B}^*$$

where $\tilde{B} = [B \quad e^{-At_1}B]$. Hence, we obtain the Lyapunov equation,

$$(A - BB^*Q^{-1})Q + Q(A - BB^*Q^{-1})^* + \tilde{B}\tilde{B}^* = 0.$$

Because $\{A, B\}$ is controllable, Lemma 6.4.2 shows that the pair $\{A - BB^*Q^{-1}, B\}$ is also controllable. This readily implies that $\{A - BB^*Q^{-1}, \tilde{B}\}$ is controllable. Since Q is strictly positive, the previous Lyapunov equation, shows that $A - BK$ is stable; see Theorem 5.4.2. By summing up this analysis we arrive at the following result.

Proposition 8.2.1 *If $\{A, B\}$ is a controllable pair, then $\{A, B\}$ is stabilizable. In this case, if Q is the strictly positive operator defined in (8.2) and $K = B^*Q^{-1}$, then K is a stabilizing gain, that is, $A - BK$ is stable.*

Remark 8.2.1 Consider any $\alpha > 0$ and any finite $t_1 > 0$. Let

$$Q_\alpha = \int_0^{t_1} e^{-2\alpha t} e^{-At} BB^* e^{-A^*t}\, dt.$$

First notice that the controllability of $\{A, B\}$ along with the PBH controllability test readily shows that the pair $\{A + \alpha I, B\}$ is controllable. Repeating the above analysis with $A + \alpha I$ replacing A shows that Q_α is positive and invertible. Moreover, the gain $K = B^*Q_\alpha^{-1}$ results in $A + \alpha I - BK$ being stable. Hence, all the eigenvalues of the closed loop operator $A - BB^*Q_\alpha^{-1}$ have real parts strictly less than $-\alpha < 0$. In particular, $A - BB^*Q_\alpha^{-1}$ is stable. Notice that by using this control algorithm, one can place the eigenvalues of the closed loop system to the left of any vertical line in the complex plane.

To complete this section, let us use Lyapunov techniques to present another method to construct a stabilizing controller. To this end, consider now any $\alpha \geq 0$ such that all the eigenvalues of $-A - \alpha I$ have nonzero negative real parts, that is, $-A - \alpha I$ is stable. For example, choosing $\alpha > \|A\|$ will suffice. (This follows because the norm of any operator is greater than or equal to the magnitude of its eigenvalues.) Then Q_α is well defined for $t_1 = \infty$ and is given by

$$Q_\alpha = \Omega = \int_0^\infty e^{-2\alpha t} e^{-At} BB^* e^{-A^*t}\, dt.$$

Since Q_α is strictly positive for all finite t_1 and is a non-decreasing function of t_1, it follows that Ω is positive and invertible. Because $-A - \alpha I$ is stable, Ω is the unique solution to the Lyapunov equation

$$(-A - \alpha I)\Omega + \Omega(-A - \alpha I)^* + BB^* = 0. \tag{8.3}$$

As expected, the state feedback gain $K = B^*\Omega^{-1}$ yields a stable closed loop state operator $A - BK$. To see this, notice that

$$(A - BB^*\Omega^{-1} + \alpha I)\Omega + \Omega(A - BB^*\Omega^{-1} + \alpha I)^* + BB^* = 0.$$

Since the pair $\{-A - \alpha I, B\}$ is controllable, it follows that $\{A - BB^*\Omega^{-1} + \alpha I, B\}$ is also controllable. The above Lyapunov equation and the strict positivity of Ω imply that the operator $A - BB^*\Omega^{-1} + \alpha I$ is stable. Hence, all the eigenvalues of the closed loop operator $A - BB^*\Omega^{-1}$ have real parts less than $-\alpha \leq 0$. In particular, $A - BB^*\Omega^{-1}$ is stable. Finally, notice that the state feedback controller with gain $K = B^*\Omega^{-1}$ can be readily computed by solving the Lyapunov equation in (8.3).

8.3   Stabilizability and uncontrollable eigenvalues

As before, let A be an operator on a finite dimensional space $\mathcal{X}$ and B an operator from a space $\mathcal{U}$ to $\mathcal{X}$. Recall that the controllable subspace $\mathcal{X}_c$ for $\{A, B\}$ is the invariant subspace for A defined by

$$\mathcal{X}_c = \text{span}\{A^k B \mathcal{U} : k = 0, 1, 2, \cdots\} \,. \tag{8.4}$$

The uncontrollable subspace $\mathcal{X}_{\bar{c}}$ for $\{A, B\}$ is the orthogonal complement of the controllable subspace $\mathcal{X}_c$. So, the uncontrollable subspace $\mathcal{X}_{\bar{c}}$ is an invariant subspace for A^*. Recall also that λ is an uncontrollable eigenvalue for the pair $\{A, B\}$ if there is a nonzero vector v in $\mathcal{X}_{\bar{c}}$ such that $A^* v = \bar{\lambda} v$. Thus, λ is an uncontrollable eigenvalue for the pair $\{A, B\}$ if and only if λ is an eigenvalue for the operator $A_{\bar{c}}$ on $\mathcal{X}_{\bar{c}}$ defined by $A_{\bar{c}}^* x = A^* x$ for $x \in \mathcal{X}_{\bar{c}}$. We say that an eigenvalue λ for A is *stable* if $\Re(\lambda) < 0$. Notice that α is an uncontrollable eigenvalue of the pair $\{A, B\}$ in Example 8.1.1, and α is also an eigenvalue of the matrix $A - BK$ for every gain K. Motivated by this example, we present the following result.

Lemma 8.3.1 *If λ is an uncontrollable eigenvalue for $\{A, B\}$, then λ is an eigenvalue of $A - BK$ for every gain operator K.*

Proof. To see this, recall that if λ is an uncontrollable eigenvalue of $\{A, B\}$, then there exists a nonzero vector v in $\mathcal{X}_{\bar{c}}$ such that $A^* v = \bar{\lambda} v$. Since $\mathcal{X}_{\bar{c}}$ is orthogonal to the range of B, we must have $B^* v = 0$. So, for any gain matrix K, we obtain

$$(A - BK)^* v = \bar{\lambda} v$$

that is, $\bar{\lambda}$ is an eigenvalue of $(A - BK)^*$. Hence, λ is an eigenvalue of $A - BK$. So, regardless of the state feedback gain, λ is always an eigenvalue of $A - BK$. Therefore, we cannot alter this eigenvalue by feedback.   ∎

If λ is an uncontrollable, unstable eigenvalue for the pair $\{A, B\}$, then Lemma 8.3.1 shows that $\{A, B\}$ is not stabilizable. In other words, if $\{A, B\}$ is stabilizable, then all of its uncontrollable eigenvalues are stable. This proves part of the following result.

Theorem 8.3.2 *A pair $\{A, B\}$ is stabilizable if and only if all of its uncontrollable eigenvalues are stable.*

Proof. We present a proof of this fact based on the decomposition of the state space $\mathcal{X}$ into the controllable subspace $\mathcal{X}_c$ and the uncontrollable subspace $\mathcal{X}_{\bar{c}}$. To this end, recall that A admits a matrix representation of the form (see Section 6.2)

$$\begin{bmatrix} A_c & * \\ 0 & A_{\bar{c}} \end{bmatrix} : \begin{bmatrix} \mathcal{X}_c \\ \mathcal{X}_{\bar{c}} \end{bmatrix} \longrightarrow \begin{bmatrix} \mathcal{X}_c \\ \mathcal{X}_{\bar{c}} \end{bmatrix} . \tag{8.5}$$

Here A_c is the operator on $\mathcal{X}_c$ defined by $A_c = A|\mathcal{X}_c$ and $A_{\bar{c}}$ is the compression of A to $\mathcal{X}_{\bar{c}}$. Clearly, the uncontrollable eigenvalues for the pair $\{A, B\}$ are precisely the eigenvalues of $A_{\bar{c}}$. Furthermore, the operator B admits a matrix representation of the form

$$\begin{bmatrix} B_c \\ 0 \end{bmatrix} : \mathcal{U} \longrightarrow \begin{bmatrix} \mathcal{X}_c \\ \mathcal{X}_{\bar{c}} \end{bmatrix} . \tag{8.6}$$

Recall that the pair $\{A_c, B_c\}$ is controllable. If K is any gain operator, then it has a matrix representation of the form

$$\begin{bmatrix} K_c & K_{\bar{c}} \end{bmatrix} : \begin{bmatrix} \mathcal{X}_c \\ \mathcal{X}_{\bar{c}} \end{bmatrix} \longrightarrow \mathcal{U}. \tag{8.7}$$

Hence, $A - BK$ has the following matrix representation

$$A - BK = \begin{bmatrix} A_c - B_c K_c & * \\ 0 & A_{\bar{c}} \end{bmatrix}. \tag{8.8}$$

From this representation, we see that the eigenvalues of $A - BK$ are the union of the eigenvalues of $A_c - B_c K_c$ and $A_{\bar{c}}$. Hence, regardless of K, the eigenvalues of $A_{\bar{c}}$ are always contained in the eigenvalues of $A - BK$.

Furthermore, since the pair $\{A_c, B_c\}$ is controllable, there exists a state feedback K_c such that $A_c - B_c K_c$ is stable; see Proposition 8.2.1. Hence, except for the eigenvalues of $A_{\bar{c}}$, all eigenvalues of $A - BK$ can be placed somewhere in the open left half plane. In particular, if all the eigenvalues of $A_{\bar{c}}$ are stable, then K can be chosen so that all the eigenvalues of $A - BK$ are also stable. This completes the proof. ∎

Using Lemma 5.2.1, we obtain the following useful result.

Corollary 1 *(PBH stabilizability result.) The pair $\{A$ on $\mathcal{X}, B\}$ is stabilizable if and only if*

$$rank \begin{bmatrix} A - \lambda I & B \end{bmatrix} = \dim \mathcal{X}$$

for every complex number λ in the closed right half plane.

8.4 Eigenvalue placement

Recall that a system $\dot{x} = Ax + Bu$ or the pair $\{A, B\}$ is a *single input* system if its input space $\mathcal{U}$ is one dimensional. The system $\dot{x} = Ax + Bu$ or the pair $\{A, B\}$ is *scalar input* if $\mathcal{U} = \mathbb{C}^1$. Obviously, any single input system can be identified with a scalar input system. In this section, for single input systems, we use controllability of the pair $\{A, B\}$ to develop a procedure to compute a gain K, which places the eigenvalues of $A - BK$ at any n specified locations in the complex plane where $n = \dim \mathcal{X}$. This will provide us with a computational method for stabilization and solve a classical pole placement problem.

8.4.1 An eigenvalue placement procedure

Consider first any pair $\{A, B\}$ where A is on $\mathcal{X}$ while B maps $\mathcal{U}$ into $\mathcal{X}$. The spaces $\mathcal{X}$ and $\mathcal{U}$ are finite dimensional and of arbitrary dimension. If K is any operator from $\mathcal{X}$ into $\mathcal{U}$, we claim that

$$\det[sI - A + BK] = \det[sI - A]\det[I + K(sI - A)^{-1}B] \tag{8.9}$$

where det denotes determinant. To verify this recall that if M and N are two finite dimensional operators acting between the appropriate spaces, then $\det[I + MN] = \det[I + NM]$.

Using this, we obtain

$$\begin{aligned} \det[sI - A + BK] &= \det[(sI - A)(I + (sI - A)^{-1}BK] \\ &= \det[sI - A]\det[I + (sI - A)^{-1}BK] \\ &= \det[sI - A]\det[I + K(sI - A)^{-1}B] \, . \end{aligned}$$

Therefore, (8.9) holds.

Finally, it is noted that if the input space $\mathcal{U}$ is one dimensional, then (8.9) reduces to

$$\det[sI - A + BK] = \det[sI - A](1 + K(sI - A)^{-1}B) \, . \tag{8.10}$$

Equation (8.15) shows that, *for a single input pair $\{A, B\}$, the coefficients of the character-istic polynomial for $A - BK$ are affine in K.*

Let $\{A, B\}$ be a controllable single input system where n is the dimension of the state space $\mathcal{X}$. To complete this section we will develop a method to compute a gain K which places the eigenvalues of $A - BK$ at any n specified location in the complex plane. In particular, this shows that any single input controllable pair is stabilizable.

Theorem 8.4.1 *Let $\{A, B\}$ be a single input controllable pair where n is the dimension of the state. Let $\hat{d}$ be any monic polynomial of degree n. Then there exists a state gain K such that $\hat{d}$ is the characteristic polynomial for $A - BK$. In particular, there exists a gain K which places the eigenvalues $A - BK$ at any specified n locations in the complex plane.*

PROOF. If $\hat{d}$ is any monic polynomial of degree n, we need to show that there exists a gain K such that $\hat{d}$ is the characteristic polynomial for $A - BK$, that is,

$$\hat{d}(s) = \det[sI - A + BK] \, .$$

Recalling (8.10) we need to show that there exists a gain K such that

$$K(sI - A)^{-1}B = \frac{\hat{d}(s) - d(s)}{d(s)} \tag{8.11}$$

where $d(s) = \det[sI - A]$ is the characteristic polynomial of A. Let

$$\begin{aligned} d(s) &= a_0 + a_1 s + \cdots + a_{n-1}s^{n-1} + s^n \\ \hat{d}(s) &= \hat{a}_0 + \hat{a}_1 s + \cdots + \hat{a}_{n-1}s^{n-1} + s^n \, . \end{aligned}$$

It now follows from Lemma 1.3.5 that (8.11) holds if and only if

$$\left[\begin{array}{cccc} KB & KAB & \cdots & KA^{n-1}B \end{array} \right] \Upsilon = \left[\begin{array}{cccc} \hat{a}_0 - a_0 & \hat{a}_1 - a_1 & \cdots & \hat{a}_{n-1} - a_{n-1} \end{array} \right] \tag{8.12}$$

where Υ is the invertible Hankel matrix given by

$$\Upsilon = \left[\begin{array}{cccccc} a_1 & a_2 & \cdots & a_{n-2} & a_{n-1} & 1 \\ a_2 & a_3 & \cdots & a_{n-1} & 1 & 0 \\ a_3 & a_4 & \cdots & 1 & 0 & 0 \\ \vdots & \vdots & & \vdots & \vdots & \vdots \\ a_{n-1} & 1 & \cdots & 0 & 0 & 0 \\ 1 & 0 & \ldots & 0 & 0 & 0 \end{array} \right] \, . \tag{8.13}$$

Without loss of generality we can assume that $\mathcal{U} = \mathbb{C}^1$. Let W be the controllability operator from $\mathbb{C}^n$ into $\mathcal{X}$ defined by

$$W = \begin{bmatrix} B & AB & \cdots & A^{n-1}B \end{bmatrix}. \tag{8.14}$$

Then condition (8.12) on K reduces to

$$KW\Upsilon = \begin{bmatrix} \hat{a}_0 - a_0 & \hat{a}_1 - a_1 & \cdots & \hat{a}_{n-1} - a_{n-1} \end{bmatrix}.$$

Because the pair $\{A, B\}$ is controllable, W is invertible. This readily implies that the gain K is uniquely given by

$$K = \begin{bmatrix} \hat{a}_0 - a_0 & \hat{a}_1 - a_1 & \cdots & \hat{a}_{n-1} - a_{n-1} \end{bmatrix} \Upsilon^{-1} W^{-1}. \tag{8.15}$$

Notice that the gain K is uniquely determined by the coefficients $\{\hat{a}_i\}_0^{n-1}$ of the desired characteristic polynomial for $A - BK$. So, if $\hat{d}$ is any monic polynomial of the form $\hat{d}(s) = \hat{a}_0 + \hat{a}_1 s + \cdots + \hat{a}_{n-1} s^{n-1} + s^n$, and the gain K is given by (8.15), then $\hat{d}$ is the characteristic polynomial for $A - BK$. ∎

As before, let $\{A, B\}$ be a single input controllable pair with state space dimension n. Let $\{\lambda_i\}_1^n$ be any n specified numbers in the complex plane and $\hat{d} = \hat{a}_0 + \hat{a}_1 s + \cdots + \hat{a}_{n-1} s^{n-1} + s^n$ be the polynomial uniquely determined by $\hat{d}(s) = \prod_1^n (s - \lambda_i)$. If K is the gain computed in (8.15), then $\hat{d}$ is the characteristic polynomial for $A - BK$ and $\{\lambda_i\}_1^n$ are the eigenvalues for $A - BK$.

8.4.2 Ackermann's Formula

The following lemma provides another method for computing a state feedback gain K to assign prescribed eigenvalues to $A - BK$.

Lemma 8.4.2 (Ackermann's Formula [1]) *Suppose $\{A, B\}$ is a scalar input controllable pair with state dimension n and $\hat{d}$ is any monic polynomial of degree n. Let K be the gain matrix defined by*

$$K = \begin{bmatrix} 0 & 0 & \cdots & 0 & 1 \end{bmatrix} W^{-1} \hat{d}(A). \tag{8.16}$$

where W is the controllability matrix defined in (8.14). Then $\hat{d}$ is the characteristic polynomial for $A - BK$.

PROOF. Let $v = \begin{bmatrix} 0 & 0 & \cdots & 0 & 1 \end{bmatrix} W^{-1}$. Then $vW = \begin{bmatrix} 0 & 0 & \cdots & 0 & 1 \end{bmatrix}$, that is,

$$vA^j B = 0 \quad \text{for} \quad j = 0, 1, \cdots, n-2$$
$$vA^{n-1} B = 1.$$

It now follows from Lemma 1.3.1 that $v(sI - A)^{-1}B = 1/d(s)$ where d is the characteristic polynomial for A and

$$vA^j(sI - A)^{-1}B = s^j/d(s) \quad \text{for} \quad j = 0, 1, \cdots, n-1 \tag{8.17a}$$
$$vA^n(sI - A)^{-1}B = s^n/d(s) - 1. \tag{8.17b}$$

Letting $\hat{d}(s) = \hat{a}_0 + \hat{a}_1 s + \cdots + \hat{a}_{n-1}s^{n-1} + s^n$ and noting that $K = v\hat{d}(A)$, relationships (8.17) imply that

$$
\begin{aligned}
K(sI - A)^{-1}B &= v\hat{d}(A)(sI - A)^{-1}B \\
&= \sum_{j=0}^{n-1} \hat{a}_j v A^j (sI - A)^{-1}B + vA^n(sI - A)^{-1}B \\
&= \sum_{j=0}^{n-1} \hat{a}_j s^j / d(s) + s^n/d(s) - 1 \\
&= \hat{d}(s)/d(s) - 1
\end{aligned}
$$

that is, (8.11) holds. By the proof of Theorem 8.4.1, this is equivalent to $\det[sI - A + BK] = \hat{d}(s)$, that is, $\hat{d}$ is the characteristic polynomial for $A - BK$. ∎

Exercise 22 Consider the system described by

$$
\begin{aligned}
\dot{x}_1 &= x_2 + u \\
\dot{x}_2 &= x_1 + u.
\end{aligned}
$$

(a) Is this system stabilizable?

(b) Does there exist a state feedback controller K which results in closed loop eigenvalues $\{-1, -2\}$? If so find one.

(c) Does there exist a linear state feedback controller which results in closed loop eigenvalues $\{-2, -3\}$? If so find one.

8.5   Controllable canonical form

This section is devoted to the controllable canonical form for scalar input systems. We say that the pair $\{A, B\}$ or the system $\dot{x} = Ax + Bu$ is in *controllable canonical form* if A and B are matrices with the following structure:

$$
A = \begin{bmatrix}
0 & 1 & 0 & \cdots & 0 & 0 \\
0 & 0 & 1 & \cdots & 0 & 0 \\
\vdots & \vdots & & \ddots & & \vdots \\
\vdots & \vdots & & & \ddots & \vdots \\
0 & 0 & 0 & \cdots & 0 & 1 \\
-a_0 & -a_1 & -a_2 & \cdots & -a_{n-2} & -a_{n-1}
\end{bmatrix}
\quad \text{and} \quad
B = \begin{bmatrix} 0 \\ 0 \\ \vdots \\ \vdots \\ 0 \\ 1 \end{bmatrix}.
\qquad (8.18)
$$

In particular, if the pair $\{A, B\}$ is in controllable canonical form, then A is a companion matrix. We previously encountered the controllable canonical form when we constructed state space realizations for scalar valued transfer functions; see (1.48). Finally, recall that the characteristic polynomial for A is given by

$$\det[sI - A] = a_0 + a_1 s + \cdots + a_{n-1}s^{n-1} + s^n . \tag{8.19}$$

So, given any monic polynomial $d(s) = a_0 + a_1 s + \cdots + a_{n-1}s^{n-1} + s^n$ there exists a unique controllable canonical pair $\{A, B\}$ such that d is the characteristic polynomial for A. In this case we say that $\{A, B\}$ is the controllable canonical pair determined by d. Obviously, there is a one to one correspondence between the set of all controllable canonical pairs and the set of all monic polynomials.

We claim that any pair $\{A, B\}$ in controllable canonical form is controllable. To see this, notice that the controllability operator for the pair $\{A, B\}$ in controllable canonical form is given by

$$W = \begin{bmatrix} B & AB & \cdots & A^{n-1}B \end{bmatrix} = \begin{bmatrix} 0 & 0 & \cdots & 0 & 1 \\ 0 & 0 & \cdots & 1 & * \\ \vdots & \vdots & \ddots & * & * \\ 0 & 1 & \cdots & * & * \\ 1 & * & \cdots & * & * \end{bmatrix} \tag{8.20}$$

where the $*$ entries are unspecified. Since this matrix has rank n, the pair $\{A, B\}$ is controllable.

Recall that, for a controllable single input pair $\{A, B\}$, one can arbitrarily assign the eigenvalues of $A - BK$ by appropriate choice of the gain matrix K. We now show that this can easily be demonstrated when $\{A, B\}$ is in controllable canonical form. This will play a major role in developing another algorithm to arbitrarily assign the eigenvalues of $A - BK$ for any controllable single input pair.

Let $\{A \text{ on } \mathbb{C}^n, B\}$ be in controllable canonical form, and consider any n specified complex numbers $\{\lambda_j\}_1^n$. Now consider the problem of finding a gain matrix K from $\mathbb{C}^n$ into $\mathbb{C}^1$ such that $\{\lambda_j\}_1^n$ are the eigenvalues of $A - BK$. To solve this eigenvalue placement problem, let

$$\hat{d}(s) = s^n + \hat{a}_{n-1}s^{n-1} + \cdots + \hat{a}_1 s + \hat{a}_0 = \prod_{j=1}^{n}(s - \lambda_j) \tag{8.21}$$

be the unique monic polynomial whose roots are the desired closed loop eigenvalues $\{\lambda_j\}_1^n$. Introducing the $1 \times n$ gain matrix

$$K = \begin{bmatrix} \hat{a}_0 - a_0 & \hat{a}_1 - a_1 & \cdots & \hat{a}_{n-1} - a_{n-1} \end{bmatrix} \tag{8.22}$$

where $a_0, a_1, \cdots, a_{n-1}$ are given by the last row of A, we obtain

$$A - BK = \begin{bmatrix} 0 & 1 & 0 & \cdots & 0 & 0 \\ 0 & 0 & 1 & \cdots & 0 & 0 \\ \vdots & \vdots & & \ddots & & \vdots \\ \vdots & \vdots & & & \ddots & \vdots \\ 0 & 0 & 0 & \cdots & 0 & 1 \\ -\hat{a}_0 & -\hat{a}_1 & -\hat{a}_2 & \cdots & -\hat{a}_{n-2} & -\hat{a}_{n-1} \end{bmatrix} . \tag{8.23}$$

Because $A - BK$ is a companion matrix, $\hat{d}$ is the characteristic polynomial for $A - BK$. Hence, $\{\lambda_j\}_1^n$ are the eigenvalues for the closed loop system matrix $A - BK$. By this method, we can arbitrarily assign the eigenvalues of $A - BK$.

8.5.1   Transformation to controllable canonical form

In the last section, we saw that if a scalar input pair $\{A, B\}$ is in controllable canonical form, then one can easily place the eigenvalues of $A - BK$ by appropriate choice of a gain matrix K. Now consider any single input pair $\{A \text{ on } \mathcal{X}, B\}$ which is controllable and let n be the dimension of $\mathcal{X}$. We will show that $\{A, B\}$ is similar to a pair $\{\tilde{A} \text{ on } \mathbb{C}^n, \tilde{B}\}$ in controllable canonical form, that is, there is an invertible operator T from $\mathbb{C}^n$ onto $\mathcal{X}$ satisfying

$$AT = T\tilde{A} \qquad \text{and} \qquad B = T\tilde{B}. \tag{8.24}$$

In this case, we say that T intertwines $\{\tilde{A}, \tilde{B}\}$ with $\{A, B\}$. From this we will obtain an alternative proof of the fact that, for any single input controllable pair $\{A, B\}$, one can arbitrarily place the eigenvalues of $A - BK$ by the appropriate choice of a gain matrix K.

First consider any single input pair $\{A, B\}$ with the property that it is similar to a pair $\{\tilde{A}, \tilde{B}\}$ which is in controllable canonical form, that is, (8.24) is satisfied with some invertible operator T. Using the first equation in (8.24) one can readily show that, for all integers $k \geq 0$, one has $A^k T = T\tilde{A}^k$, and hence, $A^k B = A^k T\tilde{B} = T\tilde{A}^k \tilde{B}$. This implies that

$$W = T\tilde{W} \tag{8.25}$$

where $\tilde{W}$ and W are the controllability matrices associated with $\{\tilde{A}, \tilde{B}\}$ and $\{A, B\}$, respectively, that is,

$$\tilde{W} = \begin{bmatrix} \tilde{B} & \tilde{A}\tilde{B} & \cdots & \tilde{A}^{n-1}\tilde{B} \end{bmatrix} \quad \text{and} \quad W = \begin{bmatrix} B & AB & \cdots & A^{n-1}B \end{bmatrix}. \tag{8.26}$$

Since $\{\tilde{A}, \tilde{B}\}$ is controllable, the operator $\tilde{W}$ is invertible. The equation $W = T\tilde{W}$ also shows that W is invertible, and thus, the pair $\{A, B\}$ is controllable. Furthermore, this equation also shows that

$$T = W\tilde{W}^{-1} \tag{8.27}$$

is the only operator which intertwines $\{\tilde{A}, \tilde{B}\}$ with $\{A, B\}$. Finally, since A and $\tilde{A}$ are similar, they have the same the same characteristic polynomial.

Now consider any single input controllable pair $\{A \text{ on } \mathcal{X}, B\}$ whose state dimension is n. Let $\{\tilde{A} \text{ on } \mathbb{C}^n, \tilde{B}\}$ be the unique pair in controllable canonical form such that $\tilde{A}$ has the same characteristic polynomial as A. Since, $\{A, B\}$ and $\{\tilde{A}, \tilde{B}\}$ are controllable, their respective controllability operators W and $\tilde{W}$ are invertible. So, the equation $W = T\tilde{W}$ uniquely defines an invertible operator T. We now show that T intertwines $\{\tilde{A}, \tilde{B}\}$ with $\{A, B\}$, that is, (8.24) holds. First notice that $W = T\tilde{W}$ and the structure of $\tilde{W}$ and W imply that $B = T\tilde{B}$ and

$$A^k B = T\tilde{A}^k \tilde{B} \qquad (\text{for} \quad k = 0, 1, \cdots, n - 1).$$

Since A and $\tilde{A}$ have the same characteristic polynomial, it follows from the Cayley-Hamilton Theorem that $A^n B = T\tilde{A}^n \tilde{B}$; hence $AW = T\tilde{A}\tilde{W}$. Recalling the definition of T, we now obtain that $ATW = AW = T\tilde{A}\tilde{W}$. Since $\tilde{W}$ is invertible, we have that $AT = T\tilde{A}$. Therefore, $\{\tilde{A}, \tilde{B}\}$ and $\{A, B\}$ are similar. Actually, we have just proven the following result.

Theorem 8.5.1 *Let $\{A, B\}$ be a scalar input controllable pair with state dimension n and let*

$$\det[sI - A] = s^n + a_{n-1}s^{n-1} + \cdots + a_1 s + a_0 \tag{8.28}$$

be the characteristic polynomial of A. Suppose $\{\tilde{A}, \tilde{B}\}$ is the unique pair in controllable canonical form determined by the characteristic polynomial of A. Then $\{\tilde{A}, \tilde{B}\}$ is similar to $\{A, B\}$. Furthermore, the similarity transformation T intertwining $\{\tilde{A}, \tilde{B}\}$ with $\{A, B\}$ is uniquely given by $T = W\tilde{W}^{-1}$ where $\tilde{W}$ and W are the controllability matrices in (8.26) associated with $\{\tilde{A}, \tilde{B}\}$ and $\{A, B\}$, respectively.

A recursive algorithm. As before, suppose that $\{A \text{ on } \mathcal{X}, B\}$ is a scalar input controllable pair with state dimension n. Let $\{\tilde{A}, \tilde{B}\}$ be the unique pair in controllable canonical form which is similar to $\{A, B\}$, that is

$$\tilde{A} = \begin{bmatrix} 0 & 1 & 0 & \cdots & 0 & 0 \\ 0 & 0 & 1 & \cdots & 0 & 0 \\ \vdots & \vdots & & \ddots & & \vdots \\ \vdots & \vdots & & & \ddots & \vdots \\ 0 & 0 & 0 & \cdots & 0 & 1 \\ -a_0 & -a_1 & -a_2 & \cdots & -a_{n-2} & -a_{n-1} \end{bmatrix} \quad \text{and} \quad \tilde{B} = \begin{bmatrix} 0 \\ 0 \\ \vdots \\ \vdots \\ 0 \\ 1 \end{bmatrix} \tag{8.29}$$

where $d(s) = s^n + a_{n-1}s^{n-1} + \cdots + a_1 s + a_0$ is the characteristic polynomial of A. Due to the structure of $\{\tilde{A}, \tilde{B}\}$ one can obtain a simple recursive algorithm to compute the state transformation $T : \mathbb{C}^n \to \mathcal{X}$ which intertwines $\{\tilde{A}, \tilde{B}\}$ with $\{A, B\}$.

To this end, consider the following decomposition:

$$T = \begin{bmatrix} \phi_1 & \phi_2 & \cdots & \phi_n \end{bmatrix} \tag{8.30}$$

where $\{\phi_i\}_1^n$ are vectors in $\mathcal{X}$. Let $\{e_i\}_1^n$ be the standard basis for $\mathbb{C}^n$, that is, the i-th component of e_i is one while all the other components are zero; then $\phi_i = Te_i$ for $i = 1, \cdots, n$.

Using $\tilde{B}1 = e_n$ and $T\tilde{B} = B$, we have $\phi_n = Te_n = T\tilde{B}1 = B1$. Hence, $\phi_n = b$ where $b = B1$. By employing the companion structure of $\tilde{A}$, we obtain that $\tilde{A}e_n = e_{n-1} - a_{n-1}e_n$. Since $AT = T\tilde{A}$, it follows that

$$A\phi_n = ATe_n = T\tilde{A}e_n = Te_{n-1} - a_{n-1}Te_n = \phi_{n-1} - a_{n-1}\phi_n.$$

Therefore, $\phi_{n-1} = A\phi_n + a_{n-1}b$. In a similar fashion, we have

$$A\phi_{n-1} = ATe_{n-1} = T\tilde{A}e_{n-1} = Te_{n-2} - a_{n-2}Te_n = \phi_{n-2} - a_{n-2}\phi_n.$$

Thus, $\phi_{n-2} = A\phi_{n-1} + a_{n-2}b$. An inductive argument shows that

$$\begin{aligned} \phi_n &= b \\ \phi_{i-1} &= A\phi_i + a_{i-1}b \quad \text{(for } i = n, n-1, \cdots, 2) . \end{aligned} \tag{8.31}$$

Therefore, the invertible operator T intertwining $\{\tilde{A}, \tilde{B}\}$ with $\{A, B\}$ is given by (8.30) where the "columns" are recursively computed by (8.31).

8.5.2   Eigenvalue placement by state feedback

We are now in a position to use the controllable canonical form to develop an algorithm which arbitrarily places the eigenvalues of $A - BK$ for a scalar input controllable pair $\{A \text{ on } \mathcal{X}, B\}$. To this end, n be the dimension of $\mathcal{X}$ and let $\{\lambda_i\}_1^n$ be n specified complex numbers. As before, let

$$\hat{d}(s) = s^n + \hat{a}_{n-1}s^{n-1} + \cdots + \hat{a}_1 s + \hat{a}_0 = \prod_{i=1}^{n}(s - \lambda_i) \tag{8.32}$$

be the unique monic polynomial whose roots are the desired closed loop eigenvalues $\{\lambda_i\}_1^n$. Let $T = W\tilde{W}^{-1}$ be the invertible operator specified in Theorem 8.5.1 which intertwines $\{\tilde{A}, \tilde{B}\}$ with $\{A, B\}$. Finally, consider the $1 \times n$ real gain matrix

$$K = \begin{bmatrix} \hat{a}_0 - a_0 & \hat{a}_1 - a_1 & \cdots & \hat{a}_{n-1} - a_{n-1} \end{bmatrix} T^{-1} \tag{8.33}$$

where the characteristic polynomial for A is given in (8.28). Notice that $K = \tilde{K}T^{-1}$ where $\tilde{K}$ is the gain matrix which guarantees that $\hat{d}$ is the characteristic polynomial for $\tilde{A} - \tilde{B}\tilde{K}$. Using $K = \tilde{K}T^{-1}$, we have $(A - BK)T = T(\tilde{A} - \tilde{B}\tilde{K})$. Hence, the operators $A - BK$ and $\tilde{A} - \tilde{B}\tilde{K}$ are similar, and thus, have the same characteristic polynomial $\hat{d}$. In other words, $\{\lambda_i\}_1^n$ are the eigenvalues of $A - BK$. We have just presented another proof of the result that one can arbitrarily place the eigenvalues of $A - BK$ for a scalar input controllable pair $\{A, B\}$. Summing up this analysis readily yields the following eigenvalue placement algorithm for scalar input controllable systems.

Theorem 8.5.2 *Let $\{A \text{ on } \mathcal{X}, B\}$ be a scalar input controllable pair with state dimension n. Let T be the similarity transformation intertwining the controllable canonical pair $\{\tilde{A}, \tilde{B}\}$ with $\{A, B\}$ computed according to Theorem 8.5.1. Let $\{\lambda_i\}_1^n$ be a specified set of complex numbers and K the operator from $\mathcal{X}$ into $\mathbb{C}^1$ determined by (8.28), (8.32) and (8.33). Then $\{\lambda_i\}_1^n$ are the eigenvalues of $A - BK$.*

8.6   Multivariable eigenvalue placement

This section uses an upper triangular matrix representation to place the eigenvalues of a finite dimensional controllable system at a specified location in the complex plane. We begin with the following multi-input version of Theorem 8.5.2.

Theorem 8.6.1 *(Eigenvalue placement theorem.) Assume that $\{A \text{ on } \mathcal{X}, B\}$ is a controllable pair with state dimension n. Let $\{\lambda_j\}_1^n$ be a set of n complex numbers. Then there exists a gain K such that*

$$\det[sI - A + BK] = \prod_{j=1}^{n}(s - \lambda_j). \tag{8.34}$$

In this case, $\{\lambda_j\}_1^n$ are the eigenvalues of $A - BK$. In particular, if $\hat{d}$ is any monic polynomial of degree n, then there exists a gain K such that $\hat{d}$ is the characteristic polynomial for $A - BK$.

It follows from the above theorem that controllability implies stabilizability.

We also claim that if one can use state feedback to place all the eigenvalues of $A - BK$ at any $\dim \mathcal{X}$ points in the complex plane, then the pair $\{A, B\}$ is controllable. To see this simply recall that if $\{A, B\}$ is not controllable, then the uncontrollable eigenvalues of $\{A, B\}$ are not changed by state feedback. In this case, state feedback cannot be used to arbitrarily place all the eigenvalues of $A - BK$, which proves our claim. Combining this with Theorem 8.6.1, yields the following result.

Theorem 8.6.2 *Consider the pair $\{A, B\}$ with state dimension n. Then $\{A, B\}$ is controllable if and only if for every set $\{\lambda_j\}_1^n$ of n complex numbers there exists a state feedback K such that the eigenvalues of $A - BK$ are $\{\lambda_j\}_1^n$*

PROOF OF THEOREM 8.6.1. Let $\mathcal{U}$ be the input space for the controllable pair $\{A, B\}$. Let u_1 be any vector in $\mathcal{U}$ such that $b_1 = Bu_1$ is nonzero. Let $\mathcal{X}_1$ be the subspace spanned by $\{A^k b_1\}_0^\infty$. Clearly, $\mathcal{X}_1$ is an invariant subspace for A. Let $\tilde{\mathcal{X}}_1$ be the orthogonal complement of $\mathcal{X}_1$. Then A admits a matrix representation of the form:

$$A = \begin{bmatrix} A_1 & * \\ 0 & \tilde{A}_1 \end{bmatrix} \quad \text{on} \quad \begin{bmatrix} \mathcal{X}_1 \\ \tilde{\mathcal{X}}_1 \end{bmatrix}.$$

Here A_1 is the operator on $\mathcal{X}_1$ defined by $A_1 = A|\mathcal{X}_1$. The operator $\tilde{A}_1$ is the compression of A to $\tilde{\mathcal{X}}_1$, that is, $\tilde{A}_1 = \tilde{P}_1 A|\tilde{\mathcal{X}}_1$ where $\tilde{P}_1$ is the orthogonal projection onto $\tilde{\mathcal{X}}_1$. Let $\mathcal{U}_1$ be the one dimensional space spanned by u_1. Let B_1 be the operator from $\mathcal{U}_1$ into $\mathcal{X}_1$ defined by $B_1 = B|\mathcal{U}_1$. Since Bu_1 is cyclic for A_1, it follows that the pair $\{A_1, B_1\}$ is controllable. (Recall that a vector b is cyclic for an operator T on $\mathcal{X}$ if $\{T^k b\}_0^\infty$ spans all of $\mathcal{X}$.) According to this decomposition the operator B admits a matrix representation of the form:

$$B = \begin{bmatrix} B_1 & * \\ 0 & \tilde{B}_1 \end{bmatrix} : \begin{bmatrix} \mathcal{U}_1 \\ \tilde{\mathcal{U}}_1 \end{bmatrix} \longrightarrow \begin{bmatrix} \mathcal{X}_1 \\ \tilde{\mathcal{X}}_1 \end{bmatrix}$$

where $\tilde{\mathcal{U}}_1 = \mathcal{U} \ominus \mathcal{U}_1$. Since $\{A, B\}$ is controllable and

$$A^k B = \begin{bmatrix} A_1^k B_1 & * \\ 0 & \tilde{A}_1^k \tilde{B}_1 \end{bmatrix}$$

for all integers $k \geq 0$, it follows that the range of $\tilde{B}_1$ is cyclic for $\tilde{A}_1$. (As expected, a subspace $\mathcal{B}$ is cyclic for an operator T on $\mathcal{X}$ if $\{T^k \mathcal{B}\}_0^\infty$ spans all of $\mathcal{X}$.) Therefore, the pair $\{\tilde{A}_1, \tilde{B}_1\}$ is also controllable.

Now assume that $\tilde{\mathcal{X}}_1$ is nonzero. If $\tilde{\mathcal{X}}_1$ is zero, then $\{A, B_1\}$ is a single input controllable pair. In this case, Theorems 8.4.1 and 8.5.2 show that there exists a state feedback gain K from $\mathcal{X}$ into $\mathcal{U}_1$ which places the eigenvalues of $A - B_1 K$ at any n specified locations in the complex plane. Since $B_1 = B|\mathcal{U}_1$, it follows that this feedback K can also be used to place the eigenvalues of $A - BK$ at the same n locations in the complex plane. Hence, Theorem 8.6.1 follows from Theorem 8.5.2 when $\tilde{\mathcal{X}}_1$ equals zero. So, assume that $\tilde{\mathcal{X}}_1$ is nonzero. Then we can apply the previous procedure to decompose the controllable pair $\{\tilde{A}_1, \tilde{B}_1\}$ into upper triangular block matrices. To this end, let u_2 be a vector in $\tilde{\mathcal{U}}_1$ such that $b_2 = \tilde{B}_1 u_2$ is nonzero. Let $\mathcal{X}_2$ be the subspace of $\tilde{\mathcal{X}}_1$ generated by the span of $\{\tilde{A}_1^k b_2\}_0^\infty$. Let A_2 be the operator on $\mathcal{X}_2$ defined by $A_2 = A|\mathcal{X}_2$. Then $\tilde{A}_1$ admits a matrix representation of the form:

$$\tilde{A}_1 = \begin{bmatrix} A_2 & * \\ 0 & \tilde{A}_2 \end{bmatrix} \text{ on } \begin{bmatrix} \mathcal{X}_2 \\ \tilde{\mathcal{X}}_2 \end{bmatrix}.$$

Here $\tilde{\mathcal{X}}_2 = \tilde{\mathcal{X}}_1 \ominus \mathcal{X}_2$ and $\tilde{A}_2$ is the compression of $\tilde{A}_1$ to $\tilde{\mathcal{X}}_2$. Let $\mathcal{U}_2$ be the one dimensional space spanned by u_2, and $\tilde{\mathcal{U}}_2$ be the orthogonal complement of $\mathcal{U}_2$ in $\tilde{\mathcal{U}}_1$. According to this decomposition the operator $\tilde{B}_1$ admits a matrix representation of the form:

$$\tilde{B}_1 = \begin{bmatrix} B_2 & * \\ 0 & \tilde{B}_2 \end{bmatrix} : \begin{bmatrix} \mathcal{U}_2 \\ \tilde{\mathcal{U}}_2 \end{bmatrix} \longrightarrow \begin{bmatrix} \mathcal{X}_2 \\ \tilde{\mathcal{X}}_2 \end{bmatrix}.$$

Notice that B_2 is the operator from $\mathcal{U}_2$ into $\mathcal{X}_2$ defined by $B_2 = \tilde{B}_1|\mathcal{U}_2$, or equivalently, $B_2 = P_2 B|\mathcal{U}_2$ where P_2 is the orthogonal projection onto $\mathcal{X}_2$. Since b_2 is cyclic for A_2, the pair $\{A_2, B_2\}$ is controllable. Because $\{\tilde{A}_1, \tilde{B}_1\}$ is controllable, it follows that the pair $\{\tilde{A}_2, \tilde{B}_2\}$ is also controllable.

By substituting the matrix representations for $\{\tilde{A}_1, \tilde{B}_1\}$ into our previous matrix representations for $\{A, B\}$, we arrive at the following matrix representation for A:

$$\begin{bmatrix} A_1 & * & * \\ 0 & A_2 & * \\ 0 & 0 & \tilde{A}_2 \end{bmatrix} \text{ on } \begin{bmatrix} \mathcal{X}_1 \\ \mathcal{X}_2 \\ \tilde{\mathcal{X}}_2 \end{bmatrix}.$$

The corresponding matrix representation for B is

$$\begin{bmatrix} B_1 & * & * \\ 0 & B_2 & * \\ 0 & 0 & \tilde{B}_2 \end{bmatrix} : \begin{bmatrix} \mathcal{U}_1 \\ \mathcal{U}_2 \\ \tilde{\mathcal{U}}_2 \end{bmatrix} \longrightarrow \begin{bmatrix} \mathcal{X}_1 \\ \mathcal{X}_2 \\ \tilde{\mathcal{X}}_2 \end{bmatrix}.$$

By construction $\{A_1, B_1\}$ and $\{A_2, B_2\}$ are two single input controllable pairs. If $\tilde{\mathcal{X}}_2$ is not equal to zero, then we can apply the above procedure to the controllable pair $\{\tilde{A}_2, \tilde{B}_2\}$. By using the fact that the state space is finite dimensional, a repeated application of the above procedure shows that A admits an upper triangular matrix representation of the form:

$$A = \begin{bmatrix} A_1 & * & * & \cdots \\ 0 & A_2 & * & \cdots \\ \vdots & & \ddots & \\ 0 & 0 & \cdots & A_m \end{bmatrix} \text{ on } \begin{bmatrix} \mathcal{X}_1 \\ \mathcal{X}_2 \\ \vdots \\ \mathcal{X}_m \end{bmatrix}. \tag{8.35}$$

Here m is the first integer such that $\tilde{\mathcal{X}}_m$ is zero. In this setting $\mathcal{U} = \oplus_{j=1}^m \mathcal{U}_j \oplus \tilde{\mathcal{U}}_m$ where $\mathcal{U}_j$ is the one dimensional space spanned by the corresponding input vector u_j. The vector u_j is chosen such that b_j is nonzero where b_j is the orthogonal projection of Bu_j onto $\mathcal{X}_j$. The vectors $\{u_j\}_1^m$ are orthogonal, and b_j is cyclic for A_j. In this case, B has an upper triangular matrix representation of the form:

$$B = \begin{bmatrix} B_1 & * & * & \cdots & * \\ 0 & B_2 & * & \cdots & * \\ \vdots & & \ddots & & \vdots \\ 0 & 0 & \cdots & B_m & * \end{bmatrix} : \begin{bmatrix} \mathcal{U}_1 \\ \mathcal{U}_2 \\ \vdots \\ \mathcal{U}_m \\ \tilde{\mathcal{U}}_m \end{bmatrix} \longrightarrow \begin{bmatrix} \mathcal{X}_1 \\ \mathcal{X}_2 \\ \vdots \\ \mathcal{X}_m \end{bmatrix}. \tag{8.36}$$

By construction the pair $\{A_j$ on $\mathcal{X}_j, B_j\}$ is a controllable single input system whose input space is $\mathcal{U}_j$ for $j = 1, 2, \cdots, m$. Obviously, m is less than or equal to the dimension of $\mathcal{U}$.

Recall that n is the dimension of $\mathcal{X}$. We are now ready to show that there exists a state feedback gain K which places the eigenvalues of $A - BK$ at any n specified locations in the complex plane, that is, (8.34) holds. To see this, without loss of generality, we can assume that the controllable pair $\{A, B\}$ is given by its matrix representation in equations (8.35) and (8.36). Recall that for every $j = 1, 2, \cdots, m$ the pair $\{A_j, B_j\}$ is a single input controllable system whose input space is $\mathcal{U}_j$. So, if n_j is the dimension of $\mathcal{X}_j$, then the dimension of the state $\mathcal{X}$ is $n = \sum n_j$. According to Theorems 8.4.1 and 8.5.2, there exists a state feedback operator K_j from $\mathcal{X}_j$ into $\mathcal{U}_j$ which places the eigenvalues of $A_j - B_j K_j$ at any n_j specified locations in the complex plane. Now let K be the block diagonal matrix from $\mathcal{X} = \oplus_1^m \mathcal{X}_j$ into $\mathcal{U} = \oplus_1^m \mathcal{U}_j \oplus \tilde{\mathcal{U}}_m$ defined by

$$K = \begin{bmatrix} K_1 & 0 & \cdots & 0 \\ 0 & K_2 & \cdots & 0 \\ \vdots & & \ddots & \\ 0 & 0 & \cdots & K_m \\ 0 & 0 & \cdots & 0 \end{bmatrix}. \tag{8.37}$$

Then substituting this K into the upper triangular matrix representation for $\{A, B\}$ in (8.35) and (8.36), we have the following matrix representation:

$$A - BK = \begin{bmatrix} A_1 - B_1 K_1 & * & * & \cdots \\ 0 & A_2 - B_2 K_2 & * & \cdots \\ \vdots & & \ddots & \\ 0 & 0 & \cdots & A_m - B_m K_m \end{bmatrix}. \tag{8.38}$$

Because $A - BK$ is an upper triangular block matrix whose diagonal entries are $A_j - B_j K_j$, the eigenvalues of $A - BK$ is the union of the eigenvalues of $\{A_j - B_j K_j\}$ for $j = 1, \cdots, m$. Since we can place the eigenvalues of $A_j - B_j K_j$ at any n_j locations, it follows that we can choose K to place the poles of $A - BK$ at n specified locations in the complex plane. Therefore, (8.34) holds. ■

Remark 8.6.1 Consider the controllable decomposition for the pair $\{A$ on $\mathcal{X}, B\}$ given in (8.5) and (8.6) where $\mathcal{X} = \mathcal{X}_c \oplus \mathcal{X}_{\bar{c}}$. Since the pair $\{A_c$ on $\mathcal{X}_c, B_c\}$ is controllable, one can use state feedback to place the eigenvalues of $A_c - B_c K_c$ at any dim $\mathcal{X}_c$ specified locations in the complex plane. However, state feedback does not change the uncontrollable eigenvalues of A. So, by consulting the matrix representation for $A - BK$ in (8.7) and (8.8), it readily follows that one can use state feedback to place the eigenvalues of $A - BK$ at any dim $\mathcal{X}_c$ specified locations in the complex plane, while the remaining dim $\mathcal{X}_{\bar{c}}$ uncontrollable eigenvalues are unchanged.

Finally, it is noted that one can use the proof of Theorem 8.6.1 to construct an algorithm to compute a gain K to place the eigenvalues of $A - BK$ at n specified location in the complex plane. Let us complete this section with the following classical result in Wonham [129].

Proposition 8.6.3 *Let $\{A \text{ on } \mathcal{X}, B\}$ be a controllable system and consider any nonzero vector b in the range of B. Let B_0 the operator from $\mathbb{C}^1$ into $\mathcal{X}$ defined by $B_0\,\alpha = \alpha b$ Then there exists an operator K such that the scalar input system $\{A - BK, B_0\}$ is controllable.*

PROOF. Without loss of generality we can assume that A and B are the upper triangular block matrices given by (8.35) and (8.36), respectively, and $b = b_1$. Recall that $\{A_j, B_j\}$ is a single input controllable pair for $j = 1, 2, \cdots, m$. As before, let n_j be the dimension of $\mathcal{X}_j$ and $b_j = B_j u_j$ where u_j is any nonzero vector in $\mathcal{U}_j$. Then $\{A_j^k b_j : k = 0, 1, \cdots, n_j - 1\}$ forms a basis for $\mathcal{X}_j$. For $j = 1, 2, \cdots, m - 1$, let Φ_j be the operator from $\mathcal{X}_j$ into $\mathcal{U}_{j+1}$ defined by

$$\Phi_j A_j^{n_j - 1} b_j = u_{j+1} \quad \text{and} \quad \Phi_j A_j^k b_j = 0 \quad (\text{for } k = 0, 1, \cdots, n_j - 2). \tag{8.39}$$

Let K be the block matrix from $\mathcal{X} = \oplus_1^m \mathcal{X}_j$ into $\mathcal{U} = \oplus_1^m \mathcal{U}_j \oplus \tilde{\mathcal{U}}_m$ obtained by placing the operators $-\Phi_1, \cdots, -\Phi_{m-1}$ immediately below the main diagonal and zeros elsewhere, that is,

$$K = \begin{bmatrix} 0 & 0 & \cdots & 0 & 0 \\ -\Phi_1 & 0 & \cdots & 0 & 0 \\ 0 & -\Phi_2 & \ddots & \vdots & \vdots \\ \vdots & & \ddots & 0 & 0 \\ 0 & 0 & \cdots & -\Phi_{m-1} & 0 \\ 0 & 0 & \cdots & 0 & 0 \end{bmatrix}. \tag{8.40}$$

We claim that $\{A - BK, B_1\}$ is a controllable pair. To verify this it is sufficient to show that $\mathcal{X} = \mathcal{X}_c$ where $\mathcal{X}_c$ is the controllable subspace spanned by $\{(A - BK)^k b_1\}_0^\infty$.

The upper triangular form of A and the structure of K along with (8.39), yields

$$(A - BK)b_1 = Ab_1 + B\Phi_1 b_1 = A_1 b_1$$

when $n_1 \geq 2$. If $n_1 \geq 3$, then we obtain

$$(A - BK)^2 b_1 = (A - BK)A_1 b_1 = AA_1 b_1 + B\Phi_1 A_1 b_1 = A_1^2 b_1.$$

By continuing in this fashion we arrive at $(A - BK)^k b_1 = A_1^k b_1$ for $k = 0, 1, \cdots, n_1 - 1$. Since b_1 is cyclic for A_1, it follows that $\mathcal{X}_1$ is a subspace of $\mathcal{X}_c$. Using (8.39), we see that

$$(A - BK)^{n_1} b_1 = (A - BK)A_1^{n_1 - 1} b_1 = A_1^{n_1} b_1 + B\Phi_1 A^{n_1 - 1} b_1 = x_1 \oplus B_2 u_2 = x_1 \oplus b_2$$

where x_1 is a vector in $\mathcal{X}_1$. Because $\mathcal{X}_1$ is a subspace of $\mathcal{X}_c$, it follows that b_2 is also in $\mathcal{X}_c$.

The upper triangular form of A and the structure of K along with (8.39) implies that $(A - BK)^k b_2 = x_1^k \oplus A_2^k b_2$ where x_1^k is some vector in $\mathcal{X}_1$ for $k = 0, 1, \cdots, n_2 - 1$. Since $\mathcal{X}_1$ is a subspace of $\mathcal{X}_c$ all the vectors $\{A_2^k b_2\}_0^{n_2 - 1}$ are contained in $\mathcal{X}_c$. Therefore, $\mathcal{X}_1 \oplus \mathcal{X}_2$ is also a subspace of $\mathcal{X}_c$. By consulting (8.39) the vector $(A - BK)^{n_2} b_2$ is of the form $x_1 \oplus x_2 \oplus B_3 u_3$ where $x_1 \oplus x_2$ is some vector in $\mathcal{X}_1 \oplus \mathcal{X}_2$. Hence, $b_3 = B_3 u_3$ is also a vector in $\mathcal{X}_c$. By continuing in this fashion, it follows that $\oplus_1^m \mathcal{X}_j$ is a subspace of $\mathcal{X}_c$. Therefore, $\mathcal{X}_c$ equals $\mathcal{X}$ and $\{A - BK, B_1\}$ is a single input controllable pair. ∎

Finally, it is noted that Theorem 8.6.1 also follows from Proposition 8.6.3 and Theorem 8.4.1.

8.7 Two canonical forms

In this section, we refine the previous matrix representations to present some classical canonical forms for a controllable pair. The results in this section are self contained. The observable companion matrix M generated by the scalars $a_0, a_1, \cdots, a_{n-1}$ is the $n \times n$ matrix defined by

$$M = \begin{bmatrix} 0 & 0 & \cdots & 0 & -a_0 \\ 1 & 0 & \cdots & 0 & -a_1 \\ 0 & 1 & \cdots & 0 & -a_2 \\ \vdots & \vdots & \ddots & \vdots & \vdots \\ 0 & 0 & \cdots & 1 & -a_{n-1} \end{bmatrix}. \tag{8.41}$$

Notice that the transpose of M is the companion matrix $\tilde{A}$ in (8.29). So, the characteristic polynomial for M is given by $\det[sI - M] = a_0 + a_1 s + \cdots + a_{n-1} s^{n-1} + s^n$. Let $\{A, B\}$ be a controllable scalar input pair with n-dimensional state space $\mathcal{X}$. Because the pair $\{A, B\}$ is controllable, the vector $b = B1$ is cyclic for A. Moreover, $\{A^k b\}_0^{n-1}$ is a basis for $\mathcal{X}$. Hence,

$$P = \begin{bmatrix} b & Ab & A^2 b & \cdots & A^{n-1} b \end{bmatrix} \tag{8.42}$$

is a similarity transform from $\mathbb{C}^n$ onto $\mathcal{X}$. By consulting Example 7.5.3 or performing a direct calculation, it follows that $PM = AP$ where M is the observable companion matrix generated by the characteristic polynomial for A. Let $\tilde{b}$ be the vector in $\mathbb{C}^n$ defined by $\tilde{b} = [1\ 0\ 0\ \cdots\ 0]^{tr}$, that is, the first component of $\tilde{b}$ is one and all the other components are zero. Then $P\tilde{b} = b$. Therefore, any controllable single input pair $\{A, B\}$ is similar to a pair of the form $\{M, \tilde{b}\}$ where M is the observable companion matrix determined by the characteristic polynomial for A. According to Theorem 8.5.1 the pair $\{A, B\}$ is also similar to the controllable canonical pair $\{\tilde{A}, \tilde{B}\}$ in (8.29). In this section we will generalize both of these canonical forms to the multivariable setting.

Now assume that $\{A, B\}$ is a controllable pair where the state space $\mathcal{X}$ is finite dimensional and the input space $\mathcal{U} = \mathbb{C}^p$. Throughout $\{e_j\}_1^p$ is the standard orthonormal basis for $\mathbb{C}^p$. Let r_1 be the first integer such that Be_{r_1} is nonzero, that is, $Be_j = 0$ for $j = 1, 2, \cdots, r_1 - 1$ and $Be_{r_1} \neq 0$. In most applications $r_1 = 1$. Let $b_1 = Be_{r_1}$ and n_1 be the largest integer such that

$$\{b_1, Ab_1, A^2 b_1, \cdots, A^{n_1 - 1} b_1\}$$

is a linearly independent set. Let $\mathcal{M}_1$ be the space spanned by this set. With $\phi_j = A^{j-1} b_1$ for $j = 1, 2, \cdots, n_1$ the set $\{\phi_j\}_1^{n_1}$ is a basis for $\mathcal{M}_1$. Notice that n_1 is the smallest integer such that $A^{n_1} b_1$ is contained in the linear span of $\{A^k b_1\}_0^{n_1 - 1}$. According to Lemma 7.6.1, the space $\mathcal{M}_1$ equals the span of $\{A^k b_1\}_0^\infty$. In particular, $\mathcal{M}_1$ is an invariant subspace for A. Furthermore,

$$A\phi_{n_1} = A^{n_1} b_1 = \sum_{j=0}^{n_1 - 1} -a_{1j} A^j b_1 = \sum_{j=1}^{n_1} -a_{1j-1} \phi_j,$$

where a_{1j} for $j = 0, 1, \cdots, n_1 - 1$, are uniquely determined scalars. Now let P_1 be the similarity transformation from $\mathbb{C}^{n_1}$ onto $\mathcal{M}_1$ defined by

$$P_1 = \begin{bmatrix} \phi_1 & \phi_2 & \cdots & \phi_{n_1} \end{bmatrix}.$$

Then using $A\phi_j = \phi_{j+1}$ for $j = 1, 2, \cdots, n_1 - 1$ along with our previous calculation of $A\phi_{n_1}$, we see that $(A|\mathcal{M}_1)P_1 = P_1 A_1$ where A_1 is the $n_1 \times n_1$ observable companion matrix generated by a_{1j} for $j = 0, 1, \cdots, n_1 - 1$. Finally, it is noted that $P_1[1\ 0\ 0\ \cdots\ 0]^{tr} = b_1 = Be_{r_1}$.

Now let r_2 be the first integer such that Be_{r_2} is not contained in $\mathcal{M}_1$, that is, $Be_i \in \mathcal{M}_1$ for $1 \leq i < r_2$ and Be_{r_2} is not in $\mathcal{M}_1$. Set $b_2 = Be_{r_2}$. Let n_2 be the largest integer such that

$$\{\phi_1, \phi_2, \cdots, \phi_{n_1}, b_2, Ab_2, A^2 b_2, \cdots, A^{n_2-1} b_2\}$$

is a linearly independent set, and let $\mathcal{M}_2$ be the subspace spanned by this set of vectors. Set $\phi_{n_1+1} = b_2, \phi_{n_1+2} = Ab_2, \cdots, \phi_{n_1+n_2} = A^{n_2-1} b_2$. Then $\{\phi_j : j = 1, 2, \cdots, n_1 + n_2\}$ is a basis for $\mathcal{M}_2$. Moreover, because n_2 is the largest integer which makes this set linearly independent, the space $\mathcal{M}_2$ is given by

$$\mathcal{M}_2 = \bigvee_{k=0}^{\infty} \{A^k b_i : i = 1, 2\} = \bigvee_{k=0}^{\infty} \{A^k Be_i : i = 1, 2, \cdots, r_2\}.$$

In particular, $\mathcal{M}_2$ is an invariant subspace for A. Furthermore,

$$A\phi_{n_1+n_2} = A^{n_2} b_{r_2} = \sum_{j=1}^{n_1} \beta_{2j}\phi_j + \sum_{j=0}^{n_2-1} -a_{2j}A^j b_{r_2},$$

where β_{2j} for $j = 1, 2, \cdots, n_1$ and a_{2j} for $j = 0, 1, \cdots, n_2 - 1$ are uniquely determined scalars. Let P_2 be the similarity transformation from $\mathbb{C}^{n_1+n_2}$ onto the subspace $\mathcal{M}_2$ defined by

$$P_2 = \begin{bmatrix} \phi_1 & \phi_2 & \cdots & \phi_{n_1+n_2} \end{bmatrix}.$$

Let C_2 be the $1 \times n_2$ row vector defined by $C_2 = [0\ 0\ \cdots\ 0\ 1]$, that is, all the entries of C_2 are zero except the last entry which is one. Let f_{12} be the vector in $\mathbb{C}^{n_1}$ defined by $f_{12} = [\beta_{21}\ \beta_{22}\ \cdots\ \beta_{2n_1}]^{tr}$. Then using $A\phi_j = \phi_{j+1}$ for $j = n_1 + 1, \cdots, n_1 + n_2 - 1$ along with our previous calculation of $A\phi_{n_1+n_2}$ and $P_1 = P_2|\mathbb{C}^{n_1}$, we see that

$$(A|\mathcal{M}_2)P_2 = P_2 \begin{bmatrix} A_1 & f_{12}C_2 \\ 0 & A_2 \end{bmatrix}.$$

Here A_2 is the $n_2 \times n_2$ observable companion matrix generated by a_{2j} for $j = 0, 1, \cdots, n_2 - 1$.

Let $\tilde{b}_1$ be the vector in $\mathbb{C}^{n_1}$ defined by $\tilde{b}_1 = [1\ 0\ 0\ \cdots\ 0]^{tr}$, that is, the first entry of $\tilde{b}_1$ is one and all the other entries of $\tilde{b}_1$ are zero. Observe that $P_2(\tilde{b}_1 \oplus 0) = b_1 = Be_{r_1}$. Let $\tilde{b}_2$ be the vector in $\mathbb{C}^{n_2}$ defined by $\tilde{b}_2 = [1\ 0\ 0\ \cdots\ 0]^{tr}$, that is, the first entry of $\tilde{b}_2$ is one and all the other entries are zero. Then $P_2(0 \oplus \tilde{b}) = b_2 = Be_{r_2}$. Recall that $Be_i \in \mathcal{M}_1$ for $1 \leq i < r_2$. Hence, for $1 \leq i < r_2$, we have $Be_i = P_2(v_i \oplus 0)$ for some v_i in $\mathbb{C}^{n_1}$. Therefore, the first r_2 columns of B can be factored into an upper triangular matrix times P_2, that is,

$$B\begin{bmatrix} e_1 & e_2 \cdots & e_{r_2} \end{bmatrix} = P_2 \begin{bmatrix} 0 & \tilde{b}_1 & * & \cdots & * & 0 \\ 0 & 0 & 0 & \cdots & 0 & \tilde{b}_2 \end{bmatrix}.$$

Notice that the first column in the matrix on the right is a block column of zeros, and this column is not present if Be_1 is nonzero. Obviously, $\{A_j, \tilde{b}_j\}$ is a controllable scalar input pair for $j = 1, 2$.

Continuing in this fashion, let r_3 be the first integer such that Be_{r_3} is not contained in $\mathcal{M}_2$, that is, $Be_i \in \mathcal{M}_2$ for $1 \le i < r_3$ and Be_{r_3} is not in $\mathcal{M}_2$. Set $b_3 = Be_{r_3}$. Let n_3 be the largest integer such that

$$\{\phi_1, \phi_2, \cdots, \phi_{n_1+n_2}, b_3, Ab_3, A^2 b_3, \cdots, A^{n_3-1} b_3\}$$

is a linearly independent set, and let $\mathcal{M}_3$ be the subspace spanned by this set of vectors. Set $\phi_{n_1+n_2+1} = b_3, \cdots, \phi_{n_1+n_2+n_3} = A^{n_3-1} b_3$. Then $\{\phi_j : j = 1, 2, \cdots, n_1 + n_2 + n_3\}$ is a basis for $\mathcal{M}_3$. Moreover, $\mathcal{M}_3$ is the invariant subspace for A given by

$$\mathcal{M}_3 = \bigvee_{k=0}^{\infty} \{A^k b_i : i = 1, 2, 3\} = \bigvee_{k=0}^{\infty} \{A^k Be_i : i = 1, 2, \cdots, r_3\}.$$

Furthermore,

$$A\phi_{n_1+n_2+n_3} = A^{n_3} b_{r_3} = \sum_{j=1}^{n_1+n_2} \beta_{3j} \phi_j + \sum_{j=0}^{n_3-1} -a_{3j} A^j b_{r_3},$$

where β_{3j} for $j = 1, 2, \cdots, n_1 + n_2$ and a_{3j} for $j = 0, 1, \cdots, n_3 - 1$ are uniquely determined scalars. Let P_3 be the similarity transformation from $\mathbb{C}^{n_1+n_2+n_3}$ onto the subspace $\mathcal{M}_3$ defined by $P_3 = [\phi_1 \ \phi_2 \ \cdots \ \phi_{n_1+n_2+n_3}]$. Let C_3 be the $1 \times n_3$ row vector defined by $C_3 = [0 \ 0 \ \cdots \ 0 \ 1]$. Let f_{13} be the vector in $\mathbb{C}^{n_1}$ and f_{23} be the vector in $\mathbb{C}^{n_2}$ defined by

$$\begin{aligned} f_{13} &= [\ \beta_{31} \ \ \beta_{32} \ \ \cdots \ \ \beta_{3n_1} \ ]^{tr} \\ f_{23} &= [\ \beta_{3n_1+1} \ \ \beta_{3n_1+2} \ \ \cdots \ \ \beta_{3n_1+n_2} \ ]^{tr}. \end{aligned}$$

Then using $A\phi_j = \phi_{j+1}$ for $j = n_1 + n_2 + 1, \cdots, n_1 + n_2 + n_3 - 1$ along with our previous calculation of $A\phi_{n_1+n_2+n_3}$, we see that

$$(A|\mathcal{M}_3) P_3 = P_3 \begin{bmatrix} A_1 & f_{12}C_2 & f_{13}C_3 \\ 0 & A_2 & f_{23}C_3 \\ 0 & 0 & A_3 \end{bmatrix},$$

where A_3 is the $n_3 \times n_3$ observable companion matrix generated by a_{3j} for $j = 0, 1, \cdots, n_3 - 1$.

Since P_2 is an invertible transformation from $\mathbb{C}^{n_1+n_2}$ onto $\mathcal{M}_2$, the equation $P_2(b_{1i} \oplus b_{2i}) = Be_i$ has a unique solution $b_{1i} \oplus b_{2i}$ in $\mathbb{C}^{n_1} \oplus \mathbb{C}^{n_2}$ for $r_2 + 1 \le i < r_3$. Let $\tilde{b}_3$ be the vector in $\mathbb{C}^{n_3}$ defined by $\tilde{b}_3 = [1 \ 0 \ 0 \ \cdots \ 0]^{tr}$, that is, the first entry is one and all the other entries are zero. Then $P_3(0 \oplus 0 \oplus \tilde{b}_3) = Be_{r_3}$. Notice that $P_2 = P_3|(\mathbb{C}^{n_1} \oplus \mathbb{C}^{n_2})$. Therefore, the first r_3 columns of B admit a factorization of the form

$$B \begin{bmatrix} e_1 & e_2 \cdots & e_{r_3} \end{bmatrix} = P_3 \begin{bmatrix} 0 & \tilde{b}_1 & * & \cdots & * & 0 & * & \cdots & * & 0 \\ 0 & 0 & 0 & \cdots & 0 & \tilde{b}_2 & * & \cdots & * & 0 \\ 0 & 0 & 0 & \cdots & 0 & 0 & 0 & \cdots & 0 & \tilde{b}_3 \end{bmatrix}.$$

Notice that the first column in the matrix on the right is a block column of zeros, and this column is not present if Be_1 is nonzero. Clearly, $\{A_j, \tilde{b}_j\}$ is a controllable scalar input pair for $j = 1, 2, 3$.

By continuing in this fashion, we obtain a set of integers $\{n_j\}_1^m$ and $\{r_j\}_1^m$ such that

$$\{b_1, \cdots, A^{n_1-1}b_1, b_2, \cdots\cdots, b_m, Ab_m, \cdots, A^{n_m-1}b_m\}$$

is a linearly independent set for the space

$$\mathcal{M}_m = \bigvee_{k=0}^{\infty}\{A^k b_i : i = 1, 2, \cdots, m\} = \bigvee_{k=0}^{\infty}\{A^k B e_i : i = 1, 2, \cdots, p\}.$$

Moreover, $b_m = Be_{r_m}$ where r_m is the first integer such that Be_{r_m} is not in $\mathcal{M}_{m-1}$. Because the state space is finite dimensional and the pair $\{A, B\}$ is controllable, this recursive procedure must eventually end with $\mathcal{M}_m = \mathcal{X}$ for some integer m. So, the dimension of $\mathcal{X}$ is given by $n = \sum_1^m n_j$. The integers $\{n_j\}_1^m$ are uniquely determined and are called the *controllability indices* for the pair $\{A, B\}$. The operator P_m defined by

$$P_m = \begin{bmatrix} b_1 & \cdots & A^{n_1-1}b_1 & b_2 & \cdots & \cdots b_m & Ab_m & \cdots & A^{n_m-1}b_m \end{bmatrix} \tag{8.43}$$

is a similarity transform from $\mathbb{C}^n$ onto $\mathcal{X} = \mathcal{M}_m$. Let C_j be the $1 \times n_j$ row vector defined by $C_j = [0\ 0\ \cdots\ 0\ 1]$ for $j = 2, 3, \cdots, m$. Then $AP_m = P_m A_o$ where A_o on $\oplus_1^m \mathbb{C}^{n_j}$ is an upper triangular block matrix of the form:

$$A_o = \begin{bmatrix} A_1 & f_{12}C_2 & \cdots & f_{1m}C_m \\ 0 & A_2 & \cdots & f_{2m}C_m \\ \vdots & \ddots & \ddots & \vdots \\ 0 & 0 & \cdots & A_m \end{bmatrix}. \tag{8.44}$$

The block diagonal matrices $\{A_j\}_1^m$ are all observable companion matrices. Finally, $P_m B_o = B$ where B_o is the upper triangular block matrix mapping $\mathbb{C}^p$ into $\oplus_1^m \mathbb{C}^{n_j}$ of the form:

$$B_o = \begin{bmatrix} 0 & \tilde{b}_1 & * & 0 & * & 0 & * & \cdots & 0 & * \\ 0 & 0 & 0 & \tilde{b}_2 & * & 0 & * & \cdots & 0 & * \\ 0 & 0 & 0 & 0 & 0 & \tilde{b}_3 & * & \cdots & 0 & * \\ 0 & 0 & 0 & 0 & 0 & 0 & 0 & \cdots & 0 & * \\ \vdots & \vdots & \vdots & \vdots & \vdots & \vdots & \vdots & \ddots & \vdots & \vdots \\ 0 & 0 & 0 & 0 & 0 & 0 & 0 & \cdots & \tilde{b}_m & * \end{bmatrix}. \tag{8.45}$$

Here $\tilde{b}_j = [1\ 0\ 0\ \cdots\ 0]^{tr}$ is the vector in $\mathbb{C}^{n_j}$ with one in the first entry and all the other entries are zero. Obviously, $\{A_j, \tilde{b}_j\}$ is a controllable scalar input pair for $j = 1, 2, \cdots, m$. As before, the first column is a block column of zeros, and this column is not present if Be_1 is nonzero. Finally, it is noted that $m \leq p$ where the input space is $\mathcal{U} = \mathbb{C}^p$. Summing up this analysis we obtain the following result.

Proposition 8.7.1 *Let $\{A, B\}$ be a controllable pair where the input space $\mathcal{U} = \mathbb{C}^p$. Then $\{A, B\}$ is similar to the pair $\{A_o, B_o\}$ whose canonical form is given in (8.44) and (8.45). In fact, $AP_m = P_m A_o$ and $P_m B_o = B$ where P_m is the similarity transform defined in (8.43).*

Recall that the companion matrix $\tilde{A}$ generated by $\{a_j\}_0^{k-1}$ is the $k \times k$ matrix defined by

$$
\tilde{A} = \begin{bmatrix}
0 & 1 & 0 & \cdots & 0 \\
0 & 0 & 1 & \cdots & 0 \\
\vdots & \vdots & \vdots & \ddots & \vdots \\
0 & 0 & 0 & \cdots & 1 \\
-a_0 & -a_1 & -a_2 & \cdots & -a_{k-1}
\end{bmatrix}. \tag{8.46}
$$

Recall also that $a_0 + a_1 s + \cdots + a_{k-1}s^{k-1} + s^k$ is the characteristic polynomial for $\tilde{A}$. Let $\tilde{B}$ be the vector in $\mathbb{C}^k$ defined by $\tilde{B} = [0\ 0\ \cdots\ 0\ 1]^{tr}$, that is, all the components of $\tilde{B}$ are zero except the last one which is one. Let Q be the similarity transform defined by (8.42) with $\tilde{A}$ replacing A and $\tilde{B}$ replacing b, that is,

$$
Q = \begin{bmatrix} \tilde{B} & \tilde{A}\tilde{B} & \tilde{A}^2\tilde{B} & \cdots & \tilde{A}^{k-1}\tilde{B} \end{bmatrix} \text{ on } \mathbb{C}^k. \tag{8.47}
$$

Then $QM = \tilde{A}Q$ where M is the observable companion matrix generated by $\{a_j\}_0^{k-1}$; see (8.41) with $k = n$. Notice that in this case Q is a matrix with one on the off diagonal and zeros above the off diagonal, that is, Q is a matrix of the form:

$$
Q = \begin{bmatrix}
0 & 0 & \cdots & 0 & 1 \\
0 & 0 & \cdots & 1 & * \\
\vdots & \vdots & & \vdots & \vdots \\
0 & 1 & \cdots & * & * \\
1 & * & \cdots & * & *
\end{bmatrix}.
$$

Using this it follows that $Q\tilde{b} = \tilde{B}$ where $\tilde{b} = [1\ 0\ \cdots\ 0\ 0]^{tr}$. So, the similarity transform Q intertwines $\{M, \tilde{b}\}$ with the controllable canonical pair $\{\tilde{A}, \tilde{B}\}$. The pair $\{\tilde{A}, \tilde{B}\}$ is precisely the one used for state feedback. Moreover, the inverse of Q has ones on the off diagonal and zeros below the off diagonal, that is, the inverse of Q is a matrix of the form:

$$
Q^{-1} = \begin{bmatrix}
* & * & \cdots & * & 1 \\
* & * & \cdots & 1 & 0 \\
\vdots & \vdots & & \vdots & \vdots \\
* & 1 & \cdots & 0 & 0 \\
1 & 0 & \cdots & 0 & 0
\end{bmatrix}.
$$

Due to the triangular structure one can recursively compute the inverse of Q. Finally, it is noted that the inverse of Q can also be computed by using the recursive algorithm presented in the paragraph following Theorem 8.5.1, that is, $Q^{-1} = T$ where T in (8.30) is the operator intertwining $\{\tilde{A}, \tilde{B}\}$ with $\{M, \tilde{b}\}$.

Now let us return to the canonical form $\{A_o, B_o\}$ in (8.44) and (8.45) computed from the controllable pair $\{A, B\}$. Let $\tilde{A}_j$ be the transpose of A_j for $j = 1, 2, \cdots, m$. Obviously, $\tilde{A}_j$ and A_j have the same characteristic polynomial. Let $\tilde{B}_{jr_j}$ be the vector in $\mathbb{C}^{n_j}$ whose last component is one and all the other components are zero. Finally, let Q_j be the similarity transform on $\mathbb{C}^{n_j}$ defined by replacing $\tilde{A}$ by $\tilde{A}_j$ and $\tilde{B}$ by $\tilde{B}_{jr_j}$ in (8.47). Then Q_j intertwines the pair $\{A_j, b_{jr_j}\}$ with $\{\tilde{A}_j, \tilde{B}_{jr_j}\}$. Moreover, let R_j be the $1 \times n_j$ row vector whose first component is one and all the other components are zero. Using the special form of the inverse of Q_j, we see that $C_j Q_j^{-1} = R_j$. Now let Q_c be the similarity transform on $\oplus_1^m \mathbb{C}^{n_j}$ defined by the block diagonal matrix

$$Q_c = \text{diag } [Q_1 \; Q_2 \; \cdots \; Q_m]. \qquad (8.48)$$

Then $\tilde{A}_c = Q_c A_o Q_c^{-1}$ is the upper triangular matrix given by

$$\tilde{A}_c = \begin{bmatrix} \tilde{A}_1 & Q_1 f_{12} R_2 & \cdots & Q_1 f_{1m} R_m \\ 0 & \tilde{A}_2 & \cdots & Q_2 f_{2m} R_m \\ \vdots & & \ddots & \vdots \\ 0 & 0 & \cdots & \tilde{A}_m \end{bmatrix}. \qquad (8.49)$$

The block diagonal matrices $\{\tilde{A}_j\}_1^m$ are all companion matrices. Finally, $Q_c B_o = \tilde{B}_c$ where $\tilde{B}_c$ is the upper triangular block matrix mapping $\mathbb{C}^p$ into $\oplus_1^m \mathbb{C}^{n_j}$ of the form:

$$\tilde{B}_c = \begin{bmatrix} 0 & \tilde{B}_{1r_1} & * & 0 & * & 0 & * & \cdots & 0 & * \\ 0 & 0 & 0 & \tilde{B}_{2r_2} & * & 0 & * & \cdots & 0 & * \\ 0 & 0 & 0 & 0 & 0 & \tilde{B}_{3r_3} & * & \cdots & 0 & * \\ 0 & 0 & 0 & 0 & 0 & 0 & 0 & \cdots & 0 & * \\ \vdots & \vdots & \vdots & \vdots & \vdots & \vdots & \vdots & \ddots & \vdots & \vdots \\ 0 & 0 & 0 & 0 & 0 & 0 & 0 & \cdots & \tilde{B}_{mr_m} & * \end{bmatrix}. \qquad (8.50)$$

By construction the pairs $\{\tilde{A}_j, \tilde{B}_{jr_j}\}_1^m$ are in controllable canonical form. As before, the first column in $\tilde{B}_c$ is a block column of zeros and this column is not present if Be_1 is nonzero. Clearly the pair $\{A, B\}$ is similar to the pair $\{\tilde{A}_c, \tilde{B}_c\}$. This readily proves the following result.

Proposition 8.7.2 *Let $\{A, B\}$ be a controllable pair where the input space $\mathcal{U} = \mathbb{C}^p$. Then $\{A, B\}$ is similar to the pair $\{\tilde{A}_c, \tilde{B}_c\}$ whose controllable canonical form is given in (8.49) and (8.50). In fact, $P_m Q_c^{-1}$ is the similarity transformation intertwining $\{\tilde{A}_c, \tilde{B}_c\}$ with $\{A, B\}$ where Q_c is defined in (8.48) while P_m is defined in (8.43).*

One can use the canonical form $\{\tilde{A}_c, \tilde{B}_c\}$ for $\{A, B\}$ to develop an algorithm to place the eigenvalues of $A - BK$ at any n specified locations in the complex plane. Since $\{\tilde{A}_j, \tilde{B}_{jr_j}\}$ is in controllable conical form, it is easy to construct a state feedback gain $\hat{K}_j$ from $\mathbb{C}^{n_j}$ into $\mathbb{C}^1$ to place the eigenvalues of $\tilde{A}_j - \tilde{B}_{jr_j} \hat{K}_j$ at any n_j locations in the complex plane; see (8.21), (8.22) and (8.23) in Section 8.5. Now let $\hat{K}$ be the block matrix from $\oplus_1^m \mathbb{C}^{n_j}$ into $\mathbb{C}^p$ whose block entries $\hat{K}_{i,k}$ from $\mathbb{C}^{n_k}$ into $\mathbb{C}^1$ are defined by

$$\hat{K}_{i,k} = \delta(i - r_j, k - j)\hat{K}_j \qquad (\text{for } i = 1, 2, \cdots, p \text{ and } j, k = 1, 2, \cdots, m).$$

Here $\delta(i, k)$ is the Kronecker delta, that is, $\delta(i, k) = 1$ if $i = k = 0$, otherwise $\delta(i, k) = 0$. In this case, $\tilde{A}_c - \tilde{B}_c \hat{K}$ is an upper triangular block matrix with $\{\tilde{A}_j - \tilde{B}_{jr_j} \hat{K}_j\}_1^m$ for its diagonal entries. So, the eigenvalues of $\tilde{A}_c - \tilde{B}_c\hat{K}$ is the union of the eigenvalues of $\{\tilde{A}_j - \tilde{B}_{jr_j}\hat{K}_j\}$. In particular, since $n = \sum n_j$, we can compute a state feedback $\hat{K}$ to place the eigenvalues of $\tilde{A}_c - \tilde{B}_c\hat{K}$ any n specified locations in the complex plane. Now let K be the operator form $\mathcal{X}$ into $\mathbb{C}^p$ defined by $K = \hat{K}Q_cP_m^{-1}$. Then $A - BK$ is similar to $\tilde{A}_c - \tilde{B}_c\hat{K}$. In particular, they have the same eigenvalues. Therefore, the canonical form $\{\tilde{A}_c, \tilde{B}_c\}$ can be used to compute a gain K to place the eigenvalues of $A - BK$ any n specified locations in the complex plane.

Finally, it is noted that one can obtain the canonical form $\{\tilde{A}_c, \tilde{B}_c\}$ for $\{A, B\}$ without first computing the canonical form $\{A_o, B_o\}$. To accomplish this one simply uses the vectors constructed in the section following Theorem 8.5.1 as basis for the subspaces $\mathcal{M}_j$. The details are left as an exercise.

8.8 Transfer functions and feedback

To complete this chapter, let us show how state feedback affects the transfer function for certain feedback control systems. To this end, consider the open loop system $\{A, B, C, D\}$ given by

$$\dot{x} = Ax + Bu \quad \text{and} \quad y = Cx + Du. \tag{8.51}$$

As before, A is an operator on $\mathcal{X}$ and B maps $\mathcal{U}$ into $\mathcal{X}$, while C maps $\mathcal{X}$ into $\mathcal{Y}$ and D maps $\mathcal{U}$ into $\mathcal{Y}$. Now consider a state feedback controller of the form

$$u = -Kx + w \tag{8.52}$$

where the gain K is an operator from $\mathcal{X}$ into $\mathcal{U}$ and w is a function with values in $\mathcal{U}$. The function w is referred to as the reference input for the above controller. Substituting (8.52) into (8.51) results in

$$\dot{x} = (A - BK)x + Bw \quad \text{and} \quad y = (C - DK)x + Dw. \tag{8.53}$$

In this setting the transfer function $\mathbf{F}$ from reference input w to the output y is given by

$$\mathbf{F}(s) = D + (C - DK)(sI - A + BK)^{-1}B. \tag{8.54}$$

In other words, $\{A-BK, B, C-DK, D\}$ is a realization of $\mathbf{F}$. The system in (8.53) is called the *closed loop system* corresponding to controller (8.52) and $\mathbf{F}$ in (8.54) is the corresponding closed loop transfer function. Obviously, if λ is a pole of $\mathbf{F}$, then λ is an eigenvalue of $A-BK$. In particular, if $A - BK$ is stable, then the closed loop transfer function $\mathbf{F}$ is also stable.

Lemma 6.4.2 shows that the open loop system $\{A, B, C, D\}$ is controllable if and only if the closed loop system in (8.53) is controllable. However, the closed loop system may not be observable even if the open loop system is observable. For example, the open loop system $\{1, 1, 1, 1\}$ is controllable and observable. If $u = -x + w$, then the corresponding closed loop system $\{0, 1, 0, 1\}$ is not observable.

Let $\{A, B, C, D\}$ be a realization of a transfer function $\mathbf{G}$. Then we claim that the closed loop transfer function $\mathbf{F}$ in (8.54) is also given by

$$\mathbf{F}(s) = [D + C(sI - A)^{-1}B][I + K(sI - A)^{-1}B]^{-1}. \tag{8.55}$$

So, if $\mathbf{H}$ is the transfer function for the system $\{A, B, K, I\}$, then the closed loop transfer function $\mathbf{F}$ is given by

$$\mathbf{F} = \mathbf{GH}^{-1}.$$

To verify that (8.55) holds, let $\Phi(s) = (sI - A)^{-1}$. Using the operator identity

$$(I + RT)^{-1}R = R(I + TR)^{-1},$$

we obtain

$$
\begin{aligned}
\mathbf{F}(s) &= D + (C - DK)(sI - A + BK)^{-1}B \\
&= D + (C - DK)(I + \Phi(s)BK)^{-1}\Phi(s)B \\
&= D + (C - DK)\Phi(s)B(I + K\Phi(s)B)^{-1} \\
&= [D(I + K\Phi(s)B) + (C - DK)\Phi(s)B][I + K\Phi(s)B]^{-1} \\
&= (D + C\Phi(s)B)(I + K\Phi(s)B)^{-1} \\
&= \mathbf{G}(s)[I + K(sI - A)^{-1}B]^{-1}.
\end{aligned}
$$

This yields (8.55).

Recalling (8.9) we see that

$$\det[\mathbf{H}(s)] = \det[sI - A + BK]/\det[sI - A]. \tag{8.56}$$

When the input space $\mathcal{U}$ is one dimensional, we obtain

$$\mathbf{H}(s) = \det[sI - A + BK]/\det[sI - A]. \tag{8.57}$$

In this case, Proposition 7.1.3 shows that $\{A, B, K, 1\}$ is a minimal realization if and only if the characteristic polynomials for A and $A - BK$ have no common zeros.

Now let $\{A \text{ on } \mathcal{X}, B, C, D\}$ be a realization for a scalar valued transfer function $\mathbf{G}$. Then $\mathbf{G}$ admits a decomposition of the form $\mathbf{G} = p/d$ where d is the characteristic polynomial for A and p is a polynomial satisfying $\deg p \leq \deg d$. Moreover, $\{A, B, C, D\}$ is a minimal realization if and only if the polynomials p and d have no common zeros; see Proposition 7.1.3. In the scalar case, the closed loop transfer function $\mathbf{F}$ in (8.55) reduces to

$$\mathbf{F} = \frac{D + C(sI - A)^{-1}B}{1 + K(sI - A)^{-1}B}. \tag{8.58}$$

Recalling (8.9), we obtain that

$$1 + K(sI - A)^{-1}B = \det[sI - A + BK]/\det[sI - A] = q(s)/d(s)$$

where q is the characteristic polynomial for $A - BK$. Using $D + C(sI - A)^{-1}B = p(s)/d(s)$ yields $\mathbf{F} = p/q$, that is,

$$\mathbf{F}(s) = \frac{p(s)}{\det[sI - A + BK]} \quad \text{where} \quad \mathbf{G}(s) = \frac{p(s)}{\det[sI - A]}. \tag{8.59}$$

This relationship implies that the zeros of the closed loop transfer function $\mathbf{F}$ are contained in the zeros of the open loop transfer function $\mathbf{G}$. Thus state feedback does not introduce new zeros. However, p and q can have common roots. So, the closed loop transfer function $\mathbf{F}$ can have fewer zeros than the open loop transfer function $\mathbf{G}$. This pole-zero cancellation also explains why the open loop realization can be minimal while the closed loop realization $\{A - BK, B, C - DK, D\}$ for $\mathbf{F}$ may only be controllable. If $\{A, B, C, D\}$ is minimal, then p and d have no common zeros and the McMillan degree of $\mathbf{G}$ equals the degree of d. If there is a pole-zero cancellation in $\mathbf{F} = p/q$, then the McMillan degree of $\mathbf{F}$ is strictly less then the dimension of $\mathcal{X}$, and thus, the realization $\Sigma = \{A - BK, B, C - DK, D\}$ of $\mathbf{F}$ is not minimal; see Proposition 7.1.3. So, the realization Σ of $\mathbf{F} = p/q$ is minimal if and only if p and q have no common zeros. In other words, the realization Σ is minimal if and only if the state feedback K does not place any eigenvalues of $A - BK$ at the zeros of p. In particular, if $p(s) \equiv \gamma$ where γ is a constant, then Σ is a minimal realization for all state feedback gain operators K.

For the moment assume that the pair $\{A, B\}$ is controllable, and recall that $\{A, B, K, I\}$ is a realization for $\mathbf{H} = q/d$. The characteristic polynomials q and d may have some common zeros. According to Proposition 7.1.3, the polynomials q and d have no common zeros if and only if $\{A, B, K, I\}$ is observable. If $\{A, B, K, I\}$ is observable, then q and $\mathbf{H}$ have the same zeros.

8.9 Notes

All the results in this chapter are classical; see Kailath [68] and Rugh [110] for a history of this subject area. Corollary 8.4.2 is due to Ackermann [1]. For some further results on state variable feedback see Brockett [21], Chen [26], Kailath [68], Rugh [110] and Wonham [129].

Chapter 9

State Estimators and Detectability

This chapter is devoted to asymptotic state estimation. First we introduce the concept of detectability. It is shown that detectability is the dual of stabilizability. Then we construct asymptotic state estimators for detectable systems.

9.1 Detectability

Consider a state space system of the form:

$$\begin{aligned} \dot{x} &= Ax + Bu \\ y &= Cx + Du \end{aligned} \tag{9.1}$$

where A is an operator on $\mathcal{X}$ and B maps $\mathcal{U}$ into $\mathcal{X}$ while C maps $\mathcal{X}$ into $\mathcal{Y}$ and D maps $\mathcal{U}$ into $\mathcal{Y}$. As before, all spaces are finite dimensional. The system in (9.1) will be referred to as the plant. Suppose that at each instant of time t we can measure the output $y(t)$ and the input $u(t)$ and we wish to estimate the state $x(t)$. Recall that this system is observable if given the input $u(t)$ and output $y(t)$ over an interval $[0, t_1]$ (with $t_1 > 0$), one can uniquely determine the state $x(t)$ on this interval. In many applications, it suffices to obtain an asymptotic estimate of $x(t)$ as t tends to infinity. With this in mind, we say that $\hat{x}$ is an *asymptotic estimate* of x if

$$\lim_{t \to \infty} x(t) - \hat{x}(t) = 0. \tag{9.2}$$

Since we wish to obtain an asymptotic estimate of x using only information on the plant input and output, we define an asymptotic *state estimator* to be any system with input $\{u, y\}$, output $\hat{x}$ and which has the following property: If $\{u, y\}$ is any input-output pair for plant (9.1), then the corresponding output $\hat{x}$ of the estimator is an asymptotic estimate of the corresponding plant state x. Thus, for any initial plant state, an asymptotic state estimator, using only knowledge of plant input and output, produces an asymptotic estimate $\hat{x}$ of the plant state, that is, (9.2) holds. We say that the system $\{A, B, C, D\}$ in (9.1) is *detectable* if there exists an asymptotic state estimator for the system. If system (9.1) is observable, then one can find an estimator such that $\hat{x}(t) = x(t)$ for all $t > 0$. So, if the pair $\{C, A\}$ is observable, then $\{A, B, C, D\}$ is detectable.

145

As expected, detectability of the system $\{A, B, C, D\}$ depends only on the pair $\{C, A\}$. To see this, recall that all solutions to the differential equation in (9.1) satisfy

$$x(t) = e^{At}x(0) + \int_0^t e^{A(t-\tau)}Bu(\tau)\,d\tau\,.$$

Since u is known, estimating x is equivalent to estimating $e^{At}x(0)$. Because u and y are known,

$$g(t) = Ce^{At}x(0) = y(t) - \int_0^t Ce^{A(t-\tau)}Bu(\tau)\,d\tau - Du(t)$$

is known. Thus, the system in (9.1) is detectable if and only if given the function $g(t) = Ce^{At}x(0)$ over $[0, \infty)$ one can determine an asymptotic estimate $\hat{x}(t)$ of $e^{At}x(0)$ from g. Therefore, detectability of system (9.1) depends only on the pair $\{C, A\}$ and is independent of the operators B and D. Hence, we say that the pair $\{C, A\}$ is detectable if system (9.1) is detectable.

Recall that the unobservable subspace for the pair $\{C, A\}$ is the invariant subspace for A defined by

$$\mathcal{X}_{\bar{o}} = \{x : CA^k x = 0 \text{ for } k = 0, 1, 2, \cdots\}\,. \tag{9.3}$$

Recall also that a complex number λ is an unobservable eigenvalue for the pair $\{C, A\}$ if there is a nonzero vector v in the unobservable subspace $\mathcal{X}_{\bar{o}}$ satisfying $Av = \lambda v$. In this case, we say that v is an unobservable eigenvector corresponding to λ. Notice λ is an unobservable eigenvalue if and only if λ is an eigenvalue of the operator $A_{\bar{o}}$ on $\mathcal{X}_{\bar{o}}$ defined by $A_{\bar{o}} = A|\mathcal{X}_{\bar{o}}$. In other words, the unobservable eigenvalues for the pair $\{C, A\}$ are precisely the eigenvalues of $A_{\bar{o}}$. An unobservable eigenvalue λ is stable if $\Re(\lambda) < 0$. If $\{\lambda, v\}$ is an unobservable eigenvalue eigenvector pair for $\{C, A\}$, then $Ce^{At}v = e^{\lambda t}Cv = 0$. In particular, $x(t) = e^{\lambda t}v$ is a solution of

$$\dot{x} = Ax \qquad \text{and} \qquad y = Cx$$

with $y(t) = 0$. Hence, one cannot distinguish the solution $x(t) = e^{\lambda t}v$ from the zero solution $x(t) = 0$. If λ is not stable, then $e^{\lambda t}v$ does not asymptotically approach zero. Since $e^{\lambda t}v$ and the zero solution produce the same output (that is, zero) and their difference does not asymptotically converge to zero, one cannot construct an asymptotic state estimator for a system with an unobservable eigenvalue which is not stable. So, if λ is not stable, then the pair $\{C, A\}$ is not detectable. Therefore, the detectability of $\{C, A\}$ implies that all the unobservable eigenvalues of $\{C, A\}$ are stable. This proves part of the following result.

Theorem 9.1.1 *A pair $\{C, A\}$ is detectable if and only if all of its unobservable eigenvalues are stable.*

PROOF. To complete the proof it remains to show that if all the unobservable eigenvalues of $\{C, A\}$ are stable, then $\{C, A\}$ is detectable. Let $\mathcal{X}$ be the state space for the pair $\{C, A\}$, and $\mathcal{X}_o = \mathcal{X}_{\bar{o}}^{\perp}$ be the corresponding observable subspace. Recall that $\mathcal{X}_o$ is an invariant subspace for A^* and A admits a matrix representation of the form:

$$A = \begin{bmatrix} A_{\bar{o}} & A_{\bar{o}o} \\ 0 & A_o \end{bmatrix} : \begin{bmatrix} \mathcal{X}_{\bar{o}} \\ \mathcal{X}_o \end{bmatrix} \longrightarrow \begin{bmatrix} \mathcal{X}_{\bar{o}} \\ \mathcal{X}_o \end{bmatrix} ; \tag{9.4}$$

see (6.15) and (6.16) in Section 6.2. The operator A_o is the compression of A to $\mathcal{X}_o$, that is, A_o is the operator on $\mathcal{X}_o$ defined by $A_o = P_o A | \mathcal{X}_o$ where P_o is the orthogonal projection onto $\mathcal{X}_o$. The operator $A_{\bar{o}}$ on $\mathcal{X}_{\bar{o}}$ is given by the restriction of A to $\mathcal{X}_{\bar{o}}$. Furthermore, the operator C admits a matrix representation of the form

$$C = [\, 0 \;\; C_o \,] : \begin{bmatrix} \mathcal{X}_{\bar{o}} \\ \mathcal{X}_o \end{bmatrix} \longrightarrow \mathcal{Y}, \tag{9.5}$$

where $C_o = C|\mathcal{X}_o$. Since the kernel of C contains $\mathcal{X}_{\bar{o}}$, the first entry of C is zero, that is, $C_{\bar{o}} = C|\mathcal{X}_{\bar{o}} = 0$. Recall also that the pair $\{C_o, A_o\}$ is observable. Let $x_{\bar{o}} \oplus x_o$ be the decomposition of the state x with respect to $\mathcal{X} = \mathcal{X}_{\bar{o}} \oplus \mathcal{X}_o$. Thus,

$$\begin{aligned} \dot{x}_{\bar{o}} &= A_{\bar{o}} x_{\bar{o}} + A_{\bar{o}o} x_o \\ \dot{x}_o &= A_o x_o \\ y &= C_o x_o . \end{aligned}$$

Since the pair $\{C_o, A_o\}$ is observable, one can uniquely determine $x_o(t)$ for all $t > 0$ from the knowledge of $y(t)$ on $[0, \infty)$. It follows from the first differential equation above that

$$x_{\bar{o}}(t) = e^{A_{\bar{o}} t} x_{\bar{o}}(0) + \int_0^t e^{A_{\bar{o}}(t-\tau)} A_{\bar{o}o} x_o(\tau)\, d\tau .$$

Now assume that all the unobservable eigenvalues of $\{C, A\}$ are stable. Then $A_{\bar{o}}$ is stable. Considering the known function $\hat{x}_{\bar{o}}$ given by $\hat{x}_{\bar{o}}(t) = \int_0^t e^{A_{\bar{o}}(t-\tau)} A_{\bar{o}o} x_o(\tau)\, d\tau$, it follows that as t approaches infinity, $x_{\bar{o}}(t) - \hat{x}_{\bar{o}}(t) = e^{A_{\bar{o}} t} x_{\bar{o}}(0)$ approaches zero. Therefore, $\hat{x}_{\bar{o}} \oplus x_o$ is an asymptotic estimate of x. ∎

Obviously, if A is stable, then $\{C, A\}$ is detectable regardless of C. By consulting the PBH characterization of unobservable eigenvalues in Lemma 4.2.1, we obtain the following corollary.

Lemma 9.1.2 *(PBH detectability lemma.)* *The pair $\{C, A$ on $\mathcal{X}\}$ is detectable if and only if the kernel of*

$$\Gamma_\lambda = \begin{bmatrix} A - \lambda I \\ C \end{bmatrix} \tag{9.6}$$

is zero for every complex number λ in the closed right half plane.

Since the kernel of Γ_λ is zero when λ is not an eigenvalue of A, one only has to check the kernel of Γ_λ when λ is an eigenvalue of A which is not stable. The following result presents some duality between stabilizability and detectability.

Theorem 9.1.3 *Consider a pair $\{C, A\}$ where A is on $\mathcal{X}$ and C maps $\mathcal{X}$ into $\mathcal{Y}$. Then the following statements are equivalent.*

(i) The pair $\{C, A\}$ is detectable.

(ii) The pair $\{A^*, C^*\}$ is stabilizable.

(iii) There exists an operator L from $\mathcal{Y}$ into $\mathcal{X}$ such that $A - LC$ is stable.

PROOF. Recall that a complex number λ is an unobservable eigenvalue for $\{C, A\}$ if and only if $\bar{\lambda}$ is an uncontrollable eigenvalue for the pair $\{A^*, C^*\}$; see Section 5.2. Since $\{A^*, C^*\}$ is stabilizable if and only if all its uncontrollable eigenvalues are stable, Theorem 9.1.1 shows that $\{C, A\}$ is detectable if and only if $\{A^*, C^*\}$ is stabilizable. So, Parts (i) and (ii) are equivalent. By definition, the pair $\{A^*, C^*\}$ is stabilizable if there is an operator K such that $A^* - C^*K$ is stable. Now let $L = K^*$. Obviously, $A^* - C^*K$ is stable if and only if its adjoint $A - LC$ is stable. Therefore, $\{A^*, C^*\}$ is stabilizable if and only if there is an operator L such that $A - LC$ is stable. Hence, Parts (ii) and (iii) are equivalent. ∎

Example 9.1.1 Consider the following system corresponding to an unattached mass with velocity measurement, that is, $m\ddot{x}_1 = 0$ and $y = \dot{x}_1$ where x_1 is the inertial displacement of the mass along its line of motion. A state space representation for this system is given by

$$\dot{x} = \begin{bmatrix} 0 & 1 \\ 0 & 0 \end{bmatrix} x \quad \text{and} \quad y = \begin{bmatrix} 0 & 1 \end{bmatrix} x$$

where $x_2 = \dot{x}_1$ and $x = x_1 \oplus x_2$. In this case, the operator Γ_λ in Lemma 9.1.2 becomes

$$\Gamma_\lambda = \begin{bmatrix} A - \lambda I \\ C \end{bmatrix} = \begin{bmatrix} -\lambda & 1 \\ 0 & -\lambda \\ 0 & 1 \end{bmatrix}.$$

Obviously, the kernel of Γ_λ is nonzero for $\lambda = 0$. Hence, $\lambda = 0$ is an unobservable eigenvalue and this system in not detectable. Note that for any 2×1 matrix $L = \begin{bmatrix} a & b \end{bmatrix}^{tr}$, we have

$$A - LC = \begin{bmatrix} 0 & 1 - a \\ 0 & -b \end{bmatrix}.$$

In this case, the characteristic polynomial for $A - LC$ is given by

$$\det[sI - A + LC] = s(s + b).$$

So, regardless of the choice of L, zero is an eigenvalue of $A - LC$. This happens in general, that is, the unobservable eigenvalues of $\{C, A\}$ are contained in the eigenvalues of $A - LC$.

9.2   State estimators

In this section we will present an asymptotic state estimator (sometimes called an *observer*) for the system $\{A, B, C, D\}$ in (9.1). To obtain this observer assume that the pair $\{C, A\}$ is detectable. Motivated by Theorem 9.1.3, we propose the following simple state estimator by adding a "correction term" to a copy of the original system whose state is to be estimated, that is,

$$\begin{aligned} \dot{\hat{x}} &= A\hat{x} + Bu + L(y - \hat{y}) \\ \hat{y} &= C\hat{x} + Du \end{aligned} \tag{9.7}$$

where the vector $\hat{x}(t)$ in $\mathcal{X}$ is the estimated state. The operator L, called the *observer gain*, is chosen so that $A - LC$ is stable. The initial state $\hat{x}(0)$ for the observer is arbitrary. In practice, one chooses $\hat{x}(0)$ to be the best available estimate of $x(0)$. Substituting $\hat{y} = C\hat{x} + Du$ into the first equation in (9.7), yields the following description of the observer

$$\dot{\hat{x}} = (A - LC)\hat{x} + (B - LD)u + Ly. \tag{9.8}$$

It should be clear that we can regard the observer as a linear system whose input is $u \oplus y$ and whose output is $\hat{x}$. Finally, it is noted that substituting $y = Cx + Du$ into equation (9.8), gives

$$\dot{\hat{x}} = (A - LC)\hat{x} + LCx + Bu. \tag{9.9}$$

We now introduce the *state estimation error*

$$\tilde{x}(t) = x(t) - \hat{x}(t).$$

Subtracting the differential equation in (9.9) from $\dot{x} = Ax + Bu$, we see that the evolution of the estimation error is governed by

$$\dot{\tilde{x}} = (A - LC)\tilde{x}. \tag{9.10}$$

Since, by construction, the operator $A - LC$ is stable, $\tilde{x}(t)$ approaches zero as t tends towards infinity, that is, (9.2) holds. In other words the estimation error goes to zero as t goes to infinity. So, the dynamical system in (9.7) yields an asymptotic state estimate $\hat{x}$ for x. Finally, it is noted that if the initial observer state $\hat{x}(0) = x(0)$, then $\tilde{x}(t) = 0$ for all t, and thus, the estimate $\hat{x}(t) = x(t)$ for all t. Summing up this analysis yields the following result.

Proposition 9.2.1 *Let $\{A, B, C, D\}$ in (9.1) be detectable and let L be any observer gain such that $A - LC$ is stable. Then the state space system in (9.8) yields an asymptotic state estimate $\hat{x}$ of x, that is, $x(t) - \hat{x}(t)$ approaches zero as t tends to infinity.*

If A is stable, $\{C, A\}$ is detectable regardless of C and one can choose L to be zero. In practice, $L = 0$ may not result in satisfactory behavior or performance of the error dynamics. In addition to stability, one usually wants to meet additional behavior or performance criteria. The problem of choosing a operator L so that $A - LC$ is stable is equivalent to the stabilizability problem of choosing K so that $A^* - C^*K$ is stable. If one solves the stabilizability problem for K, then $L = K^*$ solves the original state estimation problem.

9.3 Eigenvalue placement for estimation error

In the previous section, we saw that if the pair $\{C, A\}$ is detectable, then one can asymptotically estimate the state of system $\{A, B, C, D\}$. To accomplish this one needs to find an operator L such that $A - LC$ is stable. Then the dynamical system in (9.7) yields a asymptotic state estimate $\hat{x}$ of x. In this section we will use duality results to show how one can construct an observer gain L such that $A - LC$ is stable.

If the pair $\{C, A\}$ is observable, then one can use the Lyapunov techniques in Section 8.2 to compute an observer gain such that $A - LC$ is stable. To see this simply notice that the pair $\{A^*, C^*\}$ is controllable. Then the Lyapunov techniques in Section 8.2 provide a computational method to construct a controller gain K such that $A^* - C^*K$ is stable. So, if $L = K^*$, then $A - LC$ is also stable.

Now assume that the pair $\{C, A\}$ is observable with state dimension n. Then the results in Sections 8.4, 8.5 and 8.6 can also be used to construct an observer gain which places the eigenvalues of $A - LC$ at any n specified locations $\{\lambda_j\}_1^n$ in the complex plane. Since $\{A^*, C^*\}$ is controllable, there exists a controller gain K which places the eigenvalues of $A^* - C^*K$ at $\{\bar{\lambda}_j\}_1^n$; see Theorem 8.6.1. (Sections 8.4, 8.5 and 8.6 provide some algorithms to compute the gain K.) So, if $L = K^*$, then $\{\lambda_j\}_1^n$ are the eigenvalues of $A - LC$. This yields the following result.

Theorem 9.3.1 *Let $\{C, A\}$ be an observable pair with state dimension n. Let $\{\lambda_j\}_1^n$ be a set of specified complex numbers. Then there exists an observer gain L such that*

$$\det[sI - A + LC] = \prod_{j=1}^{n}(s - \lambda_i).$$

In particular, if p is any monic polynomial of degree n, then there exists an observer gain L such that p is the characteristic polynomial of $A - LC$.

By using the fact that $\{C, A\}$ is observable if and only if $\{A^*, C^*\}$ is controllable along with Theorem 8.6.2 we obtain the following result.

Theorem 9.3.2 *Consider the pair $\{C, A\}$ with state dimension n. Then $\{C, A\}$ is observable if and only if there exists an observer gain L which places the eigenvalues of $A - LC$ at any n points in the complex plane.*

As before, let $\mathcal{X} = \mathcal{X}_{\bar{o}} \oplus \mathcal{X}_o$ be the observable decomposition for the pair $\{C, A \text{ on } \mathcal{X}\}$ where $\mathcal{X}_{\bar{o}}$ is the unobservable subspace defined in (9.3). The corresponding matrix representations for A and C are given in (9.4) and (9.5). So, if L from $\mathcal{Y}$ into $\mathcal{X}$ is any observer gain, then L admits a matrix representation of the form

$$L = \begin{bmatrix} L_{\bar{o}} \\ L_o \end{bmatrix} : \mathcal{Y} \longrightarrow \begin{bmatrix} \mathcal{X}_{\bar{o}} \\ \mathcal{X}_o \end{bmatrix}. \tag{9.11}$$

Using these decompositions we see that $A - LC$ admits a matrix representation of the form

$$A - LC = \begin{bmatrix} A_{\bar{o}} & * \\ 0 & A_o - L_o C_o \end{bmatrix} : \begin{bmatrix} \mathcal{X}_{\bar{o}} \\ \mathcal{X}_o \end{bmatrix} \longrightarrow \begin{bmatrix} \mathcal{X}_{\bar{o}} \\ \mathcal{X}_o \end{bmatrix}. \tag{9.12}$$

Notice that if λ is an eigenvalue of $A_{\bar{o}}$, then λ is also an eigenvalue of $A - LC$. In other words, the unobservable eigenvalues of A are contained in the eigenvalues of $A - LC$. Hence, the observer gain does not alter the unobservable eigenvalues.

Recall that the pair $\{C, A\}$ is detectable if and only if all of its unobservable eigenvalues are stable. Now assume that $\{C, A\}$ is detectable. Then all the eigenvalues of $A_{\bar{o}}$ are stable. Since the pair $\{C_o, A_o\}$ is observable, there exists an observer gain L_o from $\mathcal{Y}$ into $\mathcal{X}_o$ such that $A_o - L_o C_o$ is stable. In fact, one can choose L_o to place the eigenvalues of $A_o - L_o C_o$ at any $\dim \mathcal{X}_o$ specified locations in the complex plane. In this case (9.12) shows that $A - LC$ is stable. So, if $\{C, A\}$ is detectable, then one can use the decompositions in (9.4), (9.5) and (9.11) to compute an observer gain such that $A - LC$ is stable.

9.4 Notes

The Kalman filter [69] is an optimal stochastic observer or state estimator. The notation of a deterministic observer is due to Luenberger [84]. Our presentation of observers is standard. For some further results in this direction see Kailath [68] and Rugh [110].

Chapter 10

Output Feedback Controllers

In this chapter we consider the problem of obtaining stabilizing output feedback controllers.

10.1 Static output feedback

To implement a state feedback controller one requires knowledge of the state. In many practical problems, it is not feasible to access the complete state. However, one can usually obtain a portion of the state which we call the *measured output*. Here we investigate the problem of designing stabilizing controllers which are based only on a measured output. To this end, consider the state space system described by

$$\begin{aligned} \dot{x} &= Ax + Bu \\ y &= Cx \end{aligned} \tag{10.1}$$

with state $x(t) \in \mathcal{X}$, control input $u(t) \in \mathcal{U}$, and measured output $y(t) \in \mathcal{Y}$. As before, all spaces are finite dimensional. To eliminate some minor technical problems we have assumed that the direct transmission term $D = 0$. We will refer to $\{A, B, C, 0\}$ in (10.1) as the plant. In Section 10.3 we obtain a controller based on the measured output y to stabilize a stabilizable and detectable plant.

The simplest type of controller is a memoryless or static linear output feedback controller of the form

$$u(t) = -Zy(t) \tag{10.2}$$

where the gain Z is an operator from $\mathcal{Y}$ to $\mathcal{U}$. At each instant of time the current control input $u(t)$ is a linear function of the current output $y(t)$. Because the gain Z acts on the output y, the controller $u(t) = -Zy(t)$ is called an *output feedback controller*. Applying this controller to the plant in (10.1) yields the following *closed loop system*:

$$\dot{x} = (A - BZC)x. \tag{10.3}$$

If the *open loop system* $\dot{x} = Ax$ is not stable, a natural question is whether one can choose Z so that the closed loop system is stable. For full state feedback ($C = I$), we have seen that it is possible to do this if $\{A, B\}$ is controllable, or less restrictively, if $\{A, B\}$ is stabilizable. If $\{C, A\}$ is observable or detectable, we might expect to be able to stabilize the plant with static output feedback. This is not the case as the following example illustrates.

153

Example 10.1.1 Consider the rectilinear motion of a unit mass which is subject to a single input force u, that is, $\ddot{y} = u$. We assume that one can measure y, the inertial displacement of the mass along its line of motion. If we set $x = y \oplus \dot{y}$, then this system is described by

$$\dot{x} = \begin{bmatrix} 0 & 1 \\ 0 & 0 \end{bmatrix} x + \begin{bmatrix} 0 \\ 1 \end{bmatrix} u \quad \text{and} \quad y = [\, 1 \quad 0 \,] x . \tag{10.4}$$

Clearly, this system is both controllable and observable. In this case, all linear static output feedback controllers are given by $u = -zy$ where z is a scalar. This controller yields the following the closed loop system

$$\dot{x} = \begin{bmatrix} 0 & 1 \\ -z & 0 \end{bmatrix} x . \tag{10.5}$$

Notice that the eigenvalues of the above 2×2 state matrix are $\pm\sqrt{-z}$. So, regardless of z, the system in (10.5) is not stable. Finally, it is noted that if the plant has some damping in it, that is,

$$\dot{x} = \begin{bmatrix} 0 & 1 \\ 0 & -d \end{bmatrix} x + \begin{bmatrix} 0 \\ 1 \end{bmatrix} u \quad \text{and} \quad y = [\, 1 \quad 0 \,] x$$

where $d > 0$, then the closed loop system is given by

$$\dot{x} = \begin{bmatrix} 0 & 1 \\ -z & -d \end{bmatrix} x .$$

In this case, the closed loop system is stable if and only if $z < 0$.

10.1.1   Transfer function considerations

Recall that a transfer function is *stable* if all of its poles are in the open left half part of the complex plane, $\{s \in \mathbb{C} : \Re(s) < 0\}$. Let $\mathbf{G}$ be the transfer function for the plant $\{A, B, C, 0\}$ in (10.1). Then with all the initial conditions set equal to zero, $\mathbf{y} = \mathbf{G}\mathbf{u}$ where u is the control input and y is the output. (Recall that $\mathbf{f}$ denotes the Laplace transform of a function f.) Consider the static output feedback controller

$$u(t) = -Zy(t) + w(t) \tag{10.6}$$

where Z is an operator from $\mathcal{Y}$ into $\mathcal{U}$ and w is a function with values in $\mathcal{U}$. The function w is referred to as the reference signal Obviously, $\mathbf{u} = -Z\mathbf{y} + \mathbf{w}$ is the Laplace transform of u. Substituting this into $\mathbf{y} = \mathbf{G}\mathbf{u}$, yields $\mathbf{y} + \mathbf{G}Z\mathbf{y} = \mathbf{G}\mathbf{w}$. Hence, $\mathbf{y} = (I + \mathbf{G}Z)^{-1}\mathbf{G}\mathbf{w}$. Recall that the transfer function from w to y is the function $\mathbf{F}$ defined by $\mathbf{y} = \mathbf{F}\mathbf{w}$. Therefore, the transfer function from w to y is given by

$$\mathbf{F} = (I + \mathbf{G}Z)^{-1}\mathbf{G} . \tag{10.7}$$

The function $\mathbf{F}$ is called the transfer function for the closed loop system with the controller (10.6). So, a classical problem in control systems is to determine a gain Z such that the closed loop system is stable, that is, find a gain Z such that all the poles of $\mathbf{F} = (I - \mathbf{G}Z)^{-1}\mathbf{G}$ live in the open left half plane.

Substitution of (10.6) into (10.1) yields the following state space description of the closed loop system:

$$\dot{x} = (A - BZC)x + Bw \qquad \text{and} \qquad y = Cx. \qquad (10.8)$$

Hence, $\{A - BZC, B, C, 0\}$ is a realization for $\mathbf{F}$. Now suppose that $\{A, B, C, 0\}$ is a controllable and observable realization of $\mathbf{G}$. Theorem 6.3.1 shows that the transfer function $\mathbf{G}$ is stable if and only if A is stable. According to Lemma 6.4.2, the closed loop system $\{A - BZC, B, C, 0\}$ is also controllable and observable. By Theorem 6.3.1, we see that λ is a pole of $(I + \mathbf{G}Z)^{-1}\mathbf{G}$ if and only if λ is an eigenvalue of $A - BZC$. In particular, the feedback transfer function $\mathbf{F}$ is stable if and only if $A - BZC$ is stable. So, the problem of finding a gain Z such that all the poles of the feedback transfer function $(I + \mathbf{G}Z)^{-1}\mathbf{G}$ are in the open left half plane is equivalent to finding an operator Z such that $A - BZC$ is stable. Summing up this analysis readily yields the following result.

Proposition 10.1.1 *Let $\{A, B, C, 0\}$ be a realization of the transfer function $\mathbf{G}$. Let Z be an operator from $\mathcal{Y}$ into $\mathcal{U}$. Then $\{A - BZC, B, C, 0\}$ is a realization for the closed loop transfer function $(I + \mathbf{G}Z)^{-1}\mathbf{G}$. Moreover, $\{A, B, C, 0\}$ is controllable and observable if and only if $\{A - BZC, B, C, 0\}$ is controllable and observable. In this case, λ is a pole of $(I + \mathbf{G}Z)^{-1}\mathbf{G}$ if and only if λ is an eigenvalue of $A - BZC$.*

Example 10.1.1 shows that it is not always possible to stabilize a transfer function by static output feedback. In this example the open loop transfer function for the state space system in (10.4) is given by $\mathbf{G}(s) = 1/s^2$. Since zero is a pole for $\mathbf{G}$, it follows that this transfer function is unstable. In this case, the feedback transfer function is given by

$$\mathbf{F} = \frac{\mathbf{G}}{1 + \mathbf{G}z} = \frac{1}{s^2 + z}$$

where z is a scalar. Clearly, the closed loop transfer function $\mathbf{F}$ is unstable for all scalars z. If $\{A, B, C, 0\}$ is any minimal realization of $\mathbf{G} = 1/s^2$, then $\{A - BZC, B, C, 0\}$ is a minimal realization of the closed loop transfer function $1/(s^2 + z)$. So, according to Proposition 10.1.1, the operator $A - BZC$ is unstable for any operator Z on $\mathbb{C}^1$.

To see why static output feedback is not sufficient to stabilize a transfer function, let $\mathbf{G} = p/d$ be any scalar valued proper rational transfer function where p and d are polynomials with no common zeros. Then the closed loop transfer function is given by

$$\mathbf{F} = \frac{\mathbf{G}}{1 + \mathbf{G}z} = \frac{p}{d + zp} \qquad (10.9)$$

where z is a scalar. Because p and d have no common zeros, the polynomials p and $d + zp$ have no common zeros. So, $\mathbf{F}$ is stable if and only if all the zeros of $d + zp$ lie in the open left half plane. Since $d + zp$ has $\deg d$ roots, it is not always possible vary the single parameter z to guarantee that all the $\deg d$ roots of $d + zp$ lie in the open left half plane. In other words, varying one parameter z is not enough to guarantee that all the poles of the closed loop system lie in the open left half plane. In the scalar case, the classical root locus provides a graphical method to determine the roots of $d + zp$ as z varies in $(-\infty, \infty)$. So, using root locus techniques one can design output feedback controllers to stabilize certain scalar valued transfer functions. Finally, it is noted that equation (10.9) also shows that output feedback does not change the zeros of the transfer function. In other words, $\mathbf{G} = p/d$ and $\mathbf{F}$ have the same zeros. In this case, the roots of p are precisely the zeros of both $\mathbf{G}$ and $\mathbf{F}$.

For multivariable systems there are currently are no easily verifiable necessary and sufficient conditions for the existence of a stabilizing static output feedback controller. This leads us to consider dynamic output feedback controllers, that is, controllers which are dynamic systems whose input is the measurement y and whose output is the control u.

Exercise 23 Let $\mathbf{G}$ be the transfer function for the state space system

$$\dot{x} = Ax + Bu \quad \text{and} \quad y = Cx + Du. \tag{10.10}$$

Consider the static output feedback controller $u(t) = -Zy(t) + w(t)$ where Z is an operator from $\mathcal{Y}$ into $\mathcal{U}$ and w is the reference signal. Then the closed loop transfer function from the reference signal w to y is given by $\mathbf{F} = (I - \mathbf{G}Z)^{-1}\mathbf{G}$. Assume that -1 is not an eigenvalue of ZD.

(i) Show that the control u is given by

$$u = -(I + ZD)^{-1}ZCx + (I + ZD)^{-1}w. \tag{10.11}$$

(ii) Show that a state space realization for the closed loop transfer function $\mathbf{F}$ is given by

$$\begin{aligned} \dot{x} &= (A - B(I + ZD)^{-1}ZC)x + B(I + ZD)^{-1}w \\ y &= (I + DZ)^{-1}Cx + (I + DZ)^{-1}Dw. \end{aligned} \tag{10.12}$$

(iii) Show that $\{A, B, C, D\}$ is controllable, respectively observable if and only if the system in (10.12) is controllable, respectively observable.

(iv) If $\{A, B, C, D\}$ is minimal, then show that λ is a pole of $\mathbf{F}$ if and only if λ is an eigenvalue of $A - B(I + ZD)^{-1}ZC$. In particular, the closed loop transfer function $\mathbf{F}$ is stable if and only if $A - B(I + ZD)^{-1}ZC$ is stable.

10.2 Dynamic output feedback

In this section we discuss dynamic output feedback controllers for state space systems of the form

$$\begin{aligned} \dot{x} &= Ax + Bu \\ y &= Cx. \end{aligned} \tag{10.13}$$

As before, A is an operator on $\mathcal{X}$ and B maps $\mathcal{U}$ into $\mathcal{X}$ while C maps $\mathcal{X}$ into $\mathcal{Y}$. In general, a *linear dynamic output feedback controller* is described by a state space system of the form

$$\begin{aligned} \dot{f} &= A_c f + B_c y \\ u &= -C_c f - D_c y \end{aligned} \tag{10.14}$$

where the *controller state* $f(t)$ lies in some finite dimensional space $\mathcal{F}$. The dimension of the controller state space $\mathcal{F}$ is called the *order of the controller*. In particular, if $\{A_c, B_c, C_c, D_c\}$ is controllable and observable, then the order of the controller equals the McMillan degree of $\{A_c, B_c, C_c, D_c\}$. This controller is a linear time invariant system whose input is the measured output y of the plant in (10.13), and whose output u is the control input to the plant. If C_c is zero then the controller in (10.14) is precisely a static feedback controller as discussed in Section 10.1.

Applying the controller (10.14) to the plant in (10.13) results in the following closed loop system

$$\begin{bmatrix} \dot{x} \\ \dot{f} \end{bmatrix} = \begin{bmatrix} A - BD_cC & -BC_c \\ B_cC & A_c \end{bmatrix} \begin{bmatrix} x \\ f \end{bmatrix}. \tag{10.15}$$

This is a linear time invariant system whose state is $x \oplus f$ and state space matrix is given by

$$\mathcal{A} = \begin{bmatrix} A - BD_cC & -BC_c \\ B_cC & A_c \end{bmatrix} \text{ on } \begin{bmatrix} \mathcal{X} \\ \mathcal{F} \end{bmatrix}. \tag{10.16}$$

So, the system in (10.15) has state dimension $n + n_c$ where n the dimension of the plant state and n_c is the dimension of the controller state. Finally, we say that the plant $\{A, B, C, 0\}$ is *stabilizable by a linear dynamic output feedback controller* if there exists a controller $\{A_c, B_c, C_c, D_c\}$ such that the state operator $\mathcal{A}$ in (10.16) is stable.

In classical control of scalar input scalar output systems, a widely used controller is the PI (proportional integral) controller described by

$$u(t) = -\alpha y(t) - \beta \int_0^t y(\tau) \, d\tau \tag{10.17}$$

where α and β are scalars. Let f be the function defined by $\dot{f} = y$. Then it is easy to verify that this PI controller can be represented by the following first order dynamical system

$$\begin{aligned} \dot{f} &= y \\ u &= -\beta f - \alpha y. \end{aligned} \tag{10.18}$$

We will show that stabilizability of $\{A, B\}$ and detectability of $\{C, A\}$ are necessary and sufficient conditions to stabilize the plant $\{A, B, C, 0\}$ by a linear output feedback controller. The following lemma states the necessity of stabilizability and detectability. In the next section, we show that these conditions are sufficient for output feedback stabilizability, by constructing specific stabilizing controllers.

Lemma 10.2.1 *If the plant $\{A, B, C, 0\}$ is stabilizable by a linear dynamic output feedback controller, then $\{A, B\}$ is stabilizable and $\{C, A\}$ is detectable.*

PROOF. Consider the plant in (10.13) subject to any controller of the form (10.14) and let $\mathcal{A}$ be the state matrix of the resulting closed loop system given in (10.16). We first show that if λ is an unobservable eigenvalue of $\{C, A\}$, then λ is an eigenvalue of $\mathcal{A}$. To see this, suppose that $\{\lambda, v\}$ is an unobservable eigenvalue eigenvector pair for $\{C, A\}$. Then $Av = \lambda v$ and $Cv = 0$. Using this, it should be clear that

$$\mathcal{A} \begin{bmatrix} v \\ 0 \end{bmatrix} = \lambda \begin{bmatrix} v \\ 0 \end{bmatrix}.$$

Since v is nonzero, it follows that λ is an eigenvalue of $\mathcal{A}$. So, all the unobservable eigenvalues of $\{C, A\}$ are contained in the eigenvalues of $\mathcal{A}$. If the plant is stabilizable by dynamic output feedback, then there exists a controller such that all the eigenvalues of $\mathcal{A}$ are stable. In particular, all the unobservable eigenvalues of $\{C, A\}$ must also be stable. Therefore, if the plant is stabilizable by a linear dynamic output feedback controller, then $\{C, A\}$ is detectable.

We now claim that if λ is an uncontrollable eigenvalue of $\{A, B\}$, then λ is an eigenvalue of $\mathcal{A}$. Suppose that λ is an uncontrollable eigenvalue of $\{A, B\}$. Then $A^*v = \bar{\lambda}v$ and $B^*v = 0$ for some nonzero vector v. Using the structure of $\mathcal{A}$, it readily follows that $\bar{\lambda}$ is an eigenvalue of $\mathcal{A}^*$ with eigenvector $[v, 0]^{tr}$. Hence, λ is an eigenvalue of $\mathcal{A}$. So, all the uncontrollable eigenvalues of $\{A, B\}$ are contained in the eigenvalues of $\mathcal{A}$. Therefore, if the plant is stabilizable by a linear dynamic output feedback controller, then $\{A, B\}$ is stabilizable. ∎

Exercise 24 Show that the unattached mass with position feedback in Example 10.1.1 can be stabilized with a first order dynamic output feedback controller. What controller parameters place all the eigenvalues of the closed loop system at -1?

10.3   Observer based controllers

As before, consider the plant in (10.13). Recall that if $\{A, B, C, 0\}$ is stabilizable by dynamic output feedback, then $\{A, B\}$ is stabilizable and $\{C, A\}$ is detectable. We now demonstrate that if these conditions are satisfied, then closed loop stability can be achieved with a controller of order no more than the state dimension of the plant. To achieve this, we combine our previous results on asymptotic state estimation and stabilization via state feedback. We consider controllers which obtain an asymptotic estimate of the plant state and then use this estimate for the state in a stabilizing state feedback controller. These controllers are called observer based controllers and have the following structure:

$$\begin{aligned} \dot{\hat{x}} &= A\hat{x} + Bu + L(y - C\hat{x}) \\ u &= -K\hat{x}. \end{aligned} \qquad (10.19)$$

Notice that this controller is completely determined by specifying the state feedback gain K and the observer gain L. Moreover, using $u = -K\hat{x}$ this controller can be written as

$$\begin{aligned} \dot{\hat{x}} &= (A - BK - LC)\hat{x} + Ly \\ u &= -K\hat{x}. \end{aligned} \qquad (10.20)$$

This is a dynamic output feedback controller with controller state $f = \hat{x}$, the state estimate of x; see (10.14). Hence, $n_c = n$, that is, the controller and the plant have the same dimension. In this setting, $A_c = A - BK - LC$ and $B_c = L$. Moreover, $C_c = K$ and $D_c = 0$.

Combining the plant in (10.13) with the controller description (10.20), yields the closed loop system (see (10.15)) described by

$$\begin{bmatrix} \dot{x} \\ \dot{\hat{x}} \end{bmatrix} = \begin{bmatrix} A & -BK \\ LC & A - BK - LC \end{bmatrix} \begin{bmatrix} x \\ \hat{x} \end{bmatrix}. \qquad (10.21)$$

This is a linear time invariant system with state $x \oplus \hat{x}$ and state matrix given by

$$\mathcal{A} = \begin{bmatrix} A & -BK \\ LC & A - BK - LC \end{bmatrix} \quad \text{on} \quad \begin{bmatrix} \mathcal{X} \\ \mathcal{X} \end{bmatrix}. \tag{10.22}$$

Let $\tilde{x}$ be the estimation error, that is $\tilde{x} = x - \hat{x}$. Then subtracting $\dot{\hat{x}}$ from $\dot{x}$ in (10.21) shows that $x \oplus \tilde{x}$ satisfies the following state space equation

$$\begin{bmatrix} \dot{x} \\ \dot{\tilde{x}} \end{bmatrix} = \begin{bmatrix} A - BK & BK \\ 0 & A - LC \end{bmatrix} \begin{bmatrix} x \\ \tilde{x} \end{bmatrix}. \tag{10.23}$$

This system has state $x \oplus \tilde{x}$ and state matrix

$$\tilde{\mathcal{A}} = \begin{bmatrix} A - BK & BK \\ 0 & A - LC \end{bmatrix} \quad \text{on} \quad \begin{bmatrix} \mathcal{X} \\ \mathcal{X} \end{bmatrix}. \tag{10.24}$$

Now consider the invertible operator T on $\mathcal{X} \oplus \mathcal{X}$ defined by

$$T = \begin{bmatrix} I & 0 \\ I & -I \end{bmatrix}. \tag{10.25}$$

Then it is easy to show that $\tilde{\mathcal{A}}T = T\mathcal{A}$. Hence, $\mathcal{A}$ is similar to $\tilde{\mathcal{A}}$. In particular, $\mathcal{A}$ and $\tilde{\mathcal{A}}$ have the same eigenvalues including their multiplicity. Since $\tilde{\mathcal{A}}$ is upper triangular, it follows that the set of eigenvalues of $\mathcal{A}$ are simply the union of the eigenvalues of $A - BK$ and $A - LC$. One way to see this is to notice that

$$\det[sI - \tilde{\mathcal{A}}] = \det[sI - A + BK] \det[sI - A + LC].$$

In other words, the characteristic polynomial of $\tilde{\mathcal{A}}$ is the product of the characteristic polynomials of $A - BK$ and $A - LC$. Since the eigenvalues of any finite dimensional operator are the roots of its characteristic polynomial, the set of eigenvalues of $\tilde{\mathcal{A}}$, or equivalently $\mathcal{A}$, are the union of the eigenvalues of $A - BK$ and $A - LC$.

It now follows that if both $A - BK$ and $A - LC$ are stable, then the closed loop system (10.21) is stable. If $\{A, B\}$ is stabilizable, one can choose a controller gain K so that $A - BK$ is stable. If $\{C, A\}$ is detectable, one can choose an observer gain L such that $A - LC$ is stable. Combining these observations with Lemma 10.2.1 leads to the following result.

Theorem 10.3.1 *Consider the plant $\{A, B, C, 0\}$. Then the following statements are equivalent.*

(a) *The pair $\{A, B\}$ is stabilizable and $\{C, A\}$ is detectable.*

(b) *The plant $\{A, B, C, 0\}$ is stabilizable by a linear dynamic output feedback controller.*

(c) *The plant $\{A, B, C, 0\}$ is stabilizable by a linear dynamic output feedback controller whose order is less than or equal to the state dimension of the plant.*

Remark 10.3.1 The above analysis provides a method for constructing a stabilizing dynamic controller for a stabilizable and detectable plant of the form (10.13). To accomplish this, simply use any method to compute the gains K and L which ensure that both $A - BK$ and $A - LC$ are stable. Then a stabilizing controller is given by (10.19). In particular, if the plant is controllable and observable, then one can choose K and L to arbitrarily place the eigenvalues of $A - BK$ and $A - LC$, respectively. In this case, one can arbitrarily place the eigenvalues of the closed system in (10.21) by the appropriate choice of K and L.

Example 10.3.1 Let us use the above results to construct a stabilizing dynamic controller for the simple mechanical system presented in Example 10.1.1, that is,

$$\dot{x} = \begin{bmatrix} 0 & 1 \\ 0 & 0 \end{bmatrix} x + \begin{bmatrix} 0 \\ 1 \end{bmatrix} u \quad \text{and} \quad y = \begin{bmatrix} 1 & 0 \end{bmatrix} x .$$

Clearly, this system is controllable and observable. For this system, the observer based controllers in (10.19) are given by

$$\begin{aligned} \dot{\hat{x}} &= \begin{bmatrix} 0 & 1 \\ 0 & 0 \end{bmatrix} \hat{x} + \begin{bmatrix} 0 \\ 1 \end{bmatrix} u + \begin{bmatrix} l_1 \\ l_2 \end{bmatrix} (x_1 - \hat{x}_1) \\ u &= -k_1\hat{x}_1 - k_2\hat{x}_2 \end{aligned}$$

where $K = \begin{bmatrix} k_1 & k_2 \end{bmatrix}$ and $L = \begin{bmatrix} l_1 & l_2 \end{bmatrix}^{tr}$. The closed loop system in (10.21) is described by

$$\begin{bmatrix} \dot{x}_1 \\ \dot{x}_2 \\ \dot{\hat{x}}_1 \\ \dot{\hat{x}}_2 \end{bmatrix} = \begin{bmatrix} 0 & 1 & 0 & 0 \\ 0 & 0 & -k_1 & -k_2 \\ l_1 & 0 & -l_1 & 1 \\ l_2 & 0 & -(l_2 + k_1) & -k_2 \end{bmatrix} \begin{bmatrix} x_1 \\ x_2 \\ \hat{x}_1 \\ \hat{x}_2 \end{bmatrix} . \tag{10.26}$$

Let $\mathcal{A}$ be the 4×4 matrix given in the previous equation. In this case, the state estimation error $\tilde{x}_j = x_j - \hat{x}_j$ for $j = 1, 2$. The state estimation error equation in (10.23) is now given by

$$\begin{bmatrix} \dot{x}_1 \\ \dot{x}_2 \\ \dot{\tilde{x}}_1 \\ \dot{\tilde{x}}_2 \end{bmatrix} = \begin{bmatrix} 0 & 1 & 0 & 0 \\ -k_1 & -k_2 & k_1 & k_2 \\ 0 & 0 & -l_1 & 1 \\ 0 & 0 & -l_2 & 0 \end{bmatrix} \begin{bmatrix} x_1 \\ x_2 \\ \tilde{x}_1 \\ \tilde{x}_2 \end{bmatrix} .$$

Let $\tilde{\mathcal{A}}$ be the 4×4 matrix given in the previous equation. Recall that $\tilde{\mathcal{A}}$ is similar to $\mathcal{A}$. It is easy to show that the characteristic polynomial for $\tilde{\mathcal{A}}$ is given by

$$\det[sI - \tilde{\mathcal{A}}] = (s^2 + k_2 s + k_1)(s^2 + l_1 s + l_2) .$$

Clearly, one can arbitrarily assign the roots of this polynomial by choice of the appropriate choice of the scalar gains $\{k_1, k_2, l_1, l_2\}$. In particular, the closed loop systems is stable if and only if all the these scalar gains are positive.

Exercise 25 Let $\{A, B, C, D\}$ be the state space system given by

$$\dot{x} = Ax + Bu \quad \text{and} \quad y = Cx + Du . \tag{10.27}$$

Consider the dynamic output feedback controller

$$\begin{aligned}
\dot{f} &= A_c f + B_c y \\
u &= -C_c f - D_c y
\end{aligned} \tag{10.28}$$

where the controller state f lies in some finite dimensional space $\mathcal{F}$. Assume that -1 is not an eigenvalue of $D_c D$.

(i) Show that the control u and output y are given by

$$\begin{aligned}
u &= -(I + D_c D)^{-1} D_c C x - (I + D_c D)^{-1} C_c f \\
y &= (I + D D_c)^{-1} C x - (I + D D_c)^{-1} D C_c f .
\end{aligned} \tag{10.29}$$

(ii) Show that a state space realization for the closed loop system is given by

$$\begin{bmatrix} \dot{x} \\ \dot{f} \end{bmatrix} = \begin{bmatrix} A - B(I + D_c D)^{-1} D_c C & -B(I + D_c D)^{-1} C_c \\ B_c(I + D D_c)^{-1} C & A_c - B_c(I + D D_c)^{-1} D C_c \end{bmatrix} \begin{bmatrix} x \\ f \end{bmatrix} . \tag{10.30}$$

Let $\mathcal{A}_c$ be the 2×2 block matrix on $\mathcal{X} \oplus \mathcal{F}$ in (10.30). Notice that the block matrices $\mathcal{A}_c$ and $\mathcal{A}$ in (10.16) are closely related. By replacing B by $B(I + D_c D)^{-1}$ and B_c by $B_c(I + D D_c)^{-1}$ and A_c by $A_c - B_c(I + D D_c)^{-1} D C_c$ in $\mathcal{A}$, we obtain $\mathcal{A}_c$.

(iii) The system $\{A, B, C, D\}$ in (10.27) is stabilizable by dynamic output feedback if there exists a controller of the form (10.28) such that $\mathcal{A}_c$ is stable. Show that $\{A, B, C, D\}$ is stabilizable by dynamic output feedback if and only if the pair $\{A, B\}$ is stabilizable and $\{C, A\}$ is detectable.

(iv) Consider the observer based controller

$$\begin{aligned}
\dot{\hat{x}} &= A\hat{x} + Bu + L(y + (DK - C)\hat{x}) \\
u &= -K\hat{x} .
\end{aligned} \tag{10.31}$$

For this controller show that the closed loop system in (10.30) is given by

$$\begin{bmatrix} \dot{x} \\ \dot{\hat{x}} \end{bmatrix} = \begin{bmatrix} A & -BK \\ LC & A - BK - LC \end{bmatrix} \begin{bmatrix} x \\ \hat{x} \end{bmatrix} . \tag{10.32}$$

(v) Let $\tilde{x} = x - \hat{x}$ be the state estimation error. Then shows that $x \oplus \tilde{x}$ satisfies the following state space equation

$$\begin{bmatrix} \dot{x} \\ \dot{\tilde{x}} \end{bmatrix} = \begin{bmatrix} A - BK & BK \\ 0 & A - LC \end{bmatrix} \begin{bmatrix} x \\ \tilde{x} \end{bmatrix} . \tag{10.33}$$

Let $\mathcal{A}$ be the block matrix in (10.32) and $\tilde{\mathcal{A}}$ the block matrix in (10.33). Then $T\mathcal{A} = \tilde{\mathcal{A}}T$ where T is the similarity transformation defined in (10.25). So, $\mathcal{A}$ is similar to $\tilde{\mathcal{A}}$. In particular, if $\{A, B, C, D\}$ is stabilizable and detectable, then one can compute operator gains K and L such that $A - BK$ and $A - LC$ are stable. In this case, the state space system in (10.28) is a dynamic output stabilizing controller for the plant in (10.27).

10.4 Notes

The results in this chapter are classical. For some further results on feedback controllers see
Kailath [68] and Rugh [110].

Chapter 11

Zeros of Transfer Functions

This chapter is devoted to a state space interpretation of the zeros of a transfer function.

11.1 Zeros

This section introduces the concept of a zero for a rational transfer function. To begin, consider a scalar proper rational function $\mathbf{g}$. We say that a complex number λ is a zero of $\mathbf{g}$ if $\mathbf{g}(\lambda) = 0$. So, λ is a zero of $\mathbf{g}$ if and only if the rational function given by $\mathbf{g}(s)/(s - \lambda)$ is analytic at λ. We now obtain a state space interpretation of zeros. The dimension of the state space for any minimal realization of a proper transfer function $\mathbf{G}$ is called the McMillan degree of $\mathbf{G}$ and is denoted by $\operatorname{mdeg} \mathbf{G}$. If a proper rational function $\mathbf{g}$ equals n/d where n and d are two scalar polynomials with no common zeros, then the zeros of $\mathbf{g}$ are precisely the zeros of n. Furthermore, the McMillan degree of $\mathbf{g}$ is simply the degree of d, that is, $\operatorname{mdeg} \mathbf{g} = \deg d$. From this it readily follows that if λ is not a zero of $\mathbf{g}$, then $\operatorname{mdeg}(\mathbf{g}/(s - \lambda)) = \operatorname{mdeg}(\mathbf{g}) + 1$. However, $\operatorname{mdeg}(\mathbf{g}/(s - \lambda)) = \operatorname{mdeg}(\mathbf{g})$ when λ is a zero of $\mathbf{g}$.

Consider any rational function F. Then the *normal rank* of F, denoted by $\operatorname{nrank} F$, is defined by

$$\operatorname{nrank} F = \sup\{\operatorname{rank} F(s) : s \text{ is not a pole of } F\}. \tag{11.1}$$

If P is an operator valued polynomial acting between two finite dimensional vector spaces, then the rank of $P(\lambda)$ equals the normal rank of P except at a finite number of points. To see this, assume, without loss of generality, that P is matrix valued. Then the rank of $P(\lambda)$ is the largest of the orders of the nonzero minors of $P(\lambda)$. Since each minor of P is a scalar-valued polynomial, it is either identically zero or has a finite number of zeros. Because P has a finite number of minors, there is only a finite number of points for which the rank of $P(\lambda)$ is below that of the normal rank of P. If F is an operator valued rational function acting between two finite dimensional vector spaces, then $F = P/d$ where P is a operator valued polynomial and d is a scalar valued polynomial. Therefore, the rank of $F(\lambda)$ equals the normal rank of F except at a finite number of points.

If a complex number λ is not a pole of F, we say that it is a *zero* of F if the rank of $F(\lambda)$ is strictly less than the normal rank of F. Motivated by the following lemma, we will generalize the definition of a zero to include the possibility of pole being a zero.

Lemma 11.1.1 *If a complex number λ is not a pole of a proper rational transfer function* **G**, *then*

$$mdeg\left(\mathbf{G}/(s-\lambda)\right) = mdeg\,\mathbf{G} + rank\,\mathbf{G}(\lambda).$$

PROOF. First notice that

$$\frac{\mathbf{G}(s)}{s-\lambda} = \frac{\mathbf{G}(s) - \mathbf{G}(\lambda)}{s-\lambda} + \frac{\mathbf{G}(\lambda)}{s-\lambda}. \tag{11.2}$$

Let $\{A \text{ on } \mathcal{X}, B, C, D\}$ be a minimal realization of **G**. Then $\mathbf{G}(s) = C(sI - A)^{-1}B + D$. Since λ is not a pole of **G** and the realization is minimal, λ is not an eigenvalue of A. Hence, $\lambda I - A$ is invertible and

$$
\begin{aligned}
(sI - A)^{-1} - (\lambda I - A)^{-1} &= (sI - A)^{-1}[\lambda I - A - (sI - A)](\lambda I - A)^{-1}\\
&= (s - \lambda)(sI - A)^{-1}(A - \lambda I)^{-1}.
\end{aligned}
$$

We now have that

$$\frac{\mathbf{G}(s) - \mathbf{G}(\lambda)}{s - \lambda} = C(sI - A)^{-1}(A - \lambda I)^{-1}B.$$

Since $(A - \lambda I)^{-1}$ is invertible and commutes with A, it follows from the controllability of $\{A, B\}$ that $\mathcal{X}$ is spanned by $A^i(A - \lambda I)^{-1}B\mathcal{U}$ for $i = 0, 1, 2, \cdots$ where $\mathcal{U}$ is the input space. Hence, $\{A \text{ on } \mathcal{X}, (A - \lambda I)^{-1}B, C, 0\}$ is a controllable and observable realization of the transfer function $(\mathbf{G}(s) - \mathbf{G}(\lambda))/(s - \lambda)$. Therefore mdeg $((\mathbf{G}(s) - \mathbf{G}(\lambda))/(s - \lambda))$ is the dimension of $\mathcal{X}$ which is the same as mdeg **G**.

To obtain a realization of $\mathbf{G}(\lambda)/(s - \lambda)$, let $\mathcal{X}_2$ be the range of $\mathbf{G}(\lambda)$ and A_2 the operator on $\mathcal{X}_2$ defined by $A_2 x_2 = \lambda x_2$. A simple calculation shows that $\{A_2, \mathbf{G}(\lambda), C_2, 0\}$, with $C_2 = I|\mathcal{X}_2$, is a minimal realization of $\mathbf{G}(\lambda)/(s - \lambda)$. Hence, mdeg $\mathbf{G}/(\mathbf{s} - \lambda)$ is the dimension of $\mathcal{X}_2$ which is the rank of $\mathbf{G}(\lambda)$.

Since the transfer functions $(\mathbf{G}(s) - \mathbf{G}(\lambda))/(s - \lambda)$ and $\mathbf{G}(\lambda)/(s - \lambda)$ have no common poles, it follows from equation (11.2) that

$$\text{mdeg}\left(\frac{\mathbf{G}(s)}{s-\lambda}\right) = \text{mdeg}\left(\frac{\mathbf{G}(s) - \mathbf{G}(\lambda)}{s-\lambda}\right) + \text{mdeg}\left(\frac{\mathbf{G}(\lambda)}{s-\lambda}\right) = \text{mdeg}\,\mathbf{G} + \text{rank}\,\mathbf{G}(\lambda).$$

The second equality follows from Exercise 26 below.   ∎

If λ is not a pole of a proper rational function **G**, then the above lemma shows that

$$\text{mdeg}\left(\mathbf{G}/(s-\lambda)\right) \leq \text{mdeg}\,\mathbf{G} + \text{nrank}\,\mathbf{G}. \tag{11.3}$$

Moreover, λ is a zero of **G** if and only if

$$\text{mdeg}\left(\mathbf{G}/(s-\lambda)\right) < \text{mdeg}\,\mathbf{G} + \text{nrank}\,\mathbf{G}. \tag{11.4}$$

Later we will see that, as expected, the inequality (11.3) also holds when λ is a pole of **G**. Motivated by this discussion, we say that a complex number λ is a *zero* of **G** if the inequality (11.4) holds. For example, the normal rank of the transfer function

$$\mathbf{G}(s) = \begin{bmatrix} \frac{1}{s+1} & 0 \\ 0 & \frac{s+1}{s+2} \end{bmatrix}$$

is two. The McMillan degree of $\mathbf{G}$ is two. Clearly, if λ is not a pole of $\mathbf{G}$, that is, $\lambda \neq -1, -2$, then $\mathbf{G}(\lambda)$ has rank two, and thus, the normal rank of $\mathbf{G}$ is two. So, if $\lambda \neq -1, -2$, then λ is not a zero of $\mathbf{G}$. Since the McMillan degree of $\mathbf{G}/(s+2)$ is four, it follows that -2 is not a zero. However, the McMillan degree of $\mathbf{G}/(s+1)$ is three; hence $\mathbf{G}$ has a zero at -1.

Exercise 26 Let $\Sigma_1 = \{A_1 \text{ on } \mathcal{X}_1, B_1, C_1, D_1\}$ and $\Sigma_2 = \{A_2 \text{ on } \mathcal{X}_2, B_2, C_2, D_2\}$ be respectively realizations for $\mathbf{G}_1$ and $\mathbf{G}_2$, where $\mathbf{G}_1$ and $\mathbf{G}_2$ have values in $\mathcal{L}(\mathcal{U}, \mathcal{Y})$. Let $\Sigma = \{A \text{ on } \mathcal{X}, B, C, D\}$ be the state space system determined by

$$A = \begin{bmatrix} A_1 & 0 \\ 0 & A_2 \end{bmatrix} \quad \text{and} \quad B = \begin{bmatrix} B_1 \\ B_2 \end{bmatrix}$$

$$C = \begin{bmatrix} C_1 & C_2 \end{bmatrix} \quad \text{and} \quad D = D_1 + D_2$$

where the state $\mathcal{X} = \mathcal{X}_1 \oplus \mathcal{X}_2$.

(i) Show that Σ is a realization for $\mathbf{G}_1 + \mathbf{G}_2$.

(ii) Now assume that Σ_1 and Σ_2 are both minimal realizations. Moreover, assume that A_1 and A_2 have no common eigenvalues. Then show that Σ is a minimal realization for $\mathbf{G}_1 + \mathbf{G}_2$. In particular, the McMillan degree of $\mathbf{G}_1 + \mathbf{G}_2$ equals $\dim \mathcal{X}_1 + \dim \mathcal{X}_2$.

(iii) If $\mathbf{G}_1$ and $\mathbf{G}_2$ are two transfer functions with no common poles, then show that $\mathrm{mdeg}\,(\mathbf{G}_1 + \mathbf{G}_2) = \mathrm{mdeg}\,\mathbf{G}_1 + \mathrm{mdeg}\,\mathbf{G}_2$.

11.2   The system matrix

In this section we introduce and study the system matrix for a linear system. A simple calculation readily proves the following useful result.

Lemma 11.2.1 *Let T be an operator mapping $\mathcal{X} \oplus \mathcal{U}$ into $\mathcal{X} \oplus \mathcal{Y}$ defined by*

$$T = \begin{bmatrix} X & Y \\ W & Z \end{bmatrix}. \tag{11.5}$$

If X is invertible, then T admits a factorization of the form

$$T = \begin{bmatrix} I & 0 \\ WX^{-1} & I \end{bmatrix} \begin{bmatrix} X & 0 \\ 0 & Z - WX^{-1}Y \end{bmatrix} \begin{bmatrix} I & X^{-1}Y \\ 0 & I \end{bmatrix}. \tag{11.6}$$

As before, assume T is given by (11.5) with X is invertible. The operator $Z - WX^{-1}Y$ is called the *Schur complement* of T. The above factorization of T shows that T is invertible if and only if its Schur complement is invertible. Moreover, the rank of T equals the dimension of $\mathcal{X}$ plus the rank of its Schur complement.

Let $\{A \text{ on } \mathcal{X}, B, C, D\}$ be a realization for a transfer function $\mathbf{G}$. Then, for any complex number λ, let T_λ be the system matrix from $\mathcal{X} \oplus \mathcal{U}$ into $\mathcal{X} \oplus \mathcal{Y}$ given by

$$T_\lambda = \begin{bmatrix} A - \lambda I & B \\ C & D \end{bmatrix}. \tag{11.7}$$

If λ is not an eigenvalue of A, then $\lambda I - A$ is invertible, and the Schur complement of T_λ is $\mathbf{G}(\lambda)$. Moreover, the rank of T_λ equals the dimension of $\mathcal{X}$ plus the rank of $\mathbf{G}(\lambda)$, that is,

$$\operatorname{rank} T_\lambda = \dim \mathcal{X} + \operatorname{rank}\mathbf{G}(\lambda) \,. \tag{11.8}$$

Since $T(s) := T_s$ defines a polynomial and $\mathbf{G}$ is a rational function, they achieve their normal rank everywhere except at a finite number of points. Therefore,

$$\operatorname{nrank} T = \dim \mathcal{X} + \operatorname{nrank}\mathbf{G} \,. \tag{11.9}$$

In particular, if $\{A \text{ on } \mathcal{X}, B, C, D\}$ is a minimal realization of $\mathbf{G}$ and λ is not a pole of $\mathbf{G}$, then λ is not an eigenvalue of A and the dimension of $\mathcal{X}$ is the McMillan degree of $\mathbf{G}$. Hence, $\operatorname{rank} T_\lambda = \operatorname{mdeg}\mathbf{G} + \operatorname{rank}\mathbf{G}(\lambda)$. Using Lemma 11.1.1, we obtain that the rank of T_λ is the McMillan degree of $\mathbf{G}(s)/(s - \lambda)$. The following theorem, whose proof is independent of the previous analysis, states that this result also holds when λ is a pole of $\mathbf{G}$.

Theorem 11.2.2 *Let $\{A, B, C, D\}$ be a minimal realization of a transfer function $\mathbf{G}$. Then, the rank of the operator T_λ given in (11.7) equals the McMillan degree of the transfer function $\mathbf{G}(s)/(s - \lambda)$.*

PROOF. First, consider the case $\lambda = 0$. Let $\mathbf{G}(s) = \sum_0^\infty G_i/s^i$ be the power series expansion for $\mathbf{G}$. The power series expansion for $\mathbf{G}/s$ is given by $\mathbf{G}(s)/s = \sum_0^\infty G_i/s^{i+1}$. It follows that the McMillan degree of $\mathbf{G}/s$ is the rank of the Hankel matrix

$$H = \begin{bmatrix} G_0 & G_1 & G_2 & \cdots \\ G_1 & G_2 & G_3 & \cdots \\ G_2 & G_3 & G_4 & \cdots \\ \vdots & \vdots & \vdots & \end{bmatrix} \,.$$

Since $\{A, B, C, D\}$ is a realization of $\mathbf{G}$, we have $G_0 = D$ and $G_i = CA^{i-1}B$ for all integers $i \geq 1$. Hence,

$$H = \begin{bmatrix} D & CB & CAB & \cdots \\ CB & CAB & CA^2B & \cdots \\ CAB & CA^2B & CA^3B & \cdots \\ \vdots & \vdots & \vdots & \end{bmatrix} = \begin{bmatrix} D & CW_c \\ W_oB & W_oAW_c \end{bmatrix} \,,$$

where W_o and W_c are the observability and controllability operators defined by

$$W_o = \begin{bmatrix} C \\ CA \\ CA^2 \\ \vdots \end{bmatrix} \quad \text{and} \quad W_c = \begin{bmatrix} B & AB & A^2B & \cdots \end{bmatrix} \,,$$

respectively. Therefore, H admits a factorization of the form

$$H = \begin{bmatrix} I & 0 \\ 0 & W_o \end{bmatrix} \begin{bmatrix} D & C \\ B & A \end{bmatrix} \begin{bmatrix} I & 0 \\ 0 & W_c \end{bmatrix} \,.$$

Since the realization $\{A, B, C, D\}$ is minimal, the pair $\{C, A\}$ is observable, and thus, the matrix corresponding to the operator $I \oplus W_o$ is one-to-one. Likewise, the pair $\{A, B\}$ is controllable and thus, the matrix corresponding to the operator $I \oplus W_c$ is onto. Therefore, the rank of H is precisely the rank of

$$\begin{bmatrix} A & B \\ C & D \end{bmatrix}.$$

This proves the theorem when λ is zero.

Consider now any complex number λ. Let $\{\tilde{A}, \tilde{B}, \tilde{C}, \tilde{D}\}$ be a minimal realization of a transfer function $\mathbf{F}$. Then, $\{\tilde{A} - \lambda I, \tilde{B}, \tilde{C}, \tilde{D}\}$ is realization of the transfer function $\mathbf{F}(s+\lambda)$. A simple application of the PBH tests for controllability and observability shows that this realization is minimal. Therefore $\mathbf{F}$ and its translation $\mathbf{F}(s + \lambda)$ have the same McMillan degree, that is, mdeg $\mathbf{F}$ = mdeg $\mathbf{F}(s + \lambda)$. Now notice that $\mathbf{G}(s + \lambda)/s$ is a translation of $\mathbf{G}(s)/(s - \lambda)$; thus, these two transfer functions have the same McMillan degree. Also, since $\{A, B, C, D\}$ is minimal realization of $\mathbf{G}$, the realization $\{A - \lambda I, B, C, D\}$ is minimal for $\mathbf{G}(s + \lambda)$. By our previous analysis, the McMillan degree of $\mathbf{G}(s + \lambda)/s$ is the rank of T_λ; hence the McMillan degree of $\mathbf{G}(s)/(s - \lambda)$ is the rank of T_λ. ∎

If $\{A$ on $\mathcal{X}, B, C, D\}$ is minimal realization of a rational transfer function $\mathbf{G}$, then $\dim \mathcal{X} = $ mdeg $\mathbf{G}$. It now follows from (11.8) that

$$\text{nrank}\, T = \text{mdeg}\, \mathbf{G} + \text{nrank}\mathbf{G}. \tag{11.10}$$

By employing Theorem 11.2.2, we now see that inequality (11.3) holds for all λ. Also, a complex number λ is a zero of $\mathbf{G}$ if and only if rank $T_\lambda <$ nrank T. This yields the following result.

Corollary 11.2.3 *Let T_λ be the system matrix in (11.7) associated with a minimal realization $\{A, B, C, D\}$ of a transfer function $\mathbf{G}$. Then, a complex number λ is a zero of $\mathbf{G}$ if and only if*

$$rank\, T_\lambda < nrank\, T.$$

In particular, every transfer function has at most a finite number of zeros.

Since the normal rank of a nonzero scalar transfer function is one, the previous corollary readily yields the following result.

Corollary 11.2.4 *Let T_λ be the system matrix in (11.7) associated with a minimal realization $\{A$ on $\mathcal{X}, B, C, D\}$ of a nonzero scalar transfer function $\mathbf{G}$. Then, a complex number λ is a zero of $\mathbf{G}$ if and only if rank $T_\lambda \le \dim \mathcal{X}$.*

Corollary 11.2.5 *Let T_λ be the system matrix in (11.7) corresponding to the realization $\{A, B, C, D\}$ of a transfer function $\mathbf{G}$. If*

$$rank\, T_\lambda < nrank\, T, \tag{11.11}$$

then λ is a zero of $\mathbf{G}$ or λ is an uncontrollable eigenvalue of $\{A, B\}$ or λ is an unobservable eigenvalue of $\{C, A\}$. On the other hand, if λ is a zero of $\mathbf{G}$, then (11.11) holds.

PROOF. Let $\mathcal{X} = \mathcal{X}_{c\bar{o}} \oplus \mathcal{X}_{co} \oplus \mathcal{X}_{\bar{c}}$ be the controllable/observable decomposition of the state space $\mathcal{X}$ for the system $\{A, B, C, D\}$. Here $\mathcal{X}_{c\bar{o}}$ is the controllable/unobservable subspace while $\mathcal{X}_{co}$ is the controllable/observable subspace and $\mathcal{X}_{\bar{c}}$ is the uncontrollable subspace. With respect to this decomposition, A, B and C have the following matrix representations:

$$
A = \begin{bmatrix} A_{c\bar{o}} & * & * \\ 0 & A_{co} & * \\ 0 & 0 & A_{\bar{c}} \end{bmatrix} \quad \text{and} \quad B = \begin{bmatrix} B_{c\bar{o}} \\ B_{co} \\ 0 \end{bmatrix} \quad \text{and} \quad C = \begin{bmatrix} 0 & C_{co} & C_{\bar{c}} \end{bmatrix}.
$$

Using this decomposition, T_λ is given by

$$
T_\lambda = \begin{bmatrix} A_{c\bar{o}} - \lambda I & * & * & B_{c\bar{o}} \\ 0 & A_{co} - \lambda I & * & B_{co} \\ 0 & 0 & A_{\bar{c}} - \lambda I & 0 \\ 0 & C_{co} & C_{\bar{c}} & D \end{bmatrix}. \tag{11.12}
$$

If λ is neither an uncontrollable eigenvalue of $\{A, B\}$ nor an unobservable eigenvalue of $\{C, A\}$, then both $A_{c\bar{o}} - \lambda I$ and $A_{\bar{c}} - \lambda I$ are invertible. It now follows from the location of the zeros in the structure of T_λ that the rank of T_λ equals the dimension of $\mathcal{X}_{c\bar{o}} \oplus \mathcal{X}_{\bar{c}}$ plus the rank of the matrix

$$
\tilde{T}_\lambda = \begin{bmatrix} A_{co} - \lambda I & B_{co} \\ C_{co} & D \end{bmatrix}.
$$

Because an operator valued polynomial achieves its normal rank everywhere except at a finite number of points, the normal rank of T equals the dimension of $\mathcal{X}_{c\bar{o}} \oplus \mathcal{X}_{\bar{c}}$ plus the normal rank of $\tilde{T}$.

If the rank of T_λ is less than the normal rank of T, and λ is neither an uncontrollable or unobservable eigenvalue, then the rank of $\tilde{T}_\lambda$ is less than the normal rank of $\tilde{T}$. Since $\tilde{T}_\lambda$ is the system matrix associated with the minimal realization $\{A_{co}, B_{co}, C_{co} D\}$ of $\mathbf{G}$, it follows from the above corollary that λ is a zero of $\mathbf{G}$.

Suppose, on the other hand, that λ is a zero of $\mathbf{G}$. From the structure of T_λ it should be clear that

$$
\operatorname{rank} T_\lambda \leq \dim \mathcal{X}_{c\bar{o}} + \operatorname{rank} \begin{bmatrix} A_{co} - \lambda I & * & B_{co} \\ 0 & A_{\bar{c}} - \lambda I & 0 \\ C_{co} & C_{\bar{c}} & D \end{bmatrix}.
$$

Since the rank of the matrix on the right hand side of the above inequality is less than or equal to $\dim \mathcal{X}_{\bar{c}}$ plus the rank of $\tilde{T}_\lambda$, we obtain that

$$
\operatorname{rank} T_\lambda \leq \dim (\mathcal{X}_{c\bar{o}} \oplus \mathcal{X}_{\bar{c}}) + \operatorname{rank} \tilde{T}_\lambda.
$$

Since λ is a zero of $\mathbf{G}$ and $\tilde{T}_\lambda$ is the system matrix associated with a minimal realization of $\mathbf{G}$, it follows from the above theorem that $\operatorname{rank} \tilde{T}_\lambda < \operatorname{nrank} \tilde{T}$. Hence,

$$
\operatorname{rank} T_\lambda < \dim (\mathcal{X}_{c\bar{o}} \oplus \mathcal{X}_{\bar{c}}) + \operatorname{nrank} \tilde{T} = \operatorname{nrank} T.
$$

Therefore (11.11) holds.   ∎

To see that an uncontrollable or unobservable eigenvalue λ does not necessarily result in $\operatorname{rank} T_\lambda < \operatorname{nrank} T$, consider the system with $A = 0$ on $\mathbb{C}^2$, with $B = I$ on $\mathbb{C}^2$ while $C = [\ 0\ \ 1\ ]$ and $D = 0$ on $\mathbb{C}$. The corresponding system matrix

$$T_\lambda = \begin{bmatrix} -\lambda & 0 & 1 & 0 \\ 0 & -\lambda & 0 & 1 \\ 0 & 1 & 0 & 0 \end{bmatrix}$$

has rank three for all λ. Clearly $\lambda = 0$ is an unobservable eigenvalue while the rank of T_0 equals the normal rank of T.

Remark 11.2.1 (Zeros and static state feedback) Recall that for a scalar system, static state feedback does not create new zeros. It can eliminate a zero, however to do so it must create an unobservable eigenvalue at the same location. This is sometimes referred to as pole/zero cancellation by state feedback.

We now demonstrate that this holds in general. To see this, consider the controllable and observable system given by

$$\begin{aligned} \dot{x} &= Ax + Bu \\ y &= Cx + Du. \end{aligned} \tag{11.13}$$

We say that λ is a zero of $\{A, B, C, D\}$ if λ is a zero of its transfer function. Suppose that $u = Kx + v$ where K is a state feedback operator. Then, the resulting closed-loop system is described by

$$\begin{aligned} \dot{x} &= (A + BK)x + Bv \\ y &= (C + DK)x + Dv. \end{aligned} \tag{11.14}$$

Recall, that a simple application of a PBH test shows that the closed-loop system is controllable. Notice that for all λ the two system matrices

$$\begin{bmatrix} A - \lambda I & B \\ C & D \end{bmatrix} \quad \text{and} \quad \begin{bmatrix} A + BK - \lambda I & B \\ C + DK & D \end{bmatrix}$$

have the same rank. Hence, they have the same normal rank. If the closed loop system (11.14) is observable, then it is minimal, and by Corollary 11.2.3, the two systems (11.13) and (11.14) have the same zeros.

Every zero of the closed loop system is a zero of the open loop system. To see this, let λ be a zero of the closed loop system. By the previous corollary, the rank of the system matrix for the closed loop system is less than its normal rank. Hence, the rank of the system matrix for the open loop system is less than its normal rank. Since the open loop system is minimal, λ is a zero of the open loop system.

Every zero of the open loop system is either a zero or an unobservable eigenvalue of the closed loop system. To see this, let λ be a zero of the open loop system. By the previous corollary, the rank of the system matrix for the open loop system is less than its normal rank. Hence, the rank of the system matrix for the closed loop system is less than its normal rank. Since the closed loop system is controllable, it follows from the previous corollary that λ is either a zero or an unobservable eigenvalue of the closed loop system.

Remark 11.2.2 (A dynamical interpretation of zeroes.) Suppose that

$$\dot{x} = Ax + Bu$$
$$y = Cx + Du$$

is a minimal realization of a transfer function $\mathbf{G}$. Assume that the normal rank of $\mathbf{G}$ is the same as the dimension of the input space $\mathcal{U}$. Then λ is a zero of $\mathbf{G}$ if and only if there exists a nonzero input $u(t) = u_0 e^{\lambda t}$ with $u_0 \in \mathcal{U}$ and an initial condition $x(0) = x_0$ so that $y(t) = 0$ for all t.

To see this, first notice that the normal rank of the system matrix T corresponding to $\{A, B, C, D\}$ equals the dimension of the state space $\mathcal{X}$ plus the dimension of $\mathcal{U}$, that is, $\operatorname{nrank} T = \dim \mathcal{X} + \dim \mathcal{U}$. So, if λ is a zero of $\mathbf{G}$, then there exists a nonzero vector $x_0 \oplus u_0$ in the kernel of T_λ, that is,

$$(A - \lambda I)x_0 + Bu_0 = 0$$
$$Cx_0 + Du_0 = 0. \tag{11.15}$$

Now consider the initial condition $x(0) = x_0$ and input $u(t) = u_0 e^{\lambda t}$. It readily follows from the above two equations that $x(t) = x_0 e^{\lambda t}$ is the unique solution to $\dot{x} = Ax + Bu$ and $y(t) = 0$ for all t. Notice that u_0 must be nonzero. If u_0 is zero, then $0 = y(t) = Ce^{At}x_0$ for all t. By observability of the system, we have $x_0 = 0$. This contradicts the fact that $x_0 \oplus u_0$ is nonzero.

Suppose, on the other hand, that the output y of the system under consideration is zero for some initial condition $x(0) = x_0$ and nonzero input $u(t) = u_0 e^{\lambda t}$. Then,

$$y(t) = Cx(t) + Du_0 e^{\lambda t} = 0 \tag{11.16}$$

for all t, and in particular,

$$Cx_0 + Du_0 = 0. \tag{11.17}$$

Differentiating (11.16) and setting $Du_0 e^{\lambda t} = -Cx(t)$, yields

$$\dot{y}(t) = C(A - \lambda I)x(t) + CBu_0 e^{\lambda t} = 0.$$

By differentiating the above expression and using induction, one may show that

$$\frac{d^k y}{dt^k}(t) = CA^{k-1}(A - \lambda I)x(t) + CA^{k-1}Bu_0 e^{\lambda t} = 0$$

for all integers $k \geq 1$. In particular,

$$CA^{k-1}[(A - \lambda I)x_0 + Bu_0] = 0 \qquad (\text{for } k = 1, 2, \cdots).$$

Since $\{C, A\}$ is observable, the above implies that $(A - \lambda I)x_0 + Bu_0 = 0$. Combining this with (11.17) results in (11.15), that is, the non-zero vector $x_0 \oplus u_0$ is in the kernel of T_λ. This means the rank of T_λ is less than the normal rank of T. Hence, λ is a zero of $\mathbf{G}$.

11.3   Notes

The results in this chapter are standard; see Kailath [68] and Rugh [110]. For a Smith-McMillan interpretation of the zeros of a transfer function see Kailath [68].

Chapter 12

Linear Quadratic Regulators

This chapter uses least squares optimization techniques to solve the linear quadratic regular problem. Two different methods are presented to solve a linear quadratic tracking problem. A special outer factorization is used to develop a connection between the classical root locus and the linear quadratic regular problem for single input single output systems.

12.1 The finite horizon problem

There has been considerable research on linear quadratic optimal control. For simplicity of presentation, we will not attempt to develop the general theory. Instead we will demonstrate how operator techniques can be used to solve a simple linear quadratic regulator problem. Throughout this chapter, A is an operator on $\mathcal{X}$ and B maps $\mathcal{U}$ into $\mathcal{X}$, while C is an operator mapping $\mathcal{X}$ into $\mathcal{Y}$. The spaces $\mathcal{U}$, $\mathcal{X}$ and $\mathcal{Y}$ are all finite dimensional. In this section we will solve the following classical linear quadratic regulator problem.

For each initial state x_0 in $\mathcal{X}$, find the optimal cost $\varepsilon(x_0)$ in the optimization problem:

$$\varepsilon(x_0) = \inf\left\{ \int_{t_0}^{t_1} (\|y(\sigma)\|^2 + \|u(\sigma)\|^2)\, d\sigma : u \in L^2([t_0, t_1], \mathcal{U}) \right\}$$

subject to $\quad \dot{x} = Ax + Bu \ \text{and} \ y = Cx \ \text{and} \ x(t_0) = x_0 \,.$ $\qquad$ (12.1)

In addition, when a minimum exists, find an optimal input $\hat{u}$ which achieves this minimum, that is,

$$\varepsilon(x_0) = \int_{t_0}^{t_1} (\|\hat{y}(\sigma)\|^2 + \|\hat{u}(\sigma)\|^2)\, d\sigma \qquad (12.2)$$

where the optimal state $\hat{x}$ and output $\hat{y}$ are given by $\dot{\hat{x}} = A\hat{x} + B\hat{u}$ and $\hat{y} = C\hat{x}$ with $\hat{x}(t_0) = x_0$.

It turns out that the minimum exists for each initial state x_0. Moreover, for each initial state x_0, there exists a unique optimal input $\hat{u}$ in $L^2([t_0, t_1], \mathcal{U})$ which achieves the minimum. We demonstrate these facts and obtain the optimal cost and input by using the following *Riccati differential equation:*

$$\dot{P} = -A^*P - PA - C^*C + PBB^*P \qquad (P(t_1) = 0)\,. \qquad (12.3)$$

171

Later we shall see that $P(t)$ is a well defined self-adjoint operator on $\mathcal{X}$ for $t \leq t_1$. By completion of squares, it is easy to verify that the optimal solution to the linear quadratic regulator problem in (12.1) is given by $\hat{u}(t) = -B^*P(t)\hat{x}(t)$. To see this, notice that the Riccati differential equation for P in (12.3) along with $\dot{x} = Ax + Bu$ gives

$$
\begin{aligned}
\frac{d}{dt}(Px, x) &= (\dot{P}x, x) + (P\dot{x}, x) + (Px, \dot{x}) \\
&= (\dot{P}x, x) + (PAx, x) + (PBu, x) + (Px, Ax) + (Px, Bu) \\
&= ((\dot{P} + PA + A^*P)x, x) + (u, B^*Px) + (B^*Px, u) \\
&= ||B^*Px||^2 - ||Cx||^2 + (u, B^*Px) + (B^*Px, u) \\
&= -||y||^2 - ||u||^2 + ||B^*Px + u||^2 .
\end{aligned}
\tag{12.4}
$$

By integrating from t_0 to t_1 and using the fact that $P(t_1) = 0$ and $x_0 = x(t_0)$, we have

$$
(P(t_0)x_0, x_0) + \int_{t_0}^{t_1} ||B^*P(\sigma)x(\sigma) + u(\sigma)||^2 \, d\sigma = \int_{t_0}^{t_1} \left(||y(\sigma)||^2 + ||u(\sigma)||^2 \right) d\sigma .
\tag{12.5}
$$

This readily implies that

$$
(P(t_0)x_0, x_0) \leq \int_{t_0}^{t_1} \left(||y(\sigma)||^2 + ||u(\sigma)||^2 \right) d\sigma
$$

for every input u. Moreover, we have equality if and only if $u(t) = -B^*P(t)x(t)$. Hence, the optimal solution $\hat{u}$ to the linear quadratic regulator problem in (12.1) is given by the feedback law $\hat{u}(t) = -B^*P\hat{x}(t)$. Since $\dot{\hat{x}} = A\hat{x} + B\hat{u}$, the optimal state trajectory satisfies $\dot{\hat{x}} = (A - BB^*P)\hat{x}$ with the initial condition $\hat{x}(t_0) = x_0$. Furthermore, the optimal cost is given by $\varepsilon(x_0) = (P(t_0)x_0, x_0)$. In particular, the optimal cost is a quadratic function of the initial state. So, obtaining the optimal solution is rather easy once we have the Riccati differential equation. In the next section we will use some operator techniques to derive the Riccati differential equation, and thus, the optimal feedback law $\hat{u} = -B^*P\hat{x}$.

We now show that the Riccati differential equation in (12.3) has a unique solution P defined on the interval $[t_0, t_1]$ and this solution is positive (≥ 0). Clearly, this Riccati differential equation is locally Lipschitz in P. Hence, this differential equation has a unique solution over some interval $(t_2, t_1]$ for t_2 sufficiently close to t_1. To verify that this solution can be extended over the interval $(-\infty, t_1]$, it is sufficient to show that over any interval $(t_2, t_1]$ on which P is defined, there is a bound M such that $||P(t)|| \leq M$ for $t_2 < t \leq t_1$. Considering $x(t) = x_0$ and replacing t_0 with t in (12.5), yields

$$
(P(t)x_0, x_0) = -\int_t^{t_1} ||B^*P(\sigma)x(\sigma) + u(\sigma)||^2 \, d\sigma + \int_t^{t_1} \left(||y(\sigma)||^2 + ||u(\sigma)||^2 \right) d\sigma .
$$

By setting $u = -B^*Px$, we see that $(P(t)x_0, x_0) \geq 0$. Since this holds for all x_0 in $\mathcal{X}$, the operator $P(t)$ is positive. On the other hand, considering $u = 0$, results in

$$
(P(t)x_0, x_0) \leq \int_t^{t_1} ||y(\sigma)||^2 \, d\sigma = \int_t^{t_1} ||Ce^{A(\sigma - t)}x_0||^2 \, d\sigma \leq M||x_0||^2
$$

where M is the norm of $\int_0^{t_1-t_2} e^{A^*\sigma} C^* C e^{A\sigma}\,d\sigma$. Since $P(t)$ is positive, we have established that $||P(t)|| \leq M$ over any interval $(t_2, t_1]$. Therefore, the solution to the Riccati equation can be extended over $[t_0, t_1]$, In fact, this solution can be extended over $(-\infty, t_1]$. Summing up this analysis yields the following fundamental result in linear quadratic optimal control.

Theorem 12.1.1 *Consider the linear quadratic regulator problem in (12.1). Then the corresponding Riccati differential equation in (12.3) has a unique solution P defined on the interval $[t_0, t_1]$ and this solution is positive. For any initial state x_0 in $\mathcal{X}$, the optimal cost in (12.1) is given by*

$$\varepsilon(x_0) = (P(t_0)x_0, x_0)\,. \tag{12.6}$$

Moreover, this cost is uniquely attained by the optimal input

$$\hat{u}(t) = -B^* P(t)\hat{x}(t) \tag{12.7}$$

where the optimal state trajectory $\hat{x}$ is uniquely determined by

$$\dot{\hat{x}} = (A - BB^* P(t))\hat{x} \qquad and \qquad \hat{x}(t_0) = x_0\,. \tag{12.8}$$

Remark 12.1.1 It is emphasized that one must integrate the Riccati differential equation in (12.3) backwards in time to find $P(t)$. However, one can easily convert this equation to a Riccati differential equation moving forward in time. To see this let $\Omega(\tau) = P(t_1 - \tau)$. Then (12.3) gives

$$\dot{\Omega} = A^*\Omega + \Omega A + C^* C - \Omega BB^*\Omega \qquad (\text{with} \quad \Omega(0) = 0)\,. \tag{12.9}$$

Therefore, one can obtain P by solving forward for Ω in the Riccati differential equation (12.9). Then $P(t) = \Omega(t_1 - t)$.

Let Ω be the solution to the Riccati differential equation in (12.9). Then $\{\Omega(\tau)\}$ forms an increasing sequence of positive operators, that is, if $0 \leq \tau_1 \leq \tau_2$, then $\Omega(\tau_1) \leq \Omega(\tau_2)$ where $\Omega(0) = 0$. To see this first notice that, due to the time invariant nature of the system under consideration, it follows from (12.1) and (12.6) that

$$(P(t_0)x_0, x_0) = \inf\left\{ \int_0^{t_1-t_0} (||y(\sigma)||^2 + ||u(\sigma)||^2)\,d\sigma : u \in L^2([t_0, t_1], \mathcal{U}) \right\}$$
$$\text{subject to} \qquad \dot{x} = Ax + Bu \text{ and } y = Cx \text{ and } x(0) = x_0\,.$$

Hence,

$$(\Omega(\tau)x_0, x_0) = \inf\left\{ \int_0^\tau (||y(\sigma)||^2 + ||u(\sigma)||^2)\,d\sigma : u \in L^2([t_0, t_1], \mathcal{U}) \right\}$$
$$\text{subject to} \qquad \dot{x} = Ax + Bu \text{ and } y = Cx \text{ and } x(0) = x_0\,. \tag{12.10}$$

If $\tau_1 \leq \tau_2$, then obviously

$$\int_0^{\tau_1} (||y(\sigma)||^2 + ||u(\sigma)||^2)d\sigma \leq \int_0^{\tau_2} (||y(\sigma)||^2 + ||u(\sigma)||^2)\,d\sigma\,.$$

By taking the infimum, this readily implies that $\Omega(\tau_1) \leq \Omega(\tau_2)$. Hence, $\{\Omega(\tau)\}$ is an increasing sequence of positive operators.

Remark 12.1.2 If the pair $\{C, A\}$ is observable, then $\Omega(\tau)$ is strictly positive for $\tau > 0$. We have already seen that $\Omega(\tau)$ is positive. Now consider any $\tau > 0$ and suppose that $(\Omega(\tau)x_0, x_0) = 0$ for some initial state x_0. We have shown that

$$(\Omega(\tau)x_0, x_0) = \int_0^\tau (\|\hat{y}(\sigma)\|^2 + \|\hat{u}(\sigma)\|^2)\, d\sigma$$

where $\hat{u}$ is the optimal input and $\hat{y}$ is the optimal output for the initial state x_0. Since $(\Omega(\tau)x_0, x_0) = 0$, it follows that both $\hat{u}$ and $\hat{y}$ are zero. Thus, $\hat{y}(\sigma) = Ce^{A\sigma}x_0$ is zero for all σ in $[0, \tau]$. The observability of $\{C, A\}$ now implies that x_0 is zero. Hence, $\Omega(\tau)$ is strictly positive. In particular, if the pair $\{C, A\}$ is observable, then $P(t)$ is strictly positive for $t < t_1$.

12.1.1 Problems with control weights

In many control applications one considers linear quadratic regular problems of the form:

$$\varepsilon(x_0) = \inf\left\{\int_{t_0}^{t_1} (\|y(\sigma)\|^2 + (Ru(\sigma), u(\sigma)))\, d\sigma : u \in L^2([t_0, t_1], \mathcal{U})\right\}$$

$$\text{subject to} \qquad \dot{x} = Ax + Bu \text{ and } y = Cx \text{ and } x(t_0) = x_0. \tag{12.11}$$

Here R is a strictly positive operator on $\mathcal{U}$. As expected, the minimum exists and there exists a unique optimal input $\hat{u}$ in $L^2([t_0, t_1], \mathcal{U})$ which achieves this minimum. To obtain the optimal input $\hat{u}$, let P be the solution, over the interval $[t_0, t_1]$, to the following Riccati differential equation,

$$\dot{P} = -A^*P - PA - C^*C + PBR^{-1}B^*P \qquad (P(t_1) = 0). \tag{12.12}$$

The solution P to this Riccati equation is well defined. Furthermore, for any initial state x_0 in $\mathcal{X}$, the optimal cost in (12.11) is given by $\varepsilon(x_0) = (P(t_0)x_0, x_0)$. Moreover, this cost is uniquely attained by the optimal input

$$\hat{u}(t) = -R^{-1}B^*P(t)\hat{x}(t), \tag{12.13}$$

where the optimal state trajectory $\hat{x}$ is uniquely determined by

$$\dot{\hat{x}} = (A - BR^{-1}B^*P(t))\hat{x} \qquad (\hat{x}(t_0) = x_0). \tag{12.14}$$

To prove this, we simply convert the weighted linear quadratic regular problem in (12.11) to the linear quadratic regular problem considered in (12.1). To this end, introduce a new input $v = R^{1/2}u$, where $R^{1/2}$ is the positive square root of R. Let $\tilde{B}$ be the operator from $\mathcal{U}$ into $\mathcal{X}$ defined by $\tilde{B} = BR^{-1/2}$. Then the linear quadratic regular problem in (12.11) is equivalent to the linear quadratic regular problem in (12.1) with u and B replaced with v and $\tilde{B}$, respectively. Substituting $\tilde{B}$ into our previous Riccati differential equation (12.3), we obtain the Riccati differential equation in (12.12). In particular, this shows that the solution P to this Riccati differential equation is well defined and positive over the interval $[t_0, t_1]$. Recall that $\hat{v} = -\tilde{B}^*P\hat{x}$ is the optimal input which uniquely solves the linear quadratic regular problem in (12.1) with u and B replaced with v and $\tilde{B}$, respectively. Therefore, $\hat{u} = R^{-1/2}\hat{v} = -R^{-1}B^*P\hat{x}$ is the solution to the linear quadratic regular problem in (12.11). Finally, since both linear quadratic regular problems (12.1) and (12.11) have the same cost, it follows that $\varepsilon(x_0) = (P(t_0)x_0, x_0)$. This proves our claim.

Exercise 27 Let P_1 be a positive operator on $\mathcal{X}$. Obviously, Theorem 12.1.1 can be extended to the time varying case. Consider the following linear quadratic regular problem:

$$\varepsilon(x_0) = \inf\left\{(P_1 x(t_1), x(t_1)) + \int_{t_0}^{t_1}(\|y(\sigma)\|^2 + (R(\sigma)u(\sigma), u(\sigma))\,d\sigma : u \in L^2([t_0, t_1], \mathcal{U}\right\}$$

subject to $\quad \dot{x} = A(t)x + B(t)u$ and $y = C(t)x$ and $x(t_0) = x_0$. $\quad\quad$ (12.15)

Here A is a continuous function with values in $\mathcal{L}(\mathcal{X}, \mathcal{X})$ and B is a continuous function with values in $\mathcal{L}(\mathcal{U}, \mathcal{X})$ while C is a continuous function with values in $\mathcal{L}(\mathcal{X}, \mathcal{Y})$. Let R be a continuous function with values in $\mathcal{L}(\mathcal{U}, \mathcal{U})$ satisfying $R(t) \geq \epsilon I$ for some $\epsilon > 0$. The Riccati differential equation associated with this linear quadratic optimization problem is given by

$$\dot{P} = -A^*P - PA - C^*C + PBR^{-1}B^*P \quad\quad (P(t_1) = P_1). \quad\quad (12.16)$$

Show that this Riccati differential equation has a unique solution over any finite interval. Moreover, show that there exists a unique optimal input $\hat{u}$ in $L^2([t_0, t_1], \mathcal{U})$ which achieves the minimum in (12.15) and $\varepsilon(x_0) = (P(t_0)x_0, x_0)$. Finally, show that $\hat{u} = -R^{-1}B^*P\hat{x}$ where the optimal state trajectory $\hat{x}$ satisfies $\dot{\hat{x}} = (A - BR^{-1}B^*P)\hat{x}$ with $\hat{x}(t_0) = x_0$.

12.2 An operator approach

In this section, we use operator techniques to gain further insight into the linear quadratic regulator problem and the role of the Riccati equation in its solution.

12.2.1 An operator based solution

To solve the linear quadratic regulator problem via operator methods, let $\mathcal{H}$ be the Hilbert space defined by $\mathcal{H} = L^2([t_0, t_1], \mathcal{U} \oplus \mathcal{Y})$, that is, $[f\ g]^{tr}$ is in $\mathcal{H}$ if and only if f is in $L^2([t_0, t_1], \mathcal{U})$ and g is in $L^2([t_0, t_1], \mathcal{Y})$. (Recall that tr denotes transpose.) The inner product on $\mathcal{H}$ is defined by

$$([f_1\ g_1]^{tr}, [f_2\ g_2]^{tr}) = \int_{t_0}^{t_1}((f_1(\sigma), f_2(\sigma))_\mathcal{U} + (g_1(\sigma), g_2(\sigma))_\mathcal{Y})\,d\sigma.$$

Using this inner product, the linear quadratic regular problem in (12.1) is equivalent to

$$\varepsilon(x_0) = \inf\left\{\left\|\begin{bmatrix} u \\ y \end{bmatrix}\right\|_\mathcal{H}^2 : \dot{x} = Ax + Bu \text{ and } y = Cx \text{ with } x(t_0) = x_0\right\}. \quad\quad (12.17)$$

To solve this problem, let F be the input output operator mapping $L^2([t_0, t_1], \mathcal{U})$ into $L^2([t_0, t_1], \mathcal{Y})$ defined by

$$(Fu)(t) = \int_{t_0}^t Ce^{A(t-\tau)}Bu(\tau)\,d\tau \quad\quad (u \in L^2([t_0,\ t_1], \mathcal{U})). \quad\quad (12.18)$$

Let C_o be the observability operator from $\mathcal{X}$ into $L^2([t_0, t_1], \mathcal{Y})$ defined by

$$(C_o x)(t) = Ce^{A(t-t_0)}x \qquad (x \in \mathcal{X}). \tag{12.19}$$

Clearly, $y = C_o x_0 + Fu$. This readily implies that

$$\left\| \begin{bmatrix} u \\ y \end{bmatrix} \right\|_{\mathcal{H}} = \left\| \begin{bmatrix} 0 \\ C_o x_0 \end{bmatrix} + \begin{bmatrix} u \\ Fu \end{bmatrix} \right\|_{\mathcal{H}}.$$

Now let T be the linear operator from $L^2([t_0, t_1], \mathcal{U})$ into $\mathcal{H}$ defined by

$$Tu = -\begin{bmatrix} u \\ Fu \end{bmatrix} \qquad (u \in L^2([t_0, t_1], \mathcal{U})). \tag{12.20}$$

Then the linear quadratic regulator problem in (12.1) is equivalent to the following least squares optimization problem:

$$\varepsilon(x_0) = \inf \left\{ \left\| \begin{bmatrix} 0 \\ C_o x_0 \end{bmatrix} - Tu \right\|_{\mathcal{H}}^2 \; : \; u \in L^2([t_0, t_1], \mathcal{U}) \right\}. \tag{12.21}$$

Notice that $T^* = -[I \; F^*]$, and thus, $T^*T = I + F^*F$. Since $T^*T = I + F^*F \geq I$, it follows that T^*T is invertible. So, according to the solution of the least squares optimization problem in Theorem 16.2.4 in the Appendix, the optimal input $\hat{u}$ solving the linear quadratic regular problem in (12.1) or (12.21) is given by

$$\hat{u} = (T^*T)^{-1}T^* \begin{bmatrix} 0 \\ C_o x_0 \end{bmatrix}.$$

Thus, the optimal input $\hat{u}$ solving the quadratic regular problem is unique and is given by

$$\hat{u} = -(I + F^*F)^{-1}F^*C_o x_0. \tag{12.22}$$

The corresponding optimal output trajectory $\hat{y}$ is given by $\hat{y} = C_o x_0 + F\hat{u}$. Since $(I + F^*F)\hat{u} = -F^*C_o x_0$, it follows that

$$\hat{u} = -F^*(C_o x_0 + F\hat{u}) = -F^*\hat{y}.$$

We now show that for any x_0 in $\mathcal{X}$, the optimal cost is given by

$$\varepsilon(x_0) = (C_o^*(I + FF^*)^{-1}C_o x_0, \; x_0). \tag{12.23}$$

In particular, the optimal cost is a quadratic function of the initial state. Since $\hat{y} = C_o x_0 + F\hat{u}$ and $\hat{u} = -F^*\hat{y}$, it follows that $\hat{y} = (I + FF^*)^{-1}C_o x_0$. Hence,

$$\begin{aligned} \varepsilon(x_0) &= \|\hat{y}\|^2 + \|\hat{u}\|^2 = \|\hat{y}\|^2 + \|F^*\hat{y}\|^2 \\ &= ((I + FF^*)\hat{y}, \hat{y}) = (C_o x_0, \; (I + FF^*)^{-1}C_o x_0) \\ &= (C_o^*(I + FF^*)^{-1}C_o x_0, \; x_0). \end{aligned}$$

Therefore, (12.23) holds. Summarizing the above results, we obtain the following operator based solution to the linear quadratic problem.

Theorem 12.2.1 *Consider the linear quadratic regulator problem in (12.1). Let F be the input output operator defined in (12.18) and C_o the observability operator in (12.19). Then for any initial state x_0 in $\mathcal{X}$, the optimal cost in (12.1) is given by*

$$\varepsilon(x_0) = (C_o^*(I + FF^*)^{-1}C_o x_0, \; x_0). \tag{12.24}$$

This cost is uniquely attained by the optimal input

$$\hat{u} = -(I + F^*F)^{-1}F^*C_o x_0 . \tag{12.25}$$

Moreover, if $\hat{y}$ is the optimal output trajectory associated with $\hat{u}$, that is, $\hat{y} = C_o x_0 + F\hat{u}$, then

$$\hat{u} = -F^*\hat{y} . \tag{12.26}$$

Finally, it is noted that the elementary operator equation $\hat{u} = -F^*\hat{y}$ plays a fundamental role in our approach to solving the linear quadratic regular problem.

12.2.2 The adjoint system

Here we use the above operator based results to obtain another characterization of the solution to the linear quadratic problem in (12.1). From this characterization, we will naturally arrive at the Riccati equation. This characterization utilizes the adjoint system associated with a linear system $\{A, B, C, 0\}$. First we need an explicit expression for the adjoint of an input output operator defined by (12.18). To achieve this, consider any operator Γ from $L^2([t_0, t_1], \mathcal{U})$ into $L^2([t_0, t_1], \mathcal{Y})$ defined by

$$(\Gamma h)(t) = \int_{t_0}^{t} G(t - \tau)h(\tau)d\tau \qquad (h \in L^2([t_0, t_1], \mathcal{U})) \tag{12.27}$$

where G is a continuous function with values in $\mathcal{L}(\mathcal{U}, \mathcal{Y})$. We claim that the adjoint Γ^* of Γ is the linear operator from $L^2([t_0, t_1], \mathcal{Y})$ into $L^2([t_0, t_1], \mathcal{U})$ defined by

$$(\Gamma^* g)(t) = \int_{t}^{t_1} G^*(\tau - t)g(\tau) \, d\tau \qquad (g \in L^2([t_0, t_1], \mathcal{Y})). \tag{12.28}$$

To prove this, notice that for g and h in the appropriate L^2 spaces, we have

$$
\begin{aligned}
(\Gamma h, g) &= \int_{t_0}^{t_1} \left(\int_{t_0}^{t} G(t - \tau)h(\tau) \, d\tau, \; g(t) \right) dt = \int_{t_0}^{t_1} \int_{t_0}^{t} (G(t - \tau)h(\tau), \; g(t)) \, d\tau dt \\
&= \int_{t_0}^{t_1} \int_{t_0}^{t} (h(\tau), \; G^*(t - \tau)g(t)) \, d\tau dt = \int_{t_0}^{t_1} \int_{\tau}^{t_1} (h(\tau), \; G^*(t - \tau)g(t)) \, dt d\tau \\
&= \int_{t_0}^{t_1} \left(h(\tau), \; \int_{\tau}^{t_1} G^*(t - \tau)g(t) \, dt \right) d\tau = (h, \; \Gamma^* g).
\end{aligned}
$$

Therefore, the adjoint Γ^* of Γ is given by (12.28).

As before, let F be the input output operator given by (12.18). Then with $G(t) = Ce^{At}B$, the adjoint F^* of F is given by

$$(F^*g)(t) = \int_t^{t_1} B^*e^{A^*(\tau-t)}C^*g(\tau)d\tau \,. \tag{12.29}$$

To obtain a state space realization of the adjoint map F^*, let λ be the function defined by

$$\lambda(t) = \int_t^{t_1} e^{A^*(\tau-t)}C^*g(\tau)\,d\tau \,. \tag{12.30}$$

Recall Leibnitz's rule for any differentiable function ζ

$$\frac{d}{dt}\int_{\alpha(t)}^{\beta(t)} \zeta(t,\tau)\,d\tau = \zeta(t,\beta(t))\dot{\beta} - \zeta(t,\alpha(t))\dot{\alpha} + \int_{\alpha(t)}^{\beta(t)} \frac{\partial}{\partial t}\zeta(t,\tau)\,d\tau \,. \tag{12.31}$$

Using Leibnitz's rule in (12.30), we obtain $\dot{\lambda} = -A^*\lambda - C^*g$ which yields the following state space representation of $h = F^*g$

$$\begin{aligned} \dot{\lambda} &= -A^*\lambda - C^*g \qquad \text{with} \qquad \lambda(t_1) = 0 \\ h &= B^*\lambda \,. \end{aligned} \tag{12.32}$$

To obtain h for a specified g, one must integrate the above differential equation backward in time from t_1. To see that (12.32) is a realization of F^*, notice that

$$(F^*g)(t) = -\int_{t_1}^t B^*e^{-A^*(t-\tau)}C^*g(\tau)d\tau \,. $$

From this we see that $h = F^*g$ can be viewed as a linear input output system (running backward in time) with impulse response given by $-B^*e^{-A^*t}C^*$. This impulse response has state space realization $\{-A^*, -C^*, B^*, 0\}$. So, by a slight abuse of notation, we call (12.32) a state space representation of F^*. Motivated by this, we call λ the *adjoint state*.

Now let us use the adjoint system to obtain a solution to the linear quadratic regulator problem in (12.1). Recall from Theorem 12.2.1 that this problem has a unique optimal input $\hat{u}$ and $\hat{u} = -F^*\hat{y}$ where $\hat{y}$ is the corresponding optimal output trajectory, that is, $\hat{y}$ is given by

$$\begin{aligned} \dot{\hat{x}} &= A\hat{x} + B\hat{u} \qquad \text{with} \quad \hat{x}(t_0) = x_0 \\ \hat{y} &= C\hat{x} \end{aligned} \tag{12.33}$$

and $\hat{x}$ is the optimal state trajectory. Using the above state space realization of F^*, we now see that $\hat{u}$ is given by

$$\begin{aligned} \dot{\lambda} &= -A^*\lambda - C^*\hat{y} \qquad \text{with} \qquad \lambda(t_1) = 0 \\ \hat{u} &= -B^*\lambda \,. \end{aligned}$$

Combining these equations with (12.33), the optimal input is given by $\hat{u} = -B^*\lambda$ where $[\hat{x} \ \lambda]^{tr}$ solves the following two point boundary problem

$$
\begin{aligned}
\dot{\hat{x}} &= A\hat{x} - BB^*\lambda &&\text{with } \hat{x}(t_0) = x_0 \\
\dot{\lambda} &= -C^*C\hat{x} - A^*\lambda &&\text{with } \lambda(t_1) = 0.
\end{aligned}
\tag{12.34}
$$

Since we have shown that the optimization problem (12.1) must have a solution, it follows that the two point value problem in (12.34) must have a solution. Moreover, if $[\hat{x} \ \lambda]^{tr}$ is any solution to (12.34), then $B^*\lambda = F^*C\hat{x} = F^*(F(-B^*\lambda) + C_o x_0)$ where C_o is the observability defined in (12.19). Hence, $B^*\lambda = (I + F^*F)^{-1}F^*C_o x_0$, that is, $-B^*\lambda$ is the optimal input. We now claim that (12.34) has only one solution. To see this, we first note that, for any x_0, the two point boundary problem (12.34) has a unique solution if and only if $[\hat{x} \ \lambda]^{tr} = [0 \ 0]$ is the only solution of (12.34) when $x_0 = 0$. With $x_0 = 0$, one can readily see from the problem statement (12.1) that the optimal input is $\hat{u} = 0$. Hence $-B^*\lambda = 0$, from which it follows that the only solution to (12.34) with $x_0 = 0$ is the zero solution.

We now claim that, if $[\hat{x} \ \lambda]^{tr}$ is the solution to (12.34), then the optimal cost in (12.1) is given by $\varepsilon(x_0) = (\lambda(t_0), x_0)$. To this end, we note that

$$
\begin{aligned}
\frac{d(\lambda, \hat{x})}{dt} &= (\dot{\lambda}, \hat{x}) + (\lambda, \dot{\hat{x}}) = (-C^*C\hat{x} - A^*\lambda, \ \hat{x}) + (\lambda, \ A\hat{x} - BB^*\lambda) \\
&= -\|B^*\lambda\|^2 - \|C^*\hat{x}\|^2 = -(\|\hat{u}\|^2 + \|\hat{y}\|^2).
\end{aligned}
$$

Integrating from t_0 to t_1 and using the boundary conditions in (12.34), we obtain that

$$
(\lambda(t_0), x_0) = \int_{t_0}^{t_1} (\|\hat{u}(t)\|^2 + \|\hat{y}(t)\|^2) \, dt.
$$

Hence, $\varepsilon(x_0) = (\lambda(t_0), x_0)$. We have just demonstrated the following result.

Theorem 12.2.2 *Consider the linear quadratic regulator problem in (12.1). Then the corresponding two point boundary value problem in (12.34) has a unique solution for $[\hat{x} \ \lambda]^{tr}$ on the interval $[t_0, t_1]$. The optimal cost in (12.1) is given by*

$$
\varepsilon(x_0) = (\lambda(t_0), x_0)
\tag{12.35}
$$

and this cost is uniquely attained by the optimal input

$$
\hat{u}(t) = -B^*\lambda(t).
\tag{12.36}
$$

12.2.3 The Riccati equation

Here we use the results of the previous section to arrive at the Riccati equation. More explicitly, we show that $\lambda = P\hat{x}$ where P is the solution of the Riccati differential equation in (12.3). To this end, we introduce the so called *Hamiltonian matrix* associated with the linear quadratic regulator problem in (12.1):

$$
H = \begin{bmatrix} A & -BB^* \\ -C^*C & -A^* \end{bmatrix}.
\tag{12.37}
$$

The Hamiltonian matrix is simply the state space matrix for the two point boundary value problem in (12.34).

We have seen that for each $t_0 < t_1$ and x_0 in $\mathcal{X}$ the two point boundary value problem in (12.34) has a solution for $[\hat{x} \ \lambda]^{tr}$ on the interval $[t_0, t_1]$. Hence, using Lemma 13.6.1 in the next chapter, it follows that the Riccati differential equation in (12.3) has a solution P for all $t \leq t_1$. Moreover, it follows from Remark 13.6.1 that

$$\lambda(t) = P(t)\hat{x}(t) \quad \text{and} \quad P(t) = \Phi_{21}(t - t_1)\Phi_{11}(t - t_1)^{-1} \tag{12.38}$$

where Φ_{11} and Φ_{21} are obtained from the following matrix partition of e^{Ht}:

$$e^{Ht} = \begin{bmatrix} \Phi_{11}(t) & \Phi_{12}(t) \\ \Phi_{21}(t) & \Phi_{22}(t) \end{bmatrix} \text{ on } \begin{bmatrix} \mathcal{X} \\ \mathcal{X} \end{bmatrix}. \tag{12.39}$$

So, one can obtain P by using either $P(t) = \Phi_{21}(t - t_1)\Phi_{11}(t - t_1)^{-1}$ or solving the Riccati differential equation in (12.3).

Equation (12.38) along with $\hat{u} = -B^*\lambda$, shows that the optimal input $\hat{u} = -B^*P\hat{x}$ where $\hat{x}$ is the optimal state trajectory. This yields another proof of Theorem 12.1.1.

Remark 12.2.1 Recall that $\Omega(t_1 - t) = P(t)$. By combining Theorems 12.1.1 and 12.2.1, we readily arrive at the following operator formula for $P(t_0)$

$$\Omega(t_1 - t_0) = P(t_0) = C_o^*(I + FF^*)^{-1}C_o. \tag{12.40}$$

Recall that the pair $\{C, A\}$ is observable if and only if the operator C_o is one to one. Hence, equation (12.40) shows that the pair $\{C, A\}$ is observable if and only if $P(t)$ is strictly positive for any $t < t_1$, or equivalently, $\Omega(t)$ is strictly positive for any $t > 0$. In particular, $P(t)$ is strictly positive for any $t < t_1$ if and only if $P(t)$ is strictly positive for all $t < t_1$. Likewise $\Omega(t)$ is strictly positive for any $t > 0$ if and only if $\Omega(t)$ is strictly positive for all $t > 0$.

12.3   An operator quadratic regulator problem

In this section, we present an operator version of the linear quadratic regulator problem. To this end, let F be an operator from $\mathcal{F}$ into $\mathcal{G}$ and g a vector in $\mathcal{G}$. Then the following optimization problem is an operator generalization of the linear quadratic regulator problem:

$$\rho(g) = \inf\{\|g + Fu\|^2 + \|u\|^2 : u \in \mathcal{F}\}. \tag{12.41}$$

If $g = C_0 x_0$ and F is the operator from $L^2([t_0, t_1], \mathcal{U})$ into $L^2([t_0, t_1], \mathcal{Y})$ defined in (12.18), then this optimization problem reduces to the linear quadratic regulator problem in (12.1) with $\rho(g) = \varepsilon(x_0)$.

Lemma 12.3.1 *Let F be an operator from $\mathcal{F}$ into $\mathcal{G}$ and g a vector in $\mathcal{G}$. Then*

$$((I + FF^*)^{-1}g, g) = \inf\{\|g + Fu\|^2 + \|u\|^2 : u \in \mathcal{F}\}. \tag{12.42}$$

Moreover, the optimal $\hat{u}$ in $\mathcal{F}$ solving this minimization problem is unique and given by

$$\hat{u} = -(I + F^*F)^{-1}F^*g = -F^*(I + FF^*)^{-1}g. \tag{12.43}$$

PROOF. To obtain an optimal $\hat{u}$ we simply convert (12.41) to a least squares optimization problem. To this end, let h be the vector in $\mathcal{F} \oplus \mathcal{G}$ and T the operator from $\mathcal{F}$ into $\mathcal{F} \oplus \mathcal{G}$ defined

$$h = \begin{bmatrix} 0 \\ g \end{bmatrix} \quad \text{and} \quad T = \begin{bmatrix} -I \\ -F \end{bmatrix},$$

respectively. Then the optimization problem in (12.41) is equivalent to the following least squares optimization problem:

$$\rho(g) = \inf \{ \| h - Tu \| : u \in \mathcal{F} \}.$$

Because $T^*T = I + F^*F$ is invertible, the solution $\hat{u}$ to this least squares problem is unique and is given by (see Theorem 16.2.4)

$$\hat{u} = (T^*T)^{-1}T^*h = -(I + F^*F)^{-1}F^*g.$$

Hence, the first equality in (12.43) holds. The second equality in (12.43) follows from the identity $(I + MN)^{-1}M = M(I + NM)^{-1}$ where M and N are operators acting between the appropriate spaces.

The Projection Theorem yields

$$\rho(g) = \|h\|^2 - \|P_{\mathcal{R}}h\|^2,$$

where $P_{\mathcal{R}} = T(T^*T)^{-1}T^*$ is the orthogonal projection onto the range $\mathcal{R}$ of T; see the Appendix for a review of the Projection Theorem. Since $\|h\|^2 = \|g\|^2$ and

$$\begin{aligned} \|P_{\mathcal{R}}h\|^2 &= (P_{\mathcal{R}}h, h) = ((T^*T)^{-1}T^*h, T^*h)) \\ &= (F(I + F^*F)^{-1}F^*g, g) \\ &= ((I + FF^*)^{-1}FF^*g, g), \end{aligned}$$

we obtain that

$$\rho(g) = \|g\|^2 - ((I + FF^*)^{-1}FF^*g, g).$$

Using the identity $I - (I + FF^*)^{-1}FF^* = (I + FF^*)^{-1}$ in the last equation, we arrive at $\rho(g) = ((I + FF^*)^{-1}g, g)$ which completes the proof. ∎

Remark 12.3.1 As before, let F be an operator mapping $\mathcal{F}$ into $\mathcal{G}$. Let $y = g + Fu$ where g is a specified vector in $\mathcal{G}$. Then the abstract linear quadratic regulator problem in (12.41) is equivalent to

$$\rho(g) = \inf\{\|y\|^2 + \|u\|^2 : u \in \mathcal{F}\}. \tag{12.44}$$

Hence, $\rho(g) = ((I + FF^*)^{-1}g, g)$. The optimal output $\hat{y} = g + F\hat{u}$ where $\hat{u} = -(I + F^*F)^{-1}F^*g$ is the optimal input. This readily implies that $(I + F^*F)\hat{u} = -F^*g$, or equivalently, $\hat{u} = -F^*(g + F\hat{u})$. Therefore, the optimal input $\hat{u}$ is given by $\hat{u} = -F^*\hat{y}$ where $\hat{y}$ is the optimal output.

12.4   A linear quadratic tracking problem

In this section, we solve a linear quadratic tracking problem. To this end, let r be any function in $L^2([t_0, t_1], \mathcal{Y})$. For each initial state x_0 in $\mathcal{X}$, find the optimal input $\hat{u}$ which solves the following linear quadratic tracking problem:

$$\varepsilon(x_0) = \inf \left\{ \int_{t_0}^{t_1} (\|y(\sigma) - r(\sigma)\|^2 + \|u(\sigma)\|^2) \, d\sigma : u \in L^2([t_0, t_1], \mathcal{U}) \right\}$$

subject to $\dot{x} = Ax + Bu$ and $y = Cx$ and $x(t_0) = x_0$.   (12.45)

The following result uses the Riccati equation to compute an optimal input which solves this tracking problem.

Theorem 12.4.1 *Consider the linear quadratic tracking problem in (12.45). Then the minimum exists, and there is a unique optimal input $\hat{u}$ in $L^2([t_0, t_1], \mathcal{U})$ which achieves this minimum. To compute $\hat{u}$, let P be the unique solution on the interval $[t_0, t_1]$ to the Riccati differential equation in (12.3). Let φ be the solution to the following differential equation moving backwards in time:*

$$\dot{\varphi} = -(A^* - P(t)BB^*)\varphi + C^*r(t) \qquad (\varphi(t_1) = 0). \tag{12.46}$$

Then the optimal input $\hat{u}$ is given by

$$\hat{u}(t) = -B^*P(t)\hat{x}(t) - B^*\varphi(t) \tag{12.47}$$

where the optimal state trajectory $\hat{x}$ is uniquely determined by

$$\dot{\hat{x}} = (A - BB^*P(t))\hat{x} - BB^*\varphi(t) \qquad with \qquad \hat{x}(t_0) = x_0. \tag{12.48}$$

PROOF. The proof is a minor modification of the operator proof of Theorem 12.1.1. As before, let F be the input output operator from $L^2([t_0, t_1], \mathcal{U})$ into $L^2([t_0, t_1], \mathcal{Y})$ defined in (12.18), and C_o the observability operator from $\mathcal{X}$ into $L^2([t_0, t_1], \mathcal{Y})$ defined in (12.19). Clearly, $y = C_o x_0 + Fu$. So, the optimization problem in (12.45) is equivalent to the following optimization problem

$$\varepsilon(x_0) = \inf\{\|C_o x_0 - r + Fu\|^2 + \|u\|^2 : u \in L^2([t_0, t_1], \mathcal{U})\}.$$

By consulting Lemma 12.3.1 with $g = C_o x_0 - r$, we see that the unique optimal input $\hat{u}$ solving the quadratic tracking problem is given by

$$\hat{u} = -(I + F^*F)^{-1}F^*(C_o x_0 - r). \tag{12.49}$$

Notice that this is precisely the solution to the linear quadratic regulator problem with $C_o x_o$ replaced by $C_o x_0 - r$. The corresponding optimal output is $\hat{y} = C_o x_0 + F\hat{u}$. In other words, $\hat{y}$ is given by

$$\begin{aligned} \dot{\hat{x}} &= A\hat{x} + B\hat{u} \qquad with \quad \hat{x}(t_0) = x_0 \\ \hat{y} &= C\hat{x} \end{aligned} \tag{12.50}$$

and $\hat{x}$ is the optimal state trajectory. Since $(I + F^*F)\hat{u} = -F^*(C_o x_0 - r)$, it follows that

$$\hat{u} = -F^*(C_o x_0 + F\hat{u} - r) = -F^*(\hat{y} - r).\tag{12.51}$$

Recall that the adjoint F^* of F is given by (12.29). Recall also the following state space representation of $h = F^*g$

$$\begin{aligned} \dot{\lambda} &= -A^*\lambda - C^*g & \text{with} \quad \lambda(t_1) = 0 \\ h &= B^*\lambda. \end{aligned}$$

Using $\hat{u} = -F^*(\hat{y} - r)$ in the above realization with $g = \hat{y} - r$ and $\hat{u} = -h$, we see that the optimal input $\hat{u}$ is given by

$$\begin{aligned} \dot{\lambda} &= -A^*\lambda - C^*C\hat{x} + C^*r & \text{with} \quad \lambda(t_1) = 0 \\ \hat{u} &= -B^*\lambda. \end{aligned}\tag{12.52}$$

Combining this with the state equation in (12.50) readily yields the following two point boundary value problem:

$$\begin{bmatrix} \dot{\hat{x}} \\ \dot{\lambda} \end{bmatrix} = \begin{bmatrix} A & -BB^* \\ -C^*C & -A^* \end{bmatrix} \begin{bmatrix} \hat{x} \\ \lambda \end{bmatrix} + \begin{bmatrix} 0 \\ C^*r \end{bmatrix}\tag{12.53}$$

where $\hat{x}(t_0) = x_0$ and $\lambda(t_1) = 0$.

As before, let H be Hamiltonian matrix given in (12.37). Then,

$$\begin{bmatrix} \hat{x}(t) \\ \lambda(t) \end{bmatrix} = e^{H(t-t_1)} \begin{bmatrix} \hat{x}(t_1) \\ 0 \end{bmatrix} + \int_{t_1}^t e^{H(t-\tau)} \begin{bmatrix} 0 \\ C^*r(\tau) \end{bmatrix} d\tau.\tag{12.54}$$

Using the matrix partition in (12.39) of e^{Ht} on $\mathcal{X} \oplus \mathcal{X}$, yields

$$\begin{aligned} \hat{x}(t) &= \Phi_{11}(t - t_1)\hat{x}(t_1) + \int_{t_1}^t \Phi_{12}(t - \tau)C^*r(\tau)\, d\tau \\ \lambda(t) &= \Phi_{21}(t - t_1)\hat{x}(t_1) + \int_{t_1}^t \Phi_{22}(t - \tau)C^*r(\tau)\, d\tau. \end{aligned}$$

Recall that $\Phi_{11}(t - t_1)$ is invertible and $P(t) = \Phi_{21}(t - t_1)\Phi_{11}(t - t_1)^{-1}$ satisfies the Riccati differential equation in (12.3). Eliminating $\hat{x}(t_1)$ in the previous equations, shows that $\lambda(t) = P(t)\hat{x}(t) + \varphi(t)$ where

$$\varphi(t) = \int_{t_1}^t \Theta(t, \tau)C^*r(\tau)\, d\tau \quad \text{with} \quad \Theta(t, \tau) = \Phi_{22}(t-\tau) - P(t)\Phi_{12}(t-\tau).\tag{12.55}$$

Recall that the optimal input $\hat{u} = -B^*\lambda$. So, to complete the proof it remains to show that φ is the solution to the initial value problem specified in (12.46). Clearly, $\varphi(t_1) = 0$. Notice that

$$\begin{bmatrix} \dot{\Phi}_{11} & \dot{\Phi}_{12} \\ \dot{\Phi}_{21} & \dot{\Phi}_{22} \end{bmatrix} = \begin{bmatrix} A & -BB^* \\ -C^*C & -A^* \end{bmatrix} \begin{bmatrix} \Phi_{11} & \Phi_{12} \\ \Phi_{21} & \Phi_{22} \end{bmatrix}.$$

By employing the Riccati equation in (12.3), we have

$$
\begin{aligned}
\frac{\partial \Theta}{\partial t}(t, \tau) &= \dot{\Phi}_{22}(t-\tau) - \dot{P}(t)\Phi_{12}(t-\tau) - P(t)\dot{\Phi}_{12}(t-\tau) \\
&= -C^*C\Phi_{12} - A^*\Phi_{22} - \dot{P}\Phi_{12} - PA\Phi_{12} + PBB^*\Phi_{22} \\
&= -(A^* - PBB^*)\Phi_{22} + A^*P\Phi_{12} - PBB^*P\Phi_{12} \\
&= -(A^* - PBB^*)\Theta(t, \tau) .
\end{aligned}
$$

Using this with Leibnitz's rule in (12.55), we arrive at the differential equation for φ in (12.46). Finally, $\lambda = P\hat{x} + \varphi$ and $\hat{u} = -B^*\lambda$ yields $\hat{u} = -B^*P\hat{x} - B^*\varphi$. ∎

12.5 A spectral factorization

In this section we will follow some of the ideas in Porter [101] and use the Riccati differential equation in (12.3) to compute a special factorization for $I + F^*F$ where F is the causal operator defined in (12.18). Then we will use these results to solve a general tracking problem. To establish some terminology, let Ξ be any invertible positive operator on $L^2([t_0, t_1], \mathcal{U})$. Then we say that Θ is a *spectral factor* of Ξ, if $\Xi = \Theta^*\Theta$ where Θ a causal operator on $L^2([t_0, t_1], \mathcal{U})$. Because Ξ is invertible and $\Xi = \Theta^*\Theta$, it follows that Θ is bounded below, that is, $\|\Theta f\| \geq \delta\|f\|$ for all f in $L^2([t_0, t_1], \mathcal{U})$ and some scalar $\delta > 0$. In particular, Θ is one to one. We say that Θ is a *finite time outer spectral factor* of Ξ, if Θ is an invertible spectral factor of Ξ and its inverse Θ^{-1} is causal.

Let L be the linear operator on $L^2([t_0, t_1], \mathcal{X})$ defined by

$$
(Lf)(t) = \int_{t_0}^{t} e^{A(t-\tau)} f(\tau)\, d\tau \qquad (f \in L^2([t_0, t_1], \mathcal{X})). \tag{12.56}
$$

Notice that $g = Lf$ for some f in $L^2([t_0, t_1], \mathcal{X})$ if and only if g is the unique solution to the following differential equation

$$
\dot{g} = Ag + f \qquad \text{with} \qquad g(t_0) = 0. \tag{12.57}
$$

It should also be clear that the operator $F = CLB$.

Lemma 12.5.1 *Let F be the causal operator from $L^2([t_0, t_1], \mathcal{U})$ into $L^2([t_0, t_1], \mathcal{Y})$ defined in (12.18), and P the solution to the Riccati differential equation in (12.3). Then*

$$
\Theta = I + B^*PLB \tag{12.58}
$$

is a finite time outer spectral factor for $I + F^*F$.

PROOF. The adjoint L^* of L is the operator on $L^2([t_0, t_1], \mathcal{X})$ defined by

$$
L^*\phi = \int_{t}^{t_1} e^{-A^*(t-\tau)} \phi(\tau)\, d\tau \qquad (\phi \in L^2([t_0, t_1], \mathcal{X})). \tag{12.59}
$$

By applying Leibnitz's rule to (12.59), it follows that $\xi = L^*\phi$ for some ϕ in $L^2([t_0, t_1], \mathcal{X})$ if and only if ξ is the unique solution to the following differential equation

$$\dot{\xi} = -A^*\xi - \phi \qquad \text{with} \qquad \xi(t_1) = 0. \qquad (12.60)$$

Let g be any differentiable function in $L^2([t_0, t_1], \mathcal{X})$. Using the Riccati differential equation in (12.3), we have

$$\frac{d}{dt}(Pg) = P\dot{g} + \dot{P}g = P\dot{g} - A^*Pg - PAg - C^*Cg + PBB^*Pg.$$

In other words,

$$\frac{d}{dt}(Pg) = -A^*(Pg) - C^*Cg + PBB^*Pg + P(\dot{g} - Ag).$$

Since $P(t_1)g(t_1) = 0$, it now follows from the characterization of $\xi = L^*\phi$ in (12.60) that

$$Pg = L^*(C^*Cg - PBB^*Pg - P\dot{g} + PAg).$$

This yields the following relationship:

$$(P + L^*PBB^*P - L^*C^*C)g = -L^*P(\dot{g} - Ag) \qquad (12.61)$$

for any differentiable function g in $L^2([t_0, t_1], \mathcal{X})$. Now let $g = Lf$ where f is any function in $L^2([t_0, t_1], \mathcal{X})$. Then using $\dot{g} = Ag + f$ in (12.61), we arrive at the following algebraic Riccati equation

$$L^*P + PL + L^*PBB^*PL - L^*C^*CL = 0. \qquad (12.62)$$

By applying B^* to the left and B on the right, rearranging terms and using $F = CLB$, we obtain

$$
\begin{aligned}
I + F^*F &= I + B^*L^*C^*CLB \\
&= I + B^*L^*PB + B^*PLB + B^*L^*PBB^*PLB \\
&= (I + B^*PLB)^*(I + B^*PLB).
\end{aligned}
$$

So, if $\Theta = I + B^*PLB$, then clearly $\Theta^*\Theta = I + F^*F$.

Obviously Θ is causal. In fact, Θ can be viewed as the input output map for a state space system. To be precise, if $h = \Theta v$ for some v in $L^2([t_0, t_1], \mathcal{U})$, then we claim that h is the output of the following linear system

$$\dot{q} = Aq + Bv \quad \text{and} \quad h = B^*Pq + v \qquad (q(t_0) = 0). \qquad (12.63)$$

To verify that $h = \Theta v$, simply notice that because $q(t_0) = 0$, the state $q = LBv$. Hence, $h = B^*PLBv + v = \Theta v$, which verifies our claim. By setting $\tilde{B} = B$ and $\tilde{C} = B^*P$ with $D = I$ in Lemma 12.5.2 below, it follows that Θ is an invertible causal operator. Therefore, Θ is a finite time outer spectral factor for $I + F^*F$. ∎

Lemma 12.5.2 *Let Υ be the operator from $L^2([t_0, t_1], \mathcal{U})$ into $L^2([t_0, t_1], \mathcal{Y})$ defined by*

$$(\Upsilon v)(t) = \int_{t_0}^{t} \tilde{C}(t) e^{A(t-\tau)} \tilde{B}(\tau) v(\tau) \, d\tau + D v(t) \qquad (v \in L^2([t_0, t_1], \mathcal{U})) \tag{12.64}$$

where $\tilde{B}$ and $\tilde{C}$ are continuous functions with values in $\mathcal{L}(\mathcal{U}, \mathcal{X})$ and $\mathcal{L}(\mathcal{X}, \mathcal{Y})$, respectively. Moreover, assume that D is an invertible operator from $\mathcal{U}$ into $\mathcal{Y}$. Then Υ is invertible and

$$(\Upsilon^{-1} y)(t) = -D^{-1} \int_{t_0}^{t} \tilde{C}(t) \Psi(t, \tau) \tilde{B}(\tau) D^{-1} y(\tau) \, d\tau + D^{-1} y(t) \qquad (y \in L^2([t_0, t_1], \mathcal{Y})) \tag{12.65}$$

where $\Psi(t, \tau)$ is the state transition matrix for $A - \tilde{B} D^{-1} \tilde{C}$.

PROOF. Notice that Υ is the input output map for the following state space system

$$\dot{q} = Aq + \tilde{B}v \quad \text{and} \quad y = \tilde{C}q + Dv \qquad (q(t_0) = 0). \tag{12.66}$$

In other words, $y = \Upsilon v$ if and only if v is the input and y is the output for the state space system in (12.66). Substituting $v = D^{-1} y - D^{-1} \tilde{C} q$ into the first equation in (12.66), gives

$$\dot{q} = (A - \tilde{B} D^{-1} \tilde{C}) q + \tilde{B} D^{-1} y \quad \text{and} \quad v = -D^{-1} \tilde{C} q + D^{-1} y \qquad (q(t_0) = 0). \tag{12.67}$$

This is the state space system for the operator Λ from $L^2([t_0, t_1], \mathcal{Y})$ into $L^2([t_0, t_1], \mathcal{U})$ defined by

$$(\Lambda y)(t) = -D^{-1} \int_{t_0}^{t} \tilde{C}(t) \Psi(t, \tau) \tilde{B}(\tau) D^{-1} y(\tau) \, d\tau + D^{-1} y(t) \qquad (y \in L^2([t_0, t_1], \mathcal{Y})).$$

In particular, equation (12.67) shows that if $y = \Upsilon v$, then $v = \Lambda y = \Lambda \Upsilon v$. Since this holds for all v in $L^2([t_0, t_1], \mathcal{U})$, we obtain $\Lambda \Upsilon = I$. On the other hand, if $v = \Lambda y$ for any y in $L^2([t_0, t_1], \mathcal{Y})$, then substituting $y = \tilde{C} q + D v$ into the first equation in (12.67), yields the state space system in (12.66), that is, $y = \Upsilon v$. Hence, $y = \Upsilon v = \Upsilon \Lambda y$. This readily implies that $\Upsilon \Lambda = I$. Therefore, Λ is the inverse of Υ and (12.65) holds.   ∎

12.5.1   A general tracking problem

In this section we will use some operator techniques to solve a generalization of the previous tracking problem. To this end, let $\mathcal{W}$ be a finite dimensional space. Let E be an operator from $\mathcal{W}$ into $\mathcal{X}$ and D an operator from $\mathcal{W}$ into $\mathcal{Y}$. Consider the system

$$\begin{aligned} \dot{x} &= Ax + Bu + Ew \\ y &= Cx + Dw. \end{aligned} \tag{12.68}$$

Here w is some signal or a disturbance in $L([t_0, t_1], \mathcal{W})$. This leads to the following linear quadratic optimization problem

$$\varepsilon(x_0) = \inf \left\{ \int_{t_0}^{t_1} (\|y(\sigma)\|^2 + \|u(\sigma)\|^2) d\sigma : u \in L^2([t_0, t_1], \mathcal{U}) \right\}$$

subject to the system in (12.68) and $x(t_0) = x_0$. \tag{12.69}

If E is zero, $D = -I$ and $w = r$, then this optimization problem reduces to the linear quadratic tracking problem (12.45) discussed in Section 12.4. The following presents a solution to this optimization problem.

Theorem 12.5.3 *Consider the linear quadratic tracking problem in (12.69). Then the minimum exists and there exists a unique optimal input $\hat{u}$ in $L^2([t_0, t_1], \mathcal{U})$ which achieves this minimum. To compute $\hat{u}$, let P be the unique solution on the interval $[t_0, t_1]$ to the Riccati differential equation in (12.3). Let φ be the solution to the following differential equation moving backwards in time:*

$$\dot{\varphi} = -(A - BB^*P(t))^*\varphi - (PE + C^*D)w(t) \qquad with \qquad \varphi(t_1) = 0. \tag{12.70}$$

Then the optimal input $\hat{u}$ is given by

$$\hat{u}(t) = -B^*P(t)\hat{x}(t) - B^*\varphi(t) \tag{12.71}$$

where the optimal state trajectory $\hat{x}$ is uniquely determined by

$$\dot{\hat{x}} = (A - BB^*P(t))\hat{x} - BB^*\varphi(t) + Ew(t) \qquad (\hat{x}(t_0) = x_0). \tag{12.72}$$

PROOF. One can obtain a proof of this result by following the techniques in the proof of Theorem 12.4.1. Now let us use the finite time outer spectral factor Θ to obtain an alternative proof. Recall that the input output operator F in (12.18) is given by $F = CLB$. Moreover, if we set $G = CLE + D$, then the output y in (12.68) is given by

$$y = C_o x_0 + Gw + Fu. \tag{12.73}$$

So, the optimization problem in (12.69) is equivalent to the following optimization problem

$$\varepsilon(x_0) = \inf\{\|C_o x_0 + Gw + Fu\|^2 + \|u\|^2 : u \in L^2([t_0, t_1], \mathcal{U})\}. $$

By consulting Lemma 12.3.1 with $g = C_o x_0 + Gw$, we see that the unique optimal input $\hat{u}$ solving the quadratic optimization problem in (12.69) is given by

$$\hat{u} = -(I + F^*F)^{-1}F^*(C_o x_0 + Gw). \tag{12.74}$$

Notice that this is precisely the solution to the linear quadratic regular problem with $C_o x_o$ replaced by $C_o x_0 + Gw$. Moreover, $\hat{u} + F^*F\hat{u} = -F^*(C_o x_0 + Gw)$. By employing (12.73), the optimal output $\hat{y}$ corresponding to the optimal input $\hat{u}$ is given by $\hat{y} = C_o x_0 + Gw + F\hat{u}$. Recalling the expression for $\hat{u}$, we now obtain $\hat{u} = -F^*(C_o x_0 + Gw + F\hat{u}) = -F^*\hat{y}$. As expected, the optimal control input satisfies $\hat{u} = -F^*\hat{y}$; see Remark 12.3.1. Finally, it is noted that the optimal state $\hat{x}$ is given by

$$\dot{\hat{x}} = A\hat{x} + B\hat{u} + Ew. \tag{12.75}$$

Using $g = \hat{x}$ in equation (12.61) along with (12.75), we obtain

$$(P + L^*PBB^*P - L^*C^*C)\hat{x} = -L^*P(B\hat{u} + Ew).$$

Recall that $\Theta^* = I + B^*L^*PB$. Multiplying the above equation by B^* on the left and rearranging terms, we obtain

$$
\begin{aligned}
-B^*L^*C^*C\hat{x} + B^*L^*PB\hat{u} &= -(I + B^*L^*PB)B^*P\hat{x} - B^*L^*PEw \\
&= -\Theta^*B^*P\hat{x} - B^*L^*PEw\,.
\end{aligned}
$$

By employing the optimal output $\hat{y} = C\hat{x} + Dw$ and the optimal input $\hat{u} = -F^*\hat{y} = -B^*L^*C^*\hat{y}$ in the previous equation, yields

$$
\begin{aligned}
\Theta^*\hat{u} &= (I + B^*L^*PB)\hat{u} = -B^*L^*C^*\hat{y} + B^*L^*PB\hat{u} \\
&= -B^*L^*C^*(C\hat{x} + Dw) + B^*L^*PB\hat{u} \\
&= -\Theta^*B^*P\hat{x} - B^*L^*(PE + C^*D)w\,.
\end{aligned}
$$

According to Lemma 12.5.1, the operator $\Theta^* = I + B^*L^*PB$ is invertible. This along with the identity $(I + NM)^{-1}N = N(I + MN)^{-1}$, yields

$$
\begin{aligned}
\hat{u} &= -B^*P\hat{x} - (I + B^*L^*PB)^{-1}B^*L^*(PE + C^*D)w \\
&= -B^*P\hat{x} - B^*(I + L^*PBB^*)^{-1}L^*(PE + C^*D)w \\
&= -B^*P\hat{x} - B^*\varphi
\end{aligned}
$$

where $\varphi := (I + L^*PB^*B)^{-1}L^*(PE + C^*D)w$. Using $(I + L^*PB^*B)\varphi = L^*(PE + C^*D)w$, we obtain

$$
\varphi = L^*\left(-PB^*B\varphi + (PE + C^*D)w\right)\,.
$$

Recall that if $\xi = L^*\phi$, then ξ satisfies the differential equation in (12.60). Hence,

$$
\dot{\varphi} = -(A - BB^*P)^*\varphi - (PE + C^*D)w \qquad (\varphi(t_1) = 0)\,.
$$

This is precisely the differential equation in (12.70). Since $\hat{u} = -B^*P\hat{x} - B^*\varphi$, equation (12.71) holds. Substituting this into (12.75), yields (12.72).   ∎

Problems with control weights. Now consider the following linear quadratic tracking problem associated with the linear quadratic regular problem discussed in Section 12.1.1:

$$
\varepsilon(x_0) = \inf\left\{\int_{t_0}^{t_1} (\|y(\sigma)\|^2 + (Ru(\sigma), u(\sigma))\,d\sigma : u \in L^2([t_0, t_1], \mathcal{U}\right\}
$$

subject to the system in (12.68) and $x(t_0) = x_0$.   (12.76)

As before, R is a strictly positive operator on $\mathcal{U}$. The minimum exists and there exists a unique optimal input $\hat{u}$ in $L^2([t_0, t_1], \mathcal{U})$ which achieves the minimum. To obtain the optimal input, let P be the solution to the following Riccati differential equation

$$
\dot{P} = -A^*P - PA - C^*C + PBR^{-1}B^*P \qquad (P(t_1) = 0)\,.   \tag{12.77}
$$

Let φ be the solution to the following differential equation moving backwards in time:

$$
\dot{\varphi} = -(A - BR^{-1}B^*P)^*\varphi - (PE + C^*D)w \qquad (\varphi(t_1) = 0)\,.   \tag{12.78}
$$

Then for any initial state x_0 in $\mathcal{X}$, the optimal input

$$\hat{u}(t) = -R^{-1}B^*(P(t)\hat{x}(t) + \varphi(t))$$

where the optimal state trajectory $\hat{x}$ is uniquely determined by

$$\dot{\hat{x}} = (A - BR^{-1}B^*P)\hat{x} - BR^{-1}B^*\varphi + Ew \quad \text{and} \quad \hat{x}(t_0) = x_0. \tag{12.79}$$

Finally, it is noted that this tracking result also holds in the time varying case, that is, when $\{A(t), B(t), C(t), D(t), E(t), R(t)\}$ are continuous function with values in the appropriate $\mathcal{L}(\cdot, \cdot)$ space and $R(t) \geq \epsilon I$ for some $\epsilon > 0$ and all t. Because the proof of this result is almost identical to the proof in Section 12.1.1, the details are left as an exercise.

12.6 The infinite horizon problem

This section is concerned with the following infinite horizon linear quadratic regulator problem. For each initial state x_0 in $\mathcal{X}$, find the optimal cost $\varepsilon(x_0)$ and an optimal input $\hat{u}$ which solves the optimization problem:

$$\varepsilon(x_0) = \inf\left\{ \int_0^\infty (\|y(\sigma)\|^2 + \|u(\sigma)\|^2)\, d\sigma : u \in L^2([0,\infty), \mathcal{U}) \right\}$$

subject to $\quad \dot{x} = Ax + Bu$ and $y = Cx$ and $x(0) = x_0$. $\tag{12.80}$

As before, A is an operator on $\mathcal{X}$ and B maps $\mathcal{U}$ into $\mathcal{X}$ while C is an operator mapping $\mathcal{X}$ into $\mathcal{Y}$. The spaces $\mathcal{X}$, $\mathcal{Y}$ and $\mathcal{U}$ are finite dimensional.

Throughout this section it is assumed that the pair $\{A, B\}$ is stabilizable. If $\{A, B\}$ is not stabilizable, then for some initial states x_0, the cost in (12.80) is infinite for every input; hence the infimum is infinite and the optimization problem is trivial. For example, consider the system,

$$\dot{x} = \begin{bmatrix} 1 & 0 \\ 0 & 0 \end{bmatrix} x + \begin{bmatrix} 0 \\ 1 \end{bmatrix} u \quad \text{and} \quad y = \begin{bmatrix} 1 & 0 \end{bmatrix} x,$$

which is not stabilizable. Here $y(t) = e^t x_{10}$ where x_{10} is the first component of x_0. So, for a nonzero x_{10} and any input, we have

$$\int_0^\infty (\|y(\sigma)\|^2 + \|u(\sigma)\|^2)\, d\sigma \geq \int_0^\infty e^{2\sigma} |x_{10}|^2\, d\sigma = +\infty.$$

Therefore, the infimum in (12.80) is infinite for all nonzero x_{10}, and the optimization is trivial. On the other hand, if $\{A, B\}$ is stabilizable, then there exists a feedback gain K from $\mathcal{X}$ into $\mathcal{U}$ such that $A - BK$ is stable. In other words, if $u = -Kx$, then all solutions of the resulting closed loop system decays exponentially to zero. Specifically, there are constants α and β for the closed loop system such that $\|x(t)\| \leq \beta \|x_0\| e^{-\alpha t}$ for all $t \geq 0$. Since $y = Cx$ and $u = -Kx$, it follows that

$$\int_0^\infty (\|y(\sigma)\|^2 + \|u(\sigma)\|^2)\, d\sigma \leq M\|x_0\|^2 \tag{12.81}$$

for some finite scalar M. So, in this case the linear quadratic optimization problem in (12.80) is non-trivial.

12.7   The algebraic Riccati equation

Solving an infinite horizon problem involves the algebraic Riccati equation associated with
the Riccati differential equation in (12.3). In this section, we establish some properties of
this algebraic Riccati equation.

As in Remark 12.1.1, let $\Omega(t) = P(t_1-t)$ where P is the solution to the Riccati differential
equation in (12.3). Remark 12.1.1 states that $\{\Omega(t)\}$ is an increasing sequence of positive
operators which satisfy the following Riccati differential equation

$$\dot{\Omega} = A^*\Omega + \Omega A + C^*C - \Omega BB^*\Omega \qquad (\Omega(0) = 0). \qquad (12.82)$$

Moreover,

$$(\Omega(t_1)x_0, x_0) = \inf \left\{ \int_0^{t_1} (||y(\sigma)||^2 + ||u(\sigma)||^2)\, d\sigma : u \in L^2([0, t_1], \mathcal{U}) \right\}$$

$$\text{subject to} \qquad \dot{x} = Ax + Bu \text{ and } y = Cx \text{ and } x(0) = x_0. \qquad (12.83)$$

Now assume that the pair $\{A, B\}$ is stabilizable. Then there is a finite scalar M such
that for any initial state x_0, there is an input u such that (12.81) holds. Hence, for any $t \geq 0$,

$$(\Omega(t)x_0, x_0) \leq \int_0^t (||y(\sigma)||^2 + ||u(\sigma)||^2)\, d\sigma$$

$$\leq \int_0^\infty (||y(\sigma)||^2 + ||u(\sigma)||^2)\, d\sigma \leq M||x_0||^2. \qquad (12.84)$$

Since $\Omega(t) \geq 0$, this shows that $\Omega(t)$ is uniformly bounded for all t, that is, $||\Omega(t)|| \leq M$
where M is a finite positive scalar. Recall that a sequence of positive increasing uniformly
bounded operators converges strongly to a positive operator; see Halmos [59]. Therefore,
$\Omega(t)$ converges to a positive operator Q as $t \to \infty$, that is,

$$Q = \lim_{t \to \infty} \Omega(t). \qquad (12.85)$$

Equations (12.84) and (12.85) show that, for any initial state x_0, the scalar (Qx_0, x_0) is a
lower bound on the optimal cost, that is,

$$(Qx_0, x_0) \leq \inf_u \int_0^\infty (||y(\sigma)||^2 + ||u(\sigma)||^2)\, d\sigma. \qquad (12.86)$$

Later we will show that, when $\{C, A\}$ is detectable, $(Qx_0, x_0) = \varepsilon(x_0)$ where $\varepsilon(x_0)$ is the
optimal cost for the infinite time horizon optimization problem in (12.80).

Since $\Omega(t)$ approaches a constant Q as t tends to infinity, our intuition tells us that Q is a
constant solution to the Riccati differential equation (12.82). In other words, Q is a solution
of the following *algebraic Riccati equation*:

$$A^*Q + QA + C^*C - QBB^*Q = 0. \qquad (12.87)$$

Our intuitive result can be rigorously justified by using Lemma 12.7.4 at the end of this
section. If the pair $\{C, A\}$ is observable, then Q is strictly positive. In this case, Remark
12.1.2 or 12.2.1 shows that $\Omega(t)$ is strictly positive for all $t > 0$. Because $\{\Omega(t)\}$ is increasing
and converge to Q, it follows that Q is strictly positive. This proves the following result.

Lemma 12.7.1 *Let $\{A, B\}$ be a stabilizable pair. Then the Riccati differential equation in (12.82) has a unique solution Ω for all $t \geq 0$. Moreover, $\Omega(t)$ converges to a positive operator Q as t tends to infinity and Q satisfies the algebraic Riccati equation in (12.87). In addition, if the pair $\{C, A\}$ is observable, then Q is a strictly positive solution to (12.87).*

We say that Q is a *stabilizing solution* to the algebraic Riccati equation in (12.87) if Q is a self-adjoint operator satisfying (12.87) and the operator $A - BB^*Q$ is stable. If the algebraic Riccati equation in (12.87) admits a stabilizing solution Q, then it is unique. To see this let Z be another stabilizing solution, that is, assume that

$$A^*Z + ZA + C^*C - ZBB^*Z = 0 \tag{12.88}$$

and $A_1 = A - BB^*Z$ is stable. Let $A_c = A - BB^*Q$ and set $\Delta = Q - Z$. Using (12.87) and (12.88), we obtain

$$
\begin{aligned}
A_c^*\Delta + \Delta A_1 &= (A^* - QBB^*)(Q - Z) + (Q - Z)(A - BB^*Z) \\
&= A^*Q + QA - QBB^*Q + QBB^*Z - A^*Z - ZA + ZBB^*Z - QBB^*Z \\
&= -C^*C + C^*C \\
&= 0.
\end{aligned} \tag{12.89}
$$

Therefore, $A_c^*\Delta + \Delta A_1 = 0$. Since A_c and A_1 are both stable, the only solution to this Lyapunov equation is $\Delta = 0$. To demonstrate this simply notice that

$$
\begin{aligned}
0 &= \int_0^\infty e^{A_c^*t} 0 e^{A_1 t} dt = \int_0^\infty e^{A_c^*t}(A_c^*\Delta + \Delta A_1)e^{A_1 t} dt \\
&= \int_0^\infty \frac{d\left(e^{A_c^*t}\Delta e^{A_1 t}\right)}{dt} dt = e^{A_c^*t}\Delta e^{A_1 t}\Big|_0^\infty = -\Delta.
\end{aligned} \tag{12.90}
$$

Thus, Δ equals zero and $Q = Z$. In other words, Q is the only stabilizing solution to the algebraic Riccati equation in (12.87).

Recall that a pair $\{C, A\}$ is detectable if all the unobservable eigenvalues for $\{C, A\}$ are stable, that is, if $Af = \lambda f$ and $Cf = 0$ for some eigenvector f with eigenvalue λ, then $\Re(\lambda) < 0$. The following result shows that if $\{C, A\}$ is detectable, then the algebraic Riccati equation has a unique stabilizing solution and $\Omega(t)$ converges to this solution.

Lemma 12.7.2 *Let $\{A, B, C\}$ be a stabilizable and detectable system. Let $Q = \lim_{t\to\infty} \Omega(t)$ where Ω is the solution to the Riccati differential equation in (12.82). Then Q is positive and is a stabilizing solution to the algebraic Riccati equation in (12.87). Moreover, Q is the only positive solution and the only stabilizing solution to this algebraic Riccati equation.*

PROOF. According to Lemma 12.7.1, the operator $Q = \lim_{t\to\infty} \Omega(t)$ is positive and satisfies the algebraic Riccati equation in (12.87). Now let us show that $A - BB^*Q$ is stable and Q is the only positive solution to this algebraic Riccati equation. To this end, notice that the algebraic Riccati equation in (12.87) is equivalent to

$$(A - BB^*Q)^*Q + Q(A - BB^*Q) + \begin{bmatrix} C \\ B^*Q \end{bmatrix}^* \begin{bmatrix} C \\ B^*Q \end{bmatrix} = 0.$$

If we set $\tilde{C} = [C \quad B^*Q]^{tr}$ and $A_c = A - BB^*Q$, then it follows that the positive Q satisfies the Lyapunov equation

$$A_c^*Q + QA_c + \tilde{C}^*\tilde{C} = 0 .$$

To show that A_c is stable it is sufficient to verify that the pair $\{\tilde{C}, A_c\}$ is detectable; see Lemma 12.7.3 at the end of this section. We now claim that the kernel of

$$J(\lambda) = \begin{bmatrix} A_c - \lambda I \\ \tilde{C} \end{bmatrix} = \begin{bmatrix} A - BB^*Q - \lambda I \\ C \\ B^*Q \end{bmatrix} \tag{12.91}$$

is zero for all complex numbers λ in the closed right half plane. Then by the PBH test $\{\tilde{C}, A_c\}$ is detectable; see Lemma 9.1.2. If f is any vector in the kernel of $J(\lambda)$, then $B^*Qf = 0$. Thus, $(A - \lambda I)f = 0$ and $Cf = 0$. Since $\{C, A\}$ is detectable and $\Re(\lambda) \geq 0$, the PBH test guarantees that the kernel of $[A - \lambda I \quad C]^{tr}$ is zero. In other words, f must be zero which proves our claim. Therefore, A_c is stable.

We now demonstrate that Q is the only positive solution to the algebraic Riccati equation in (12.87). Let Z be any positive solution to this algebraic Riccati equation, that is, assume that (12.88) holds. Then by replacing Q by Z in our previous analysis, we see that $A_1 = A - BB^*Z$ is stable. Because the stabilizing solution to this algebraic Riccati equation is unique, we must have $Z = Q = \lim_{t \to \infty} \Omega(t)$.   ∎

If $\{A, B, C\}$ is stabilizable and detectable, then Lemma 12.7.2 shows that the algebraic Riccati in (12.87) admits a unique positive solution Q. Moreover, this solution is also the unique stabilizing solution. Let us complete this section with the following lemmas which were used above.

Lemma 12.7.3 *Assume that the pair $\{C, A\}$ is a detectable pair and the Lyapunov equation*

$$A^*\Xi + \Xi A + C^*C = 0 \tag{12.92}$$

has a positive solution for Ξ. Then A is stable.

PROOF. Consider any eigenvalue λ of A and let f be a corresponding eigenvector. We need to show that λ is stable, that is, $\Re(\lambda) < 0$. Using $Af = \lambda f$ in (12.92) gives

$$-\|Cf\|^2 = (A^*\Xi f, f) + (\Xi Af, f) = (\Xi f, Af) + \lambda(\Xi f, f) = 2\Re(\lambda)(\Xi f, f) .$$

Hence, $2\Re(\lambda)(\Xi f, f) = -\|Cf\|^2$. Since Ξ is positive, we have $(\Xi f, f) \geq 0$. If either $(\Xi f, f)$ or $\Re(\lambda)$ is zero, then $\|Cf\| = 0$, and thus, $Cf = 0$. Since $\{C, A\}$ is detectable, we must have $\Re(\lambda) < 0$. Now suppose that $(\Xi f, f) > 0$ and $\Re(\lambda) \neq 0$. Then, $\|Cf\| \neq 0$ and $\Re(\lambda) = -\|Cf\|/2(\Xi f, f) < 0$.   ∎

Lemma 12.7.4 *Consider the differential equation $\dot{x} = f(x)$ where f is a continuous function on a finite dimensional space $\mathcal{X}$. Suppose that x is any solution of this differential equation which converges to a constant vector x^e in $\mathcal{X}$ as t tends to infinity. Then $f(x^e) = 0$.*

PROOF. For any $t \geq 0$, we have

$$x(t+1) - x(t) = \int_t^{t+1} f(x(\sigma)) \, d\sigma = \int_t^{t+1} (f(x(\sigma)) - f(x^e)) \, d\sigma + f(x^e). \tag{12.93}$$

By the hypothesis, $x(t)$ approaches x^e as t tends towards infinity. This also implies that $x(t+1)$ approaches x^e as t tends towards infinity. Hence, $x(t+1) - x(t)$ converges to zero as t approaches infinity. Since the function f is continuous, $f(x(\sigma)) - f(x^e)$ also converges to zero as σ tends towards infinity. This readily implies that

$$\lim_{t \to \infty} \int_t^{t+1} (f(x(\sigma)) - f(x^e)) \, d\sigma = 0.$$

By taking limits as t approaches infinity in equation (12.93), we obtain $f(x^e) = 0$. ∎

12.8 Solution to the infinite horizon problem

In this section we use the stabilizing solution to the algebraic Riccati equation to obtain a state feedback solution to the infinite horizon linear quadratic regulator problem.

Theorem 12.8.1 *Consider the infinite horizon linear quadratic regulator problem in (12.80) where $\{A, B, C\}$ is a stabilizable and detectable system. Then the algebraic Riccati equation in (12.87) has a unique positive solution Q and for any initial state x_0 in $\mathcal{X}$, the optimal cost in (12.80) is given by*

$$\varepsilon(x_0) = (Qx_0, x_0). \tag{12.94}$$

Moreover, this cost is uniquely attained by the optimal input

$$\hat{u}(t) = -B^* Q \hat{x}(t) \tag{12.95}$$

where the optimal state trajectory $\hat{x}$ is uniquely determined by

$$\dot{\hat{x}} = (A - BB^* Q)\hat{x} \quad \text{and} \quad \hat{x}(0) = x_0. \tag{12.96}$$

PROOF. Recall from (12.86) that, for any initial state x_0, the scalar (Qx_0, x_0) is a lower bound on the optimal cost, that is, $(Qx_0, x_0) \leq \varepsilon(x_0)$. We show that the input given by $\hat{u} = -B^* Q \hat{x}$ achieves this lower bound, and hence, is optimal. To see this, notice that the algebraic Riccati equation for Q in (12.87) along with $\dot{x} = Ax + Bu$ and $y = Cx$ gives

$$\begin{aligned} \frac{d}{dt}(Qx, x) &= (Q\dot{x}, x) + (Qx, \dot{x}) \\ &= (QAx, x) + (QBu, x) + (Qx, Ax) + (Qx, Bu) \\ &= (QAx, x) + (A^* Qx, x) + (u, B^* Qx) + (B^* Qx, u) \tag{12.97} \\ &= -\|Cx\|^2 + \|B^* Qx\|^2 + (u, B^* Qx) + (B^* Qx, u) \\ &= -\|y\|^2 - \|u\|^2 + \|B^* Qx + u\|^2. \end{aligned}$$

By integrating from 0 to t with $x_0 = x(0)$, we obtain

$$\int_0^t (||y(\sigma)||^2 + ||u(\sigma)||^2)\, d\sigma \;=\; (Qx_0, x_0) - (Qx(t), x(t))$$

$$+ \int_0^t ||B^*Qx(\sigma) + u(\sigma)||^2\, d\sigma. \qquad (12.98)$$

Considering $u = \hat{u} = -B^*Q\hat{x}$ and $y = \hat{y}$ yields

$$\int_0^t (||\hat{y}(\sigma)||^2 + ||\hat{u}(\sigma)||^2)\, d\sigma = (Qx_0, x_0) - (Q\hat{x}(t), \hat{x}(t)).$$

Since $\dot{\hat{x}} = (A - BB^*Q)\hat{x}$ and $A - BB^*Q$ is stable, it follows that $\hat{x}(t)$ approaches zero as t tends towards infinity. Letting t approach infinity, we now obtain

$$\int_0^\infty (||\hat{y}(\sigma)||^2 + ||\hat{u}(\sigma)||^2)\, d\sigma = (Qx_0, x_0).$$

Therefore, $\hat{u} = -B^*Q\hat{x}$ is an optimal input and the cost $\varepsilon(x_0) = (Qx_0, x_0)$.

To complete the proof, it remains to show that $\hat{u} = -B^*Q\hat{x}$ is the unique optimal input. To this end, assume that v is another function in $L^2([0,\infty), \mathcal{U})$ which achieves the cost (Qx_0, x_0). The output z corresponding to v is given by $z = Cf$ where $\dot{f} = Af + Bv$ and $f(0) = x_0$. Since v achieves the optimal cost $\varepsilon(x_0)$, the optimal output z must be in $L^2([0,\infty), \mathcal{Y})$. According to Lemma 12.8.2 given at the end of this section, the state f is in $L^2([0,\infty), \mathcal{X})$. Lemma 12.8.3 presented below, shows that $f(t)$ approaches zero, as t tends to infinity. So, by letting t approach infinity in (12.98), we obtain

$$(Qx_0, x_0) = \int_0^\infty (||z(\sigma)||^2 + ||v(\sigma)||^2)\, d\sigma = (Qx_0, x_0) + \int_0^\infty ||B^*Qf(\sigma) + v(\sigma)||^2\, d\sigma.$$

This readily implies that $||B^*Qf + v||_{L^2} = 0$. Hence, $v = -B^*Qf$. Substituting this into $\dot{f} = Af + Bv$, yields $\dot{f} = (A - BB^*Q)f$ where $f(0) = x_0$. Thus, $f = \hat{x}$ and $v = -B^*Qf = \hat{u}$. Therefore, the optimal input $\hat{u}$ is unique. ∎

Remark 12.8.1 Now consider the possibly unstable system $\dot{x} = Ax + Bu$. Recall that an operator K mapping $\mathcal{X}$ into $\mathcal{U}$ is a *stabilizing gain* if the state feedback operator $A - BK$ is stable. The algebraic Riccati equation in (12.87) can be used to compute a stabilizing gain K for any stabilizable pair $\{A, B\}$. To see this simply choose any operator C such that the pair $\{C, A\}$ is detectable. For example, choose $C = I$. Let Q be the unique positive solution to the algebraic Riccati equation in (12.87). Then Lemma 12.7.2 shows that $K = B^*Q$ is a stabilizing gain for $\{A, B\}$.

Remark 12.8.2 Consider any stable system $\{A, B, C\}$ and let F be the operator from $L^2([0,\infty), \mathcal{U})$ into $L^2([0,\infty), \mathcal{Y})$ defined by

$$(Fu)(t) = \int_0^t Ce^{A(t-\tau)}Bu(\tau)d\tau \qquad (u \in L^2([0,\infty), \mathcal{U})). \qquad (12.99)$$

Let C_o be the operator from $\mathcal{X}$ into $L^2([0,\infty), \mathcal{Y})$ defined by $(C_o x_0)(t) = Ce^{At}x_0$ where x_0 is in $\mathcal{X}$. Suppose Q is the unique stabilizing solution to the algebraic Riccati equation in (12.87). By combining Lemma 12.3.1 and Theorem 12.8.1, we see that $Q = C_o^*(I + FF^*)^{-1}C_o$.

Lemma 12.8.2 *Consider the detectable system*

$$\dot{x} = Ax + Bu \quad and \quad y = Cx + Du\,. \tag{12.100}$$

If the input u is in $L^2([0,\infty),\mathcal{U})$ and the output y is in $L^2([0,\infty),\mathcal{Y})$, then the state trajectory x is in $L^2([0,\infty),\mathcal{X})$.

PROOF. Since the pair $\{C, A\}$ is detectable, there exists an operator L from $\mathcal{Y}$ into $\mathcal{X}$ such that the operator $A - LC$ is stable; see Theorem 9.1.3. So, we can rewrite the system in (12.100) as

$$\dot{x} = (A - LC)x + Ly + (B - LD)u\,.$$

This readily implies that

$$x(t) = e^{(A-LC)t}x(0) + \int_0^t e^{(A-LC)(t-\tau)}(Ly(\tau) + (B - LD)u(\tau))\,d\tau\,.$$

Because both $(B - LD)u$ and Ly are in $L^2([0,\infty),\mathcal{X})$ and $A - LC$ is stable, it follows that x is in $L^2([0,\infty),\mathcal{X})$. ∎

Lemma 12.8.3 *Consider the system $\dot{x} = Ax + Bu$ where A is an operator on $\mathcal{X}$ and B maps $\mathcal{U}$ into $\mathcal{X}$. If the input u is in $L^2([0,\infty),\mathcal{U})$ and the state x is in $L^2([0,\infty),\mathcal{X})$, then $x(t)$ approaches zero as t tends to infinity.*

PROOF. Since $\dot{x} = Ax + Bu$ and u and x are in the appropriate L^2 spaces, it follows that $\dot{x}$ is in $L^2([0,\infty),\mathcal{X})$. We now observe that

$$(\dot{x}, x) = \int_0^\infty (\dot{x}(\tau), x(\tau))\,d\tau = \lim_{t\to\infty} \int_0^t (\dot{x}(\tau), x(\tau))\,d\tau \tag{12.101}$$

and

$$\|x(t)\|^2 - \|x(0)\|^2 = \int_0^t \frac{d}{d\tau}(x(\tau), x(\tau))\,d\tau = \int_0^t (\dot{x}(\tau), x(\tau))\,d\tau + \int_0^t (x(\tau), \dot{x}(\tau))\,d\tau\,. \tag{12.102}$$

Considering limits as t approaches infinity, it follows from (12.101) and (12.102) that

$$\lim_{t\to\infty} \|x(t)\|^2 = \|x(0)\|^2 + 2\Re(\dot{x}, x)\,.$$

Thus $\|x(t)\|^2$ converges a limit as t tends to infinity. Because x is in $L^2([0,\infty),\mathcal{X})$, this limit must be zero. Hence, $x(t)$ approaches zero as t tends to infinity. ∎

12.8.1 Problems with control weights

In many control applications one considers linear quadratic regular problems of the form:

$$\varepsilon(x_0) = \inf\left\{\int_0^\infty (\|y(\sigma)\|^2 + (Ru(\sigma), u(\sigma))\,d\sigma : u \in L^2([0,\infty),\mathcal{U})\right\}$$

$$\text{subject to} \quad \dot{x} = Ax + Bu \text{ and } y = Cx \text{ and } x(0) = x_0\,. \tag{12.103}$$

Here R is a strictly positive operator on $\mathcal{U}$. By following the procedure used in Section 12.1.1, introducing a new input $v = R^{1/2}u$ and setting $\tilde{B} = BR^{-1/2}$, we readily obtain the following solution to the infinite horizon linear quadratic optimization problem in (12.103).

If $\{A, B\}$ is stabilizable and $\{C, A\}$ is detectable, then the algebraic Riccati equation

$$A^*Q + QA + C^*C - QBR^{-1}B^*Q = 0,   \tag{12.104}$$

has a unique positive solution for Q. For any initial state x_0 in $\mathcal{X}$, the optimal cost in (12.103) is given by $\varepsilon(x_0) = (Qx_0, x_0)$. Moreover, this cost is uniquely attained by the optimal input $\hat{u} = -R^{-1}B^*Q\hat{x}$, where the optimal state trajectory $\hat{x}$ is uniquely determined by

$$\dot{\hat{x}} = (A - BR^{-1}B^*Q)\hat{x}   (\hat{x}(0) = x_0).   \tag{12.105}$$

Finally, the operator $A - BR^{-1}B^*Q$ is stable. In this setting, we say that Q is a stabilizing solution to the algebraic Riccati equation in (12.104), if Q is a self-adjoint operator satisfying (12.104) and $A - BR^{-1}B^*Q$ is stable. So, if $\{A, B, C\}$ is a stabilizable and detectable system, then the algebraic Riccati equation in (12.104) admits a unique stabilizing solution and this solution is the only positive solution. In Chapter 13 we present a computational method based on the Hamiltonian matrix in (12.37) to compute the unique stabilizing solution.

12.9   An outer spectral factorization

Let $\{A, B, C, 0\}$ be a realization for a transfer function $\mathbf{G}$. In this section, we utilize the algebraic Riccati equation in (12.104) to obtain a special factorization of $R + G(-\bar{s})^*G(s)$. We begin with the following result which is useful in the inversion of transfer functions.

Lemma 12.9.1 *Let $\{A, B, C, D\}$ be a state space realization of a transfer function $\mathbf{H}$ and assume that D is invertible. Then the inverse of $\mathbf{H}$ exists and is the proper rational function given by*

$$\mathbf{H}(s)^{-1} = D^{-1} - D^{-1}C(sI - A + BD^{-1}C)^{-1}BD^{-1}.   \tag{12.106}$$

In other words, $\{A - BD^{-1}C, BD^{-1}, -D^{-1}C, D^{-1}\}$ is a state space realization of $\mathbf{H}^{-1}$. Furthermore, if $\mathbf{H}$ has values in $\mathcal{L}(\mathcal{U}, \mathcal{U})$, then

$$\det[\mathbf{H}(s)] = \frac{\det[D]\det[sI - A + BD^{-1}C]}{\det[sI - A]}.   \tag{12.107}$$

PROOF. Proposition 6.4.1 shows that the inverse of $\mathbf{H}$ is given by (12.106). To complete the proof it remains to establish (12.107). Recall that if M and N are two finite dimensional operators acting between the appropriate spaces, then $\det[I + MN] = \det[I + NM]$. Using this we obtain

$$
\begin{aligned}
\det[\mathbf{H}(s)] &= \det[D + C(sI - A)^{-1}B] \\
&= \det[D]\det[I + D^{-1}C(sI - A)^{-1}B] \\
&= \det[D]\det[I + (sI - A)^{-1}BD^{-1}C] \\
&= \det[D]\det[(sI - A)^{-1}]\det[sI - A + BD^{-1}C] \\
&= \det[D]\det[sI - A + BD^{-1}C]/\det[sI - A].
\end{aligned}
$$

Therefore, (12.107) holds.   ∎

If $\{A, B, C, D\}$ is a realization of a scalar valued transfer function $\mathbf{H}$ and D is nonzero, then (12.107) reduces to

$$\mathbf{H}(s) = \frac{D \det[sI - A + BD^{-1}C]}{\det[sI - A]}. \tag{12.108}$$

Let J be any rational function with values in $\mathcal{L}(\mathcal{U}, \mathcal{Y})$. Then $J^{\sharp}$ is the rational function with values in $\mathcal{L}(\mathcal{Y}, \mathcal{U})$ defined by $(J^{\sharp})(s) = J(-\bar{s})^*$. The following result uses the algebraic Riccati equation to compute a spectral factorization. This factorization will play a fundamental role in connecting the classical root locus to the solution of the infinite horizon linear quadratic regular problem.

Theorem 12.9.2 *Suppose $\{A, B, C, 0\}$ is a stabilizable and detectable realization of a transfer function $\mathbf{G}$. Let Q be any self-adjoint solution to the algebraic Riccati equation (12.104) where R is a self-adjoint invertible operator. Finally, let Θ be the transfer function defined by*

$$\Theta(s) = I + K(sI - A)^{-1}B \qquad where \qquad K = R^{-1}B^*Q. \tag{12.109}$$

Then $\Theta^{\sharp}R\Theta$ is a factorization of $R + \mathbf{G}^{\sharp}\mathbf{G}$, that is,

$$\Theta^{\sharp}R\Theta = R + \mathbf{G}^{\sharp}\mathbf{G}. \tag{12.110}$$

Moreover, the inverse of Θ is given by

$$\Theta(s)^{-1} = I - K(sI - A + BK)^{-1}B. \tag{12.111}$$

If R is strictly positive and Q is the unique stabilizing solution to the algebraic Riccati equation in (12.104), then Θ^{-1} is stable.

PROOF. Using $K = R^{-1}B^*Q$ in the algebraic Riccati equation (12.104) yields

$$C^*C = K^*RK - A^*Q - QA.$$

For any complex number s this implies that

$$C^*C = K^*RK + (-sI - A^*)Q + Q(sI - A).$$

Now let $\Phi(s)$ be the inverse of $sI - A$. Multiplying by $B^*\Phi^{\sharp}$ on the left and by ΦB on the right, we obtain

$$
\begin{aligned}
B^*\Phi^{\sharp}C^*C\Phi B &= B^*\Phi^{\sharp}K^*RK\Phi B + B^*Q\Phi B + B^*\Phi^{\sharp}QB \\
&= B^*\Phi^{\sharp}K^*RK\Phi B + RK\Phi B + B^*\Phi^{\sharp}K^*R.
\end{aligned}
$$

Using $\mathbf{G} = C\Phi B$ and $\Theta = I + K\Phi B$, we arrive at

$$R + \mathbf{G}^{\sharp}\mathbf{G} = (I + K\Phi B)^{\sharp}R(I + K\Phi B) = \Theta^{\sharp}R\Theta.$$

The expression for the inverse of Θ follows from Lemma 12.9.1. If R is strictly positive and Q is the stabilizing solution, then obviously $A - BK$ is stable, and thus, Θ^{-1} is stable. ∎

A rational function Ψ with values in $\mathcal{L}(\mathcal{U}, \mathcal{Y})$ is called an *invertible outer function* if Ψ is a stable proper rational function, its inverse Ψ^{-1} exists and is also a stable proper rational function. If $\{A, B, C, D\}$ is a stable realization for a transfer function $\mathbf{H}$ where D is invertible and $A - BD^{-1}C$ is stable, then Lemma 12.9.1 shows that $\mathbf{H}$ is an invertible outer function. If Ξ is a rational function with values in $\mathcal{L}(\mathcal{U}, \mathcal{U})$, then we say that Ψ is an *invertible outer spectral factor* of Ξ if Ψ is an invertible outer function and $\Psi^\sharp \Psi = \Xi$. If Ξ admits an invertible outer spectral factor, then $\Xi^\sharp = \Xi$. Obviously, not every rational function admits an invertible outer spectral factor. If $\{A, B, C, 0\}$ is a stable realization of $\mathbf{G}$ and $R > 0$ and Q is the stabilizing solution to the algebraic Riccati equation in (12.104), then $R^{1/2}\Theta$ is an invertible outer spectral factor for $R + \mathbf{G}^\sharp \mathbf{G}$.

12.10   The root locus and the quadratic regulator

In this section we use the factorization in Theorem 12.9.2 to obtain a root locus interpretation for the parameter R in the weighted single input single output infinite horizon regulator problem in (12.103). To this end, let $\{A \text{ on } \mathcal{X}, B, C, 0\}$ be a stabilizable and detectable realization of a scalar valued transfer function $\mathbf{G}$, that is, $\mathbf{G}(s) = C(sI - A)^{-1}B$ where $\mathcal{U} = \mathcal{Y} = \mathbb{C}^1$. Let $r > 0$ be a scalar, and Q the unique stabilizing solution to the algebraic Riccati equation

$$A^*Q + QA + C^*C - r^{-1}QBB^*Q = 0. \tag{12.112}$$

Recall that the optimal input $\hat{u}$ to the weighted infinite time horizon linear quadratic regulator problem in (12.103), is given by $\hat{u} = -r^{-1}B^*Q\hat{x}$ where $\hat{x}$ is the optimal state trajectory. In this case, $\dot{\hat{x}} = A_c\hat{x}$ where A_c is the stable closed loop state operator $A_c = A - r^{-1}BB^*Q$.

Let $\{A, B, C, D\}$ be a realization for a scalar valued transfer function $\mathbf{G}$. Then using $(sI - A)^{-1} = \text{adj}(sI - A)/\det[sI - A]$, it follows that $\mathbf{G} = p/d$ where p is a polynomial and $d(s) = \det[sI - A]$ is the characteristic polynomial for A. If $\{A, B, C, D\}$ is a minimal realization of $\mathbf{G} = p/d$ where p and d are co-prime polynomials, then Proposition 7.1.3 shows that d equals the characteristic polynomial of A up to a nonzero constant, that is, $d = \gamma \det[sI - A]$ where γ is a constant. Realizations where the characteristic polynomial of A equals the denominator polynomial d are used in the following result.

Lemma 12.10.1 *Let $\{A, B, C, 0\}$ be a realization of a scalar valued transfer function $\mathbf{G} = p/d$ where p is a polynomial and d is the characteristic polynomial for A. Let Q be any self-adjoint solution to the algebraic Riccati equation in (12.112) where r is a nonzero scalar and let Δ be the characteristic polynomial for $A - BK$, that is,*

$$\Delta(s) = \det[sI - A + BK] \qquad where \qquad K = r^{-1}B^*Q. \tag{12.113}$$

Then we have the following factorization:

$$\Delta^\sharp \Delta = d^\sharp d + r^{-1}p^\sharp p. \tag{12.114}$$

PROOF. From Theorem 12.9.2, we have $\Theta^\sharp r \Theta = r + \mathbf{G}^\sharp \mathbf{G}$ where $\Theta(s) = I + K(sI - A)^{-1}B$. Applying relationship (12.107) of Lemma 12.9.1 with $\mathbf{H} = \Theta$, we obtain

$$\det[\Theta(s)] = \det[sI - A + BK]/\det[sI - A].$$

This readily implies that

$$\det[sI - A + BK] = \det[sI - A]\det[\Theta(s)].$$

Since $\Theta(s)$ is a scalar, $\det[\Theta(s)] = \Theta(s)$, and hence, $\Delta(s) = d(s)\Theta(s)$. If we rewrite $\Theta^\sharp r \Theta = r + \mathbf{G}^\sharp \mathbf{G}$ as

$$d^\sharp \Theta^\sharp \Theta d = d^\sharp d + r^{-1}p^\sharp p.$$

Applying $\Theta = \Delta/d$, we obtain the factorization in (12.114). ∎

Now assume that $\{A, B, C, 0\}$ is stabilizable and detectable. Let Q be the stabilizing solution to the weighted algebraic Riccati equation in (12.112) where $r > 0$. Clearly, the operator $A - BK$ is stable where $K = r^{-1}B^*Q$. Hence, all the roots of Δ are in the open left half complex plane. Notice that λ is a root of a polynomial q if and only if $-\bar\lambda$ is a root of the polynomial $q^\sharp$. Therefore, all the roots of $\Delta^\sharp$ are contained in the open right half plane. In particular, Δ and $\Delta^\sharp$ have no common roots and their product $\Delta^\sharp \Delta$ has no roots on the imaginary axis. It now follows from (12.114) that the roots of Δ are the left half plane roots of $d^\sharp d + r^{-1}p^\sharp p$. This immediately yields the following result.

Theorem 12.10.2 *Suppose that $\{A, B, C, 0\}$ is a stabilizable and detectable realization of a scalar valued transfer function $\mathbf{G} = p/d$ where p is a polynomial and $d(s) = \det[sI - A]$. Let $K = r^{-1}B^*Q$ where Q is the stabilizing solution to the algebraic Riccati equation in (12.112) with $r > 0$. Then the roots of the characteristic polynomial of $A - BK$ are precisely the left half plane roots of the polynomial $d^\sharp d + r^{-1}p^\sharp p$.*

To complete this section, we use root locus techniques to see how the weight r affects the eigenvalues of the closed loop state space operator $A - BK$. First let us recall some classical root locus results; for further details see Ogata [95]. Consider a parameter dependent polynomial of the form $q + k\eta$ where η and q are two polynomials with real coefficients and real parameter $k \geq 0$. Furthermore, it is assumed that the degree of η is less than or equal to the degree of q. Recall that the root locus for $q + k\eta$ is the graph of the zeros of $q + k\eta$ as k varies from zero to infinity. Moreover, for $k = 0$ the $\deg q$ branches of the root locus of $q + k\eta$ start at the zeros of q. As k tends to infinity, $\deg \eta$ branches of the root locus approach the zeros of η. If $\delta = \deg q - \deg \eta > 0$, then the remaining δ branches tend to infinity as k approaches infinity. Each of these δ branches asymptotically approach one of the asymptotes of the root locus. Assuming q is monic and α is the coefficient of the highest order term of η, these asymptotes are half lines whose angles are given by

$$\frac{\pi + 2\pi j}{\delta} \quad \text{for } j = 0, 1, 2, \cdots, \delta - 1 \qquad \text{if } \alpha > 0$$

$$\frac{2\pi j}{\delta} \quad \text{for } j = 0, 1, 2, \cdots, \delta - 1 \qquad \text{if } \alpha < 0.$$

Recall that a monic polynomial is a polynomial of the form $s^j + a_{j-1}s^{j-1} + \cdots a_1 s + a_0$.

Throughout the rest of this section, we assume that $\{A, B, C, 0\}$ is a stabilizable and detectable realization for a scalar valued transfer function $\mathbf{G} = p/d$ where A, B, C are real matrices and p, d are polynomials whose coefficients are real while d is the characteristic polynomial for A. Moreover, Q is the unique stabilizing solution of the algebraic Riccati equation in (12.112) where $r > 0$ is a scalar. In particular, this implies that $d^\sharp(s) = d(-s)$ and $p^\sharp(s) = p(-s)$. Using (12.114), it follows that

$$\Delta(-s)\Delta(s) = d(-s)d(s) + r^{-1}p(-s)p(s)$$

where Δ is the characteristic polynomial for $A - BK$. So, if we let $k = 1/r$, then the root locus of $d(-s)d(s) + kp(-s)p(s)$ is a graph of the eigenvalues of $A - BK$ and $-(A - BK)^*$ as r varies from infinity to zero. In particular, the left half plane root locus of $d(-s)d(s) + r^{-1}p(-s)p(s)$ is precisely the eigenvalues of $A - BK$ as r varies from infinity to zero. Notice that for any polynomial q with real coefficients, the zeros of $q(-s)q(s)$ are symmetric about the real and imaginary axis. Hence, the root locus of $d(-s)d(s) + kp(-s)p(s) = \Delta(-s)\Delta(s)$ is also symmetric about the real and imaginary axis.

As $k = 1/r$ varies from zero to infinity, the branches of the root locus of the polynomial $d(-s)d(s) + kp(-s)p(s)$ move from the zeros of $d(-s)d(s)$ to the zeros of $p(-s)p(s)$ and the appropriate asymptotes. The asymptotes of $d(-s)d(s) + kp(-s)p(s)$ are determined by $m = \deg d - \deg p$. To be precise, if m is even, then $d(-s)d(s) + kp(-s)p(s) = 0$ can be expressed as $b(s) + k\alpha a(s) = 0$ where $\alpha > 0$ and a and b are monic polynomials. Since α is the coefficient of the highest order term of αa, the angles ϕ_{ej} for the asymptotes are given by

$$\phi_{ej} = \frac{\pi + 2\pi j}{2m} \qquad \text{for } j = 0, 1, 2, \cdots, 2m - 1. \tag{12.115}$$

However, if m is odd, then $d(-s)d(s) + kp(-s)p(s) = 0$ can be expressed as $b(s) + k\alpha a(s) = 0$ where $\alpha < 0$ and a and b are monic polynomials. In this case, the angles ϕ_{oj} for the asymptotes are given by

$$\phi_{oj} = \frac{\pi j}{m} \qquad \text{for } j = 0, 1, 2, \cdots, 2m - 1. \tag{12.116}$$

Moreover, because $d(-s)d(s) + kp(-s)p(s)$ is symmetric about the real and imaginary axis, the origin of these asymptotes is zero. Notice that none of these asymptotes lie on the imaginary axis. Therefore, as $r = 1/k$ varies from infinity to zero the eigenvalues of $A - BK$ start at the zeros of $d(-s)d(s)$ in the open left half plane and move towards the zeros of $p(-s)p(s)$ in the open left half plane and the appropriate asymptotes in the open left half plane. For m even, the asymptote angles ϕ_{ej} are given by (12.115). For m odd, the asymptote angles ϕ_{oj} are given by (12.116).

Recall that d is the characteristic polynomial of A. Hence, the eigenvalues of A are the zeros of d including their multiplicity. Let $\{\lambda_1, \lambda_2, \cdots, \lambda_\mu\}$ be the eigenvalues of A including their multiplicity in the closed left half plane, and let $\{\lambda_{\mu+1}, \lambda_{\mu+2}, \cdots, \lambda_n\}$ be the remaining eigenvalues of A including multiplicity. Obviously,

$$\Lambda = \{\lambda_1, \lambda_2, \cdots, \lambda_\mu, -\lambda_{\mu+1}, -\lambda_{\mu+2}, \cdots, -\lambda_n\} \tag{12.117}$$

are the zeros of $d(-s)d(s)$ in the closed left half plane. Moreover, $-\Lambda \bigcup \Lambda$ are all the zeros of $d(-s)d(s)$ including their multiplicity. Let $\{z_1, z_2, \cdots, z_\nu\}$ be the zeros of p in the closed left half plane including their multiplicity, and let $\{z_{\nu+1}, z_{\nu+2}, \cdots, z_m\}$ be the remaining zeros of p including multiplicity. Clearly,

$$\mathcal{Z} = \{z_1, z_2, \cdots, z_\nu, -z_{\nu+1}, -z_{\nu+2}, \cdots, -z_m\} \tag{12.118}$$

are the closed left half plane zeros of $p(-s)p(s)$. Furthermore, $-\mathcal{Z} \bigcup \mathcal{Z}$ are all the zeros of $p(-s)p(s)$ including their multiplicity. Therefore, as $r = 1/k$ varies from infinity to zero the eigenvalues of $A - BK$ start at the roots Λ given in (12.117) and then follow the branches of the root locus for $d(-s)d(s) + kp(-s)p(s)$ and move towards the roots $\mathcal{Z}$ given in (12.118) and the corresponding asymptotes with angles ϕ_{ej} or ϕ_{oj} in the open left half plane.

In the scalar case, the cost in the weighted linear quadratic regulator problem is given by

$$\varepsilon(x_0, r) = \inf \left\{ \int_0^\infty (||y(\sigma)||^2 + r||u(\sigma)||^2) d\sigma : u \in L^2[0, \infty) \right\}. \tag{12.119}$$

If $r = 1/k$ is large, then the term $r||u(t)||^2$ is heavily weighted in computing the cost, and hence, any input effort used to move poles is "expensive". Moreover, in this case the root locus shows that the eigenvalues of $A - BK$ are in a neighborhood of the roots Λ given in (12.117). If any one of the roots Λ in (12.117) lie on the imaginary axis, then a large r will place an eigenvalue of $A - BK$ in the open left half plane in some neighborhood of that root. In other words, for large r, the optimal input $\hat{u} = -r^{-1}B^*Q\hat{x}$ barely moves the stable eigenvalues of A, moves the eigenvalues of A on the imaginary axis slightly into the open left half plane, and moves the unstable eigenvalues of A close to $\{-\lambda_{\mu+1}, -\lambda_{\mu+2}, \cdots, -\lambda_n\}$. Because a large r penalizes the input, the result is that the optimal input only moves the eigenvalues of A necessary to stabilize the closed loop state operator $A - BK$.

On the other hand, if r is small, then the term $r||u(t)||^2$ is barely weighted in computing the cost $\varepsilon(x_0, r)$, and thus, input effort is "cheap". In other words, a large input effort has little effect on the cost. In this case the root locus shows that m of eigenvalues of $A - BK$ are in a neighborhood of the points $\mathcal{Z}$, and the remaining eigenvalues of $A - BK$ move towards infinity along the asymptotes in the open left half plane corresponding to the angles ϕ_{ej} in (12.115) for even m, respectively ϕ_{oj} in (12.116) for odd m. So, if r is small, then the optimal controller moves m of the eigenvalues of A to $\mathcal{Z}$ and the places remaining eigenvalues of A as far as possible in the left half plane.

Example 12.10.1 To complete this section we demonstrate the above results with a simple example. To this end, let $\{A, B, C, 0\}$ be any minimal realization of

$$\mathbf{G}(s) = \frac{s+3}{(s+1)(s-2)(s+2+\imath)(s+2-\imath)}.$$

Clearly, the McMillan degree of $\mathbf{G}$ is four and A is unstable. The root locus for $d(-s)d(s) + kp(-s)p(s)$ is given in Figure 1. In this case, $m = 3$. According to (12.116) the angles for the asymptotes for the root locus are given by

$$0, \frac{\pi}{3}, \frac{2\pi}{3}, \pi, \frac{4\pi}{3}, \frac{5\pi}{3}.$$

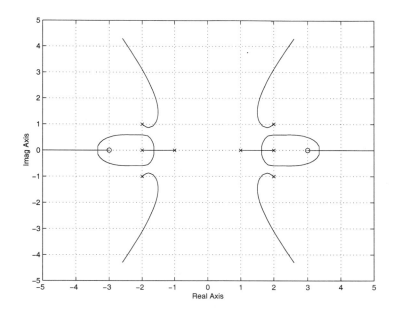

Figure 12.1: A root locus for design

In particular, the asymptotes for the root locus in the left half plane occur at the angles $2\pi/3, \pi, 4\pi/3$. So, as $r = 1/k$ moves from infinity to zero the eigenvalues of $A - BK$ move from the points $-1, -2, -2 + \imath, -2 - \imath$ along the branches of the root locus in the open left half plane to the points

$$-3, \infty e^{2\pi\imath/3}, \infty e^{\pi\imath}, \infty e^{4\pi\imath/3}. \tag{12.120}$$

In particular, if r is large the eigenvalues of $A - BK$ are in some neighborhood of the points $-1, -2, -2+\imath, -2-\imath$. So, for large r the optimal control barely moved the stable eigenvalues $-1, -2 + \imath, -2 - \imath$ for A and shifted the unstable eigenvalue 2 of A to -2. On the other hand, if r is small, then the optimal control moved the eigenvalues of A to (12.120), which is as far as they can possible go in the left half plane.

12.10.1   Some comments on the outer spectral factor

In this section we present some comments concerning the minimality of the realization $\{A, B, K, I\}$ of the spectral factor Θ defined in (12.109). Let $\{A, B, C, 0\}$ be a controllable and observable realization of a transfer function $\mathbf{G}$. Assume that R is strictly positive and let Q be the stabilizing solution to the weighted algebraic Riccati equation (12.104). Finally, let Θ be the transfer function defined by

$$\Theta(s) = I + K(sI - A)^{-1}B \qquad \text{where} \qquad K = R^{-1}B^{*}Q. \tag{12.121}$$

Theorem 12.9.2 shows that $\Theta^{\sharp}R\Theta$ is a factorization of $R + \mathbf{G}^{\sharp}\mathbf{G}$, that is, (12.110) holds. Moreover, $\{A - BK, B, -K, I\}$ is a stable realization of Θ^{-1}. In particular, if A is stable, then Θ is an invertible outer function, that is, both Θ and Θ^{-1} are stable transfer functions.

Obviously, $\{A, B, K, I\}$ is a controllable realization of Θ. However, this realization is not necessarily observable. For an example, let $\{A, B, C, 0\}$ be any minimal realization of $\mathbf{G}(s) = (2 - s)/(s + 1)(s + 2)$, and set $R = 1$. Then we claim that $\{A, B, B^*Q, I\}$ is not observable. To verify this recall that Θ is an outer spectral factor of $I + \mathbf{G}^{\sharp}\mathbf{G}$. Thus,

$$\Theta^{\sharp}(s)\Theta(s) = (2 - s^2)/(1 - s^2), \qquad (12.122)$$

where all the poles and zeros of $\Theta(s)$ are in the open left half plane. Hence, all the poles and zeros of $\Theta^{\sharp}$ are in the open right half plane. Using this along with the factorization in (12.122), it follows that $\Theta(s) = (s + \sqrt{2})/(s + 1)$. However, $\{A, B, B^*Q, I\}$ is a two dimensional controllable realization of Θ. Obviously, the state dimension of the minimal realization of $(s + \sqrt{2})/(s + 1)$ is one. So, the system $\{A, B, B^*Q, I\}$ is not observable. Since $\{A - BK, B, -B^*Q, I\}$ is a two dimensional controllable realization of $\Theta^{-1} = (s + 1)/(s + \sqrt{2})$, it also follows that $\{A - BK, B, -B^*Q, I\}$ is not observable.

Throughout the rest of this section, we assume that $\{A, B, C, 0\}$ is a minimal realization for a scalar valued strictly proper rational transfer function $\mathbf{G} = p/d$, where p and d are two polynomials with real coefficients and no common zeros. Because the coefficients of p and d are real, we have $p^{\sharp}(s) = p(-s)$ and $d^{\sharp}(s) = d(-s)$. Moreover, we also assume that $p(-s)p(s)$ and $d(-s)d(s)$ have no common zeros, Q is the stabilizing solution to the algebraic Riccati equation in (12.104) and $R = r > 0$. Then we claim that $\{A, B, K, I\}$ is a minimal realization of Θ. Because Θ and Θ^{-1} have the same McMillan degree, it follows that $\{A - BK, B, -K, I\}$ is a minimal realization of Θ^{-1}. In particular, the eigenvalues of $A - BK$ are precisely the zeros of Θ, including their multiplicity; see Proposition 7.1.3. Furthermore, the McMillan degree of $1/\Theta$ equals the degree of the polynomial d. Hence, there are $\deg d$ zeros of Θ, all of which are in the open left half plane.

To prove that $\{A$ on $\mathcal{X}, B, K, I\}$ is a minimal realization, notice that because $R = r$ is scalar valued, the factorization in (12.110) reduces to

$$\Theta(-s)\Theta(s) = 1 + \frac{\mathbf{G}(-s)\mathbf{G}(s)}{r} = \frac{rd(-s)d(s) + p(-s)p(s)}{rd(-s)d(s)}. \qquad (12.123)$$

Since $p(-s)p(s)$ and $d(-s)d(s)$ have no common zeros, the polynomials $rd(-s)d(s)$ and $rd(-s)d(s) + p(-s)p(s)$ have no common zeros. Now let us proceed by contradiction. If $\{A, B, K, I\}$ is not observable, then $\Theta = a/b$ where a and b are polynomials and $\deg b < \deg d = \dim \mathcal{X}$. Substituting this into (12.123) shows that we must have cancellation between the zeros of $rd(-s)d(s)$ and $rd(-s)d(s) + p(-s)p(s)$. This contradicts the fact that $p(-s)p(s)$ and $d(-s)d(s)$ have no common zeros. Therefore, $\{A, B, K, I\}$ is controllable and observable. Since Θ and $1/\Theta$ have the same McMillan degree, it also follows that $\{A - BK, B, -K, I\}$ is a controllable and observable realization of $1/\Theta$.

As before, assume that $p(-s)p(s)$ and $d(-s)d(s)$ are co-prime. Then $1 + \mathbf{G}(-s)\mathbf{G}(s)/r$ and $rd(-s)d(s) + p(-s)p(s)$ have the same zeros. Moreover, these zeros are symmetric about the imaginary axis. Because $rd(-s)d(s) + p(-s)p(s)$ is a polynomial of degree $2\deg d$, we see that $1 + \mathbf{G}(-s)\mathbf{G}(s)/r$ has $2\deg d$ zeros and these zeros are symmetric about the imaginary axis. Therefore, the zeros of $1 + \mathbf{G}(-s)\mathbf{G}(s)/r$ in the open left half plane are precisely the eigenvalues of $A - BK$.

12.11 Notes

All the results in this section are classical and date back to Kalman [70]. The linear quadratic regulator problem plays a fundamental role in systems and control. For some further results on linear quadratic methods and control theory see Anderson-Moore [4, 6], Kwakernaak-Sivan [77], and Dorato-Abdallah-Cerone [36]. The linear quadratic regulator problem is an important problem in optimal control theory; see Athans-Falb [7], Berkovitz [15], Bryson-Ho [22], and Lee-Markus [80]. For some results on game theory and optimal control see Basar-Bernhard [13]. There are many different ways to derive the two point boundary value problem in (12.34). For example one can use the calculus of variations or the maximum principle in optimal control theory; see, for example, Leitmann [81]. Here operator methods were used to derive the two point boundary value problem. Operator techniques have been widely used to solve the linear quadratic regulator problem and many other problems in control theory; see Balakrishnan [8], Fuhrmann [47], Naylor-Sell [93], Luenberger [85] and Porter [100]. Finally, it is noted that linear quadratic methods also play a basic role in H^∞ control theory; see Green-Limebeer [57], Mustafa-Glover [92] and Zhou-Doyle-Glover [131]. For some numerical procedures to solve the algebraic Riccati equation see Arnold-Laub [3] and Van Dooren [121].

Chapter 13

The Hamiltonian Matrix and Riccati Equations

Associated with any Riccati equation is a matrix called the Hamiltonian matrix. In this chapter we present elementary properties of Hamiltonian matrices. We show how one can compute the stabilizing solution to an algebraic Riccati equation by using the Hamiltonian matrix associated with that equation. We also show how one can compute a solution to a Riccati differential equation from the state transition matrix of the associated Hamiltonian matrix. To this end, consider the algebraic Riccati equation

$$A^*Q + QA + QRQ + S = 0 \tag{13.1}$$

where A, R and S are operators on a finite dimensional space $\mathcal{X}$ with S and R self-adjoint. When $S = C^*C$ and $R = -BB^*$, this equation is the same as the algebraic Riccati equation (12.87) encountered in the linear quadratic regulator problem. We are considering a more general Riccati equation here because the results of this chapter are also useful in the later chapters on H^∞ analysis and control. The Hamiltonian matrix associated with the algebraic Riccati equation in (13.1) is given by

$$H = \begin{bmatrix} A & R \\ -S & -A^* \end{bmatrix} \text{ on } \begin{bmatrix} \mathcal{X} \\ \mathcal{X} \end{bmatrix}. \tag{13.2}$$

13.1 The Hamiltonian matrix and stabilizing solutions

We say that Q is a *stabilizing solution* to the algebraic Riccati equation in (13.1) if Q is an operator on $\mathcal{X}$ satisfying (13.1) and $A + RQ$ is stable. As expected, if there exists a stabilizing solution Q, then it is self-adjoint and is the only stabilizing solution. To see that a stabilizing solution Q is self-adjoint, use the Riccati equation to obtain

$$(Q^* - Q)(A + RQ) = Q^*A + Q^*RQ - QA - QRQ = Q^*A + Q^*RQ + A^*Q + S.$$

Since the operator on right-hand-side of the second equality is self-adjoint, it follows that $\tilde{Q}A_c$ is self-adjoint where $\tilde{Q} = Q^* - Q$ and $A_c = A + RQ$; thus $\tilde{Q}A_c = A_c^*\tilde{Q}^*$. Since $\tilde{Q}^* = -\tilde{Q}$, we obtain the Lyapunov equation $\tilde{Q}A_c + A_c^*\tilde{Q} = 0$. Because A_c is stable, the Lyapunov equation has only one solution, namely $\tilde{Q} = 0$; hence $Q^* = Q$.

To see that the stabilizing solution is unique, consider any two stabilizing solutions Q_1 and Q_2 and now let $\tilde{Q} = Q_2 - Q_1$. With $\Lambda_1 = A + RQ_1$ and $\Lambda_2 = A + RQ_2$, we obtain

$$
\begin{aligned}
\tilde{Q}\Lambda_1 + \Lambda_2^*\tilde{Q} &= (Q_2 - Q_1)(A + RQ_1) + (A + RQ_2)^*(Q_2 - Q_1) \\
&= Q_2 A + A^* Q_2 + Q_2 RQ_2 - Q_1 A - A^* Q_1 - Q_1 RQ_1 \\
&= S - S = 0.
\end{aligned}
$$

Because Λ_1 and Λ_2 are both stable, $\tilde{Q} = 0$, that is, $Q_2 = Q_1$ and the stabilizing solution is unique.

Recall that the Hamiltonian matrix associated with the algebraic Riccati equation in (13.1) is given by (13.2). If we rearrange the Riccati equation (13.1) as

$$-S - A^*Q = Q(A + RQ)$$

and let $A_c = A + RQ$, then the Riccati equation is equivalent to

$$
\begin{aligned}
A + RQ &= A_c \\
-S - A^*Q &= QA_c
\end{aligned}
$$

or

$$H \begin{bmatrix} I \\ Q \end{bmatrix} = \begin{bmatrix} I \\ Q \end{bmatrix} A_c. \tag{13.3}$$

Thus Q is a stabilizing solution of the Riccati equation (13.1) if and only if $[I\ Q]^{tr}$ satisfies (13.3) where A_c is stable. In this case, $A_c = A + RQ$. Since the operator $[I\ Q]^{tr}$ is one to one, our new equivalent condition for a stabilizing Q is equivalent to the requirement that the range $\mathcal{R}$ of $[I\ Q]^{tr}$ is invariant for H and the restriction of H to this subspace is stable. In particular, the restriction of H to $\mathcal{R}$ is similar to $A + RQ$.

Consider now any invertible operator X which maps onto $\mathcal{X}$. Then, postmultiplying (13.3) by X, we obtain

$$H\Gamma = \Gamma\Lambda \tag{13.4}$$

where $\Gamma = [X\ Y]^{tr}$ with $Y = QX$ and Λ is the stable operator given by $\Lambda = X^{-1}A_c X$. Thus, we can say that if the Riccati equation has a stabilizing solution, then there exists an operator $\Gamma = [X\ Y]^{tr}$ mapping into $\mathcal{X} \oplus \mathcal{X}$ with X invertible such that (13.4) holds for some stable operator Λ. To demonstrate the converse, suppose there is an operator $\Gamma = [X\ Y]^{tr}$ mapping into $\mathcal{X} \oplus \mathcal{X}$ with X invertible such that (13.4) holds with Λ stable. Postmultiplying (13.4) by X^{-1}, we see that (13.3) holds with $Q = YX^{-1}$ and $A_c = X\Lambda X^{-1}$. Thus Q is a stabilizing solution to the Riccati equation (13.1) and $A + RQ = X\Lambda X^{-1}$. So $A + RQ$ and Λ are similar and have the same characteristic polynomial. We have just demonstrated the following result.

Lemma 13.1.1 *The algebraic Riccati equation (13.1) has a stabilizing solution if and only if there exists an operator $\Gamma = [X\ Y]^{tr}$ mapping into $\mathcal{X} \oplus \mathcal{X}$ with X invertible such that (13.4) holds for some stable operator Λ. In this case, the stabilizing solution is uniquely given by*

$$Q = YX^{-1} \tag{13.5}$$

and $A + RQ$ is similar to Λ.

Note that when (13.3) holds and $\Gamma = [X \; Y]^{tr}$ with X invertible, then Γ is one to one with rank n where $n = \dim[\mathcal{X}]$. Our next step is to characterize all the one to one operators of rank n which satisfy (13.4) for some stable Λ. Before carrying out this step we make some observations.

Suppose that Q is any self-adjoint solution to the algebraic Riccati equation in (13.1) and let W be the invertible block matrix defined by

$$W = \begin{bmatrix} I & 0 \\ Q & I \end{bmatrix} \quad \text{on} \quad \begin{bmatrix} \mathcal{X} \\ \mathcal{X} \end{bmatrix}. \tag{13.6}$$

Notice that the inverse of W is given by replacing Q by $-Q$ in the definition of W. A simple calculation shows that

$$W^{-1}HW = \begin{bmatrix} A + RQ & R \\ 0 & -(A + RQ)^* \end{bmatrix}. \tag{13.7}$$

Recall that if F is any operator valued rational function, then $(F^{\sharp})(s) = F(-\bar{s})^*$. We now note that, if p is the characteristic polynomial of an operator M on a space of finite dimension n, then $(-1)^n p^{\sharp}$ is the characteristic polynomial of $-M^*$. This can be shown as follows:

$$\det[sI + M^*] = (-1)^n \det[-\bar{s}I - M]^* = (-1)^n p(-\bar{s})^* = (-1)^n p^{\sharp}(s).$$

It now follows from (13.7) that

$$\det[sI - H] = (-1)^n \Delta(s)\Delta^{\sharp}(s) \tag{13.8}$$

where Δ is the characteristic polynomial of $A + RQ$.

From (13.8), we see that if the algebraic Riccati equation admits a stabilizing solution, then the corresponding Hamiltonian matrix does not have any eigenvalues on the imaginary axis. However, the converse is not necessarily true. For example, if $A = 1$, $S = 2$ and $R = 0$, then one and minus one are the eigenvalues of H. In this case, the only solution to the algebraic Riccati equation is $Q = -1$. Obviously, $1 = A + RQ$ is not stable.

It now follows from Lemma 13.1.1 that if there exists an operator $\Gamma = [X \; Y]^{tr}$ with X invertible such that (13.4) holds for some stable operator Λ, then $\det[sI - H] = (-1)^n \Delta(s)\Delta^{\sharp}(s)$ where Δ is the characteristic polynomial of Λ; hence H has no imaginary eigenvalues. Shortly, we will see that if H has no imaginary eigenvalues, then there is a one to one operator $\Gamma = [X \; Y]^{tr}$ such that (13.4) holds for some stable operator Λ and $\det[sI - H] = (-1)^n \Delta(s)\Delta^{\sharp}(s)$ where $\Delta(s) = \det[sI - \Lambda]$. However X may not be invertible.

Consider now any Hamiltonian matrix H of the form (13.2) where R and S are self-adjoint and let J be the invertible operator defined by

$$J = \begin{bmatrix} 0 & -I \\ I & 0 \end{bmatrix} \quad \text{on} \quad \begin{bmatrix} \mathcal{X} \\ \mathcal{X} \end{bmatrix}. \tag{13.9}$$

Notice that $J^{-1} = J^* = -J$. A simple calculation shows that

$$JH = \begin{bmatrix} S & A^* \\ A & R \end{bmatrix}$$

is self-adjoint; hence, $JH = H^*J^* = -H^*J$. Thus, the operators $-H^*$ and H are similar and have the same characteristic polynomial; hence the characteristic polynomial Δ_H of H must satisfy

$$\Delta_H^\sharp = \Delta_H . \tag{13.10}$$

This implies that a scalar λ is a zero of Δ_H with multiplicity m if and only if $-\bar{\lambda}$ is a zero of Δ_H with multiplicity m. Thus, λ is an eigenvalue of H, if and only if $-\bar{\lambda}$ is also an eigenvalue of H.

Now assume that the Hamiltonian matrix does not have any eigenvalues on the imaginary axis, that is, Δ_H has no imaginary zeros. Let Δ be the unique monic polynomial whose zeros (multiplicities included) are the stable (negative real part) zeros of Δ_H. Then Δ must be of degree n and $\Delta_H = (-1)^n \Delta \Delta^\sharp$ where n is the dimension of $\mathcal{X}$. Consider now the invariant subspace $\mathcal{R} = \ker \Delta(H)$ for H. We call this the *stable subspace* associated with H. We claim that it has dimension n. To see this, we use the relationship $JH = -H^*J$ to obtain

$$J\Delta(H) = \Delta(-H^*)J = \Delta^\sharp(H)^*J$$

It now follows that $\ker \Delta(H)$ and $\ker \Delta^\sharp(H)$ have the same dimension. Since the polynomials Δ and $\Delta^\sharp$ have no common zeros and $\Delta(H)\Delta(H)^\sharp = (-1)^n \Delta_H(H) = 0$, it follows from Lemma 13.1.6 at the end of this section that $\dim[\ker \Delta(H)] = n$. We have just demonstrated the following result.

Lemma 13.1.2 *Let H be the Hamiltonian matrix in (13.2) where R and S are self-adjoint operators on a finite dimensional space $\mathcal{X}$ of dimension n and assume that H has no imaginary eigenvalues. Then,*

$$\det[sI - H] = (-1)^n \Delta(s)\Delta^\sharp(s) \tag{13.11}$$

where Δ is a stable monic polynomial of degree n. Moreover the dimension of $\ker \Delta(H)$ is n.

The above result allows us to characterize all the one to one maps of rank n which satisfy (13.4) with Λ stable. To see this, suppose H is a Hamiltonian matrix with no imaginary eigenvalues. Let Γ be any one to one map whose range equals the kernel of $\Delta(H)$ where Δ is the monic polynomial whose zeros (multiplicities included) are the stable (negative real part) zeros of Δ_H. Then, according to the above lemma, the rank of Γ is n. Since $\ker \Delta(H)$ is invariant for H, there is an operator Λ such that $H\Gamma = \Gamma\Lambda$. (In fact, $\Lambda = (\Gamma^*\Gamma)^{-1}\Gamma^*H\Gamma$.) It now follows that $\Delta(H)\Gamma = \Gamma\Delta(\Lambda)$. Because the range of Γ equals the kernel of $\Delta(H)$, we obtain that $\Gamma\Delta(\Lambda) = 0$. Since Γ is one to one, we must have $\Delta(\Lambda) = 0$. Recalling that Δ is a stable polynomial we conclude that Λ is stable.

Now suppose that Γ is any one to one map of rank n which satisfies (13.4) for some stable Λ. Then the range of Γ is an invariant subspace for H and the restriction of H to the range of Γ is similar to Λ. Thus H is similar to a matrix of the form

$$\begin{bmatrix} \Lambda & * \\ 0 & \Lambda_2 \end{bmatrix} .$$

Hence $\Delta_H(s) = \det[sI - H] = d(s)d_2(s)$ where d and d_2 are the characteristic polynomials of Λ and Λ_2 respectively. Recall that if λ is a zero of multiplicity m of Δ_H, then $-\bar{\lambda}$ is also a zero of multiplicity m of Δ_H. Since d divides Δ_H and d is a monic polynomial whose roots have negative real parts, it follows that $(-1)^n d^\sharp$ (the monic polynomial whose roots are precisely the negative of the roots of d) also divides Δ_H. Since d and $d^\sharp$ are monic polynomials of order n which divide Δ_H and Δ_H is a monic polynomial of order $2n$, we must have $\Delta_H = (-1)^n dd^\sharp$. Thus $\det[sI - H] = (-1)^n d(s)d^\sharp(s)$. Hence H has no imaginary eigenvalues. Since $d(H)\Gamma = \Gamma d(\Lambda) = 0$ and $\dim[\ker d(H)] = n$, the range of Γ equals the kernel of $d(H)$. The above analysis leads to the following result.

Lemma 13.1.3 *Let H be the Hamiltonian matrix in (13.2) where R and S are self-adjoint operators on a space $\mathcal{X}$ of finite dimension n. Then there exists a one to one operator Γ of rank n satisfying $H\Gamma = \Gamma\Lambda$ for some stable Λ if and only if H has no imaginary eigenvalues. In this case, let Δ be the unique monic polynomial whose zeros are the stable (negative real part) zeros (including multiplicity) of the characteristic polynomial of H. Then Γ is a one to one operator of rank n satisfying (13.4) for some stable Λ if and only if Γ is a one to one operator whose range equals the kernel of $\Delta(H)$. Moreover, $\det[sI - \Lambda] = \Delta(s)$.*

Note that if Γ and Γ_2 are any two operators which are one to one and have the same range, then there is an invertible operator M such that $\Gamma_2 = \Gamma M$; in fact, $M = (\Gamma^*\Gamma)^{-1}\Gamma^*\Gamma_2$. Hence, we have the following corollary.

Corollary 13.1.4 *Let H be the Hamiltonian matrix in (13.2) where R and S are self-adjoint operators on a space $\mathcal{X}$ of finite dimension n. Then there exists a one to one operator Γ of rank n satisfying (13.4) for some stable Λ if and only if H has no imaginary eigenvalues. Moreover, Γ is unique up to a similarity transformation on the right, that is, if $H\Gamma_2 = \Gamma_2\Lambda_2$ where Γ_2 is one to one with rank n and Λ_2 is stable, then there exists an invertible operator M such that $\Gamma_2 = \Gamma M$ and $\Lambda_2 = M^{-1}\Lambda M$.*

By combining the preceding results of this section, we arrive at the following result.

Theorem 13.1.5 *Let R and S be self-adjoint operators on a space $\mathcal{X}$ of finite dimension n. Then the algebraic Riccati equation in (13.1) admits a stabilizing solution if and only if the following two conditions hold.*

(i) The corresponding Hamiltonian matrix H has no imaginary eigenvalues.

(ii) If $\Gamma = [X\ Y]^{tr}$ is any one to one operator of rank n satisfying

$$H\Gamma = \Gamma\Lambda \tag{13.12}$$

where Λ is a stable operator, then X is invertible.

In this case, the unique stabilizing solution to the algebraic Riccati equation in (13.1) is given by

$$Q = YX^{-1}. \tag{13.13}$$

Moreover $A + RQ$ is similar to Λ and the characteristic polynomial Δ of $A + RQ$ satisfies

$$\Delta\Delta^\sharp = (-1)^n\Delta_H \tag{13.14}$$

where Δ_H is the characteristic polynomial of H.

PROOF. Suppose the algebraic Riccati equation admits a stabilizing solution Q. We have already shown that the corresponding Hamiltonian matrix H has no imaginary eigenvalues and $\Gamma = [I\ Q]^{tr}$ satisfies (13.12) with $\Lambda = A + RQ$. Clearly Γ is a one to one operator of rank n. Because the one to one operators Γ of rank n which satisfy (13.12) for a stable Λ are unique up to a similarity transformation on the right, Part (ii) holds; see Corollary 13.1.4.

On the other hand, suppose the conditions in (i) and (ii) hold. It follows from Lemma 13.1.3 that there exists a one to one operator $\Gamma = [X\ Y]^{tr}$ of rank n which satisfies (13.12). Since X is invertible it follows from Lemma 13.1.1 that $Q = YX^{-1}$ is a stabilizing solution to the algebraic Riccati equation; also $A + RQ$ is similar to Λ. ∎

13.1.1   Computation of the stabilizing solution

The Schur decomposition provides a numerically efficient method to compute a one to one operator Γ of rank n satisfying $H\Gamma = \Gamma\Lambda$ where Λ is a stable operator. Suppose that the Hamiltonian matrix H has no eigenvalues on the imaginary axis. Then, $\det[sI - H] = (-1)\Delta(s)\Delta^\sharp(s)$ where Δ is a stable monic polynomial of order n. According to the Schur decomposition, H is unitarily equivalent to an upper triangular matrix of the form

$$\begin{bmatrix} \Lambda & * \\ 0 & \Lambda_2 \end{bmatrix}$$

where $\det[sI - \Lambda] = \Delta(s)$ and $\det[sI - \Lambda_2] = (-1)^n\Delta^\sharp(s)$. Thus, Λ is a stable operator and

$$H\begin{bmatrix} X & U \\ Y & V \end{bmatrix} = \begin{bmatrix} X & U \\ Y & V \end{bmatrix}\begin{bmatrix} \Lambda & * \\ 0 & \Lambda_{22} \end{bmatrix}. \qquad (13.15)$$

All these matrices are block matrices on $\mathcal{X} \oplus \mathcal{X}$. The 2×2 block matrix N consisting of the operators X, Y, U, V is unitary, that is, $N^*N = I$. Hence, $[X\ Y]^{tr}$ is one to one and of rank n. Finally, if X is invertible, then YX^{-1} is the unique stabilizing solution to the algebraic Riccati equation.

For completeness let us note that one can also use the generalized eigenvectors corresponding to the stable eigenvalues of H to compute Γ. In this case, Γ is the operator from $\mathbb{C}^n$ into $\mathcal{X} \oplus \mathcal{X}$ consisting of the eigenvectors and generalized eigenvectors for the stable eigenvalues of H and Λ is simply the corresponding Jordan matrix. If the eigenvalues of H are distinct, then Γ is the operator from $\mathbb{C}^n$ into $\mathcal{X} \oplus \mathcal{X}$ consisting of the eigenvectors corresponding the stable eigenvalues of H and Λ is the diagonal matrix on $\mathbb{C}^n$ consisting of the stable eigenvalues of H.

Lemma 13.1.6 *Suppose H is an operator on a finite dimensional space $\mathcal{X}$ and a and b are two polynomials with no common zeros which satisfy*

$$a(H)b(H) = 0.$$

Then every vector x in $\mathcal{X}$ can be uniquely expressed as $x = u + v$ where u is in $\ker a(H)$ and v is in $\ker b(H)$.

PROOF. First we show that $\ker a(H)$ and $\ker b(H)$ are invariant for H. To see this, consider any vector v in $\ker a(H)$. Then $a(H)Hv = Ha(H)v = 0$; hence Hv is in $\ker a(H)$. It now follows that $\ker a(H)$ is an invariant subspace for H. Similarly for $\ker b(H)$. Thus, the intersection of $\ker a(H)$ and $\ker b(H)$ is an invariant subspace for H.

We now claim that the intersection of $\ker a(H)$ and $\ker b(H)$ contain only the zero vector. Suppose, on the contrary, that the above intersection contains a nonzero vector. Since this intersection is invariant for H, it contains an eigenvector v for H. Thus, $v \neq 0$ and $Hv = \lambda v$ for some eigenvalue λ of H; also $a(H)v = 0$ and $b(H)v = 0$. Since v is an eigenvector of H corresponding to eigenvalue λ, we have $a(\lambda)v = a(H)v = 0$ and $b(\lambda)v = b(H)v = 0$. Since v is nonzero, it follows that λ is a zero of both the polynomials a and b. This contradicts the hypothesis that a and b have no common zeros. Hence, the intersection of $\ker a(H)$ and $\ker b(H)$ contains only the zero vector.

Since $a(H)b(H) = 0$, it follows that $\operatorname{ran} b(H)$ is contained in $\ker a(H)$. Hence,

$$\dim[\ker a(H)] \geq \dim[\operatorname{ran} b(H)] = n - \dim[\ker b(H)]$$

where n is the dimension of $\mathcal{X}$. This yields $\dim[\ker a(H)] + \dim[\ker b(H)] \geq n$. Since $\ker a(H)$ and $\ker b(H)$ are subspaces of $\mathcal{X}$ which only intersect at zero, it follows that $\dim[\ker a(H)] + \dim[\ker b(H)] \leq n$. Hence, $\dim[\ker a(H)] + \dim[\ker b(H)] = n$. It now follows that every vector x is $\mathcal{X}$ can be uniquely expressed as $x = u + v$ where u is in $\ker a(H)$ and v is in $\ker b(H)$. ∎

13.2 Characteristic polynomial of the Hamiltonian matrix

The following result presents an expression for the characteristic polynomial of the Hamiltonian matrix H in (13.2).

Theorem 13.2.1 *Let R and S be self-adjoint operators on a space $\mathcal{X}$ of finite dimension n. Then the characteristic polynomial Δ_H for the Hamiltonian matrix in (13.2) is given by*

$$\Delta_H = (-1)^n d d^\sharp \det[I - \Phi R \Phi^\sharp S] \qquad (13.16)$$

where d is the characteristic polynomial for A and $\Phi(s) = (sI - A)^{-1}$.

PROOF. If M_{ij} for $i, j = 1, 2$ are all operators on $\mathcal{X}$ and M_{22} is invertible, then a simple calculation shows that

$$M = \begin{bmatrix} M_{11} & M_{12} \\ M_{21} & M_{22} \end{bmatrix} = \begin{bmatrix} I & M_{12}M_{22}^{-1} \\ 0 & I \end{bmatrix} \begin{bmatrix} M_{11} - M_{12}M_{22}^{-1}M_{21} & 0 \\ 0 & M_{22} \end{bmatrix} \begin{bmatrix} I & 0 \\ M_{22}^{-1}M_{21} & I \end{bmatrix}.$$
$$(13.17)$$

Recall that $M_{11} - M_{12}M_{22}^{-1}M_{21}$ is a Schur complement for M. In particular, this shows that

$$\det[M] = \det[M_{22}]\det[M_{11} - M_{12}M_{22}^{-1}M_{21}]. \qquad (13.18)$$

Since

$$sI - H = \begin{bmatrix} sI - A & -R \\ S & sI + A^* \end{bmatrix}, \qquad (13.19)$$

we obtain

$$\begin{aligned} \det[sI - H] &= \det[sI + A^*]\det[sI - A + R(sI + A^*)^{-1}S] \\ &= \det[sI - A]\det[sI + A^*]\det[I + (sI - A)^{-1}R(sI + A^*)^{-1}S]. \end{aligned}$$

Since $\Phi(s) = (sI - A)^{-1}$ and $d(s) = \det[sI - A]$ it follows that $(sI + A^*)^{-1} = -\Phi^\sharp(s)$ and $\det[sI + A^*] = (-1)^n d^\sharp(s)$. This yields (13.16). ∎

The expression for the characteristic polynomial for H in (13.16) also shows that $\Delta_H = \Delta_H^\sharp$ where Δ_H is the characteristic polynomial for H. Equations (13.8) and (13.16) readily yield the following result.

Corollary 13.2.2 *Let R and S be self-adjoint operators on a space $\mathcal{X}$ of finite dimension n. Suppose that Q is any self-adjoint solution to the algebraic Riccati equation in (13.1) and Δ is the characteristic polynomial for $A + RQ$. Then*

$$\Delta\Delta^\sharp = dd^\sharp \det[I - \Phi R \Phi^\sharp S] \qquad (13.20)$$

where d is the characteristic polynomial for A and $\Phi(s) = (sI - A)^{-1}$.

13.3   Some special cases

We have seen that if the Riccati equation has a stabilizing solution, then the pair $\{A, R\}$ is stabilizable and the corresponding Hamiltonian matrix has no imaginary eigenvalues. The next result states that the converse is also true when R is positive ($R \geq 0$) or negative ($R \leq 0$).

Theorem 13.3.1 *Let R and S be self-adjoint operators on a space $\mathcal{X}$ of finite dimension with R positive or negative. Then the algebraic Riccati equation in (13.1) admits a stabilizing solution if and only if the pair $\{A, R\}$ is stabilizable and the corresponding Hamiltonian matrix H in (13.2) has no imaginary eigenvalues.*

PROOF. Suppose that the Riccati equation in (13.1) has a stabilizing solution Q. By definition of a stabilizing solution, $A + RQ$ is stable; hence $\{A, R\}$ is a stabilizable pair. We have already seen that the Hamiltonian matrix H has no imaginary eigenvalues when the Riccati equation has a stabilizing solution.

To complete the proof, assume that the pair $\{A, R\}$ is stabilizable and H has no imaginary eigenvalues. It now follows from Lemma 13.1.3 that there is a one to one operator $\Gamma = [X\ Y]^{tr}$ of rank n such that $H\Gamma = \Gamma\Lambda$ for some stable operator Λ where n is the dimension of $\mathcal{X}$. Thus, X and Y satisfy

$$\begin{aligned} AX + RY &= X\Lambda & (13.21a) \\ -SX - A^*Y &= Y\Lambda & (13.21b) \end{aligned}$$

where Λ is a stable operator. We first demonstrate that the operator Y^*X is self-adjoint. To see this, premultiply equations (13.21a) and (13.21b) by Y^* and $-X^*$, respectively, and add the resulting equations to obtain

$$Y^*AX + X^*A^*Y + Y^*RY + X^*SX = (Y^*X - X^*Y)\Lambda.$$

Since the left hand side of this equation defines a self-adjoint operator it follows that $(Y^*X - X^*Y)\Lambda$ is a self-adjoint operator. If we set $Z = Y^*X - X^*Y$, then $\Lambda^*Z^* = Z\Lambda$. Since $Z^* = -Z$, we obtain the Lyapunov equation $Z\Lambda + \Lambda^*Z = 0$. Because Λ is stable, $Z = 0$ is the only solution to this Lyapunov equation. Thus, $Y^*X = X^*Y$ and Y^*X is a self-adjoint operator. As a consequence, we also obtain that X and Y satisfy

$$Y^*AX + X^*A^*Y + Y^*RY + X^*SX = 0. \tag{13.22}$$

Now let us show that X is invertible. Because $\mathcal{X}$ is finite dimensional, it is sufficient to show that X is one to one. So, suppose that $Xv = 0$ for some vector v. By consulting (13.22) we see that

$$X^*A^*Yv + Y^*RYv = 0.$$

This implies that $(RYv, Yv) = (Y^*RYv, v) = -(X^*A^*Yv, v) = -(A^*Yv, Xv) = 0$. Because R is either positive or negative, $RYv = 0$. It now follows from (13.21a) that $X\Lambda v = 0$. In other words, the kernel of X is an invariant subspace for Λ. Suppose, on the contrary, that the kernel of X is nonzero. Then there is an eigenvector v for Λ in the kernel of X with eigenvalue λ, that is, $\Lambda v = \lambda v$ and $Xv = 0$. By employing (13.21), we obtain $RYv = \lambda Xv = 0$ and $-A^*Yv = \lambda Yv$. Hence, $(-\lambda I - A^*)Yv = 0$ and $RYv = 0$. Since λ is a stable eigenvalue of Λ and by hypothesis, the pair $\{R, A^*\}$ is detectable it follows that $-\lambda$ is not an unobservable eigenvalue of the pair $\{R, A^*\}$. Using the PBH observability test we must have $Yv = 0$. Because Xv is zero and $[X \ Y]^{tr}$ is one to one, v equals zero. This contradicts v being an eigenvector, and hence, nonzero. Therefore, X is one to one and invertible. Since X is invertible, it now follows from Lemma 13.1.1 that $Q = YX^{-1}$ is a stabilizing solution to the algebraic Riccati equation in (13.1). ∎

Lemma 13.3.2 *Let R and S be operators on a space $\mathcal{X}$ of finite dimension with R negative and S positive. Then an imaginary number λ is an eigenvalue of the Hamiltonian matrix in (13.2) if and only if λ is either an uncontrollable eigenvalue of $\{A, R\}$ or an unobservable eigenvalue of $\{S, A\}$. Hence, H has no imaginary eigenvalues if and only if the system $\{A, R, S, 0\}$ has no uncontrollable or unobservable eigenvalues on the imaginary axis.*

PROOF. If λ is an unobservable eigenvalue of the pair $\{S, A\}$, then there is a nonzero vector x in $\mathcal{X}$ such that $(A - \lambda I)v = 0$ and $Sv = 0$. Letting $v = x \oplus 0$, we obtain $Hv = \lambda v$; thus λ is an eigenvalue of H. Hence, if λ is an imaginary unobservable eigenvalue of the pair $\{S, A\}$, then λ is an imaginary eigenvalue of the Hamiltonian matrix H.

If λ is an uncontrollable eigenvalue of the pair $\{A, R\}$, then there is a nonzero vector y in $\mathcal{X}$ such that $(A^* - \bar{\lambda}I)y = 0$ and $Ry = 0$. Letting $v = 0 \oplus y$, we obtain $Hv = -\bar{\lambda}v$; hence $-\bar{\lambda}$ is an eigenvalue of H. Since H has the property that $-\bar{\lambda}$ is an eigenvalue of H if and only if λ is an eigenvalue of H, we obtain that λ is also an eigenvalue of H. Hence, if λ is an imaginary uncontrollable eigenvalue of the pair $\{A, R\}$, then λ is an imaginary eigenvalue of the Hamiltonian matrix H.

Now suppose that H has an imaginary eigenvalue λ. Then there is a nonzero vector eigenvector $v = x \oplus y$ in $\mathcal{X} \oplus \mathcal{X}$ such that $Hv = \lambda v$, that is,

$$(-A + \lambda I)x = Ry \tag{13.23a}$$
$$(-A^* - \lambda I)y = Sx. \tag{13.23b}$$

By taking the appropriate inner products, we obtain

$$(Ry, y) = ((-A+\lambda I)x, \, y)$$

and

$$(Sx, x) = (x, Sx) = (x, (-A^* - \lambda I)y) = ((-A - \bar{\lambda}I)x, \, y) \, .$$

Since λ is imaginary, we have $-\bar{\lambda} = \lambda$, and hence, $(Ry, y) = (Sx, x)$. Because R is negative and S is positive, we must have $(Ry, y) = 0$ and $(Sx, x) = 0$; hence $Ry = 0$ and $Sx = 0$. By consulting (13.23a), we see that $(A - \lambda I)x = 0$ and $Sx = 0$. If x is nonzero, then according to the PBH test for unobservable eigenvalues, λ is an unobservable eigenvalue of $\{S, A\}$. If x is zero, then y must be nonzero and (13.23b) implies that $(A^* + \lambda I)y = 0$ and $Ry = 0$. By the PBH test for uncontrollable eigenvalues, we obtain that $\lambda = -\bar{\lambda}$ is an uncontrollable eigenvalue of $\{A, R\}$. So we can conclude that, if λ is an imaginary eigenvalue of H, then λ is either an uncontrollable eigenvalue of $\{A, R\}$ or an unobservable eigenvalue of $\{S, A\}$. ∎

If R is negative, S is positive and the algebraic Riccati equation in (13.1) admits a stabilizing solution Q, then Q is positive. To see this simply rearrange the Riccati equation to obtain

$$(A + RQ)^*Q + Q(A + RQ) + S - QRQ = 0 \, .$$

Because $A_c = A + RQ$ is stable, we have

$$Q = \int_0^\infty e^{A_c^*\sigma} (S - QRQ) \, e^{A_c\sigma} \, d\sigma \, . \tag{13.24}$$

Since $S - QRQ$ is positive, this implies that Q is positive. Furthermore, if the pair $\{S, A\}$ is observable, then this shows that Q is strictly positive. If R is positive and there exists a stabilizing solution Q, then Q is not necessarily positive. For example, if $A = 3$, $R = 1$ and $S = 8$, then $Q = -4$ is the stabilizing solution to the corresponding algebraic Riccati equation.

13.4   The linear quadratic regulator

Recall now the algebraic Riccati equation associated with the linear quadratic regulator problem, namely

$$A^*Q + QA - QBB^*Q + C^*C = 0 \, . \tag{13.25}$$

Here the self-adjoint operators R and S are given by $R = -BB^*$ and $S = C^*C$ where B maps $\mathcal{U}$ into $\mathcal{X}$ while C maps $\mathcal{X}$ into $\mathcal{Y}$. Hence the corresponding Hamiltonian matrix is given by

$$H = \begin{bmatrix} A & -BB^* \\ -C^*C & -A^* \end{bmatrix} \, . \tag{13.26}$$

Assume that the pair $\{A, B\}$ is stabilizable. Since the pairs $\{A, B\}$ and $\{A, R\}$ have the same uncontrollable eigenvalues, it follows that $\{A, R\}$ is stabilizable and has no uncontrollable imaginary eigenvalues. Since R is negative, it follows from Lemma 13.3.1 that the above algebraic Riccati equation has a stabilizing solution Q if and only if H has no imaginary eigenvalues.

Since the pairs $\{C, A\}$ and $\{S, A\}$ have the same unobservable eigenvalues and R is negative while S is positive, it follows from Lemma 13.3.2 that H has no imaginary eigenvalues if and only if the pair $\{C, A\}$ has no unobservable eigenvalues on the imaginary axis. With these observations, the comments at the end of the last section and Theorem 13.1.5 we can deduce the following result.

Theorem 13.4.1 *Consider a system $\{A$ on $\mathcal{X}, B, C, 0\}$ and assume that the pair $\{A, B\}$ is stabilizable. Then the following statements are equivalent.*

(i) The pair $\{C, A\}$ has no imaginary unobservable eigenvalues.

(ii) The Hamiltonian matrix H in (13.26) has no imaginary eigenvalues.

(iii) There exists a unique stabilizing solution Q to the algebraic Riccati equation (13.25).

In this case, the unique stabilizing solution to the algebraic Riccati equation in (13.1) is given by

$$Q = YX^{-1} \tag{13.27}$$

where $\Gamma = [X \ Y]^{tr}$ is any one to one operator of rank $n = \dim \mathcal{X}$ satisfying

$$H\Gamma = \Gamma\Lambda \tag{13.28}$$

with Λ a stable operator. Moreover, the stabilizing solution Q is positive. Finally, if the pair $\{C, A\}$ is observable, then Q is strictly positive.

The above result is stronger than our previous result. Previously we showed that if the pair $\{A, B\}$ is stabilizable and the pair $\{C, A\}$ is detectable, then there exists a unique stabilizing solution to the algebraic Riccati equation in (13.25); see Lemma 12.7.2.

Let Δ_H be the characteristic polynomial for the Hamiltonian matrix H given in (13.26). We now demonstrate that

$$\Delta_H = (-1)^n dd^\sharp \det[I + \mathbf{F}\mathbf{F}^\sharp] \tag{13.29}$$

where n is the dimension of $\mathcal{X}$ while d is the characteristic polynomial for A and $\mathbf{F}$ is the transfer function for the system $\{A, B, C, 0\}$. To see this, apply the results of Theorem 13.2.1 with $R = -BB^*$ and $S = C^*C$ to obtain

$$\begin{aligned}
\Delta_H &= (-1)^n dd^\sharp \det[I + \Phi BB^* \Phi^\sharp C^* C] \\
&= (-1)^n dd^\sharp \det[I + C\Phi BB^* \Phi^\sharp C^*] \\
&= (-1)^n dd^\sharp \det[I + \mathbf{F}\mathbf{F}^\sharp].
\end{aligned}$$

The second equality follows from the fact that $\det[I + MN] = \det[I + NM]$. Furthermore, when the algebraic Riccati equation has a stabilizing solution Q, it now follows from (13.14) that, the characteristic polynomial Δ for $A + RQ$ satisfies

$$\Delta\Delta^\sharp = dd^\sharp \det[I + \mathbf{F}\mathbf{F}^\sharp]. \tag{13.30}$$

If the transfer function $\mathbf{F}$ is scalar valued, then the expressions in (13.29) and (13.30) reduce to

$$\Delta_H = (-1)^n dd^\sharp (1 + \mathbf{F}\mathbf{F}^\sharp) \tag{13.31}$$

and

$$\Delta\Delta^\sharp = dd^\sharp (1 + \mathbf{F}\mathbf{F}^\sharp), \tag{13.32}$$

respectively. In particular, if $\mathbf{F} = p/d$ where p is a polynomial, then (13.31) and (13.32) respectively yield

$$\Delta_H = (-1)^n (dd^\sharp + pp^\sharp). \tag{13.33}$$

and

$$\Delta\Delta^\sharp = dd^\sharp + pp^\sharp. \tag{13.34}$$

It is noted that (13.34) is precisely the formula for $\Delta\Delta^\sharp$ in (12.114) when we replace p with $p/\sqrt{r}$. Also, if λ is an imaginary number, then (13.33) implies that

$$\Delta_H(\lambda) = (-1)^n (d(\lambda)d(-\bar\lambda)^*) + p(\lambda)p(-\bar\lambda)^*) = (-1)^n (|d(\lambda)|^2 + |p(\lambda)|^2).$$

Hence λ is an eigenvalue for H if and only if λ is a common zero for d and p. This is another demonstration of the fact that λ is an imaginary eigenvalue of H if and only if λ is an uncontrollable or unobservable eigenvalue of the system $\{A, B, C, 0\}$.

13.5 H^∞ analysis and control

In H^∞ system analysis and control design, the following Riccati equation plays a major role:

$$A^*Q + QA - QBB^*Q + QE^*EQ + C^*C = 0 \tag{13.35}$$

where E is an operator from $\mathcal{W}$ into $\mathcal{X}$ and B is an operator from $\mathcal{U}$ into $\mathcal{X}$ while C is an operator from $\mathcal{X}$ into $\mathcal{Y}$. This is a special case of the general Riccati equation (13.1) where the self-adjoint operators R and S are given by $R = EE^* - BB^*$ and $S = C^*C$, respectively. In this case Q is a stabilizing solution if Q is an operator satisfying (13.35) and $A + EE^*Q - BB^*Q$ is stable. Here, the Hamiltonian matrix in (13.2) becomes

$$H = \begin{bmatrix} A & EE^* - BB^* \\ -C^*C & -A^* \end{bmatrix} \text{ on } \begin{bmatrix} \mathcal{X} \\ \mathcal{X} \end{bmatrix}. \tag{13.36}$$

The following result yields an expression for the characteristic polynomial of this Hamiltonian matrix.

Theorem 13.5.1 *Let $\mathbf{F}$ be the transfer function for the system $\{A \text{ on } \mathcal{X}, B, C, 0\}$ and $\mathbf{T}$ be the transfer function for $\{A, E, C, 0\}$. Let d be the characteristic polynomial for A and n be the dimension of $\mathcal{X}$. Then the characteristic polynomial Δ_H for the Hamiltonian matrix in (13.36) is given by*

$$\Delta_H = (-1)^n dd^\sharp \det[I + \mathbf{F}\mathbf{F}^\sharp - \mathbf{T}\mathbf{T}^\sharp]. \tag{13.37}$$

PROOF. By applying Theorem 13.2.1 with $R = -BB^* + EE^*$ and $S = C^*C$, we obtain that

$$\Delta_H = (-1)^n dd^\sharp \det[I + \Phi(BB^* - EE^*)\Phi^\sharp C^*C]$$

where $\Phi(s) = (sI - A)^{-1}$. Using the relationship, $\det[I + MN] = \det[I + MN]$ where M and N are operators acting between the appropriate spaces, we obtain

$$
\begin{aligned}
\Delta_H &= (-1)^n dd^\sharp \det[I + C\Phi(BB^* - EE^*)\Phi^\sharp C^*] \\
&= (-1)^n dd^\sharp \det[I + C\Phi BB^*\Phi^\sharp C^* - C\Phi EE^*\Phi^\sharp C^*] \\
&= (-1)^n dd^\sharp \det[I + \mathbf{F}\mathbf{F}^\sharp - \mathbf{T}\mathbf{T}^\sharp].
\end{aligned}
$$

This yields (13.37). ∎

The expression for the characteristic polynomial for H in (13.37) readily shows that λ is an eigenvalue for H if and only if $-\bar{\lambda}$ is an value for H. Equations (13.8) and (13.37) readily yield the following result.

Corollary 13.5.2 Let $\mathbf{F}$ be the transfer function for $\{A, B, C, 0\}$ and $\mathbf{T}$ be the transfer function for $\{A, E, C, 0\}$. Let d be the characteristic polynomial for A. Suppose that Q is a self-adjoint solution to the algebraic Riccati equation in (13.35) and let Δ be the characteristic polynomial for $A + EE^*Q - BB^*Q$. Then

$$\Delta\Delta^\sharp = dd^\sharp \det[I + \mathbf{F}\mathbf{F}^\sharp - \mathbf{T}\mathbf{T}^\sharp]. \tag{13.38}$$

If the transfer functions $\mathbf{T}$ and $\mathbf{F}$ are scalar valued, then the expression for the characteristic polynomial for H in (13.37) reduces to

$$\Delta_H = (-1)^n dd^\sharp(1 + \mathbf{F}\mathbf{F}^\sharp - \mathbf{T}\mathbf{T}^\sharp). \tag{13.39}$$

In particular, if $\mathbf{T} = p_1/d$ and $\mathbf{F} = p_2/d$ where p_1 and p_2 are polynomials, then (13.39) yields

$$\Delta_H = (-1)^n(dd^\sharp - p_1 p_1^\sharp + p_2 p_2^\sharp). \tag{13.40}$$

Furthermore, if Q is a self-adjoint solution to the algebraic Riccati equation in (13.35), then (13.38) shows that

$$\Delta\Delta^\sharp = dd^\sharp - p_1 p_1^\sharp + p_2 p_2^\sharp \tag{13.41}$$

where Δ is the characteristic polynomial for $A + EE^*Q - BB^*Q$.

Theorem 13.5.3 Let A be an operator on $\mathcal{X}$ with no eigenvalues on the imaginary axis. Let $\mathbf{T}$ be the transfer function for $\{A, E, C, 0\}$ and $\mathbf{F}$ be the transfer function for $\{A, B, C, 0\}$. Then the Hamiltonian matrix H in (13.36) has no eigenvalues on the imaginary axis if and only if there exists a scalar $\epsilon > 0$ such that

$$I + \mathbf{F}(\imath\omega)\mathbf{F}(\imath\omega)^* - \mathbf{T}(\imath\omega)\mathbf{T}(\imath\omega)^* \geq \epsilon I \qquad (\text{for all } -\infty < \omega < \infty). \tag{13.42}$$

PROOF. As before, let d be the characteristic polynomial for A and n the dimension of $\mathcal{X}$. Using $(\mathbf{F}^\sharp)(\imath\omega) = \mathbf{F}(\imath\omega)^*$ along with the corresponding results for $\mathbf{T}$ and d in (13.37), we arrive at

$$\Delta_H(\imath\omega) = (-1)^n |d(\imath\omega)|^2 \det[I + \mathbf{F}(\imath\omega)\mathbf{F}(\imath\omega)^* - \mathbf{T}(\imath\omega)\mathbf{T}(\imath\omega)^*]. \qquad (13.43)$$

Let $\Xi(\imath\omega) = I + \mathbf{F}(\imath\omega)\mathbf{F}(\imath\omega)^* - \mathbf{T}(\imath\omega)\mathbf{T}(\imath\omega)$. If $\Xi(\imath\omega) \geq \epsilon I$ for some $\epsilon > 0$ and all ω, then (13.43) shows that the Hamiltonian matrix H has no eigenvalues on the imaginary axis.

Now assume that H has no eigenvalues on the imaginary axis. Because d has no zeros on the imaginary axis, equation (13.43) implies that $\Xi(\imath\omega)$ is a self-adjoint invertible operator for all ω. Since $\mathbf{F}$ and $\mathbf{T}$ are strictly proper rational functions, $\Xi(\imath\omega)$ converges to the identity operator as ω tends to $\pm\infty$. Notice that the smallest eigenvalue $\lambda_{min}(\omega)$ of $\Xi(\imath\omega)$ is a continuous function of ω. Moreover, $\lambda_{min}(\omega)$ converges to one as ω tends to $\pm\infty$. We claim that $\Xi(\imath\omega) \geq \epsilon I$ for some $\epsilon > 0$ and all ω. If $\Xi(\imath\omega_1) \leq 0$ for some frequency ω_1, then $\lambda_{min}(\omega_1) \leq 0$ and $\lambda_{min}(\omega_o)$ must be zero for some frequency ω_o. In other words, $\Xi(\imath\omega_o)$ is not invertible. This contradicts the fact that $\det[\Xi(\imath\omega)]$ is nonzero for all ω. Therefore, $\Xi(\imath\omega) > 0$ and $\lambda_{min}(\omega) > 0$ for all ω. Since λ_{min} is a nonzero positive continuous function which converges to one as ω tends to $\pm\infty$, it follows that $\lambda_{min}(\omega) \geq \epsilon$ for some $\epsilon > 0$. ∎

H^∞ **analysis.** In H^∞ analysis, $B = 0$. In this case, the Hamiltonian matrix in (13.36) reduces to

$$H = \begin{bmatrix} A & EE^* \\ -C^*C & -A^* \end{bmatrix} \text{ on } \begin{bmatrix} \mathcal{X} \\ \mathcal{X} \end{bmatrix}. \qquad (13.44)$$

The corresponding algebraic Riccati equation is given by

$$A^*Q + QA + QEE^*Q + C^*C = 0. \qquad (13.45)$$

In this case Q is a stabilizing solution if Q is a solution to (13.45) and $A + EE^*Q$ is stable. Now assume that A is stable. Then obviously, the pair $\{A, EE^*\}$ is stabilizable. By combining Theorem 13.3.1 with Theorem 13.5.3, we obtain the equivalence of Parts (i), (ii) and (iii) in the following result.

Corollary 13.5.4 *Let $\mathbf{T}$ be the transfer function for the stable system $\{A, E, C, 0\}$. Then the following statements are equivalent.*

(i) The algebraic Riccati equation in (13.45) admits a stabilizing solution Q.

(ii) The Hamiltonian matrix H in (13.44) has no eigenvalues on the imaginary axis.

(iii) The H^∞ norm $\|\mathbf{T}\|_\infty < 1$.

In this case, the unique stabilizing solution Q is positive.

PROOF. To complete the proof it remains to show that the stabilizing solution Q is positive. In fact, any self-adjoint solution to this algebraic Riccati equation is positive. Because A is stable, the Lyapunov form of (13.45) implies that

$$Q = \int_0^\infty e^{A^*\sigma} \left(C^*C + QEE^*Q \right) e^{A\sigma} \, d\sigma. \qquad (13.46)$$

This readily implies that Q is positive. ∎

13.6 The Riccati differential equation

In this section, we present some useful properties of the Riccati differential equation associated with the algebraic Riccati equation in (13.1). This differential equation is given by

$$\dot{P} + A^*P + PA + PRP + S = 0 \tag{13.47}$$

where $P(t)$ is an operator on $\mathcal{X}$.

13.6.1 A two point boundary value problem

Here we show that the solution of Riccati differential equation (13.47) with terminal condition $P(t_1) = 0$ is related to the solutions of two point boundary problems associated with the Hamiltonian differential equation

$$\begin{bmatrix} \dot{x} \\ \dot{\lambda} \end{bmatrix} = H \begin{bmatrix} x \\ \lambda \end{bmatrix} \tag{13.48}$$

where x and λ are in $\mathcal{X}$ and H is the Hamiltonian matrix associated with (13.47) as given by

$$H = \begin{bmatrix} A & R \\ -S & -A^* \end{bmatrix} \text{ on } \begin{bmatrix} \mathcal{X} \\ \mathcal{X} \end{bmatrix}. \tag{13.49}$$

Specifically, we demonstrate the following result.

Lemma 13.6.1 *The Riccati differential equation (13.47) with terminal condition $P(t_1) = 0$ has a solution for P on an interval $[t_0, t_1]$ if and only if for each t' in $[t_0, t_1)$ and each x_0 in $\mathcal{X}$, the Hamiltonian differential equation in (13.48) has a solution $[x \ \lambda]^{tr}$ with*

$$x(t') = x_0 \quad \text{and} \quad \lambda(t_1) = 0. \tag{13.50}$$

In this case,

$$\lambda(t) = P(t)x(t). \tag{13.51}$$

PROOF. Suppose first that the Riccati differential equation in (13.47) has a solution on an interval $[t_0, t_1]$ with $P(t_1) = 0$. Consider any x_0 in $\mathcal{X}$ and any t' in $[t_0, t_1)$. Let x be the solution to the initial value problem

$$\dot{x} = (A + RP(t))x \quad \text{and} \quad x(t') = x_0$$

and let $\lambda(t) = P(t)x(t)$. Then $\lambda(t_1) = 0$ and $\dot{x} = Ax + R\lambda$. Using the Riccati differential equation (13.47), we obtain

$$\dot{\lambda} = \dot{P}x + P\dot{x} = \dot{P}x + PAx + PRPx = -A^*Px - Sx = -Sx - A^*\lambda.$$

Thus, we have shown that

$$\dot{x} = Ax + R\lambda$$
$$\dot{\lambda} = -Sx - A^*\lambda$$

which is equivalent to the Hamiltonian differential equation (13.48). Also, the boundary conditions in (13.50) are satisfied.

Now suppose that for each t' in $[t_0, t_1)$, and for each x_0 in $\mathcal{X}$, the Hamiltonian differential equation in (13.48) has a solution for $[x\ \lambda]^{tr}$ which satisfies the boundary conditions in (13.50). Consider the following matrix partition of e^{Ht}

$$e^{Ht} = \begin{bmatrix} \Phi_{11}(t) & \Phi_{12}(t) \\ \Phi_{21}(t) & \Phi_{22}(t) \end{bmatrix} \text{ on } \begin{bmatrix} \mathcal{X} \\ \mathcal{X} \end{bmatrix}. \tag{13.52}$$

Consider any t' in $[t_0, t_1)$, and any x_0 in $\mathcal{X}$ and let $[x\ \lambda]^{tr}$ be a solution of the Hamiltonian differential equation in (13.48) which satisfies the boundary conditions in (13.50). Then,

$$\begin{aligned} x(t) &= \Phi_{11}(t-t_1)x(t_1) \\ \lambda(t) &= \Phi_{21}(t-t_1)x(t_1). \end{aligned} \tag{13.53}$$

In particular, $x_0 = \Phi_{11}(t'-t_1)x(t_1)$. Since x_0 can be any vector in $\mathcal{X}$, it follows that the operator $\Phi_{11}(t'-t_1)$ is onto $\mathcal{X}$. Because $\mathcal{X}$ is finite dimensional, $\Phi_{11}(t'-t_1)$ is invertible. Noting that $\Phi_{11}(t_1-t_1) = I$, we have shown that the operator $\Phi_{11}(t'-t_1)$ is invertible for all t' in $[t_0, t_1]$. Eliminating $x(t_1)$ from equations (13.53) yields $\lambda(t) = P(t)x(t)$ where

$$P(t) = \Phi_{21}(t-t_1)\Phi_{11}(t-t_1)^{-1}$$

for $t_0 \leq t \leq t_1$. We claim that P satisfies the Riccati differential equation in (13.47) with $P(t_1) = 0$. Since $\Phi_{21}(t_1-t_1) = 0$, it follows that $P(t_1) = 0$. Because P satisfies

$$P(t)\Phi_{11}(t-t_1) - \Phi_{21}(t-t_1) = 0,$$

we can differentiate with respect to t to obtain

$$\dot{P}(t)\Phi_{11}(\tau) + P(t)\dot{\Phi}_{11}(\tau) - \dot{\Phi}_{21}(\tau) = 0 \tag{13.54}$$

where $\tau = t-t_1$. Obviously,

$$\begin{bmatrix} \dot{\Phi}_{11} & \dot{\Phi}_{12} \\ \dot{\Phi}_{21} & \dot{\Phi}_{22} \end{bmatrix} = \begin{bmatrix} A & R \\ -S & -A^* \end{bmatrix} \begin{bmatrix} \Phi_{11} & \Phi_{12} \\ \Phi_{21} & \Phi_{22} \end{bmatrix}.$$

In particular,

$$\begin{aligned} \dot{\Phi}_{11} &= A\Phi_{11} + R\Phi_{21} \\ \dot{\Phi}_{21} &= -S\Phi_{11} - A^*\Phi_{21}. \end{aligned}$$

Substituting this into (13.54), it now follows that

$$\dot{P}(t)\Phi_{11}(\tau) + P(t)A\Phi_{11}(\tau) + P(t)R\Phi_{21}(\tau) + S\Phi_{11}(\tau) + A^*\Phi_{21}(\tau) = 0.$$

Multiplying this equation by $\Phi_{11}(\tau)^{-1}$ on the right and recalling that $P(t) = \Phi_{21}(\tau)\Phi_{11}(\tau)^{-1}$, yields

$$\dot{P} + PA + A^*P + PRP + S = 0.$$

This is precisely the Riccati differential equation in (13.47).   ∎

Remark 13.6.1 The above proof shows that one can obtain a solution P to the Riccati differential equation (13.47) by computing e^{Ht} where H is the Hamiltonian matrix corresponding to (13.47) and setting

$$P(t) = \Phi_{21}(t-t_1)\Phi_{11}(t-t_1)^{-1} \tag{13.55}$$

where Φ_{11} and Φ_{21} are as defined in the matrix partition of e^{Ht} in (13.52). Also, a solution $[x \ \lambda]^{tr}$ to the two point boundary value problem defined by (13.48) and (13.50) is given by

$$
\begin{aligned}
x(t) &= \Phi_{11}(t-t_1)\Phi_{11}(t'-t_1)^{-1}x_0 \\
\lambda(t) &= \Phi_{21}(t-t_1)\Phi_{11}(t'-t_1)^{-1}x_0 .
\end{aligned}
$$

This solution is also given by

$$
\begin{aligned}
\dot{x} &= (A+RP)x \qquad \text{and} \qquad x(t') = x_0 \\
\lambda &= Px .
\end{aligned}
$$

13.6.2 Some properties

We now establish some fundamental properties for the solution P of the Riccati differential equation with terminal condition $P(t_1) = 0$. To this end, suppose that P is an operator valued function defined on an interval $[t_0, t_1]$ where $P(t)$ is a self-adjoint operator on $\mathcal{X}$. We say the P is a decreasing function or is *increasing backwards in time* if $P(t') \geq P(t'')$ whenever $t_0 \leq t' \leq t'' \leq t_1$. We have now the following result.

Lemma 13.6.2 *Consider the Riccati differential equation (13.47) where R and S are operators on a finite dimensional space with R self-adjoint and S positive. Suppose P is a solution on $[t_0, t_1]$ to this differential equation with terminal condition $P(t_1) = 0$. Then $P(t)$ is positive for each t in $[t_0, t_1]$ and P is a decreasing function. Moreover, if $S = C^*C$ and $\{C, A\}$ is observable, then $P(t)$ is strictly positive for each t in $[t_0, t_1]$.*

PROOF. To see that $P(t)$ is self-adjoint, take the adjoint of each term in the Riccati differential equation (13.47) to obtain

$$\dot{P}^* + A^*P^* + P^*A + P^*RP^* + S = 0 .$$

Thus P^* satisfies the Riccati differential equation (13.47) and $P(t_1)^* = 0$. Since the Riccati differential equation is locally Lipschitz in P, the solution corresponding to $P(t_1) = 0$ is unique. Hence, $P(t)^* = P(t)$ and $P(t)$ is self-adjoint.

To show that P is decreasing and $P(t)$ is positive, rewrite the Riccati differential in (13.47) as

$$\dot{P}(t) + \tilde{A}(t)^*P(t) + P(t)\tilde{A}(t) + S = 0 , \tag{13.56}$$

where $\tilde{A}(t) = A + \frac{1}{2}RP(t)$. Let $\tilde{\Phi}(t, \sigma)$ be the state transition operator for $\tilde{A}$. Multiplying the above Riccati equation on the left by $\tilde{\Phi}(t, t_1)^*$ and on the right by $\tilde{\Phi}(t, t_1)$ yields

$$
\begin{aligned}
&\tilde{\Phi}(t, t_1)^*\dot{P}(t)\tilde{\Phi}(t, t_1) + \tilde{\Phi}(t, t_1)^*\tilde{A}(t)^*P(t)\tilde{\Phi}(t, t_1) + \tilde{\Phi}(t, t_1)^*P(t)\tilde{A}(t)\tilde{\Phi}(t, t_1) \\
&+ \ \tilde{\Phi}(t, t_1)^*S\tilde{\Phi}(t, t_1) = 0 ,
\end{aligned}
$$

that is,

$$\frac{d}{dt}\left(\tilde{\Phi}(t,t_1)^*P(t)\tilde{\Phi}(t,t_1)\right) + \tilde{\Phi}(t,t_1)^*S\tilde{\Phi}(t,t_1) = 0.$$

Integrating from t' to t_1 and using the terminal condition $P(t_1) = 0$ results in

$$-\tilde{\Phi}(t',t_1)^*P(t')\tilde{\Phi}(t',t_1) + \int_{t'}^{t_1} \tilde{\Phi}(t,t_1)^*S\tilde{\Phi}(t,t_1)\,dt = 0.$$

Multiply the above equation on the right and left by $\tilde{\Phi}(t_1,t')$ and $\tilde{\Phi}(t_1,t')^*$ and use the relationships $\tilde{\Phi}(t,t_1)\tilde{\Phi}(t_1,t') = \tilde{\Phi}(t,t')$ and $\tilde{\Phi}(t',t') = I$ to obtain that

$$P(t') = \int_{t'}^{t_1} \tilde{\Phi}(t,t')^*S\tilde{\Phi}(t,t')\,dt. \tag{13.57}$$

Since S is positive, this clearly shows that $P(t')$ is positive.

To show that P is decreasing, differentiate the Riccati differential equation (13.47) to obtain

$$\ddot{P} + (A + RP)^*\dot{P} + \dot{P}(A + RP) = 0.$$

It also follows from Riccati equation (13.47) and $P(t_1) = 0$ that $\dot{P}(t_1) = -S$. From this one can readily show that

$$\dot{P}(t) = -\tilde{\Phi}(t_1,t)^*S\tilde{\Phi}(t_1,t)$$

where $\tilde{\Phi}(t,\sigma)$ is now the state transition operator for $A + RP$. Hence $\dot{P}(t)$ is negative for $t_0 \leq t \leq t_1$. It now follows that P is decreasing.

Suppose that $S = C^*C$ and the pair $\{C, A\}$ is observable. We will show that $P(t')$ is strictly positive for each t' in $[t_0, t_1]$. To this end, consider any t' in the interval $[t_0, t_1]$ and suppose v is any vector in the kernel of $P(t')$, that is, $P(t')v = 0$. It now follows from (13.57) that

$$0 = (P(t')v, v) = \int_{t'}^{t_1} (\tilde{\Phi}(t,t')^*C^*C\tilde{\Phi}(t,t')v, v)\,dt = \int_{t'}^{t_1} \|C\tilde{\Phi}(t,t')v\|^2\,dt$$

where $\tilde{\Phi}(t,\sigma)$ is the state transition operator for $A + \frac{1}{2}RP(t)$. Hence, $C\tilde{\Phi}(t,t')v = 0$ for all t in $[t', t_1]$. In particular $Cv = C\tilde{\Phi}(t',t')v = 0$. Recalling the Riccati differential equation in (13.47) with $S = C^*C$, we now obtain that

$$\dot{P}(t')v + P(t')Av = 0. \tag{13.58}$$

Hence,

$$(\dot{P}(t')v, v) = -(P(t')Av, v) = -(Av, P(t')v) = 0.$$

Since $\dot{P}(t')$ is negative, we must have $P(t')v = 0$. It now follows from (13.58) that $P(t')Av = 0$. Thus Av is in the kernel of $P(t')$. Since Av is in the kernel of $P(t')$, it follows from the above analysis that $CAv = 0$. By induction we can show that $CA^kv = 0$ for all integers $k \geq 0$. This means that v is in the unobservable subspace for $\{C, A\}$. Since $\{C, A\}$ is observable, v must be zero. Hence the kernel of $P(t')$ is zero. Because $P(t')$ is positive and its kernel is zero, $P(t')$ must be strictly positive. ∎

Remark 13.6.2 Suppose P is a solution on $[t_0, t_1]$ to the Riccati differential equation (13.47) with terminal condition $P(t_1) = 0$ and let $\Omega(\tau) = P(t_1 - \tau)$ and $\tau_1 = t_1 - t_0$. Then, $\Omega(0) = 0$ and Ω is a solution on $[0, \tau_1]$ to the following Riccati differential equation

$$\dot{\Omega} = A^*\Omega + \Omega A + \Omega R\Omega + S. \tag{13.59}$$

Moreover, it follows from the preceding lemma that $\Omega(\tau)$ is positive for each τ. Also, Ω is an increasing function, that is, $\Omega(\tau') \leq \Omega(\tau'')$ whenever $0 \leq \tau' \leq \tau'' \leq \tau_1$. Finally, if $S = C^*C$ and $\{C, A\}$ is observable, then $\Omega(\tau)$ is strictly positive for each τ in $[0, \tau_1]$.

We have now the following result.

Lemma 13.6.3 *Suppose A, R and S are operators on a finite dimensional space $\mathcal{X}$ with R self-adjoint and S positive. Then the following statements are equivalent.*

(a) The algebraic Riccati equation (13.1) has a positive solution Q.

(b) The Riccati differential equation (13.59) has a uniformly bounded solution Ω on the interval $[0, \infty)$ with $\Omega(0) = 0$.

In this case, the solution Ω converges to a limit Ω_∞, that is,

$$\Omega_\infty = \lim_{\tau \to \infty} \Omega(\tau). \tag{13.60}$$

Moreover, the limit Ω_∞ is the minimal positive solution to the algebraic Riccati equation (13.1), that is, if Q is any other positive solution to (13.1), then

$$\Omega_\infty \leq Q. \tag{13.61}$$

Finally, if $S = C^*C$ and $\{C, A\}$ is observable, then Ω_∞ is strictly positive.

PROOF. First suppose that the algebraic Riccati equation (13.1) has a positive solution Q. We will demonstrate that the Riccati differential equation (13.59) has a uniformly bounded solution Ω on the interval $[0, \infty)$ with $\Omega(0) = 0$. Clearly, this Riccati differential equation is locally Lipschitz in Ω. Hence, this differential equation has a unique solution over some interval $[0, \tau_1)$ for τ_1 sufficiently close to 0. To verify that this solution can be extended over the interval $[0, \infty)$, it is sufficient to show that over any interval $[0, \tau_1)$ on which Ω is defined, there is a bound m such that $\|\Omega(\tau)\| \leq m$ for $0 \leq \tau \leq \tau_1$. So, suppose Ω is a solution to the Riccati differential equation (13.59) on some interval $[0, \tau_1)$ and $\Omega(0) = 0$. It follows from Remark 13.6.2 that $\Omega(\tau)$ is positive. Consider any τ' in $[0, \tau_1)$ and any x_1 in $\mathcal{X}$ and let x be the solution on $[0, \tau']$ to

$$\frac{dx}{d\tau} = -(A + R(Q + \Omega)/2)x \quad \text{and} \quad x(\tau') = x_1. \tag{13.62}$$

Then

$$\frac{d}{d\tau}((Q - \Omega)x, x) = (-\dot{\Omega}x, x) + ((Q - \Omega)\dot{x}, x) + ((Q - \Omega)x, \dot{x}) = (Mx, x)$$

where

$$
\begin{aligned}
M &= -\dot{\Omega} + (\Omega - Q)(A + R(Q + \Omega)/2) + (A + R(Q + \Omega)/2)^*(\Omega - Q) \\
&= -\dot{\Omega} + \Omega A + A^*\Omega - QA - A^*Q + (\Omega - Q)R(Q + \Omega)/2 + (Q + \Omega)R(\Omega - Q)/2 \\
&= -\Omega R\Omega - S + QRQ + S + \Omega R\Omega - QRQ \\
&= 0.
\end{aligned}
$$

Hence

$$
\frac{d}{d\tau}\left((Q - \Omega)x, x\right) = 0.
$$

By integrating from 0 to τ', it follows that

$$
((Q - \Omega(\tau'))x(\tau'),\ x(\tau')) = ((Q - \Omega(0))x(0),\ x(0)).
$$

Since Q is positive and $\Omega(0) = 0$ while $x(\tau') = x_1$, we obtain that

$$
((Q - \Omega(\tau'))x_1,\ x_1) = (Qx(0),\ x(0)) \geq 0.
$$

Because the above holds for any x_1 in $\mathcal{X}$, we must have $\Omega(\tau') \leq Q$. Since $\Omega(\tau') \geq 0$, it now follows that there is a constant m such that $\|\Omega(\tau')\| \leq m$. The bound m is independent of τ'. This implies that the solution can be continued on the interval $[0, \infty)$. Moreover, the solution is uniformly bounded.

Now suppose that the Riccati differential equation (13.59) has a uniformly bounded solution Ω on the interval $[0, \infty)$ with $\Omega(0) = 0$. We will demonstrate that the algebraic Riccati equation (13.1) has a positive solution. As a consequence of Remark 13.6.2, the solution Ω is an increasing function and $\Omega(\tau)$ is positive. Recall that a uniformly bounded increasing sequence of positive operators converges strongly to a positive operator; see Halmos [59]. Therefore, $\Omega(\tau)$ converges to a positive operator Ω_∞ as $\tau \to \infty$, that is, $\lim_{\tau \to \infty} \Omega(\tau) = \Omega_\infty$. Since $\Omega(\tau)$ approaches a constant Ω_∞ as τ tends to infinity, our intuition tells us that Ω_∞ is a constant solution to the Riccati differential equation (13.59). In other words, the positive operator Ω_∞ is a solution to the algebraic Riccati equation (13.1). Our intuitive result can be rigorously justified by using Lemma 12.7.4. Consider now any other positive solution Q to the algebraic Riccati equation (13.1). It follows from our analysis in the first part of the proof that $\Omega(\tau) \leq Q$ for all $\tau \geq 0$. Hence, $\Omega_\infty \leq Q$. So, Ω_∞ is the minimal positive solution to (13.1).

Suppose now that $\{C, A\}$ is observable. Then, Remark 13.6.2 implies that $\Omega(\tau)$ is strictly positive for all $\tau \geq 0$. Since Ω is an increasing function, $\Omega(\tau) \leq \Omega_\infty$ for all $\tau \geq 0$. Hence, Ω_∞ is strictly positive. ∎

13.7   Notes

The presentation in this chapter concerning the Hamiltonian matrix and the algebraic Riccati equation is standard; see Francis [44], Kailath [68] and Zhou-Doyle-Glover [131]. The idea of using the Schur decomposition on the Hamiltonian matrix to solve the algebraic Riccati equation is due to Laub [79]. For some numerical procedures to solve the algebraic Riccati equation see Arnold-Laub [3] and Van Dooren [121]. The derivation of the Riccati differential equation from the state transition matrix for the Hamiltonian is classical; see Kalman [70] and Kwakernaak-Sivan [77].

Chapter 14

H^∞ Analysis

In this chapter we use an optimization problem to determine whether or not the norm of an input output operator T is bounded by a specified constant γ. As in the linear quadratic regulator problem, this optimization problem leads to a Riccati differential equation. It is shown that $\|T\| < \gamma$ if and only if there exists a solution to a certain Riccati differential equation.

14.1   A disturbance attenuation problem

Consider the state space system,

$$\dot{x} = Ax + Ew \qquad \text{and} \qquad z = Cx, \tag{14.1}$$

where A is an operator on $\mathcal{X}$, while E maps $\mathcal{W}$ into $\mathcal{X}$ and C is an operator mapping $\mathcal{X}$ into $\mathcal{Z}$. The spaces $\mathcal{X}$, $\mathcal{W}$, and $\mathcal{Z}$ are finite dimensional. In this setting, w is viewed as a disturbance input acting on the system while z is an output which reflects the system performance. In this section, we consider the disturbance attenuation properties of the above system over some time interval $[t_0, t_1]$ with $t_0 < t_1$. To this end, consider the system in (14.1) with zero initial state, that is, $x(t_0) = 0$. Given a specified scalar γ, we wish to determine whether or not

$$\int_{t_0}^{t_1} \|z(\sigma)\|^2 \, d\sigma \le \gamma^2 \int_{t_0}^{t_1} \|w(\sigma)\|^2 \, d\sigma \tag{14.2}$$

for every disturbance input w. Roughly speaking, the scalar γ is a measure of the ability of the system to mitigate the effect of the disturbance w on the output z.

For an operator interpretation of the above condition, let T be the input output operator from $L^2([t_0, t_1], \mathcal{W})$ into $L^2([t_0, t_1], \mathcal{Z})$ defined by

$$(Tw)(t) = \int_{t_0}^{t} Ce^{A(t-\tau)}Ew(\tau) \, d\tau. \tag{14.3}$$

Then $z = Tw$. Since (14.2) can be restated as $\|z\|^2 \le \gamma^2 \|w\|^2$, it follows that (14.2) holds if and only if the norm of T is less than or equal to γ. In general, computing the norm of an infinite rank operator is not computationally tractable. In Section 14.2 we will demonstrate how to use Riccati differential equations to determine the norm of T.

Our approach to the above disturbance attenuation problem is based on the following quadratic optimization problem:

For each initial state x_0 in $\mathcal{X}$, find the optimal cost $\delta(x_0)$ in the optimization problem:

$$\delta(x_0) = \sup\left\{\int_{t_0}^{t_1} \left(\|z(\sigma)\|^2 - \gamma^2\|w(\sigma)^2\|\right) d\sigma : w \in L^2\left([t_0, t_1], \mathcal{W}\right)\right\}$$

$$\text{subject to} \qquad \dot{x} = Ax + Ew \text{ and } z = Cx \text{ and } x(t_0) = x_0. \tag{14.4}$$

In addition, when a maximum exists, find an optimal input $\hat{w}$ which achieves this maximum, that is,

$$\delta(x_0) = \int_{t_0}^{t_1} \left(\|\hat{z}(\sigma)\|^2 - \gamma^2\|\hat{w}(\sigma)\|^2\right) d\sigma \tag{14.5}$$

where the optimal state $\hat{x}$ and output $\hat{z}$ are given by $\dot{\hat{x}} = A\hat{x} + E\hat{w}$ and $\hat{z} = C\hat{x}$ with $\hat{x}(t_0) = x_0$.

In general the supremum $\delta(x_0)$ in this optimization problem may be infinite. However, when $\|T\| < \gamma$, we show that the supremum is finite and the solution to this optimization problem is similar to the solution of the linear quadratic regular problem.

So, suppose that $\|T\| < \gamma$. Consider the system in (14.1) with initial condition $x(t_0) = x_0$ and let C_o be the observability operator from $\mathcal{X}$ into $L^2([t_0, t_1], \mathcal{Z})$ defined by

$$(C_o x)(t) = Ce^{A(t-t_0)}x \qquad (x \in \mathcal{X} \text{ and } t_0 \le t \le t_1). \tag{14.6}$$

Recall that C_o is one to one if and only if the pair $\{C, A\}$ is observable. Then the output z is given by

$$z = C_o x_0 + Tw. \tag{14.7}$$

We now show that for each initial state x_0, there exists a finite scalar $a(x_0)$ such that for every disturbance input w,

$$\int_{t_0}^{t_1} \|z(\sigma)\|^2 d\sigma \le \gamma^2 \int_{t_0}^{t_1} \|w(\sigma)\|^2 d\sigma + a(x_0), \tag{14.8}$$

or, equivalently, $\|z\|^2 \le \gamma^2\|w\|^2 + a(x_0)$. To see this, first observe that

$$\begin{aligned} \|z\|^2 &= \|Tw\|^2 + 2\Re(Tw, C_o x_0) + \|C_o x_0\|^2 \\ &\le \|Tw\|^2 + 2\|Tw\|\|C_o x_0\| + \|C_o x_0\|^2. \end{aligned}$$

Now recall that for any real numbers a and b, one has $2ab \le a^2 + b^2$. So, for any scalar $\epsilon > 0$, we obtain $2\|Tw\|\|C_o x_0\| \le \epsilon\|Tw\|^2 + \epsilon^{-1}\|C_o x_0\|^2$. Hence,

$$\|z\|^2 \le (1 + \epsilon)\|T\|^2\|w\|^2 + (1 + \epsilon^{-1})\|C_o x_0\|^2.$$

Choosing $\epsilon = \gamma^2/\|T\|^2 - 1$, yields

$$\|z\|^2 \le \gamma^2\|w\|^2 + \gamma^2\|C_o x_0\|^2/(\gamma^2 - \|T\|^2). \tag{14.9}$$

Therefore, the desired result in (14.8) holds with $a(x_0) = \gamma^2\|C_o x_0\|^2/(\gamma^2 - \|T\|^2)$.

Now consider the problem of finding the smallest $a(x_0)$ for which (14.8) holds for all w. Notice that (14.8) holds for every input w if and only if

$$\int_{t_0}^{t_1} (\|z(\sigma)\|^2 - \gamma^2 \|w(\sigma)\|^2) \, d\sigma \leq a(x_0)$$

holds for every w in $L^2([t_0, t_1], \mathcal{W})$. Hence, it follows that the smallest $a(x_0)$ is given by the solution to the quadratic optimization problem in (14.4).

Remark 14.1.1 Let T be the operator from $L^2([t_0, t_1], \mathcal{W})$ into $L^2([t_0, t_1], \mathcal{Z})$ defined in (14.3) and set $z = C_o x_0 + Tw$ where C_o is the observability operator defined in (14.6). If $\|T\| < \gamma$, then the above discussion shows that

$$\delta(x_0) = \sup_{w \in L^2} \int_{t_0}^{t_1} (\|z(\sigma)\|^2 - \gamma^2 \|w(\sigma)\|^2) \, d\sigma \leq \gamma^2 \|C_o x_0\|^2 / (\gamma^2 - \|T\|^2). \qquad (14.10)$$

In particular, if $\|T\| < \gamma$, then $\delta(x_0)$ is finite for all initial states x_0 in $\mathcal{X}$ and the optimization problem in (14.4) has a finite supremum.

14.2 A Riccati equation

Following our approach to the linear quadratic regulator problem, we seek a solution to the optimization problem in (14.4) using a Riccati differential equation. Notice that the integrand in this optimization problem is of the form $\|z\|^2 + (Rw, w)$ with $R = -\gamma^2 I$. By setting $R = -\gamma^2 I$ and $B = E$ in the Riccati differential equation (12.12) used to solve the weighted linear quadratic regulator problem, we obtain

$$\dot{P} + A^* P + PA + \gamma^{-2} PEE^* P + C^* C = 0 \qquad \text{(with } P(t_1) = 0\text{)}. \qquad (14.11)$$

Our first result states that the optimization problem in (14.4) has a finite supremum for every initial state x_0 if and only if the associated Riccati differential equation in (14.11) has a solution for P on the interval $[t_0, t_1]$. In this case, there exists a unique optimal input $\hat{w}$ in $L^2([t_0, t_1], \mathcal{W})$ which achieves the supremum, that is,

$$\delta(x_0) = \int_{t_0}^{t_1} (\|\hat{z}(\sigma)\|^2 - \gamma^{-2} \|\hat{w}(\sigma)\|^2) \, d\sigma \qquad (14.12)$$

where the optimal output is given by $\hat{z} = C\hat{x}$ while the optimal state $\hat{x}$ satisfies $\dot{\hat{x}} = A\hat{x} + E\hat{w}$ with $\hat{x}(t_0) = x_0$. Before obtaining this result we need the following observations.

Remark 14.2.1 For any time interval $[a, b]$ and any initial state x_0, let $\rho(x_0, a, b)$ be the optimal cost given by

$$\rho(x_0, a, b) = \sup \left\{ \int_a^b (\|z(\sigma)\|^2 - \gamma^2 \|w(\sigma)^2\|) \, d\sigma : w \in L^2([t_0, t_1], \mathcal{W}) \right\}$$

$$\text{subject to} \quad \dot{x} = Ax + Ew \text{ and } z = Cx \text{ and } x(t_0) = x_0. \qquad (14.13)$$

Obviously, $\delta(x_0) = \rho(x_0, t_1, t_2)$. We first claim that if $a < b < c$, then $\rho(x_0, a, b) \leq \rho(x_0, a, c)$. To see this, consider any $w \in L^2([a, b], \mathcal{W})$ and let $\tilde{w} \in L^2([a, c], \mathcal{W})$ be defined by

$$\tilde{w}(t) = \begin{cases} w(t) & \text{for} \quad a \leq t \leq b \\ 0 & \text{for} \quad b < t \leq c. \end{cases}$$

Then,

$$\int_a^b \left(\|z(\sigma)\|^2 - \gamma^2 \|w(\sigma)^2\| \right) d\sigma \leq \int_a^c \left(\|z(\sigma)\|^2 - \gamma^2 \|\tilde{w}(\sigma)^2\| \right) d\sigma \leq \rho(x_0, a, c).$$

Since the above holds for any $w \in L^2([a, b], \mathcal{W})$, it follows that $\rho(x_0, a, b) \leq \rho(x_0, a, c)$.

We now claim that if $a < b < c$, then $\rho(x_0, b, c) \leq \rho(x_0, a, c)$. First note that due to the time invariant nature of the optimization problem, it should be clear that $\rho(x_0, b+d, c+d) = \rho(x_0, b, c)$ for any real number d. The desired result now follows from

$$\rho(x_0, b, c) = \rho(x_0, a, c + a - b) \leq \rho(x_0, a, c).$$

Theorem 14.2.1 *The optimal cost $\delta(x_0)$ for the optimization problem in (14.4) is finite for every initial state x_0 in $\mathcal{X}$ if and only if the Riccati differential equation in (14.11) has a solution P on the interval $[t_0, t_1]$. In this case, In this case, the optimal cost is given by*

$$\delta(x_0) = (P(t_0)x_0, x_0) \tag{14.14}$$

and is uniquely attained by the optimal input

$$\hat{w}(t) = \gamma^{-2} E^* P(t) \hat{x}(t) \tag{14.15}$$

where the optimal state trajectory $\hat{x}$ is uniquely determined by

$$\dot{\hat{x}} = (A + \gamma^{-2} EE^* P(t)) \hat{x} \qquad \text{with} \qquad \hat{x}(t_0) = x_0. \tag{14.16}$$

PROOF. Assume that the Riccati differential equation in (14.11) has a solution P over the interval $[t_0, t_1]$. Then we claim that the supremum in the optimization problem (14.4) is attained and is given by $\delta(x_0) = (P(t_0)x_0, x_0)$. Moreover, this supremum is uniquely determined by the optimal disturbance $\hat{w} = \gamma^{-2} E^* P \hat{x}$ where the optimal state $\hat{x}$ is given in (14.16). To see this we first note that $P(t)$ is positive; see Lemma 13.6.2. This lemma also states that P is increasing backwards in time, that is, if $t' \leq t''$ then $P(t') \geq P(t'')$. We now apply the completion of squares technique and use the Riccati differential equation. Using (14.11) along with $\dot{x} = Ax + Ew$, we obtain

$$\begin{aligned} \frac{d}{dt}(Px, x) &= (\dot{P}x, x) + (P\dot{x}, x) + (Px, \dot{x}) \\ &= (\dot{P}x, x) + (PAx, x) + (PEw, x) + (Px, Ax) + (Px, Ew) \\ &= ((\dot{P} + PA + A^*P)x, x) + (w, E^*Px) + (E^*Px, w) \\ &= -\|Cx\|^2 - \gamma^{-2}\|E^*Px\|^2 + (w, E^*Px) + (E^*Px, w) \\ &= -\|z\|^2 + \gamma^2\|w\|^2 - \|\gamma^{-1}E^*Px - \gamma w\|^2. \end{aligned} \tag{14.17}$$

By integrating from t_0 to t_1 and using the facts that $P(t_1) = 0$, and $x(t_0) = x_0$, we have

$$\int_{t_0}^{t_1} \left(\|z(\sigma)\|^2 - \gamma^2 \|w(\sigma)\|^2 \right) \, d\sigma = (P(t_0)x_0, x_0) - \int_{t_0}^{t_1} \|\gamma^{-1} E^* P(\sigma)x(\sigma) - \gamma w(\sigma)\|^2 \, d\sigma \, .$$
(14.18)

This readily implies that

$$\int_{t_0}^{t_1} \left(\|z(\sigma)\|^2 - \gamma^2 \|w(\sigma)\|^2 \right) \, d\sigma \leq (P(t_0)x_0, x_0)$$
(14.19)

for every input w. Moreover, we have equality if and only if the integrand of the second integral in (14.18) is identically zero, that is, $w(t) = \gamma^{-2} E^* P(t)x(t)$. Therefore, $\hat{w} = \gamma^{-2} E^* P \hat{x}$ is the unique optimal input where the optimal state $\hat{x}$ is defined by (14.16). Hence, a maximum exists in the optimization problem (14.4) and is given by $\delta(x_0) = (P(t_0)x_0, x_0)$.

On the other hand, if $\delta(x_0)$ is finite for every initial state x_0 in $\mathcal{X}$, then we claim that the Riccati differential equation in (14.11) has a unique solution P defined on the interval $[t_0, t_1]$. To verify this, first notice that this Riccati equation is locally Lipschitz in P. Hence, this differential equation has a unique solution over some interval $(t_2, t_1]$ for t_2 sufficiently close to t_1. To verify that this solution can be extended over the interval $[t_0, t_1]$, it is sufficient to show that over any interval $(t_2, t_1]$ on which P is defined and $t_0 \leq t_2$, there is a bound M such that $\|P(t)\| \leq M$ for $t_2 < t \leq t_1$. Consider any t in the interval $(t_2, t_1]$. Then P is defined on the interval $[t, t_1]$. Recalling the analysis in the first part of the proof and Remark 14.2.1, it follows that $P(t)$ is positive and for each x_0 in $\mathcal{X}$,

$$(P(t)x_0, x_0) = \rho(x_0, t, t_1) \leq \rho(x_0, t_0, t_1) = \delta(x_0) \, .$$
(14.20)

Let $\{\psi_j\}_1^n$ be an orthonormal basis for the state space $\mathcal{X}$. Since $P(t)$ is positive,

$$\|P(t)\| \leq \operatorname{trace} P(t) = \sum_{j=1}^{n} (P(t)\psi_j, \psi_j) \leq \sum_{j=1}^{n} \delta(\psi_j) = M < \infty \, .$$

So, there exists a bound M such that $\|P(t)\| \leq M$ for $t_2 < t \leq t_1$. Therefore, the Riccati differential equation in (14.11) has a unique solution P defined on the interval $[t_0, t_1]$. ∎

Using Theorem 14.2.1 we can now obtain the following result.

Theorem 14.2.2 *Let $\{A, E, C\}$ be the linear system in (14.1) and T the corresponding input output operator from $L^2([t_0, t_1], \mathcal{W})$ into $L^2([t_0, t_1], \mathcal{Z})$ defined in (14.3). Let $\delta(x_0)$ be the optimal cost for the optimization problem in (14.4). Then the following statements are equivalent.*

(i) The norm $\|T\| < \gamma$.

(ii) The Riccati differential equation in (14.11) has a solution P on the interval $[t_0, t_1]$.

(iii) The optimal cost $\delta(x_0)$ is finite for all initial states x_0 in $\mathcal{X}$.

In this case, the supremum is attained and is given by $\delta(x_0) = (P(t_0)x_0, x_0)$.

PROOF. According to Theorem 14.2.1, Parts (ii) and (iii) are equivalent. If $\|T\| < \gamma$, then equation (14.10) in Remark 14.1.1 shows that $\delta(x_0)$ is finite for all x_0 in $\mathcal{X}$. Hence, Part (i) implies Parts (ii) and (iii).

Now we show that (ii) implies (i). Recall that if a differential equation $\dot{q} = f(q, \eta)$ has a solution on an interval $[t_0, t_1]$ where f is a continuous function and η is a parameter, then $\dot{q} = f(q, \eta - \epsilon)$ also has a solution on the interval $[t_0, t_1]$ for all ϵ sufficiently small; see [28]. Now assume that the Riccati differential equation in (14.11) has a solution on $[t_0, t_1]$. Then this Riccati differential equation also has a solution on the interval $[t_0, t_1]$ when γ is replaced by $\gamma - \epsilon$ for some $\epsilon > 0$. Theorem 14.2.1 now shows that the supremum $\delta(x_0)$ in (14.4) is finite when γ is replaced by $\gamma - \epsilon$; using $z = C_o x_0 + Tw$ we obtain

$$\sup_{w \in L^2} \left\{ \|C_o x_o + Tw\|^2 - (\gamma - \epsilon)^2 \|w\|^2 \right\} = \delta(x_0) < \infty \,.$$

Considering $x_0 = 0$ results in

$$\sup_{w \in L^2} \left\{ \|Tw\|^2 - (\gamma - \epsilon)^2 \|w\|^2 \right\} = \delta(0) < \infty \,;$$

hence, for all w, we have $\|Tw\|^2 \le (\gamma - \epsilon)^2 \|w\|^2 + \delta(0)$. Consider any integer n and replace w by nw in the above expression to obtain

$$\|Tw\|^2 \le (\gamma - \epsilon)^2 \|w\|^2 + \delta(0)/n^2 \,. \tag{14.21}$$

Since (14.21) holds for every integer n, we must have $\|Tw\|^2 \le (\gamma - \epsilon)^2 \|w\|$. Because this holds for all w we obtain that $\|T\| \le \gamma - \epsilon < \gamma$.   ∎

The following consequence of the previous theorem provides a method to compute the norm of the infinite dimensional operator T.

Corollary 14.2.3 *Let T be the operator from $L^2([t_0, t_1], \mathcal{W})$ into $L^2([t_0, t_1], \mathcal{Z})$ defined in (14.3). Then $\|T\|$ is the infimum of the set of all positive numbers γ such that the Riccati differential equation in (14.11) has a solution on the interval $[t_0, t_1]$.*

Example 14.2.1 (Simple integrator) Consider the classical problem of computing the norm of the integrator operator T on $L^2[0, 1]$ defined by

$$(Tw)(t) = \int_0^t w(\tau)\, d\tau \,;$$

see Problem 188 in Halmos [59]. This operator is the input output operator associated with the system $\dot{x} = w$ and $z = x$ where $t_0 = 0$ and $t_1 = 1$. Let us use Corollary 14.2.3 to compute the norm of T. For any $\gamma > 0$, the corresponding Riccati differential equation is given by

$$\dot{P} + \gamma^{-2} P^2 + 1 = 0 \,.$$

With the terminal condition $P(t_1) = 0$, the solution to this differential equation is given by $P(t) = \gamma \tan(\gamma^{-1}(1 - t))$ for $1 - \gamma\pi/2 < t \le 1$. Note that this solution cannot be continued beyond $t \le 1 - \gamma\pi/2$. Hence, in order for the solution to be defined on the interval $[0, 1]$, it is necessary and sufficient that $\gamma > 2/\pi$. According to Corollary 14.2.3, the norm of the operator T is $2/\pi$.

Remark 14.2.2 Assume that the supremum $\delta(x_0)$ in (14.4) is finite for all x_0 in $\mathcal{X}$, or equivalently, the Riccati differential equation in (14.11) has a solution on the interval $[t_0, t_1]$. If the pair $\{C, A\}$ is observable, then $P(t)$ is strictly positive for all $t_0 \leq t < t_1$. This follows from Lemma 13.6.2. We can also demonstrate this as follows. Consider any t in the interval $[t_0, t_1)$ and let $x(t) = x_0$ and $w = 0$. Then $z(\sigma) = Ce^{A(\sigma-t)}x_0$ and replacing t_0 with t in (14.18), we obtain that

$$(P(t)x_0, x_0) \geq \int_t^{t_1} \|Ce^{A(\sigma-t)}x_0\|^2 \, d\sigma \, .$$

Because $\{C, A\}$ is observable, $\int_t^{t_1} \|Ce^{A(\sigma-t)}x_0\|^2 \, d\sigma > 0$ for all nonzero x_0. Therefore, the operator $P(t)$ is strictly positive.

Remark 14.2.3 It is emphasized that one must integrate the Riccati differential equation in (14.11) backwards in time to find $P(t)$. However, one can easily convert this equation to a Riccati differential equation moving forward in time. To see this let $\Omega(\tau) = P(t_1-\tau)$. Then equation (14.11) gives

$$\dot{\Omega} = A^*\Omega + \Omega A + \gamma^{-2}\Omega EE^*\Omega + C^*C \qquad (\Omega(0) = 0). \tag{14.22}$$

Therefore, one can obtain P by solving forward in time for Ω in the Riccati differential equation (14.22). Then $P(t) = \Omega(t_1-t)$.

Notice that, due to the time invariant nature of the system under consideration, it follows from (14.20) and (14.13) that

$$(P(t)x_0, x_0) = \sup \left\{ \int_0^{t_1-t} (\|z(\sigma)\|^2 - \gamma^2\|w(\sigma)\|^2) \, d\sigma : w \in L^2([0, t_1-t], \mathcal{W}) \right\}$$

subject to $\quad \dot{x} = Ax + Ew$ and $z = Cx$ and $x(0) = x_0$.

Hence,

$$(\Omega(\tau)x_0, x_0) = \sup \left\{ \int_0^{\tau} (\|z(\sigma)\|^2 - \gamma^2\|w(\sigma)\|^2) \, d\sigma : w \in L^2([0, t_1-t], \mathcal{W}) \right\}$$

subject to $\quad \dot{x} = Ax + Ew$ and $z = Cx$ and $x(0) = x_0$. $\tag{14.23}$

Notice that one can also express the solution Ω to the Riccati differential equation in (14.22) as the solution to the following integral equation

$$\Omega(\tau) = \int_0^{\tau} e^{A^*(\tau-\sigma)} \left(C^*C + \gamma^{-2}\Omega(\sigma)EE^*\Omega(\sigma)\right) e^{A(\tau-\sigma)} \, d\sigma \, . \tag{14.24}$$

An application of Leibnitz's rule shows that Ω in (14.24) is indeed a solution to the Riccati differential equation in (14.22). If the pair $\{C, A\}$ is observable, then $\Omega(\tau)$ is strictly positive for all $\tau > 0$ where $\Omega(\tau)$ is defined. Equation (14.24) shows that

$$\Omega(\tau) \geq \int_0^{\tau} e^{A^*(\tau-\sigma)}C^*Ce^{A(\tau-\sigma)} \, d\sigma > 0 \, .$$

The last equality follows because the pair $\{C, A\}$ is observable. Therefore, $\Omega(\tau)$ is strictly positive. Since $P(t) = \Omega(t_1-t)$, this also shows that $P(t)$ is strictly positive for all $t_0 \leq t < t_1$ when $\{C, A\}$ is observable.

14.3   An abstract optimization problem

In this section we introduce an optimization problem which plays a fundamental role in an operator development of the Riccati differential equation in (14.11). To this end, let J be a self-adjoint operator on a Hilbert space $\mathcal{K}$ while h and ξ are fixed vectors in $\mathcal{K}$. Consider the following optimization problem:

$$\beta_{h\xi} = \sup \left\{ \|h\|^2 + 2\Re(\varphi, \xi) - (J\varphi, \varphi) : \varphi \in \mathcal{K} \right\}. \tag{14.25}$$

If J is not positive, then $\beta_{h\xi}$ is infinite. To see this, suppose that J is not positive. Then there exists a vector φ in $\mathcal{K}$ such that $(J\varphi, \varphi) < 0$. Considering any integer $n > 0$, we obtain

$$\beta_{h\xi} \geq \|h\|^2 + 2\Re(n\varphi, \xi) - (Jn\varphi, n\varphi) = \|h\|^2 + 2n\Re(\varphi, \xi) - n^2(J\varphi, \varphi).$$

Since $(J\varphi, \varphi) < 0$, it follows that $\|h\|^2 + 2n\Re(\varphi, \xi) - n^2(J\varphi, \varphi)$ approaches infinity as n tends to infinity. Hence, $\beta_{h\xi} = \infty$. The following result yields a solution to the above optimization problem when J is strictly positive.

Lemma 14.3.1 *Let J be a strictly positive operator on a Hilbert space $\mathcal{K}$ while h and ξ are fixed vectors in $\mathcal{K}$. Then the optimization problem in (14.25) has a finite supremum*

$$\beta_{h\xi} = \|h\|^2 + (J^{-1}\xi, \xi) \tag{14.26}$$

and this supremum is uniquely attained by the optimal vector $\hat{\varphi} = J^{-1}\xi$.

PROOF. The proof is based on the following completion of squares:

$$
\begin{aligned}
\|h\|^2 + 2\Re(\varphi, \xi) - (J\varphi, \varphi) &= \|h\|^2 + 2\Re(J\varphi, J^{-1}\xi) - (J\varphi, \varphi) \\
&= \|h\|^2 + (J^{-1}\xi, \xi) - (J(\varphi - J^{-1}\xi), \varphi - J^{-1}\xi). \tag{14.27}
\end{aligned}
$$

Since J is strictly positive, it follows that $(J(\varphi - J^{-1}\xi), \varphi - J^{-1}\xi) \geq 0$. Moreover, this term is zero if and only if $\varphi - J^{-1}\xi = 0$, or equivalently, $\varphi = J^{-1}\xi$. Therefore, (14.27) shows that the supremum in (14.25) is finitely given by $\|h\|^2 + (J^{-1}\xi, \xi)$ and it is uniquely attained by the optimal vector $\hat{\varphi} = J^{-1}\xi$. ■

For completeness we include the following result when J is positive. However, this result is not used to solve any control problems in this monograph and can be proven using the Riesz Representation Theorem.

Theorem 14.3.2 *Let J be a positive operator on $\mathcal{K}$ and let h and ξ be fixed vectors in $\mathcal{K}$. Let M be any operator from $\mathcal{K}$ into $\mathcal{M}$ satisfying $J = M^*M$. Then the supremum $\beta_{h\xi}$ defined in (14.25) is finite if and only if ξ is in the range of M^*. In this case,*

$$\beta_{h\xi} = \|h\|^2 + \|v\|^2 \qquad (\xi = M^*v \text{ and } v \in (\ker M^*)^\perp) \tag{14.28}$$

where v is the unique vector in the closure of the range of M satisfying $\xi = M^*v$.

14.4 An operator disturbance attenuation problem

In this section, we present and solve an operator version of the optimization problem in (14.4). To this end, let T be an operator mapping a Hilbert space $\mathcal{K}$ into a Hilbert space $\mathcal{H}$ and h a vector in $\mathcal{H}$. The following optimization problem is a generalization of (14.4). For a specified $\gamma > 0$, find $\beta(h)$ such that

$$\beta(h) = \sup\{\|h + Tw\|^2 - \gamma^2\|w\|^2 : w \in \mathcal{K}\}. \tag{14.29}$$

An optimal vector $\hat{w}$ is a vector in $\mathcal{K}$ such that $\beta(h) = \|h + T\hat{w}\|^2 - \gamma^2\|\hat{w}\|^2$. Clearly, $\beta(h) \geq 0$. Equation (14.7), along with $h = C_o x$ and $\mathcal{K} = L^2([t_0, t_1], \mathcal{W})$, readily shows that (14.4) is a special case of the optimization problem in (14.29).

We say that the norm of T is *bounded* by γ if $\|T\| \leq \gamma$. The norm of T is *strictly bounded* by γ if $\|T\| < \gamma$. Clearly, the norm of T is bounded by γ if and only if $\gamma^2 I - T^*T$ is positive. Moreover, the norm of T is strictly bounded by γ if and only if $\gamma^2 I - T^*T$ is strictly positive, or equivalently, $\gamma^2 I - T^*T$ is positive and invertible.

For any w in $\mathcal{K}$, we obtain

$$\begin{aligned}
\|h + Tw\|^2 - \gamma^2\|w\|^2 &= \|h\|^2 + 2\Re(Tw, h) + \|Tw\|^2 - \gamma^2\|w\|^2 \\
&= \|h\|^2 + 2\Re(w, T^*h) - ((\gamma^2 I - T^*T)w, w). \tag{14.30}
\end{aligned}$$

Thus, the optimization problem in (14.29) is equivalent to the following optimization problem

$$\beta(h) = \sup\{\|h\|^2 + 2\Re(w, T^*h) - ((\gamma^2 I - T^*T)w, w) : w \in \mathcal{K}\}. \tag{14.31}$$

This is precisely the optimization problem in (14.25) where $\xi = T^*h$ and $J = \gamma^2 I - T^*T$. Recall that $\beta_{h\xi} = \infty$ when J is not positive. Obviously, $\gamma^2 I - T^*T$ is not positive if and only if $\|T\| > \gamma$. So, if $\|T\| > \gamma$, then $\beta(h) = \infty$. The following result solves the optimization problem in (14.29) when the norm of T is strictly bounded by γ.

Theorem 14.4.1 *Let T be an operator mapping $\mathcal{K}$ into $\mathcal{H}$ whose norm is strictly bounded by γ and let h be a vector in $\mathcal{H}$. Then the supremum in (14.29) is attained and is given by*

$$\beta(h) = \gamma^2\left((\gamma^2 I - TT^*)^{-1}h, h\right). \tag{14.32}$$

Furthermore, optimal vector $\hat{w}$ in $\mathcal{K}$ which attains this supremum is unique and is given by

$$\hat{w} = (\gamma^2 I - T^*T)^{-1}T^*h = T^*(\gamma^2 I - TT^*)^{-1}h. \tag{14.33}$$

PROOF. Since the norm of T is strictly bounded by γ, the operator $J = \gamma^2 I - T^*T$ is strictly positive. Because the optimization problems in (14.29) and (14.31) are equivalent, Lemma 14.3.1 with $\xi = T^*h$, shows that

$$\beta(h) = \beta_{h\xi} = \|h\|^2 + ((\gamma^2 I - T^*T)^{-1}T^*h, T^*h). \tag{14.34}$$

Using $R(\gamma^2 I - NR)^{-1} = (\gamma^2 I - RN)^{-1}R$ where R and N are operators acting between the appropriate spaces, we obtain

$$\begin{aligned}
I + T(\gamma^2 I - T^*T)^{-1}T^* &= I + (\gamma^2 I - TT^*)^{-1}TT^* = (\gamma^2 I - TT^*)^{-1}(\gamma^2 I - TT^* + T^*T) \\
&= \gamma^2(\gamma^2 I - TT^*)^{-1}.
\end{aligned}$$

Employing this identity in (14.34), yields the expression for $\beta(h)$ in (14.32). Finally, Lemma 14.3.1 with $J = \gamma^2 I - T^*T$, shows that the supremum in (14.29) is uniquely attained by the vector $\hat{w} = J^{-1}\xi = (\gamma^2 I - T^*T)^{-1}T^*h$. Therefore, (14.33) holds. ∎

The following result is a generalization of the bound (14.10) in Remark 14.1.1.

Corollary 14.4.2 *Let T be an operator mapping $\mathcal{K}$ into $\mathcal{H}$ whose norm is strictly bounded by γ and let h be in $\mathcal{H}$. Then the supremum in (14.29) satisfies $\beta(h) \leq \gamma^2\|h\|^2/(\gamma^2 - \|T\|^2)$.*

PROOF. Clearly, $(\gamma^2 - \|T\|^2)I \leq \gamma^2 I - TT^*$. If N and R are two strictly positive operators satisfying $N \leq R$, then $R^{-1} \leq N^{-1}$; see Lemma 14.4.3 below. Hence,

$$(\gamma^2 I - TT^*)^{-1} \leq (\gamma^2 - \|T\|^2)^{-1}I.$$

The corollary now follows from (14.32). ∎

If T is the operator from $L^2([t_0, t_1], \mathcal{W})$ into $L^2([t_0, t_1], \mathcal{Z})$ defined in (14.3) and $h = C_o x$, then Corollary 14.4.2 readily yields the bound in (14.10).

Remark 14.4.1 Let T be an operator mapping $\mathcal{K}$ into $\mathcal{H}$ whose norm is strictly bounded by γ and let h be a vector in $\mathcal{H}$. Consider the affine map from $\mathcal{K}$ into $\mathcal{H}$ defined by $z = h + Tw$. Here z can be viewed as an output. Then the optimization problem in (14.29) is equivalent to

$$\beta(h) = \sup\{\|z\|^2 - \gamma^2\|w\|^2 : w \in \mathcal{K}\}. \tag{14.35}$$

If $\hat{w}$ is the optimal input, then the corresponding optimal output $\hat{z} = h + T\hat{w}$ depends only on h in $\mathcal{H}$. In this case the optimal input $\hat{w}$ is given by the "feedback" formula $\hat{w} = \gamma^{-2}T^*\hat{z}$. To see this, notice that (14.33) readily gives $(\gamma^2 I - T^*T)\hat{w} = T^*h$. This implies that $\gamma^2\hat{w} = T^*(h + T\hat{w})$. Using $\hat{z} = h + T\hat{w}$, we obtain $\gamma^2\hat{w} = T^*\hat{z}$, which proves our claim.

Lemma 14.4.3 *Let N and R be two strictly positive operators on $\mathcal{H}$. If $N \leq R$, then $R^{-1} \leq N^{-1}$.*

PROOF. For all f in $\mathcal{H}$, we have

$$\|N^{1/2}f\|^2 = (Nf, f) \leq (Rf, f) \leq \|R^{1/2}f\|^2.$$

By replacing f with $R^{-1/2}g$, we see that $N^{1/2}R^{-1/2}$ is a contraction. (Recall that an operator M is a contraction if $\|M\| \leq 1$.) So, its adjoint $R^{-1/2}N^{1/2}$ is also contractive. Thus, $\|R^{-1/2}N^{1/2}f\| \leq \|f\|$. Now replacing f by $N^{-1/2}h$, we see that $\|R^{-1/2}h\|^2 \leq \|N^{-1/2}h\|^2$. This implies that $R^{-1} \leq N^{-1}$ and completes the proof. ∎

14.5 The disturbance attenuation problem revisited

Now let us return to the original optimization problem posed in (14.4). Recall that we obtained a solution to this problem using the Riccati differential equation in (14.11). In this section we use operator techniques to provide some further insight into the origins of this Riccati differential equation. As in the linear quadratic regulator problem, operator techniques naturally yield a two point boundary value problem which in turn leads to the Riccati differential equation. To be more specific, recall that

$$\dot{x} = Ax + Ew \qquad \text{and} \qquad z = Cx \qquad (14.36)$$

where A is an operator on $\mathcal{X}$, while E maps $\mathcal{W}$ into $\mathcal{X}$ and C is an operator mapping $\mathcal{X}$ into $\mathcal{Z}$. The spaces $\mathcal{X}$, $\mathcal{W}$, and $\mathcal{Z}$ are all finite dimensional. In this setting T is the input output map from $L^2([t_0, t_1], \mathcal{W})$ into $L^2([t_0, t_1], \mathcal{Z})$ defined by

$$(Tw)(t) = \int_{t_0}^{t} Ce^{A(t-\tau)}Ew(\tau)\,d\tau \qquad (w \in L^2([t_0, t_1], \mathcal{W})). \qquad (14.37)$$

Recall that C_o is the observability operator from $\mathcal{X}$ into $L^2([t_0, t_1], \mathcal{Z})$ defined by

$$(C_o x)(t) = Ce^{A(t-t_0)}x \qquad (x \in \mathcal{X}). \qquad (14.38)$$

Clearly, the optimization problem in (14.4) is a special case of the general problem in (14.29) with $h = C_o x_0$. By consulting Theorem 14.4.1 and Remark 14.4.1, we readily obtain the following result.

Theorem 14.5.1 *Consider the optimization problem in (14.4). Let T be the input output operator defined in (14.37) and let C_o be the observability operator defined in (14.38). If $\|T\| < \gamma$, then the supremum is attained in (14.4) and is given by*

$$\delta(x_0) = \gamma^2 \left((\gamma^2 I - TT^*)^{-1} C_o x_0, C_o x_0 \right). \qquad (14.39)$$

Moreover, the optimal input which attains this supremum is unique and is given by

$$\hat{w} = (\gamma^2 I - T^*T)^{-1} T^* C_o x_0. \qquad (14.40)$$

Finally, the optimal input $\hat{w}$ must satisfy

$$\hat{w} = \gamma^{-2} T^* \hat{z} \qquad (14.41)$$

where the corresponding optimal output $\hat{z}$ is given by $\hat{z} = C_o x_0 + T\hat{w}$, that is,

$$\begin{aligned} \dot{\hat{x}} &= A\hat{x} + E\hat{w} \qquad \text{with} \qquad \hat{x}(t_0) = x_0 \\ \hat{z} &= C\hat{x} \end{aligned} \qquad (14.42)$$

and $\hat{x}$ is the optimal state trajectory.

14.5.1 The adjoint system

Now let us use Theorem 14.5.1 to obtain the two point boundary value problem associated with the optimization problem in (14.4). Our approach involves the adjoint system corresponding to the system $\{A, E, C\}$. As before, let T be the operator from $L^2([t_0, t_1], \mathcal{W})$ into

$L^2([t_0, t_1], \mathcal{Z})$ defined in (14.37). Then a simple calculation shows that its adjoint T^* is the operator from $L^2([t_0, t_1], \mathcal{Z})$ into $L^2([t_0, t_1], \mathcal{W})$ defined by

$$(T^*g)(t) = \int_t^{t_1} E^* e^{-A^*(t-\tau)} C^* g(\tau)\, d\tau \qquad (g \in L^2([t_0, t_1], \mathcal{Z})). \tag{14.43}$$

In fact, this formula for T^* follows from (12.27) and (12.28) with $G(t) = Ce^{At}E$. To obtain a state space realization of the adjoint map T^*, let λ be the function defined by

$$\lambda(t) = \int_t^{t_1} e^{-A^*(t-\tau)} C^* g(\tau)\, d\tau.$$

Using Leibnitz's rule, we obtain $\dot{\lambda} = -A^*\lambda - C^*g$ which yields the following state space representation of $v = T^*g$:

$$\begin{aligned} \dot{\lambda} &= -A^*\lambda - C^*g \qquad \text{with} \qquad \lambda(t_1) = 0 \\ v &= E^*\lambda. \end{aligned} \tag{14.44}$$

To obtain v for a specified g, one must integrate the above differential equation backward in time from t_1. Finally, we call λ the *adjoint state*.

Now let us use the adjoint system to characterize the solution of the optimization problem in (14.4). Suppose that $\|T\| < \gamma$. Then the supremum is attained in (14.4) for every initial state x_0 in $\mathcal{X}$. According to Theorem 14.5.1, the optimal input $\hat{w}$ which achieves this supremum must satisfy $\hat{w} = \gamma^{-2}T^*\hat{z}$ where $\hat{z}$ is the corresponding optimal output trajectory; see (14.42). Since (14.44) is the state space realization for $v = T^*g$, the state space realization for the optimal input $\hat{w} = \gamma^{-2}T^*\hat{z}$ is given by

$$\begin{aligned} \dot{\lambda} &= -A^*\lambda - C^*\hat{z} \qquad \text{with} \qquad \lambda(t_1) = 0 \\ \hat{w} &= \gamma^{-2}E^*\lambda. \end{aligned}$$

Combining these equations with (14.42), we see that the optimal input is given by

$$\hat{w} = \gamma^{-2}E^*\lambda \tag{14.45}$$

where $[\hat{x}\ \lambda]^{tr}$ solves the following two point boundary value problem:

$$\begin{aligned} \dot{\hat{x}} &= A\hat{x} + \gamma^{-2}EE^*\lambda \qquad \text{with} \qquad \hat{x}(t_0) = x_0 \\ \dot{\lambda} &= -C^*C\hat{x} - A^*\lambda \qquad \text{with} \qquad \lambda(t_1) = 0. \end{aligned} \tag{14.46}$$

Summing up the previous analysis we obtain the following result. If $\|T\| < \gamma$, then the corresponding two point boundary value problem in (14.46) has a solution for every x_0 in $\mathcal{X}$. Furthermore, the supremum is obtained in the optimization problem (14.4) for every initial state x_0 and the optimal input $\hat{w}$ which achieves this supremum is given by $\hat{w} = \gamma^{-2}E^*\lambda$.

We now claim that, if $\|T\| < \gamma$, then for every x_0 in $\mathcal{X}$ and for every t' in the interval $[t_0, t_1)$, the two point boundary value problem:

$$\begin{aligned} \dot{\hat{x}} &= A\hat{x} + \gamma^{-2}EE^*\lambda \qquad \text{with} \qquad \hat{x}(t') = x_0 \\ \dot{\lambda} &= -C^*C\hat{x} - A^*\lambda \qquad \text{with} \qquad \lambda(t_1) = 0 \end{aligned} \tag{14.47}$$

has a solution $[\hat{x} \ \lambda]^{tr}$. To see this, let T' be the operator from $L^2([t', t_1], \mathcal{W})$ into $L^2([t', t_1], \mathcal{Z})$ defined by

$$(T'w)(t) = \int_{t'}^{t} Ce^{A(t-\tau)} Ew(\tau)\, d\tau. \qquad (w \in L^2([t', t_1], \mathcal{W}). \qquad (14.48)$$

We claim that $\|T'\| \leq \|T\|$, and thus, $\|T'\| < \gamma$. To see this consider any input w in $L^2([t', t_1], \mathcal{W})$ and extend it to an input $\tilde{w}$ in $L^2([t_0, t_1], \mathcal{W})$ by setting $\tilde{w}(t) = 0$ for $t_0 \leq t < t'$. Then

$$\|T'w\| = \|T\tilde{w}\| \leq \|T\| \|\tilde{w}\| = \|T\| \|w\|.$$

This implies that $\|T'\| \leq \|T\| < \gamma$. By using $\|T'\| < \gamma$ in our previous analysis with t' and T' replacing t_0 and T, respectively, we see that the two point boundary value problem (14.47) has a solution for every x_0 in $\mathcal{X}$ and for every t' in the interval $[t_0, t_1)$.

14.5.2   The Riccati equation

As before, assume that the norm of the operator T from $L^2([t_0, t_1], \mathcal{W})$ into $L^2([t_0, t_1], \mathcal{Z})$ defined in (14.37) is strictly bounded by γ. Theorem 14.2.2 shows that the Riccati equation in (14.11) has a solution P on the interval $[t_0, t_1]$. Here we provide an independent derivation of this result. We will see that the Riccati differential equation arises naturally in the solution of the two point boundary value problem in (14.46). In particular, the solution $[\hat{x} \ \lambda]^{tr}$ to this two point boundary value problem satisfies $\lambda(t) = P(t)\hat{x}(t)$. We also obtain an explicit formula for P in terms of the elements of the state transition matrix associated with this two point boundary problem.

To this end, we introduce the *Hamiltonian matrix* associated with the optimization problem in (14.4):

$$H = \begin{bmatrix} A & \gamma^{-2}EE^* \\ -C^*C & -A^* \end{bmatrix} \quad \text{on} \quad \begin{bmatrix} \mathcal{X} \\ \mathcal{X} \end{bmatrix}. \qquad (14.49)$$

Notice that H is simply the state space matrix for the system in (14.46). Since $\|T\| < \gamma$, it follows from the previous section that the two point boundary value problem in (14.47) has a solution $[\hat{x} \ \lambda]^{tr}$ for every x_0 in $\mathcal{X}$ and t' in $[t_0, t_1)$. Hence, using Lemma 13.6.1, it follows that the Riccati differential equation in (14.11) has a solution P on the interval $[t_0, t_1]$. Moreover, it follows from Remark 13.6.1 that

$$\lambda(t) = P(t)\hat{x}(t) \quad \text{and} \quad P(t) = \Phi_{21}(t-t_1)\Phi_{11}(t-t_1)^{-1} \qquad (14.50)$$

where Φ_{11} and Φ_{21} are obtained from the following matrix partition of e^{Ht}:

$$e^{Ht} = \begin{bmatrix} \Phi_{11}(t) & \Phi_{12}(t) \\ \Phi_{21}(t) & \Phi_{22}(t) \end{bmatrix} \quad \text{on} \quad \begin{bmatrix} \mathcal{X} \\ \mathcal{X} \end{bmatrix}. \qquad (14.51)$$

So, one can obtain P by using either $P(t) = \Phi_{21}(t - t_1)\Phi_{11}(t - t_1)^{-1}$ or solving the Riccati differential equation in (14.11).

Combining the results in this section with Theorem 14.2.2 gives the following result.

Theorem 14.5.2 *Let T be the operator from $L^2([t_0, t_1], \mathcal{W})$ into $L^2([t_0, t_1], \mathcal{Z})$ defined in (14.37). Let $\delta(x_0)$ be the supremum for the optimization problem in (14.4). Then the following statements are equivalent.*

(i) The supremum $\delta(x_0)$ is finite for every $x_0 \in \mathcal{X}$.

(ii) The supremum $\delta(x_0)$ is uniquely attained for every $x_0 \in \mathcal{X}$.

(iii) The Riccati differential equation in (14.11) has a solution on the interval $[t_0, t_1]$.

(iv) The two point boundary value problem in (14.47) has a solution for every x_0 in $\mathcal{X}$ and t' in the interval $[t_0, t_1)$.

(v) The norm of the operator T is strictly bounded by γ.

Remark 14.5.1 As before, let T be the operator from $L^2([t_0, t_1], \mathcal{W})$ into $L^2([t_0, t_1], \mathcal{Z})$ defined in (14.37) and assume that $\|T\| < \gamma$. Then Theorems 14.2.1 and 14.5.1 give

$$P(t_0) = \gamma^2 C_o^* (\gamma^2 I - TT^*)^{-1} C_o. \tag{14.52}$$

Recall that the pair $\{C, A\}$ is observable if and only if the operator C_o is one to one. Hence, equation (14.52) shows that the pair $\{C, A\}$ is observable if and only if $P(t_0)$ is strictly positive. Recall that if T' is the operator from $L^2([t', t_1], \mathcal{W})$ into $L^2([t', t_1], \mathcal{Z})$ defined in (14.48) where $t_0 \le t' < t_1$, then $\|T'\| \le \|T\| < \gamma$. So, by replacing T with T' and t_0 by t', we see that $P(t')$ is strictly positive for any $t_0 \le t' < t_1$ if and only if the pair $\{C, A\}$ is observable. In particular, $P(t)$ is strictly positive for any $t_0 \le t < t_1$ if and only if $P(t)$ is strictly positive for all $t_0 \le t < t_1$. Likewise $\Omega(\tau) = P(t_1 - \tau)$ is strictly positive for any $0 < \tau \le t_1 - t_0$ if and only if $\Omega(\tau)$ is strictly positive for all $0 < \tau \le t_1 - t_0$.

Remark 14.5.2 One might conjecture that the optimization problem in (14.4) has a solution for every x_0 if the two point boundary value problem in (14.46) has a solution for every x_0 on the interval $[t_0, t_1]$. However, this is false as the following example illustrates. Recall the simple integrator system ($\dot{x} = w$ and $z = x$) in Example 14.2.1 defined on the interval $[0, 1]$. As expected, $t_0 = 0$ and $t_1 = 1$. We have already shown that the corresponding optimization problem has a solution for every x_0 if and only if $\gamma > 2/\pi$. The Hamiltonian matrix associated with this system is given by

$$H = \begin{bmatrix} 0 & \gamma^{-2} \\ -1 & 0 \end{bmatrix}.$$

It is easy to verify that

$$e^{Ht} = \begin{bmatrix} \cos(\gamma^{-1} t) & \gamma^{-1} \sin(\gamma^{-1} t) \\ -\gamma \sin(\gamma^{-1} t) & \cos(\gamma^{-1} t) \end{bmatrix}.$$

By consulting (13.53), it follows that $[\hat{x} \ \lambda]^{tr}$ is a solution of the corresponding two point boundary value problem over the interval $[0, 1]$ if and only if

$$\hat{x}(t) = \cos(\gamma^{-1}(t - 1))\hat{x}(1) \quad \text{and} \quad \lambda(t) = -\gamma \sin(\gamma^{-1}(t - 1))\hat{x}(1) \tag{14.53}$$

and $x(0) = x_0$. So, this two point boundary value problem has a solution if and only if there exists a scalar $\hat{x}(1)$ such that $x_0 = \cos(\gamma^{-1})\hat{x}(1)$. In other words, for this two point boundary value problem to have a solution for every x_0, it is necessary and sufficient that $\cos(\gamma^{-1})$ be nonzero. In this case, the solution is uniquely given by (14.53) with $\hat{x}(1) = \cos(\gamma^{-1})^{-1} x_0$. So, if $\gamma = 1/2\pi$, then $\cos(\gamma^{-1})$ is nonzero and this two point boundary value problem has a solution for every x_0. Since $\|T\| = 2/\pi$, it follows that the corresponding optimization problem with $\gamma > 1/2\pi$ does not have a solution.

14.6   A spectral factorization

In this section we obtain a finite time outer spectral factor for the operator $\gamma^2 I - T^*T$. Recall that Θ is *a finite time outer spectral factor* for a strictly positive operator Ξ if $\Xi = \Theta^*\Theta$ where Θ is a causal invertible operator with a causal inverse. Let T be the causal operator from $L^2([t_0, t_1], \mathcal{W})$ into $L^2([t_0, t_1], \mathcal{Z})$ defined in (14.3). Assume that the Riccati differential equation in (14.11) has a solution on the interval $[t_0, t_1]$, or equivalently, $\|T\| < \gamma$. In this case, we will obtain a finite time outer spectral factor Θ for $\gamma^2 I - T^*T$. To begin, let L be the linear operator on $L^2([t_0, t_1], \mathcal{X})$ defined by

$$(Lf)(t) = \int_{t_0}^{t} e^{A(t-\tau)} f(\tau)\, d\tau \qquad (f \in L^2([t_0, t_1], \mathcal{X})). \qquad (14.54)$$

Notice that $g = Lf$, for some f in $L^2([t_0, t_1], \mathcal{X})$, if and only if g satisfies the following differential equation

$$\dot{g} = Ag + f \qquad \text{with} \qquad g(t_0) = 0. \qquad (14.55)$$

It should also be clear that $T = CLE$.

Theorem 14.6.1 *Let T be the causal operator from $L^2([t_0, t_1], \mathcal{W})$ into $L^2([t_0, t_1], \mathcal{Z})$ defined in (14.3) and assume that the Riccati differential equation in (14.11) has a solution on the interval $[t_0, t_1]$. Let Θ be the causal operator on $L^2([t_0, t_1], \mathcal{W})$ defined by*

$$\Theta = \gamma I - \gamma^{-1} E^* P L E. \qquad (14.56)$$

Then Θ has a causal inverse and $\gamma^2 I - T^*T = \Theta^*\Theta$. In particular, $\|T\| < \gamma$.

PROOF. The adjoint L^* of L is the operator on $L^2([t_0, t_1], \mathcal{X})$ defined by

$$(L^*\phi)(t) = \int_{t}^{t_1} e^{-A^*(t-\tau)} \phi(\tau)\, d\tau \qquad (\phi \in L^2([t_0, t_1], \mathcal{X})). \qquad (14.57)$$

By applying Leibnitz's rule to (14.57), it follows that $\xi = L^*\phi$ for some ϕ in $L^2([t_0, t_1], \mathcal{X})$ if and only if ξ is the unique solution to the following differential equation

$$\dot{\xi} = -A^*\xi - \phi \qquad \text{with} \qquad \xi(t_1) = 0. \qquad (14.58)$$

Consider now any differentiable function g in $L^2([t_0, t_1], \mathcal{X})$. Using the Riccati differential equation in (14.11), we have

$$\frac{d}{dt}(Pg) = -A^*(Pg) - C^*Cg - \gamma^{-2} PEE^*Pg + P\dot{g} - PAg.$$

Since $P(t_1)g(t_1) = 0$, it now follows from the characterization of $\xi = L^*\phi$ in (14.58) that

$$Pg = L^*(C^*Cg + \gamma^{-2} PEE^*Pg - P\dot{g} + PAg).$$

This yields the following relationship:

$$(P - L^*C^*C - \gamma^{-2} L^*PEE^*P)g = -L^*P(\dot{g} - Ag) \qquad (14.59)$$

for any differentiable function g in $L^2([t_0, t_1], \mathcal{X})$. Consider any f in $L^2([t_0, t_1], \mathcal{X})$ and let $g = Lf$. Then $\dot{g} - Ag = f$. Substituting this into (14.59) yields

$$(P - L^*C^*C - \gamma^{-2}L^*PEE^*P)Lf = -L^*Pf.$$

Since f is arbitrary, we obtain the following operator algebraic Riccati equation

$$L^*P + PL - L^*C^*CL - \gamma^{-2}L^*PEE^*PL = 0. \tag{14.60}$$

By applying E^* to the left and E on the right, rearranging terms and using $T = CLE$, yields

$$
\begin{aligned}
\gamma^2 I - T^*T &= \gamma^2 I - E^*L^*C^*CLE \\
&= \gamma^2 I - E^*L^*PE - E^*PLE + \gamma^{-2}E^*L^*PEE^*PLE \\
&= (\gamma I - \gamma^{-1}E^*PLE)^*(\gamma I - \gamma^{-1}E^*PLE).
\end{aligned}
$$

Hence, $\gamma^2 I - T^*T = \Theta^*\Theta$ where $\Theta = \gamma I - \gamma^{-1}E^*PLE$.

Obviously, Θ is causal. In fact, Θ can be viewed as the input output map for a state space system. To be precise, if $h = \Theta w$ for some w in $L^2([t_0, t_1], \mathcal{W})$, then we claim that h is the output of the following linear system

$$\dot{q} = Aq + Ew \quad \text{and} \quad h = -\gamma^{-1}E^*Pq + \gamma w, \tag{14.61}$$

with the initial condition $q(t_0) = 0$. To verify that $h = \Theta w$, simply observe that $\dot{q} = Aq + Ew$ with $q(t_0) = 0$, yields $q = LEw$. Hence, $h = -\gamma^{-1}E^*PLEw + \gamma w = \Theta w$, which verifies our claim.

Lemma 12.5.2 shows that Θ is an invertible operator and its inverse is causal. In fact, by setting $\tilde{C} = -\gamma^{-1}E^*P$ with $\tilde{B} = E$ and $D = \gamma I$ in Lemma 12.5.2, the inverse of Θ is given by

$$(\Theta^{-1}h)(t) = \gamma^{-1}h(t) + \gamma^{-3}\int_{t_0}^{t} E^*P(t)\Psi(t, \tau)Eh(\tau)d\tau \quad (h \in L^2([t_0, t_1], \mathcal{W}))$$

where $\Psi(t, \tau)$ is the state transition matrix for $A + \gamma^{-2}EE^*P$.

Since Θ has a bounded inverse and $\gamma^2 I - T^*T = \Theta^*\Theta$, it follows that $\|T\| < \gamma$. To see this observe that $\Theta^*\Theta \geq \epsilon^2 I$ for some $\epsilon > 0$. In fact, one can choose $\epsilon = \|\Theta^{-1}\|^{-1}$. Hence, $\gamma^2 I - T^*T \geq \epsilon^2 I$, or equivalently, $T^*T \leq (\gamma^2 - \epsilon^2)I$. Therefore, $\|T\| < \gamma$. ∎

14.6.1 A general disturbance attenuation problem

The results in this section are of independent interest and are not used in our later developments. In this section we use the finite time outer spectral factor Θ in Theorem 14.6.1 to solve a generalization of the disturbance attenuation problem presented in Section 14.1. Recall that A is an operator on $\mathcal{X}$ while E is an operator from $\mathcal{W}$ into $\mathcal{X}$ and C is an operator from $\mathcal{X}$ into $\mathcal{Z}$. Let B be an operator from $\mathcal{U}$ into $\mathcal{X}$ and D an operator from $\mathcal{U}$ into $\mathcal{Z}$. Consider the system

$$
\begin{aligned}
\dot{x} &= Ax + Ew + Bu \\
z &= Cx + Du.
\end{aligned}
\tag{14.62}
$$

Here u is a signal or vector in $L([t_0, t_1], \mathcal{U})$. This leads to the following optimization problem

$$\delta(x_0) = \sup \left\{ \int_{t_0}^{t_1} (\|z(\sigma)\|^2 - \gamma^2 \|w(\sigma)\|^2) d\sigma : w \in L^2([t_0, t_1], \mathcal{W}) \right\}$$

subject to the system in (14.62) and $x(t_0) = x_0$. $\hspace{2cm}$ (14.63)

This problem can be viewed as the dual of the linear quadratic tracking problem discussed in Section 12.5.1. The following result presents a solution to this optimization problem.

Theorem 14.6.2 *Consider the optimization problem in (14.63). Assume that there exists a solution P to the Riccati differential equation in (14.11) on the interval $[t_0, t_1]$. Then the maximum exists, and there exists a unique optimal input $\hat{w}$ in $L^2([t_0, t_1], \mathcal{W})$ which achieves this maximum. To compute $\hat{w}$, let ζ be the solution over $[t_0, t_1]$ to the following differential equation:*

$$\dot{\zeta} = -(A + \gamma^{-2} EE^* P)^* \zeta - (PB + C^* D)u \hspace{1cm} (\zeta(t_1) = 0).$$ (14.64)

Then the optimal input $\hat{w}$ is given by

$$\hat{w}(t) = \gamma^{-2} E^* (P(t)\hat{x}(t) + \zeta(t))$$ (14.65)

where the optimal state trajectory $\hat{x}$ is uniquely determined by

$$\dot{\hat{x}} = (A + \gamma^{-2} EE^* P)\hat{x} + \gamma^{-2} EE^* \zeta + Bu \hspace{1cm} (\hat{x}(t_0) = x_0).$$ (14.66)

PROOF. We use the notation and results in Section 14.6 to prove this theorem. As before, let L be the operator on $L^2([t_0, t_1], \mathcal{X})$ defined in (14.54). Recall that the input output operator T in (14.3) is given by $T = CLE$. Moreover, the output z for the state space system in (14.62) can be expressed as

$$z = C_o x_0 + CLBu + Du + Tw,$$ (14.67)

where C_o is the observability operator defined in (14.6). So, if we set $h = C_o x_0 + CLBu + Du$, then the optimization problem in (14.63) is a special case of the optimization problem in (14.29). Because the Riccati differential equation has a solution over the interval $[t_0, t_1]$, Theorem 14.2.2 shows that the norm of T is strictly bounded by γ. According to Theorem 14.4.1 the supremum in (14.63) is obtained. Moreover, according to Remark 14.4.1, the unique optimal input $\hat{w}$ which attains this supremum is given by $\hat{w} = \gamma^{-2} T^* \hat{z}$ where the optimal output $\hat{z}$ is given by

$$\hat{z} = h + T\hat{w} = C_o x_0 + CLBu + Du + T\hat{w}.$$

Finally, it is noted that the optimal state trajectory $\hat{x}$ is given by

$$\dot{\hat{x}} = A\hat{x} + E\hat{w} + Bu \hspace{1cm} (\hat{x}(t_0) = x_0).$$ (14.68)

Using $g = \hat{x}$ in equation (14.59) along with (14.68), we obtain

$$(L^* C^* C + \gamma^{-2} L^* PEE^* P - P)\hat{x} = L^* P(E\hat{w} + Bu).$$

Recall that $\Theta^* = \gamma I - \gamma^{-1}E^*L^*PE$; see (14.56). Multiplying the above equation by E^* on the left and rearranging terms, we obtain

$$
\begin{aligned}
E^*L^*C^*C\hat{x} - E^*L^*PE\hat{w} &= \gamma^{-1}(\gamma I - \gamma^{-1}E^*L^*PE)E^*P\hat{x} + E^*L^*PBu \\
&= \gamma^{-1}\Theta^*E^*P\hat{x} + E^*L^*PBu.
\end{aligned}
$$

Employing $\hat{z} = C\hat{x} + Du$ and $\hat{w} = \gamma^{-2}T^*\hat{z} = \gamma^{-2}E^*L^*C^*\hat{z}$ in the previous equation yields,

$$
\begin{aligned}
\Theta^*\hat{w} &= (\gamma I - \gamma^{-1}E^*L^*PE)\hat{w} = \gamma^{-1}E^*L^*C^*\hat{z} - \gamma^{-1}E^*L^*PE\hat{w} \\
&= \gamma^{-1}E^*L^*C^*(C\hat{x} + Du) - \gamma^{-1}E^*L^*PE\hat{w} \\
&= \gamma^{-1}(E^*L^*C^*C\hat{x} - E^*L^*PE\hat{w}) + \gamma^{-1}E^*L^*C^*Du \\
&= \gamma^{-2}\Theta^*E^*P\hat{x} + \gamma^{-1}E^*L^*(PB + C^*D)u.
\end{aligned}
$$

According to Theorem 14.6.1, the operator $\Theta^* = \gamma I - \gamma^{-1}E^*L^*PE$ is invertible. This along with the identity $(I - NM)^{-1}N = N(I - MN)^{-1}$, yields

$$
\begin{aligned}
\hat{w} &= \gamma^{-2}E^*P\hat{x} + \gamma^{-1}(\gamma I - \gamma^{-1}E^*L^*PE)^{-1}E^*L^*(PB + C^*D)u \\
&= \gamma^{-2}E^*P\hat{x} + \gamma^{-2}E^*(I - \gamma^{-2}L^*PEE^*)^{-1}L^*(PB + C^*D)u \\
&= \gamma^{-2}E^*(P\hat{x} + \zeta)
\end{aligned}
$$

where $\zeta := (I - \gamma^{-2}L^*PE^*E)^{-1}L^*(PB + C^*D)u$. By removing the inverse, we obtain $(I - \gamma^{-2}L^*PE^*E)\zeta = L^*(PB + C^*D)u$. Thus,

$$
\zeta = L^*\left(\gamma^{-2}PE^*E\zeta + (PB + C^*D)u\right).
$$

Recall that if $\xi = L^*\phi$, then ξ satisfies the differential equation $\dot{\xi} = -A^*\xi - \phi$ where $\xi(t_1) = 0$; see (14.58). Hence, ζ satisfies the differential equation in (14.64). Since $\hat{w} = \gamma^{-2}E^*(P\hat{x} + \zeta)$, equation (14.65) holds. Substituting this into (14.68) yields (14.66). ■

Finally, it is noted that Theorem 14.6.2 also holds in the time varying case, that is, when A, B, C, D, E are continuous functions on $[t_0, t_1]$ with values in the appropriate $\mathcal{L}(\cdot, \cdot)$ spaces. Because the proof of this result is almost identical to the proof of Theorem 14.6.2, the details are left as an exercise.

14.7 The infinite horizon problem

This section is concerned with the following infinite horizon optimization problem. For each initial state x_0 in $\mathcal{X}$, find the supremum $\delta(x_0)$ and an optimal input $\hat{w}$ which solves the optimization problem:

$$
\delta(x_0) = \sup\left\{\int_0^\infty \left(||z(\sigma)||^2 - \gamma^2||w(\sigma)||^2\right)\,d\sigma : w \in L^2([0, \infty), \mathcal{W})\right\}
$$

subject to $\dot{x} = Ax + Ew$ and $z = Cx$ and $x(0) = x_0$.   (14.69)

As before, A is an operator on $\mathcal{X}$ and E maps $\mathcal{W}$ into $\mathcal{X}$ while C is an operator mapping $\mathcal{X}$ into $\mathcal{Z}$.

As in Remark 14.2.3, let $\Omega(\tau) = P(t_1-\tau)$ where P is the solution to the Riccati differential equation in (14.11). Remark 14.2.3 shows that Ω is an increasing function which satisfies the following Riccati differential equation

$$\dot{\Omega} = A^*\Omega + \Omega A + \gamma^{-2}\Omega EE^*\Omega + C^*C \qquad (\Omega(0) = 0). \qquad (14.70)$$

Moreover, recalling (14.23), we have

$$(\Omega(\tau)x_0, x_0) = \sup\left\{\int_0^\tau \left(\|z(\sigma)\|^2 - \gamma^2\|w(\sigma)^2\|\right) d\sigma : w \in L^2\left([0, \infty), \mathcal{W}\right)\right\}. \qquad (14.71)$$

Proceeding as in Remark 14.2.1, one can readily show that

$$\rho(x_0, 0, \tau) \leq \sup_w \int_0^\infty \left(\|z(\sigma)\|^2 - \gamma^2\|w(\sigma)^2\|\right) d\sigma = \delta(x_0). \qquad (14.72)$$

Hence,

$$(\Omega(\tau)x_0, x_0) \leq \delta(x_0) \qquad (14.73)$$

for all τ. The following result shows that $\delta(x_0)$ is the limit of $(\Omega(\tau)x_0, x_0)$ as τ approaches infinity. To this end, recall that Ω is a *uniformly bounded solution* to the Riccati differential equation in (14.70) if Ω is a solution to (14.70) for all $\tau \geq 0$ and $\|\Omega(\tau)\| \leq mI$ for some finite scalar m. If Ω is a uniformly bounded solution to this Riccati differential equation, then Lemma 13.6.3 tells us that Ω converges to a limit Ω_∞ as τ approaches infinity. In addition, $Q = \Omega_\infty$ is the minimal positive solution to the following algebraic Riccati equation

$$A^*Q + QA + \gamma^{-2}QEE^*Q + C^*C = 0. \qquad (14.74)$$

This sets the stage for the following result.

Lemma 14.7.1 *Let $\delta(x_0)$ be the optimal cost for the infinite horizon optimization problem in (14.69). Then the following statements are equivalent.*

(i) The optimal cost $\delta(x_0)$ is finite for all x_0 in $\mathcal{X}$.

(ii) There exists a uniformly bounded solution Ω to the Riccati differential equation in (14.70).

(iii) There exists a positive solution Q to the algebraic Riccati equation in (14.74).

In this case, the solution Ω converges to a limit Ω_∞, that is,

$$\Omega_\infty = \lim_{\tau \to \infty} \Omega(\tau). \qquad (14.75)$$

and the optimal cost is given by

$$\delta(x_0) = (\Omega_\infty x_0, x_0). \qquad (14.76)$$

In addition, the limit Ω_∞ is the minimal positive solution to the algebraic Riccati equation (14.74).

PROOF. We first demonstrate that (i) implies (ii). So, suppose that the optimal cost $\delta(x_0)$ is finite for all x_0 in $\mathcal{X}$. Let Ω be the solution to the Riccati differential equation in (14.70) which is defined on some interval $[0, t_1)$. We claim that Ω can be extended to a uniformly bounded solution on $[0, \infty)$. To see this it is sufficient to show that $\|\Omega(\tau)\| \le m$ for $0 \le \tau < t_1$ where m is a specified scalar independent of t_1. We first note that Lemma 13.6.2 implies that $\Omega(\tau)$ is positive for all τ. Let $\{\varphi_j\}_1^n$ be any orthonormal basis for $\mathcal{X}$. Then using $(\Omega(\tau)\varphi, \varphi) \le \delta(\varphi)$ for any φ in $\mathcal{X}$, we obtain

$$\|\Omega(\tau)\| \le \text{ trace } \Omega(\tau) = \sum_{j=1}^n (\Omega(\tau)\varphi_j, \varphi_j) \le \sum_{j=1}^n \delta(\varphi_j) < \infty.$$

So, if $m = \sum_1^n \delta(\varphi_j)$, then $\|\Omega(\tau)\| \le m$ for any $0 \le \tau < t_1$. Therefore, there exists a uniformly bounded solution Ω to the Riccati differential equation in (14.70) and Part (ii) holds.

The equivalence of Parts (ii) and (iii) follows from Lemma 13.6.3. This lemma also states that the solution Ω converges to a limit Ω_∞. Moreover, the limit Ω_∞ is the minimal positive solution to the algebraic Riccati equation (14.74). Because $(\Omega(\tau)x_0, x_0) \le \delta(x_0)$ for all $\tau \ge 0$, we have $(\Omega_\infty x_0, x_0) \le \delta(x_0)$.

We now demonstrate that (iii) implies (i). So, suppose that the algebraic Riccati equation in (14.74) admits a positive solution Q. We claim that $\delta(x_0) \le (Qx_0, x_0)$. In particular, $\delta(x_0)$ is finite for all x_0 in $\mathcal{X}$. Using the algebraic Riccati equation in (14.74), along with $\dot{x} = Ax + Ew$ and $z = Cx$, we obtain

$$
\begin{aligned}
\frac{d}{dt}(Qx, x) &= (Q\dot{x}, x) + (Qx, \dot{x}) \\
&= (QAx, x) + (QEw, x) + (Qx, Ax) + (Qx, Ew) \\
&= ((QA + A^*Q)x, x) + (w, E^*Qx) + (E^*Qx, w) \\
&= -\|Cx\|^2 - \gamma^{-2}\|E^*Qx\|^2 + (w, E^*Qx) + (E^*Qx, w) \\
&= -\|z\|^2 + \gamma^2\|w\|^2 - \|\gamma^{-1}E^*Qx - \gamma w\|^2 .
\end{aligned}
\tag{14.77}
$$

By integrating from 0 to t with $x_0 = x(0)$, we have

$$
\int_0^t \left(\|z(\sigma)\|^2 - \gamma^2\|w(\sigma)\|^2 \right) d\sigma
$$

$$
= (Qx_0, x_0) - (Qx(t), x(t)) - \int_0^t \|\gamma^{-1}E^*Qx(\sigma) - \gamma w(\sigma)\|^2 d\sigma .
\tag{14.78}
$$

This readily implies that

$$
\int_0^t \left(\|z(\sigma)\|^2 - \gamma^2\|w(\sigma)\|^2 \right) d\sigma \le (Qx_0, x_0) .
\tag{14.79}
$$

By letting t approach infinity and then taking the supremum over all w in $L^2([0, \infty), \mathcal{W})$, it follows that $\delta(x_0) \le (Qx_0, x_0)$. In particular, $\delta(x_0)$ is finite for all x_0 in $\mathcal{X}$, and thus, Part (i) holds. Therefore, Parts $(i),(ii)$ and (iii) are equivalent.

Recall that Ω_∞ is also a positive solution to the algebraic Riccati equation in (14.74). So, we obtain $\delta(x_0) \le (\Omega_\infty x_0, x_0)$. Combining this with $\delta(x_0) \ge (\Omega_\infty x_0, x_0)$, yields the equality $\delta(x_0) = (\Omega_\infty x_0, x_0)$. ∎

Remark 14.7.1 Suppose A is stable. If Q is a self-adjoint solution to the algebraic Riccati equation in (14.74), then

$$Q = \int_0^\infty e^{A^*\sigma} \left(C^*C + \gamma^{-2}QEE^*Q\right) e^{A\sigma} \, d\sigma \, .$$

Because the integrand is a positive operator, Q is positive. In other words, if Q is a self-adjoint solution to the algebraic Riccati equation in (14.74), then Q is positive.

14.7.1 Stabilizing solutions to the algebraic Riccati equation

Throughout the remainder of this chapter it is assumed that the operator A is stable. Let T be the operator from $L^2([0,\infty), \mathcal{W})$ into $L^2([0,\infty), \mathcal{Z})$ defined by

$$(Tw)(t) = \int_0^t Ce^{A(t-\tau)}Ew(\tau) \, d\tau \, . \tag{14.80}$$

Let $\mathbf{G}(s) = C(sI - A)^{-1}E$ be the transfer function for $\{A, E, C, 0\}$. Recall that the H^∞ norm of $\mathbf{G}$ equals the norm of T, that is,

$$\|T\| = \|\mathbf{G}\|_\infty := \sup\{ \|\mathbf{G}(\imath\omega)\| : -\infty < \omega < \infty\} \, . \tag{14.81}$$

As before, let C_o be the observability operator from $\mathcal{X}$ into $L^2([0,\infty), \mathcal{Z})$ defined by $C_o x_0 = Ce^{At}x_0$ where x_0 is in $\mathcal{X}$. Since A is stable, both T and C_o are bounded linear operators. Furthermore, the optimization problem in (14.69) is equivalent to the following problem

$$\delta(x_0) = \sup\left\{\|C_o x_0 + Tw\|^2 - \gamma^2\|w\|^2 : w \in L^2([0,\infty), \mathcal{W})\right\} \, . \tag{14.82}$$

If the norm of T is strictly bounded by γ, then the supremums in (14.69) and (14.82) are finite; see Theorem 14.4.1 with $h = C_o x_0$. Moreover, $\delta(x_0) = \gamma^2(C_o^*(\gamma^2 I - TT^*)^{-1}C_o x_0, x_0)$. On the other hand, if $\|T\| > \gamma$, then $\delta(x_0)$ is infinite. Finally, it is noted that even if A is stable, we can have $\|T\| > \gamma$. So, stability of A is not sufficient to guarantee a finite supremum in the optimization problem (14.69), or equivalently, (14.82).

We say that Q is a *stabilizing solution* to the algebraic Riccati equation in (14.74) if Q is a solution to (14.74) and $A + \gamma^{-2}EE^*Q$ is stable. By consulting the results in Chapter 13, it follows that if there exists a stabilizing solution, then it is self-adjoint and unique, that is, there is only one stabilizing solution. Since A is stable, any self-adjoint solution is positive. Thus, the stabilizing solution is positive. According to Corollary 13.5.4 with $\gamma^{-1}\mathbf{G} = \mathbf{T}$, there exists a stabilizing solution to the algebraic Riccati equation in (14.74) if and only if the Hamiltonian matrix H in (14.49) has no eigenvalues on the imaginary axis, or equivalently, $\|\mathbf{G}\|_\infty < \gamma$. In this case, one can use the Hamiltonian techniques in Section 13 to compute the unique stabilizing solution. This discussion proves the equivalence of Parts (*i*), (*ii*) and (*iii*) in the following result.

Theorem 14.7.2 *Let T be the operator from $L^2([0,\infty), \mathcal{W})$ into $L^2([0,\infty), \mathcal{Z})$ defined in (14.80) where $\{A, E, C\}$ is a stable system. Then the following statements are equivalent.*

(i) There exists a stabilizing solution Q to the algebraic Riccati equation in (14.74).

(ii) The Hamiltonian matrix H in (14.49) has no eigenvalues on the imaginary axis.

(iii) The norm of T is strictly bounded by γ.

(iv) The optimal cost $\delta(x_0)$ is attained for every x_0 in $\mathcal{X}$.

(v) The Riccati differential equation in (14.70) admits a uniformly bounded solution Ω and $A + \gamma^{-2}EE^*\Omega_\infty$ is stable where $\Omega_\infty = \lim_{\tau\to\infty}\Omega(\tau)$.

In this case, the unique stabilizing solution Q to the algebraic Riccati equation in (14.74) is given by $Q = \Omega_\infty$ and $\Omega_\infty = \gamma^2 C_o^*(\gamma^2 I - TT^*)^{-1}C_o$.

PROOF. Assume that Part *(iii)* holds, that is, the norm of T is strictly bounded by γ. Then Theorem 14.4.1 with $h = C_o x_0$ shows that the optimal cost $\delta(x_0)$ is uniquely attained for all x_0 in $\mathcal{X}$, that is, Part *(iv)* holds. If Part *(iv)* holds, then $\delta(x_0)$ is finite for all x_0 in $\mathcal{X}$. According to Lemma 14.7.1, there exists a uniformly bounded solution Ω to the Riccati differential equation in (14.70). Moreover, $\Omega_\infty = \lim_{\tau\to\infty}\Omega(\tau)$ is a solution to the algebraic Riccati equation in (14.74) and $(\Omega_\infty x_0, x_0) = \delta(x_0)$. Now let us show that $A + \gamma^2 EE^*\Omega_\infty$ is stable. Let $\hat{w}$ in $L^2([0,\infty),\mathcal{W})$ be the optimal disturbance which attains the optimal cost $\delta(x_0)$. In other words, $(\Omega_\infty x_0, x_0) = \delta(x_0) = \|\hat{z}\|^2 - \gamma^2\|\hat{w}\|^2$ where

$$\dot{\hat{x}} = A\hat{x} + E\hat{w} \quad\text{and}\quad \hat{z} = C\hat{x}.$$

Because A is stable, $\hat{x}$ is in $L^2([0,\infty),\mathcal{X})$. Lemma 12.8.3 shows that $\hat{x}(t)$ approaches zero as t tends to infinity. By setting $Q = \Omega_\infty$ and $w = \hat{w}$ in (14.78) and letting t approach infinity, we arrive at

$$(\Omega_\infty x_0, x_0) = \int_0^\infty \left(\|\hat{z}(\sigma)\|^2 - \gamma^2\|\hat{w}(\sigma)\|^2\right)d\sigma \tag{14.83}$$
$$= (\Omega_\infty x_0, x_0) - \int_0^\infty \|\gamma^{-1}E^*\Omega_\infty\hat{x}(\sigma) - \gamma\hat{w}(\sigma)\|^2 d\sigma.$$

So, $\|\gamma^{-1}E^*\Omega_\infty\hat{x} - \gamma\hat{w}\|^2 = 0$, and thus, $\hat{w} = \gamma^{-2}E^*\Omega_\infty\hat{x}$. Substituting this into $\dot{\hat{x}} = A\hat{x} + E\hat{w}$, we obtain $\dot{\hat{x}} = (A + \gamma^{-2}EE^*\Omega_\infty)\hat{x}$ where $\hat{x}(0) = x_0$. Because $\hat{x}(t)$ converges to zero as t tends to infinity and x_0 can be any vector in $\mathcal{X}_0$, the operator $A + \gamma^{-2}EE^*\Omega_\infty$ is stable. In particular, Part *(v)* holds. If Part *(v)* holds, then obviously Ω_∞ is a stabilizing solution to the algebraic Riccati equation in (14.74), that is, Part (i) holds.   ∎

If $\|T\| < \gamma_o$, then clearly, $\|T\| < \gamma$ for all $\gamma \geq \gamma_o$. This observation and Theorem 14.7.2 readily show that if the algebraic Riccati equation in (14.74) admits a stabilizing solution for some $\gamma = \gamma_o > 0$, then this algebraic Riccati equation admits a stabilizing solution for all $\gamma \geq \gamma_o$. In other words, if the Hamiltonian matrix H has eigenvalues on the imaginary axis for some $\gamma = \gamma_o$, then the Hamiltonian matrix has eigenvalues on the imaginary axis for all $0 < \gamma \leq \gamma_o$. Finally, it is noted that Theorem 14.7.2 also yields the following result.

Corollary 14.7.3 *Let $\mathbf{G}$ be the transfer function for the stable system $\{A, E, C, 0\}$ and let T be the corresponding input output operator from $L^2([0,\infty),\mathcal{W})$ into $L^2([0,\infty),\mathcal{Z})$ defined in (14.80). Then $\|T\| = \|\mathbf{G}\|_\infty$ and the norm of T is given by*

$$\|T\| = \inf\{\gamma: \text{ the algebraic Riccati equation in (14.74) admits a stabilizing solution}\}. \tag{14.84}$$

As before, let $\{A, E, C\}$ be a stable system and assume that the algebraic Riccati equation in (14.74) admits a stabilizing solution Q. Then Theorems 14.7.2 and 14.4.1 show that the optimization problem in (14.69) has a unique solution $\hat{w}$ which attains the supremum and $\delta(x_0) = (Qx_0, x_0)$. Moreover, by consulting the proof of Theorem 14.7.2, we obtain the following result.

Theorem 14.7.4 *Let $\{A, E, C\}$ be a stable system and assume that the algebraic Riccati equation in (14.74) admits a stabilizing solution Q where $\gamma > 0$. Then the infinite horizon optimization problem in problem in (14.69) has a unique $\hat{w}$ which attains the supremum and $\delta(x_0) = (Qx_0, x_0)$. Moreover, this optimal cost is attained by the optimal disturbance $\hat{w} = \gamma^{-2}E^*Q\hat{x}$ where the optimal state trajectory $\hat{x}$ is uniquely determined by*

$$\dot{\hat{x}} = (A + \gamma^{-2}EE^*Q)\hat{x} \qquad (\hat{x}(0) = x_0). \tag{14.85}$$

Example 14.7.1 Consider the system $\dot{x} = -x + w$ and $z = x$. Obviously, the transfer function for this system is given by $\mathbf{G}(s) = 1/(s + 1)$. Moreover, $\|\mathbf{G}\|_\infty = 1$. Therefore, the corresponding optimal cost in (14.69) is infinite ($\delta(x_0) = \infty$) when $0 \leq \gamma < 1$. The algebraic Riccati equation corresponding to this system is given by $q^2/\gamma^2 - 2q + 1 = 0$ where $q = Q$. According to the quadratic formula, the roots to this equation are given by

$$q = \gamma^2 \pm \gamma\sqrt{\gamma^2 - 1}.$$

If $\gamma > 1$, then this algebraic Riccati equation has two strictly positive roots. In this case $q = \gamma^2 - \gamma(\gamma^2 - 1)^{1/2}$ is the unique stabilizing solution. If $\gamma = 1$, then $\Omega(\tau) = \tau/(\tau + 1)$ is the unique solution to the Riccati differential equation $\dot{\Omega} = \Omega^2 - 2\Omega + 1$ with $\Omega(0) = 0$. Clearly, Ω is a uniformly bounded solution and $\lim_{\tau \to \infty} \Omega(\tau) = 1$. So, if $\gamma = 1$, then $\delta(x_0) = \|x_0\|^2$. However, $A + \gamma^{-2}EE^*\Omega_\infty = 0$ is not stable and the supremum is not attained.

14.7.2 An outer spectral factor

Recall that an operator valued rational function Ψ is an *invertible outer or minimum phase function* if Ψ is a stable proper rational function, its inverse exists and is also a stable proper rational function. If Ξ is an operator valued rational function, then we say that Ψ is an *invertible outer spectral factor* of Ξ if Ψ is an invertible outer function and $\Psi^\sharp\Psi = \Xi$.

Let Q be any self-adjoint solution to the algebraic Riccati equation in (14.74). Recall that if F is any operator valued rational function, then $(F^\sharp)(s) = F(-\bar{s})^*$. If s is any complex number, then the algebraic Riccati equation in (14.74) gives

$$-C^*C = \gamma^{-2}QEE^*Q - (-\bar{s}I - A)^*Q - Q(sI - A). \tag{14.86}$$

Now let $\Phi(s)$ be the inverse of $sI - A$. Multiplying by $E^*\Phi^\sharp$ on the left and by ΦE on the right, we obtain

$$-E^*\Phi^\sharp C^*C\Phi E = \gamma^{-2}E^*\Phi^\sharp QEE^*Q\Phi E - E^*\Phi^\sharp QE - E^*Q\Phi E. \tag{14.87}$$

Letting $\mathbf{G} = C\Phi E$ and $\Theta = \gamma I - \gamma^{-1}E^*Q\Phi E$, we arrive at

$$\gamma^2 I - \mathbf{G}^\sharp\mathbf{G} = (\gamma I - \gamma^{-1}E^*Q\Phi E)^\sharp(\gamma I - \gamma^{-1}E^*Q\Phi E) = \Theta^\sharp\Theta. \tag{14.88}$$

Since $(\mathbf{G}^\sharp)(\imath\omega) = \mathbf{G}(\imath\omega)^*$, it follows that $\|\mathbf{G}\|_\infty \leq \gamma$. By consulting Lemma 12.9.1, we obtain the following expression for the inverse of $\mathbf{\Theta}$

$$\mathbf{\Theta}^{-1} = \gamma^{-1}I + \gamma^{-3}E^*Q(sI - A - \gamma^{-2}EE^*Q)^{-1}E. \tag{14.89}$$

If Q is the stabilizing solution, then $A + \gamma^{-2}EE^*Q$ is stable. In this case $\mathbf{\Theta}^{-1}$ is stable, and thus, $\mathbf{\Theta}$ is an invertible outer function. Finally, it is noted that if the algebraic Riccati equation admits a stabilizing solution, then the factorization $\gamma^2 I - \mathbf{G}^\sharp\mathbf{G} = \mathbf{\Theta}^\sharp\mathbf{\Theta}$ readily shows that $\|\mathbf{G}\|_\infty < \gamma$. Summing up gives the following result.

Proposition 14.7.5 *Let* $\mathbf{G}$ *be the transfer function for the stable system* $\{A, E, C, 0\}$. *Assume the algebraic Riccati equation in (14.74) admits a self-adjoint solution* Q *for some* $\gamma > 0$ *and let* $\mathbf{\Theta}$ *be the transfer function for* $\{A, E, -\gamma^{-1}E^*Q, \gamma I\}$. *Then*

$$\gamma^2 I - \mathbf{G}^\sharp\mathbf{G} = \mathbf{\Theta}^\sharp\mathbf{\Theta} \tag{14.90}$$

and $\|\mathbf{G}\|_\infty \leq \gamma$. *Moreover, if* Q *is a stabilizing solution for this algebraic Riccati equation, then* $\mathbf{\Theta}$ *is an invertible outer spectral factor for* $\gamma^2 I - \mathbf{G}^\sharp\mathbf{G}$ *and* $\|\mathbf{G}\|_\infty < \gamma$.

Exercise 28 Let $\{A, E, C\}$ be a stable system. Assume that Q is a positive solution to the algebraic Riccati equation in (14.74). Then show that

$$\int_0^\infty \left(\|z(\sigma)\|^2 - \gamma^2\|w(\sigma)\|^2\right) d\sigma = (Qx_0, x_0) - \int_0^\infty \|\gamma^{-1}E^*Qx(\sigma) - \gamma w(\sigma)\|^2 d\sigma$$

for all w in $L^2([0, \infty), \mathcal{W})$ and x_0 in $\mathcal{X}$.

Exercise 29 Let $\{A, E, C\}$ be a stable system and assume that there exists a stabilizing solution Q to the algebraic Riccati equation in (14.74). Let T be the operator from $L^2([0, \infty), \mathcal{W})$ into $L^2([0, \infty), \mathcal{Z})$ defined in (14.80) and L be the operator on $L^2([0, \infty), \mathcal{X})$ defined by

$$(Lf)(t) = \int_0^t e^{A(t-\tau)}f(\tau) d\tau \qquad (f \in L^2([0, \infty), \mathcal{X})).$$

Let $\mathbf{\Theta}$ be the operator on $L^2([0, \infty), \mathcal{W})$ defined by $\mathbf{\Theta} = \gamma I - \gamma^{-1}E^*QLE$. Then show that $\mathbf{\Theta}^{-1}$ is a bounded causal operator and $\gamma^2 I - T^*T = \mathbf{\Theta}^*\mathbf{\Theta}$.

14.8   The root locus and the H^∞ norm

In this section we will use the outer spectral factorization in Proposition 14.7.5 to obtain a root locus interpretation for the parameter γ in the single input single output infinite horizon optimization problem in (14.69). To this end, let $\{A$ on $\mathcal{X}, E, C, 0\}$ be a stable realization for a scalar valued transfer function $\mathbf{G}$, that is, $\mathbf{G}(s) = C(sI - A)^{-1}E$ where $\mathcal{U} = \mathcal{Z} = \mathbb{C}$. Moreover, assume that $\gamma > \|\mathbf{G}\|_\infty$. Then there exists a unique solution $\hat{w}$ to the optimization problem in (14.69). Furthermore, there exists a unique stabilizing solution Q to the algebraic Riccati equation in (14.74). The optimal disturbance $\hat{w}$ is given by $\hat{w} = \gamma^{-2}E^*Q\hat{x}$ where $\hat{x}$ is the optimal state trajectory. Finally, the state operator $A_c = A + \gamma^{-2}EE^*Q$ is stable.

Lemma 14.8.1 *Let* $\{A, E, C, 0\}$ *be a realization for a scalar valued transfer function* $\mathbf{G} = p/d$ *where* p *is a polynomial and* $d(s) = \det[sI - A]$ *is the characteristic polynomial for* A. *Assume that* Q *is a self-adjoint solution to the algebraic Riccati equation in (14.74) and let* Δ *be the characteristic polynomial for* $A + \gamma^{-2}EE^*Q$. *Then we obtain the following factorization*

$$\Delta^\sharp \Delta = d^\sharp d - \gamma^{-2} p^\sharp p. \tag{14.91}$$

PROOF. From Proposition 14.7.5, we have $\Theta^\sharp \Theta = \gamma^2 - \mathbf{G}^\sharp \mathbf{G}$ where

$$\Theta(s) = \gamma I - \gamma^{-1} E^* Q(sI - A)^{-1} E.$$

Applying Lemma 12.9.1 with $\Theta = \det[\boldsymbol{\Theta}]$, we obtain

$$\Theta(s) = \gamma \det[sI - A - \gamma^{-2}EE^*Q] / \det[sI - A].$$

This readily implies that $\gamma \Delta(s) = d(s)\Theta(s)$. Using $\mathbf{G} = p/d$, we arrive at

$$d^\sharp \Theta^\sharp \Theta d = \gamma^2 d^\sharp d - d^\sharp \mathbf{G}^\sharp \mathbf{G} d = \gamma^2 d^\sharp d - p^\sharp p.$$

Using $\Theta = \gamma \Delta / d$, we obtain the factorization in (14.91). Finally, it is noted that equation (14.91) also follows from Corollary 13.5.2; see (13.41) with $p_1 = \gamma^{-1} p$ and $p_2 = 0$. ∎

Now assume that $\{A, E, C, 0\}$ is stable realization for a scalar valued transfer function $\mathbf{G}$ and $\gamma > \|\mathbf{G}\|_\infty$. Let Q be the stabilizing solution to the algebraic Riccati equation in (14.74). Then the operator $A + \gamma^{-2}EE^*Q$ is stable. Hence, all the roots of Δ are in the open left half complex plane. Notice that λ is a root of a polynomial q if and only if $-\bar{\lambda}$ is a root of the polynomial $q^\sharp$. Therefore, all the roots of $\Delta^\sharp$ live in the open right half plane. In particular, Δ and $\Delta^\sharp$ have no common roots and their product $\Delta^\sharp \Delta$ has no roots on the imaginary axis. It now follows from (14.91) that the roots of Δ are the left half plane roots of $d^\sharp d - \gamma^{-2} p^\sharp p$. This immediately yields the following result.

Theorem 14.8.2 *Suppose that* $\{A, E, C, 0\}$ *is a stable realization of a scalar valued transfer function* $\mathbf{G} = p/d$ *where* p *is a polynomial and* $d(s) = \det[sI - A]$. *Assume that* $\gamma > \|\mathbf{G}\|_\infty$ *and let* Q *be the stabilizing solution to the algebraic Riccati equation in (14.74). Then the roots of the characteristic polynomial of* $A + \gamma^{-2}EE^*Q$ *are precisely the left half plane roots of the polynomial* $d^\sharp d - \gamma^{-2} p^\sharp p$.

To complete this section, we will use root locus techniques to see how the parameter γ affects the eigenvalues of the closed loop state space operator $A + \gamma^{-2}EE^*Q$. To this end, assume that $\{A, E, C, 0\}$ is a stable realization for the scalar valued transfer function $\mathbf{G} = p/d$ satisfying the hypothesis of Theorem 14.8.2. To simplify some of the notation, let us also assume that $\{A, E, C\}$ are real matrices, p and d are polynomials whose coefficients are real and d is the characteristic polynomial for A. In particular, this implies that $d^\sharp(s) = d(-s)$ and $p^\sharp(s) = p(-s)$. Using (14.91), it follows that

$$\Delta(-s)\Delta(s) = d(-s)d(s) - \gamma^{-2}p(-s)p(s)$$

where Δ is the characteristic polynomial for $A + \gamma^{-2}EE^*Q$. So, if we let $k = -\gamma^{-2}$, then the root locus of $d(-s)d(s) + kp(-s)p(s)$ is a graph of the eigenvalues of $A + \gamma^{-2}EE^*Q$ and

$-(A + \gamma^{-2}EE^*Q)^*$ as γ varies from infinity to $\|\mathbf{G}\|_\infty$. In particular, the left half plane root locus of $d(-s)d(s) - \gamma^{-2}p(-s)p(s)$ corresponds precisely to the eigenvalues of $A + \gamma^{-2}EE^*Q$ as γ varies from infinity to $\|\mathbf{G}\|_\infty$. Notice that for any polynomial q with real coefficients, the zeros of $q(-s)q(s)$ are symmetric about the real and imaginary axis. Hence, the root locus of $d(-s)d(s) - \gamma^{-2}p(-s)p(s) = \Delta(-s)\Delta(s)$ is symmetric about the real and imaginary axis.

Since $\mathbf{G}$ is a strictly proper stable transfer function, $\mathbf{G}(\imath\omega)$ is continuous for $-\infty < \omega < \infty$ and $|\mathbf{G}(\imath\omega)|$ approaches zero as ω approaches $\pm\infty$. Hence, there exists a frequency ω_o such that $|\mathbf{G}(\imath\omega_o)| = \|\mathbf{G}\|_\infty$. So, if $\gamma = \|\mathbf{G}\|_\infty$, then $\imath\omega_o$ is a zero of $1 - \gamma^{-2}\mathbf{G}(-s)\mathbf{G}(s)$. Notice that the zeros of $1 - \gamma^{-2}\mathbf{G}(-s)\mathbf{G}(s)$ are contained in the zeros of $d(-s)d(s) - \gamma^{-2}p(-s)p(s)$. Thus, $\imath\omega_o$ is a zero of $d(-s)d(s) - \gamma^{-2}p(-s)p(s)$ when $\gamma = \|\mathbf{G}\|_\infty$. In particular, $\imath\omega_o$ is contained in the root locus of $d(-s)d(s) - \gamma^{-2}p(-s)p(s)$. In other words, as γ varies from infinity to $\|\mathbf{G}\|_\infty$ the root locus of $d(-s)d(s) - \gamma^{-2}p(-s)p(s)$ ends up with some points on the imaginary axis.

As γ varies from infinity to zero, the branches of the root locus of the polynomial $d(-s)d(s) - \gamma^{-2}p(-s)p(s)$ move from the zeros of $d(-s)d(s)$ to the zeros of $p(-s)p(s)$ and the appropriate asymptotes. The asymptotes of $d(-s)d(s) - \gamma^{-2}p(-s)p(s)$ are determined by $m = \deg d - \deg p$. To be precise, if m is odd, then $d(-s)d(s) - \gamma^{-2}p(-s)p(s) = 0$ can be expressed as $b(s) + \gamma^{-2}\alpha a(s) = 0$ where $\alpha > 0$ and a and b are monic polynomials. Since α is the coefficient of the highest order term of αa, the angles ϕ_{oj} for the asymptotes are given by

$$\phi_{oj} = \frac{\pi + 2\pi j}{2m} \qquad \text{for } j = 0, 1, 2, \cdots, 2m - 1. \tag{14.92}$$

If m is even, then $d(-s)d(s) - \gamma^{-2}p(-s)p(s) = 0$ can be expressed as $b(s) + \gamma^{-2}\alpha a(s) = 0$ where $\alpha < 0$ and a and b are monic polynomials. In this case, the angles ϕ_{ej} for the asymptotes are given by

$$\phi_{ej} = \frac{\pi j}{m} \qquad \text{for } j = 0, 1, 2, \cdots, 2m - 1. \tag{14.93}$$

Moreover, because $d(-s)d(s) - \gamma^{-2}p(-s)p(s)$ is symmetric about the real and imaginary axis, the origin of these asymptotes is zero. Notice that in either case two of these asymptotes lie on the imaginary axis. Therefore, as γ varies from infinity to $\|\mathbf{G}\|_\infty$ the eigenvalues of $A + \gamma^{-2}EE^*Q$ start at the eigenvalues of A (the left half plane zeros of $d(-s)d(s)$) follow the root locus and move towards the left half plane zeros of $d(-s)d(s) - \|\mathbf{G}\|_\infty^{-2}p(-s)p(s)$ and the appropriate asymptotes in the open left half plane. For m odd, the asymptote angles ϕ_{oj} are given by (14.92). For m even, the asymptote angles ϕ_{ej} are given by (14.93). Moreover, the H^∞ norm of $\mathbf{G}$ corresponds to the largest value of γ where the root locus of $d(-s)d(s) - \gamma^{-2}p(-s)p(s)$ hits the imaginary axis. In fact, the imaginary axis is contained in the root locus of $d(-s)d(s) - \gamma^{-2}p(-s)p(s)$ as γ varies from $\|\mathbf{G}\|_\infty$ to zero. To see this simply notice that $\mathbf{G}^\sharp(\imath\omega)\mathbf{G}(\imath\omega) = |\mathbf{G}(\imath\omega)|^2$ is real for all $-\infty < \omega < \infty$. Since $\mathbf{G}$ is a stable strictly proper transfer function, $|\mathbf{G}(\imath\omega)|^2$ is a continuous function whose maximum is $\|\mathbf{G}\|_\infty^2$ and infimum is zero. So, for any frequency ω there is a gain γ such that $\gamma = |\mathbf{G}(\imath\omega)|$ with $0 < \gamma \le \|\mathbf{G}\|_\infty$. In this case $\gamma^2 - |\mathbf{G}(\imath\omega)|^2 = 0$, and thus, $\imath\omega$ is on the root locus of $d(-s)d(s) - \gamma^{-2}p(-s)p(s)$. Therefore, the root locus of $d(-s)d(s) - \gamma^{-2}p(-s)p(s)$ contains the imaginary axis for $0 < \gamma \le \|\mathbf{G}\|_\infty$.

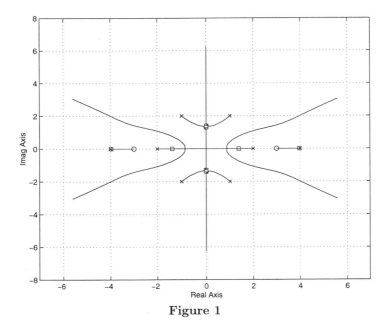

Figure 1

Example 14.8.1 To complete this section we will demonstrate how the above analysis works on a simple example. To this end, let $\{A, E, C, 0\}$ be any minimal realization of

$$\mathbf{G}(s) = \frac{10(s+3)}{(s+2)(s+4)(s+1+2i)(s+1-2i)} \, .$$

The McMillan degree of $\mathbf{G}$ is four and A is stable. Let p be the numerator and d be the denominator of $\mathbf{G}$. The root locus for $d(-s)d(s) - \gamma^{-2}p(-s)p(s)$ is given in Figure 1. In this case, $m = 3$. According to (14.92) the angles for the asymptotes for the root locus are given by

$$\phi_{odd} = \pm\pi/6, \ \pm\pi/2, \ \text{and} \ \pm 5\pi/6 \, .$$

As expected, two of the asymptotes $\pm\pi/2$ lie on the imaginary axis. Moreover, the asymptotes for this root locus in the left half plane occur at the angles $\pm 5\pi/6$. The H^∞ norm of $\mathbf{G}$ equals .7743 and this occurs at $\omega = \pm 1.3$, that is, $|\mathbf{G}(\pm 1.3i)| \approx \|\mathbf{G}\|_\infty$. So, as γ moves from infinity to $\|\mathbf{G}\|_\infty$ the eigenvalues of $A + \gamma^{-2}EE^*Q$ move from the points $-2, -4, -1 + 2i, -1 - 2i$ along the branches of the root locus in the open left half plane to the left half plane roots of $d(-s)d(s) - \|\mathbf{G}\|_\infty^{-2}p(-s)p(s)$. These roots are marked by a box in Figure 1. Finally, the root locus contains the imaginary axis as γ varies from $\|\mathbf{G}\|_\infty$ to zero.

Exercise 30 Let $\{A, E, C, 0\}$ be a minimal realization for

$$\mathbf{G}(s) = (2 - s)/(s + 1)(s + 2) \, .$$

Assume that $\gamma > \|\mathbf{G}\|_\infty$ and let Q be the corresponding stabilizing solution to the algebraic Riccati equation in (14.74). Recall that $\{A, E, -\gamma^{-1}E^*Q, \gamma I\}$ is a realization for the outer spectral factor Θ of $\gamma^2 - \mathbf{G}^\sharp\mathbf{G}$. Show that $\{A, E, -\gamma^{-1}E^*Q, \gamma I\}$ controllable and not observable.

Exercise 31 Let $\{A, E, C, 0\}$ be a minimal realization for a scalar valued stable transfer function $\mathbf{G}(s) = p/d$ where p and d are polynomials. Moreover, assume that $p^\sharp p$ and $d^\sharp d$ are co-prime. Assume that $\gamma > \|\mathbf{G}\|_\infty$ and let Q be the stabilizing solution to the corresponding algebraic Riccati equation in (14.74). Then show that $\{A, E, -\gamma^{-1}E^*Q, \gamma I\}$ is a minimal realization for the outer spectral factor Θ of $\gamma^2 - \mathbf{G}^\sharp\mathbf{G}$.

14.9 Notes

The optimization problem in (14.4) can be viewed as a linear quadratic regulator problem with an indefinite weight, and thus, the derivation of the corresponding Riccati differential equation is classical. In other words, the optimization problem in (14.4) is essentially the dual of the linear quadratic regulator problem. Hence, the derivation of the Riccati differential equation in Section 14.5.2 is almost identical to the corresponding derivation of the Riccati differential equation for the linear quadratic regulator problem. The only thing that changes is some technical issues and the sign of the PBB^*P term. For some further results on the two point boundary problem and its role in H^∞ control theory see Green-Limebeer [57] and Limebeer-Anderson-Khargonekar-Green [83]. Corollary 14.2.3 and many of the results in this section are now considered standard results in H^∞ control theory see Green-Limebeer [57], Mustafa-Glover [92] and Zhou-Doyle-Glover [131]. For some nice results on least squares optimal control problems and the algebraic Riccati equation see Willems [125].

The optimization problem in (14.25) is a basic optimization problem arising in control theory; see Porter [101]. The operator optimization problem in (14.29) plays a fundamental role in operator theory; see de Branges [17, 18] and de Branges-Rovnyak [19]. For a recent account of some of the de Branges-Rovnyak work see Sarason [113].

Many of the results in Sections 14.1 to 14.6 hold in the time varying case, that is, when A, E and C are continuous functions of time. In particular, Theorem 14.2.2 holds. Minor modifications of the techniques in this chapter show that Parts (i) and (ii) in Theorem 14.2.2 are equivalent in the time varying case. Moreover, if $\|T\| < \gamma$, then Part (iii) holds; see Remark 14.1.1. If (iii) holds, then clearly, $\|T\| \le \gamma$. To verify that $\|T\| < \gamma$, one can use Theorem 14.3.2, to show that the cost $\delta(x_0) = (Qx_0, x_0)$ for some positive operator Q on $\mathcal{X}$. In particular, $\delta(x_0)$ is continuous in x_0. Then this fact can be used to prove that Part (ii) holds. The details are left to the reader. Finally, it is noted that Theorem 14.5.2 holds in the time varying case.

Chapter 15

H^∞ Control

This chapter concentrates on a H^∞ type control problem. By combining the linear quadratic regulator problem with our previous H^∞ analysis, we present and solve a max-min optimization problem. The solution to this problem, yields a feedback controller which guarantees that the norm of the resulting closed loop system is bounded by a specified constant. Moreover, it is shown that this controller provides a tradeoff between an optimal L^2 and optimal H^∞ controller.

15.1 A H^∞ control problem

In this section, we consider systems which contain a control input in addition to a disturbance input. We examine the problem of choosing the control input to mitigate the effect of the disturbance on a specified system performance output. The systems under consideration are described by

$$\dot{x} = Ax + Ew + Bu \quad \text{and} \quad z = Cx + Du \tag{15.1}$$

which we denote by $\{A, B, C, D, E\}$. Here A is an operator on $\mathcal{X}$ and B maps $\mathcal{U}$ into $\mathcal{X}$ while C maps $\mathcal{X}$ into $\mathcal{Z}$ and E maps $\mathcal{W}$ into $\mathcal{X}$. Throughout this chapter, D is an isometry mapping $\mathcal{U}$ into $\mathcal{Z}$ whose range is orthogonal to the range of C, that is, $D^*D = I$ and $D^*C = 0$. The spaces $\mathcal{X}, \mathcal{U}, \mathcal{W},$ and $\mathcal{Z}$ are all finite dimensional. In this setting $u(t)$ is the control input, the disturbance input is $w(t)$ while $z(t)$ is an output which reflects the system performance. Finally, we define the function ϕ by $z = \phi(x_0, w, u)$ where z is the performance output due to the initial state x_0, the disturbance input w and the control input u.

Consider the system in (15.1) whose initial state at some initial time t_0 is given by $x(t_0) = x_0$. In Section 14.1, we saw that one can quantify the effect of a disturbance w on the performance output z by examining the cost

$$J(x_0, w, u) = \int_{t_0}^{t_1} \left(\|z(\sigma)\|^2 - \gamma^2 \|w(\sigma)\|^2 \right) d\sigma \tag{15.2}$$

for a fixed positive scalar γ. Note that

$$J(x_0, w, u) = \int_{t_0}^{t_1} \left(\|Cx(\sigma)\|^2 + \|u(\sigma)\|^2 - \gamma^2 \|w(\sigma)\|^2 \right) d\sigma . \tag{15.3}$$

The equality follows because $D^*D = I$ and $D^*C = 0$. For a fixed disturbance input w in $L^2([t_0, t_1], \mathcal{W})$, the best cost $\xi(x_0, w)$ that can be achieved by the control input is given by

$$\xi(x_0, w) = \inf \left\{ J(x_0, u, w) : u \in L^2([t_0, t_1], \mathcal{U}) \right\}. \tag{15.4}$$

We say u is *optimal* for w if u is a vector in $L^2([t_0, t_1], \mathcal{U})$ which attains the cost $\xi(x_0, w)$, that is, $\xi(x_0, w) = J(x_0, w, u)$. Now let $d(x_0)$ be the optimal cost that one obtains by taking the supremum over w, that is,

$$d(x_0) = \sup \left\{ \xi(x_0, w) : w \in L^2([t_0, t_1], \mathcal{W}) \right\}.$$

We denote this by

$$d(x_0) = \sup_w \inf_u \int_{t_0}^{t_1} \left(\|z(\sigma)\|^2 - \gamma^2 \|w(\sigma)\|^2 \right) d\sigma \tag{15.5}$$

where it is understood that the infimum is taken over all u in $L^2([t_0, t_1], \mathcal{U})$ while the supremum is taken over all w in $L^2([t_0, t_1], \mathcal{W})$. The significance of $d(x_0)$ is that for each disturbance input w in $L^2([t_0, t_1], \mathcal{W})$ and any $\epsilon > 0$, there exists a control input u in $L^2([t_0, t_1], \mathcal{U})$ such that

$$\|z\|^2 \leq \gamma^2 \|w\|^2 + d(x_0) + \epsilon$$

where $z = \phi(x_0, w, u)$. Whenever the infimum is attained in (15.4), then there is a control input which results in

$$\|z\|^2 \leq \gamma^2 \|w\|^2 + d(x_0).$$

For a specified initial state x_0, we say that $\{\hat{w}, \hat{u}\}$ is an optimal pair for the optimization problem in (15.5) if $\hat{w}$ is in $L^2([t_0, t_1], \mathcal{W})$ while $\hat{u}$ is in $L^2([t_0, t_1], \mathcal{U})$ and

$$J(x_0, \hat{w}, \hat{u}) = \xi(x_0, \hat{w}) = d(x_0) \tag{15.6}$$

where the optimal performance output $\hat{z}$ and optimal state $\hat{x}$ are determined by

$$\dot{\hat{x}} = A\hat{x} + E\hat{w} + B\hat{u} \quad \text{with} \quad \hat{x}(t_0) = x_0 \quad \text{and} \quad \hat{z} = C\hat{x} + D\hat{u}. \tag{15.7}$$

In other words, $\hat{u}$ is optimal for $\hat{w}$ and $\hat{w}$ maximizes the cost $\xi(x_0, \cdot)$. Finally, it is noted that if $\{w, u\}$ is any pair satisfying $J(x_0, w, u) = d(x_0)$, then $\{w, u\}$ is not necessarily an optimal pair.

15.1.1 Problem solution

As expected, the solution of the optimization problem in (15.5) involves a Riccati differential equation. Specifically,

$$\dot{P} + A^*P + PA + \gamma^{-2}PEE^*P - PBB^*P + C^*C = 0 \qquad (P(t_1) = 0). \tag{15.8}$$

Notice that, if $E = 0$, then the optimization problem in (15.5) reduces to a linear quadratic regulator problem and this Riccati differential equation becomes the corresponding linear quadratic Riccati differential equation. On the other hand, if $B = 0$, then the optimization problem in (15.5) reduces to the optimization problem in (14.4) and the Riccati differential equation in (15.8) is simply the Riccati differential equation in (14.11). Before presenting the main result of this section, we make the following observations.

Remark 15.1.1 For any time interval $[a, b]$ and any initial state x_0, let $\rho(x_0, a, b)$ be the optimal cost given by

$$\rho(x_0, a, b) = \sup_{w} \inf_{u} \int_a^b (\|z(\sigma)\|^2 - \gamma^2 \|w(\sigma)\|^2)\, d\sigma \qquad (15.9)$$

where the infimum is taken over all u in $L^2([a, b], \mathcal{U})$ while the supremum is taken over all w in $L^2([a, b], \mathcal{W})$ and $x(a) = x_0$. Obviously, $d(x_0) = \rho(x_0, t_1, t_2)$. We first claim that if $a < b < c$, then $\rho(x_0, a, b) \leq \rho(x_0, a, c)$. To see this, let w be any vector in $L^2([a, c], \mathcal{W})$ such that $w(t) = 0$ for all $b \leq t \leq c$. Then

$$\int_a^b (\|z(\sigma)\|^2 - \gamma^2 \|w(\sigma)\|^2)\, d\sigma \leq \int_a^c (\|z(\sigma)\|^2 - \gamma^2 \|w(\sigma)\|^2)\, d\sigma\,.$$

By taking the infimum over all u in $L^2([a, b], \mathcal{U})$, we obtain

$$\inf_{u} \int_a^b (\|z(\sigma)\|^2 - \gamma^2 \|w(\sigma)\|^2)\, d\sigma \leq \int_a^c (\|z(\sigma)\|^2 - \gamma^2 \|w(\sigma)\|^2)\, d\sigma\,.$$

By taking the infimum over all u in $L^2([a, c], \mathcal{U})$ in the right hand side and then taking the supremum over all w in $L^2([a, c], \mathcal{W})$ in the right hand side, we have

$$\inf_{u} \int_a^b (\|z(\sigma)\|^2 - \gamma^2 \|w(\sigma)\|^2)\, d\sigma \leq \rho(x_0, a, c)\,.$$

By taking the supremum over all w in $L^2([a, b], \mathcal{W})$, we arrive at $\rho(x_0, a, b) \leq \rho(x_0, a, c)$.

We now claim that if $a < b < c$, then $\rho(x_0, b, c) \leq \rho(x_0, a, c)$. First note that due to the time invariant nature of the optimization problem, it should be clear that $\rho(x_0, b+d, c+d) = \rho(x_0, b, c)$ for any real number d. The desired result now follows from

$$\rho(x_0, b, c) = \rho(x_0, a, c + a - b) \leq \rho(x_0, a, c)\,.$$

Theorem 15.1.1 *The optimal cost $d(x_0)$ for the optimization problem in (15.5) is finite for every initial state x_0 in $\mathcal{X}$ if and only if the Riccati differential equation in (15.8) has a solution P on the interval $[t_0, t_1]$. In this case, the optimal cost is given by*

$$d(x_0) = (P(t_0)x_0, x_0) \qquad (15.10)$$

and is uniquely attained by the optimal disturbance/control input pair

$$\hat{w}(t) = \gamma^{-2} E^* P(t)\hat{x}(t) \qquad and \qquad \hat{u}(t) = -B^* P(t)\hat{x}(t) \qquad (15.11)$$

where the optimal state trajectory $\hat{x}$ is uniquely determined by

$$\dot{\hat{x}} = (A + \gamma^{-2} EE^* P - BB^* P)\hat{x} \qquad with \qquad \hat{x}(t_0) = x_0\,. \qquad (15.12)$$

PROOF. Assume that the Riccati differential equation in (15.8) has a solution P over the interval $[t_0, t_1]$. Then we claim that, for each initial state x_0 in $\mathcal{X}$, the optimal cost $d(x_0)$ is finite and is given by $d(x_0) = (P(t_0)x_0, x_0)$. Moreover, this optimal cost is determined by the disturbance/control input pair $\hat{w} = \gamma^{-2}E^*P\hat{x}$ and $\hat{u} = -B^*P\hat{x}$ where the state trajectory $\hat{x}$ is uniquely determined by (15.12). To see this, we first note that as a consequence of Lemma 13.6.2, the operator $P(t)$ is self-adjoint for each t. Using $\dot{x} = Ax + Ew + Bu$, we obtain

$$
\begin{aligned}
\frac{d}{dt}(Px, x) &= (\dot{P}x, x) + 2\Re(P\dot{x}, x) \\
&= (\dot{P}x, x) + 2\Re(PAx, x) + 2\Re(PEw, x) + 2\Re(PBu, x) \qquad (15.13) \\
&= ((\dot{P} + PA + A^*P)x, x) + 2\Re(w, E^*Px) + 2\Re(u, B^*Px)\,.
\end{aligned}
$$

Recall that $z = Cx + Du$. Using $D^*D = I$ and $D^*C = 0$, we have $\|z\|^2 = \|Cx\|^2 + \|u\|^2$. The Riccati differential equation in (15.8) yields

$$
\begin{aligned}
((\dot{P} + PA + A^*P)x, x) &= -\gamma^{-2}\|E^*Px\|^2 + \|B^*Px\|^2 - \|Cx\|^2 \\
&= -\gamma^{-2}\|E^*Px\|^2 + \|B^*Px\|^2 + \|u\|^2 - \|z\|^2\,.
\end{aligned}
$$

By substituting this into (15.13) and completing the appropriate squares, we now obtain

$$
\frac{d}{dt}(Px, x) = -\|z\|^2 + \gamma^2\|w\|^2 - \|\gamma^{-1}E^*Px - \gamma w\|^2 + \|B^*Px + u\|^2\,. \qquad (15.14)
$$

Recall that $x(t_0) = x_0$. By integrating from t_0 to t_1, rearranging terms and using the fact that $P(t_1) = 0$, we have

$$
\begin{aligned}
\int_{t_0}^{t_1} \left(\|z(\sigma)\|^2 - \gamma^2\|w(\sigma)\|^2\right) d\sigma &= (P(t_0)x_0, x_0) - \int_{t_0}^{t_1} \|\gamma^{-1}E^*P(\sigma)x(\sigma) - \gamma w(\sigma)\|^2 \, d\sigma \\
&\quad + \int_{t_0}^{t_1} \|B^*P(\sigma)x(\sigma) + u(\sigma)\|^2 \, d\sigma\,. \qquad (15.15)
\end{aligned}
$$

It now immediately follows that for $w = \hat{w}$ and $u = \hat{u}$, where $\hat{w}$ and $\hat{u}$ are given by (15.11) and (15.12), the corresponding cost satisfies

$$
\int_{t_0}^{t_1} \left(\|\hat{z}(\sigma)\|^2 - \gamma^2\|\hat{w}(\sigma)\|^2\right) d\sigma = (P(t_0)x_0, x_0)\,. \qquad (15.16)
$$

We now demonstrate that the optimal cost $d(x_0) \leq (P(t_0)x_0, x_0)$. Consider any disturbance w in $L^2([t_0, t_1], \mathcal{W})$ and let $u(t) = -B^*P(t)x(t)$. In this case,

$$
\dot{x} = (A - BB^*P(t))x + Ew \qquad \text{with} \qquad x(t_0) = x_0\,.
$$

Then for this input u relationship (15.15) implies that

$$
J(x_0, w, u) = (P(t_0)x_0, x_0) - \|\gamma^{-1}E^*Px - \gamma w\|^2\,.
$$

Hence, the infimal cost $\xi(x_0, w)$ defined in (15.4) satisfies $\xi(x_0, w) \leq (P(t_0)x_0, x_0)$. By taking the supremum over all w in $L^2([t_0, t_1], \mathcal{W})$, we see that $d(x_0) \leq (P(t_0)x_0, x_0)$.

We now claim that $\hat{u}$ is the vector which attains the cost $\xi(x_0, \hat{w})$ in the optimization problem (15.4) with $w = \hat{w}$, that is, $J(x_0, \hat{w}, \hat{u}) = \xi(x_0, \hat{w})$. From this fact it readily follows that $(P(t_0)x_0, x_0) = \|\hat{z}\|^2 - \gamma^2\|\hat{w}\|^2 \le d(x_0)$. Hence, $d(x_0) = (P(t_0)x_0, x_0)$ and the pair $\{\hat{w}, \hat{u}\}$ is optimal. To achieve our objective, notice that for the disturbance $\hat{w} = \gamma^{-2}E^*P\hat{x}$, we obtain

$$\dot{x} = Ax + \gamma^{-2}EE^*P\hat{x} + Bu \qquad \text{with} \qquad x(t_0) = x_0.$$

Now let us introduce the new state $\tilde{x} = x - \hat{x}$ and set $\tilde{u} = u + B^*Px$. Then subtracting the differential equation in (15.12) from the previous state equation, yields

$$\dot{\tilde{x}} = (A - BB^*P(t))\tilde{x} + B\tilde{u} \qquad \text{with} \qquad \tilde{x}(t_0) = 0. \qquad (15.17)$$

Using $\hat{w} = \gamma^{-2}E^*P\hat{x}$ in (15.15) with $\tilde{x} = x - \hat{x}$, gives

$$\int_{t_0}^{t_1} \left(\|z(\sigma)\|^2 - \gamma^2\|\hat{w}(\sigma)\|^2\right) d\sigma =$$
$$(P(t_0)x_0, x_0) - \int_{t_0}^{t_1} \|\gamma^{-1}E^*P(\sigma)\tilde{x}(\sigma)\|^2 d\sigma + \int_{t_0}^{t_1} \|\tilde{u}(\sigma)\|^2 d\sigma.$$

Hence, the cost $\xi(x_0, \hat{w})$ defined in (15.4) is given by

$$\xi(x_0, \hat{w}) = (P(t_0)x_0, x_0) - \sup_{\tilde{u} \in L^2} \int_{t_0}^{t_1} \left(\|\gamma^{-1}E^*P(\sigma)\tilde{x}(\sigma)\|^2 - \|\tilde{u}(\sigma)\|^2\right) d\sigma. \qquad (15.18)$$

We claim that the supremum above is attained with $\tilde{u} = 0$. To see this notice that the Riccati differential equation in (15.8) can be rewritten as in

$$\dot{P} + (A - BB^*P)^*P + P(A - BB^*P) + PBB^*P + \gamma^{-2}PEE^*P + C^*C = 0 \qquad (15.19)$$

with $P(t_1) = 0$. This is precisely the Riccati differential equation one obtains by replacing C with $[C, \gamma^{-1}E^*P]^{tr}$ and $\gamma^{-1}E$ with B in Theorem 14.2.1. Considering any $\tilde{u}$ and computing the rate of change of $(P\tilde{x}, \tilde{x})$ and completing the appropriate square as in the proof of Theorem 14.2.1, we obtain that

$$\int_{t_0}^{t_1} \left(\|\gamma^{-1}E^*P(\sigma)\tilde{x}(\sigma)\|^2 - \|\tilde{u}(\sigma)\|^2\right) d\sigma = -\int_{t_0}^{t_1} \left(\|C\tilde{x}(\sigma)\|^2 + \|B^*P(\sigma)\tilde{x}(\sigma) - \tilde{u}(\sigma)\|^2\right) d\sigma.$$

Here we also used the facts that $P(t_1) = 0$ and $\tilde{x}(t_0) = 0$. Hence, the supremum in (15.18) is zero and this is supremum is uniquely achieved with $\tilde{u} = 0$, or equivalently, with $u = \hat{u} = -B^*P\hat{x}$. Thus,

$$d(x_0) \le (P(t_0)x_0, x_0) = \xi(x_0, \hat{w}) \le \sup_{w \in L^2} \xi(x_0, w) = d(x_0).$$

Therefore, $d(x_0) = (P(t_0)x_0, x_0)$. In particular, the optimal cost $d(x_0)$ is finite for all x_0 in $\mathcal{X}$.

Now let us show that there is only one optimal pair which attains the optimal cost $d(x_0)$. To this end, consider any pair $\{w_o, u_o\}$ which attains $d(x_0)$. Then

$$(P(t_0)x_0, x_0) = d(x_0) = \xi(x_0, w_o) = J(x_0, w_o, u_o)$$

where $z_o = \phi(x_0, w_o, u_o)$ is the output corresponding to the pair $\{w_o, u_o\}$ and u_o is optimal for w_o. Now consider the input $u = -B^*Px$ where x is the solution to

$$\dot{x} = (A - BB^*P(t))x + Ew_o \qquad \text{with} \qquad x(t_0) = x_0. \tag{15.20}$$

Since u_o is optimal for w_o, it follows from (15.15) that

$$(P(t_0)x_0, x_0) = J(x_0, w_o, u_o) \le (P(t_0)x_0, x_0) - \|\gamma^{-1}E^*Px - \gamma w_o\|^2.$$

Thus, $\|\gamma^{-1}E^*Px - \gamma w_o\|^2 = 0$, or equivalently, $w_o = \gamma^{-2}E^*Px$. Substituting this in (15.20), it now follows that $x = \hat{x}$, and hence, $w_o = \hat{w}$. Recall that $\tilde{u} = 0$ is the only function in $L^2([t_0, t_1], \mathcal{U})$ which achieves the minimal optimal cost $\xi(x_0, \hat{w})$ in (15.18). Since $\tilde{u} = u + B^*Px$, there is only one input u which achieves the minimal cost $\xi(x_0, \hat{w})$ and this $u = \hat{u} = -B^*P\hat{x}$. Therefore, $u_o = \hat{u}$ and the optimal pair $\{\hat{w}, \hat{u}\}$ is unique.

To complete the proof it remains to show that if $d(x_0)$ is finite for every initial state x_0 in $\mathcal{X}$, then the Riccati differential equation in (15.8) has a unique solution P defined on the interval $[t_0, t_1]$. To verify this, first notice that this Riccati differential equation is locally Lipschitz in P. Hence, this differential equation has a unique solution over some interval $(t_2, t_1]$ for t_2 sufficiently close to t_1. To verify that this solution can be extended over the interval $[t_0, t_1]$, it is sufficient to show that over any interval $(t_2, t_1]$ on which P is defined, there is a bound m independent of t such that $\|P(t)\| \le m$ for $t_2 < t \le t_1$.

Recall now the definition of the optimal cost $\rho(x_0, a, b)$ in (15.9). By replacing t_0 with t in our previous analysis it follows that $\rho(x_0, t, t_1) = (P(t)x_0, x_0)$. Since $t_0 \le t < t_1$, it follows from Remark 15.1.1 that $\rho(x_0, t, t_1) \le \rho(x_0, t_0, t_1)$. Hence, $(P(t)x_0, x_0) \le \rho(x_0, t_0, t_1) = d(x_0)$. In other words, $(P(t)x_0, x_0) \le d(x_0) < \infty$ for every x_0 in $\mathcal{X}$ and all t in $(t_2, t_1]$. Also, Lemma 13.6.2 states that $P(t)$ is positive. Let $\{\psi_j\}_1^n$ be an orthonormal basis for the state space $\mathcal{X}$. Then

$$\|P(t)\| \le \text{trace } P(t) = \sum_{j=1}^n (P(t)\psi_j, \psi_j) \le \sum_{j=1}^n d(\psi_j) = m < \infty.$$

So, there exists a bound m such that $\|P(t)\| \le m$ for $t_2 < t \le t_1$. Therefore, the Riccati differential equation (15.8) has a unique solution P defined on the interval $[t_0, t_1]$. ∎

Remark 15.1.2 Assume that $d(x_0)$ is finite for all x_0 in $\mathcal{X}$, or equivalently, the Riccati differential equation in (15.8) has a solution on the interval $[t_0, t_1]$. If the pair $\{C, A\}$ is observable, then $P(t)$ is strictly positive for all $t_0 \le t < t_1$. This follows from Lemma 13.6.2. We can also demonstrate this as follows. By replacing t_0 with t in (15.15) and rearranging terms, we obtain

$$(P(t)x_0, x_0) = \int_t^{t_1} \left(\|z(\sigma)\|^2 - \gamma^2 \|w(\sigma)\|^2 \right) d\sigma + \int_t^{t_1} \|\gamma^{-1}E^*P(\sigma)x(\sigma) - \gamma w(\sigma)\|^2 d\sigma$$
$$- \int_t^{t_1} \|B^*P(\sigma)x(\sigma) + u(\sigma)\|^2 d\sigma. \tag{15.21}$$

If we let $w(\sigma) = 0$ and $u(\sigma) = -B^*P(\sigma)x(\sigma)$ for $t < \sigma \le t_1$, then

$$(P(t)x_0, x_0) = \int_t^{t_1} \left(\|Cx(\sigma)\|^2 + \|u(\sigma)\|^2 \right) d\sigma + \int_t^{t_1} \|\gamma^{-1}E^*P(\sigma)x(\sigma)\|^2 d\sigma \ge 0. \tag{15.22}$$

Since $(P(t)x_0, x_0) \geq 0$ for all x_0 in $\mathcal{X}$, this also shows that the operator $P(t)$ is positive. If $(P(t)x_0, x_0) = 0$ for some initial state x_0, then $u(\sigma) = 0$ for all $t \leq \sigma \leq t_1$. In this case, $z(\sigma) \geq Ce^{A(\sigma-t)}x_0$. Hence,

$$0 = (P(t)x_0, x_0) \geq \int_t^{t_1} \|Ce^{A(\sigma-t)}x_0\|^2 \, d\sigma \, .$$

Because $\{C, A\}$ is observable, x_0 is zero. Therefore, the operator $P(t)$ is strictly positive.

Remark 15.1.3 It is emphasized that one must integrate the Riccati differential equation in (15.8) backwards in time to find $P(t)$. However, one can easily convert this equation to a Riccati differential equation moving forward in time. To see this let $\Omega(\tau) = P(t_1-\tau)$. Then equation (15.8) yields

$$\dot{\Omega} = A^*\Omega + \Omega A + \gamma^{-2}\Omega EE^*\Omega - \Omega BB^*\Omega + C^*C \qquad (\Omega(0) = 0) \,. \qquad (15.23)$$

Therefore, $P(t) = \Omega(t_1 - t)$ where Ω can be computed by solving the Riccati differential equation (15.23) forward in time. By replacing t_0 with $t_1-\tau$ in Theorem 15.1.1, we readily obtain that

$$(\Omega(\tau)x_0, x_0) = (P(t_1-\tau)x_0, x_0) = \rho(x_0, t_1-\tau, t_1) = \rho(x_0, 0, \tau) \,.$$

Hence,

$$(\Omega(\tau)x_0, x_0) = \sup_w \inf_u \int_0^\tau \left(\|z(\sigma)\|^2 - \gamma^2\|w(\sigma)^2\| \right) \, d\sigma \,. \qquad (15.24)$$

Here the infimum is taken over all u in $L^2([0, \tau], \mathcal{U})$ while the supremum is taken over all w in $L^2([0, \tau], \mathcal{W})$. Obviously, $\Omega(\tau)$ is well defined if and only if the optimal cost $\rho(x_0, 0, \tau)$ is finite for all initial states x_0 in $\mathcal{X}$.

If the pair $\{C, A\}$ is observable, then $\Omega(\tau)$ is strictly positive for all $0 < \tau$ where $\Omega(\tau)$ is defined. This follows from the fact that $P(t_1-t)$ is strictly positive when $\{C, A\}$ is observable.

Exercise 32 Consider the system $\{A, B, C, D, E\}$ and assume that the Riccati differential equation in (15.8) has a solution over the interval $[t_0, t_1]$. Consider the feedback controller $u = -B^*Px$ and let w be a vector in $L^2([t_0, t_1], \mathcal{W})$ and z be the corresponding output for this feedback system described in (15.25). Then u is not necessarily the optimal input which obtains the minimum cost $\xi(x_0, w)$, that is, for some disturbance w one can have $\xi(x_0, w) < \|z\|^2 - \gamma^2\|w\|^2$. For a counter example, consider the system $\dot{x} = w + u$ on the interval $[0, 1]$ with $x_0 = 1$, and set $\gamma = 1$. Then show that for $w = 1$, the feedback controller $u = -Px$ does not obtain the minimal cost $\xi(x_0, w)$.

15.1.2 The central controller

Assume that the Riccati differential equation in (15.8) has a solution P on the interval $[t_0, t_1]$. The control $\hat{u} = -B^*P\hat{x}$ presented in Theorem 15.1.1 is an open loop control which achieves the optimal cost $d(x_0)$ for the worst case disturbance input $\hat{w}$. For disturbances other than the worst case disturbance, there is no guaranteed performance. However, the feedback controller $u = -B^*Px$ is useful in applications.

Definition 1 Consider the system $\{A, B, C, D, E\}$ in (15.1). Assume that the Riccati differential equation in (15.8) has a solution over the interval $[t_0, t_1]$ for some specified $\gamma > 0$. The feedback controller $u = -B^*Px$ is called the *central controller* corresponding to the tolerance γ.

The closed loop system corresponding to the central controller $u = -B^*Px$ is described by

$$
\begin{aligned}
\dot{x} &= (A - BB^*P)x + Ew \\
z &= (C - DB^*P)x.
\end{aligned}
\tag{15.25}
$$

Now let $\hat{T}_\gamma$ be the operator from $L^2([t_0, t_1], \mathcal{W})$ into $L^2([t_0, t_1], \mathcal{Z})$ defined by

$$
(\hat{T}_\gamma w)(t) = \int_{t_0}^{t} (C - DB^*P(t))\Psi(t, \tau)Ew(\tau)\, d\tau \qquad (w \in L^2([t_1, t_2], \mathcal{W}))
\tag{15.26}
$$

where Ψ is the state transition operator for $A - BB^*P$. Obviously, the output z for the system in (15.25) is given by

$$
z(t) = (C - DB^*P(t))\Psi(t, t_0)x_0 + (\hat{T}_\gamma w)(t).
\tag{15.27}
$$

In particular, if $x(t_0) = 0$, then $z = \hat{T}_\gamma w$. In other words, $\hat{T}_\gamma$ is simply the input output operator from the disturbance w to the output z determined by the central controller.

A fundamental problem in control design is given a system $\{A, B, C, D, E\}$ along with a specified tolerance $\gamma > 0$, find a linear controller u such that $\|\phi(0, w, u)\| \leq \gamma\|w\|$ for all w in $L^2([t_0, t_1], \mathcal{W})$. The following result uses the central controller to solve this problem.

Corollary 15.1.2 *Consider the system $\{A, B, C, D, E\}$ and assume that the Riccati differential equation in (15.8) has a solution over the interval $[t_0, t_1]$. Let $\hat{T}_\gamma$ from $L^2([t_0, t_1], \mathcal{W})$ into $L^2([t_0, t_1], \mathcal{Z})$ be the input output operator for the closed loop system corresponding to the central controller*

$$
u = -B^*Px.
\tag{15.28}
$$

Then $\|\hat{T}_\gamma\| < \gamma$. Moreover, for any initial state $x(t_0) = x_0$, the closed loop system in (15.25) satisfies $\|z\|^2 \leq \gamma^2\|w\|^2 + d(x_0)$.

PROOF. Using $D^*D = I$ and $D^*C = 0$, we obtain

$$
(C - DB^*P)^*(C - DB^*P) = PBB^*P + C^*C.
$$

Substituting this into the Riccati differential equation in (15.19), it follows that the Riccati differential equation in (15.8) can be written as

$$
\dot{P} + (A - BB^*P^*)^*P + P(A - BB^*P) + \gamma^{-2}PEE^*P + (C - DB^*P)^*(C - DB^*P) = 0 \tag{15.29}
$$

subject to the final condition $P(t_1) = 0$. This is precisely the Riccati differential equation one obtains by replacing C and A with respectively $C - DB^*P$ and $A - BB^*P$ in Theorem 14.2.1. By using the completion of squares approach in the first part of the proof of Theorem

14.2.1, one can readily show that for any disturbance input w in $L^2([t_0, t_1], \mathcal{W})$ and for any initial state x_0, the closed loop system satisfies

$$\int_{t_0}^{t_1} \left(\|z(\sigma)\|^2 - \gamma^2 \|w(\sigma)\| \right) \, d\sigma \leq (P(t_0)x_0, x_0) = d(x_0) \,. \tag{15.30}$$

Moreover, by following the proof of Theorem 14.2.2, one can show that the norm of $\hat{T}_\gamma$ is strictly bounded by γ. ∎

15.2 Some abstract max-min problems

In this section, we develop and solve some operator max-min problems. In the next section we will apply some of these results to a standard feedback control problem. Let us begin by recalling Lemma 12.3.1 in Section 12.3 restated here for convenience.

Lemma 15.2.1 *Let F be an operator from $\mathcal{F}$ into $\mathcal{H}$ and h a vector in $\mathcal{H}$. Then*

$$((I + FF^*)^{-1}h, h) = \inf\{\|h + Fu\|^2 + \|u\|^2 : u \in \mathcal{F}\} \,. \tag{15.31}$$

Moreover, the optimal $\hat{u}$ in $\mathcal{F}$ solving this minimization problem is unique and given by

$$\hat{u} = -(I + F^*F)^{-1}F^*h = -F^*(I + FF^*)^{-1}h \,. \tag{15.32}$$

Remark 15.2.1 As before, let F be an operator mapping $\mathcal{F}$ into $\mathcal{H}$. Let $y = h + Fu$ where h is a specified vector in $\mathcal{H}$. Then the abstract linear quadratic regulator problem in (15.31) is equivalent to

$$\varepsilon(h) = \inf\{\|y\|^2 + \|u\|^2 : u \in \mathcal{F}\} \,. \tag{15.33}$$

Obviously, $\varepsilon(h) = ((I + FF^*)^{-1}h, h)$. Moreover, the optimal $\hat{u}$ which attains the optimal cost $\varepsilon(h)$ is given by $\hat{u} = -F^*\hat{y}$ where $\hat{y} = h + F\hat{u}$ is the optimal output; see Remark 12.3.1.

Now let T be an operator mapping $\mathcal{K}$ into $\mathcal{H}$ and C_o be an operator from $\mathcal{X}$ into $\mathcal{H}$. Let D be an isometry from $\mathcal{F}$ into $\mathcal{H}$ whose range is orthogonal to the range of C_o, T and F, that is $D^*C_o = 0$, $D^*T = 0$ and $D^*F = 0$. Let z be the output vector in $\mathcal{H}$ given by

$$z = \phi(x_0, w, u) = C_o x_0 + Tw + Fu + Du \tag{15.34}$$

where (the initial state) x_0 is specified. In our abstract control problem the vector u is viewed as the control and w is the disturbance. The idea is to design a control u to minimize the effect of the unknown disturbance w on the output z. This leads to the following minimization problem

$$\inf\{\|\phi(x_0, w, u)\|^2 : u \in \mathcal{F}\} = \inf\{\|C_o x_0 + Tw + Fu\|^2 + \|u\|^2 : u \in \mathcal{F}\} \,. \tag{15.35}$$

According to Lemma 15.2.1 with $h = C_o x_0 + Tw$, the optimal solution to this problem is uniquely given by $\hat{u} = -(I + F^*F)^{-1}F^*(C_o x_0 + Tw)$. To simplify some notation it is convenient to let R_* be the positive square root of $(I + FF^*)^{-1}$. Then the optimal cost in the optimization problem (15.35) is given by

$$\|R_*(C_ox_0+Tw)\|^2 = \inf\{\|C_ox_0+Tw+Fu\|^2+\|u\|^2 : u \in \mathcal{F}\} = \inf_u\{\|\phi(x_0,w,u)\|^2\}. \quad (15.36)$$

For the moment assume that the initial state $x_0 = 0$. In this case, the optimal choice for the control $\hat{u}$ gives rise to the operator R_*T from the disturbance space $\mathcal{K}$ to the output space $\mathcal{H}$. Notice that

$$\|R_*T\|^2 = \sup_{\|w\|\leq 1} \inf\{\|\phi(0,w,u)\|^2 : u \in \mathcal{F}\}. \quad (15.37)$$

In particular, this shows that if u is any vector in $\mathcal{F}$, then

$$\|R_*T\| \leq \sup\{\|\phi(0,w,u)\| : w \in \mathcal{K} \text{ and } \|w\| \leq 1\}. \quad (15.38)$$

This leads to the following design problem: Given a specified $\gamma > \|R_*T\|$ find a controller u such that $\|\phi(0,w,u)\| \leq \gamma$ whenever $\|w\| \leq 1$. Of course, the closer γ comes to $\|R_*T\|$ the closer one comes to constructing a controller to minimize the operator norm from the disturbance space $\mathcal{K}$ to the output $\mathcal{H}$.

Notice that R_*T is bounded by γ if and only if $\gamma^2I-TT^*+\gamma^2FF^*$ is positive. This follows because $\|R_*T\| \leq \gamma$ if and only if $\gamma^2I - R_*TT^*R_*$ is positive, or equivalently, $\gamma^2R_*^{-2} - TT^* = \gamma^2I - TT^* + \gamma^2FF^*$ is positive. A similar argument shows that the norm of R_*T is strictly bounded by γ if and only if $\gamma^2I - TT^* + \gamma^2FF^*$ is strictly positive.

Let Υ be an operator from $\mathcal{K}$ into $\mathcal{H}$. Let us recall the following optimization problem discussed in Section 14.4. For a specified $\gamma > 0$, find the optimal cost $\beta(h)$ defined by

$$\beta(h) = \sup\{\|h + \Upsilon w\|^2 - \gamma^2\|w\|^2 : w \in \mathcal{K}\}. \quad (15.39)$$

Recall that if $\|\Upsilon\| > \gamma$, then $\beta(h)$ is infinite and this optimization problem is undefined. If $\beta(h)$ is finite, then $\|\Upsilon\| \leq \gamma$. For convenience let us restate Theorem 14.4.1 to obtain the optimal solution when $\|\Upsilon\| < \gamma$.

Lemma 15.2.2 *Let Υ be an operator mapping $\mathcal{K}$ into $\mathcal{H}$ whose norm is strictly bounded by γ and let h be a vector in $\mathcal{H}$. Then the supremum in (15.39) is attained and is given by*

$$\beta(h) = \gamma^2 \left((\gamma^2I - \Upsilon\Upsilon^*)^{-1}h, h\right). \quad (15.40)$$

Furthermore, the optimal $\hat{w}$ in $\mathcal{K}$ which attains this supremum is unique and is given by

$$\hat{w} = (\gamma^2I - \Upsilon^*\Upsilon)^{-1}\Upsilon^*h = \Upsilon^*(\gamma^2I - \Upsilon\Upsilon^*)^{-1}h. \quad (15.41)$$

Recall that $z = C_ox_0 + Tw + Fu + Du$. By choosing $h = C_ox_0$ we can combine the optimization problems in (15.36) and (15.39) to arrive at the following max-min problem

$$d(x_0) = \sup_w\inf_u\{\|C_ox_0 + Tw + Fu\|^2 + \|u\|^2 - \gamma^2\|w\|^2\} = \sup_w\inf_u\{\|z\|^2 - \gamma^2\|w\|^2\} \quad (15.42)$$

where the infimum is taken over all u in $\mathcal{F}$ and the supremum is taken over all w in $\mathcal{K}$. Finally, let $\xi(x_0, w)$ be the cost of the infimum, that is,

$$\xi(x_0, w) = \inf_u \{\|\phi(x_0, w, u)\|^2 - \gamma^2 \|w\|^2\} = \inf_u \{\|C_o x_0 + Tw + Fu\|^2 + \|u\|^2 - \gamma^2 \|w\|^2\}. \quad (15.43)$$

Obviously, $d(x_0) = \sup_w \xi(x_0, w)$. In this setting, we say that the vector $\hat{v}$ is optimal for w if $\hat{v}$ attains the cost $\xi(x_0, w)$, that is, $\xi(x_0, w) = \|\phi(x_0, w, \hat{v})\|^2 - \gamma^2 \|w\|^2$. The pair $\{\hat{w}, \hat{u}\}$ is optimal for (15.42) if $\hat{u}$ is a vector in $\mathcal{F}$ and $\hat{w}$ is a vector in $\mathcal{K}$ satisfying

$$d(x_0) = \xi(x_0, \hat{w}) = \|\phi(x_0, \hat{w}, \hat{u})\|^2 - \gamma^2 \|\hat{w}\|^2. \quad (15.44)$$

By employing (15.36) in (15.43), it follows that for fixed w in $\mathcal{K}$

$$\xi(x_0, w) = \|R_*(C_o x_0 + Tw)\|^2 - \gamma^2 \|w\|^2. \quad (15.45)$$

Moreover, Lemma 15.2.1 shows that the cost $\xi(x_0, w)$ is uniquely attained by the vector

$$\hat{v} = -(I + F^* F)^{-1} F^* (C_o x_0 + Tw). \quad (15.46)$$

Furthermore, the optimal $\hat{v}$ can be obtained from the "feedback controller" $\hat{v} = -F^* \phi(x_0, w, \hat{v})$. To see this simply observe that (15.46) gives $(I + F^* F)\hat{v} = -F^*(C_o x_0 + Tw)$. This readily implies that $\hat{v} = -F^*(C_o x_0 + Tw + F\hat{v})$. Using $F^* D = 0$ with $\phi(x_0, w, \hat{v}) = C_o x_0 + Tw + F\hat{v} + D\hat{v}$, yields $\hat{v} = -F^* \phi(x_0, w, \hat{v})$. This proves part of the following result.

Theorem 15.2.3 *Let C_o be an operator from $\mathcal{X}$ into $\mathcal{H}$ while T is an operator from $\mathcal{K}$ into $\mathcal{H}$ and F is an operator from $\mathcal{F}$ into $\mathcal{H}$. Let $d(x_0)$ be the optimal cost in the optimization problem (15.42) and assume that $\gamma^2 I - TT^* + \gamma^2 FF^*$ is strictly positive. Then*

$$d(x_0) = \gamma^2 ((\gamma^2 I - TT^* + \gamma^2 FF^*)^{-1} C_o x_0, C_o x_0). \quad (15.47)$$

The optimal control $\hat{v}$ which uniquely attains the cost $\xi(x_0, w)$ is given by $\hat{v} = -F^* \phi(x_0, w, \hat{v})$. Furthermore, the optimal disturbance pair $\{\hat{w}, \hat{u}\}$ which attains the optimal cost $d(x_0)$ is unique and is given by $\hat{u} = -F^* \hat{z}$ and $\hat{w} = \gamma^{-2} T^* \hat{z}$ where $\hat{z} = C_o x_0 + T\hat{w} + F\hat{u} + D\hat{u}$, or equivalently, $\hat{u} = -F^* \hat{g}$ and $\hat{w} = \gamma^{-2} T^* \hat{g}$ where $\hat{g} = C_o x_0 + T\hat{w} + F\hat{u}$.

PROOF. Assume that $\gamma^2 I - TT^* + \gamma^2 FF^*$ is strictly positive, or equivalently, the norm of $R_* T$ is strictly bounded by γ. By taking the supremum in (15.45) and employing Lemma 15.2.2 with $\Upsilon = R_* T$ and $h = R_* C_o x_0$, we obtain

$$\begin{aligned} d(x_0) &= \sup_w \{\|R_* C_o x_0 + R_* Tw\|^2 - \gamma^2 \|w\|^2\} \\ &= \gamma^2 ((\gamma^2 I - R_* TT^* R_*)^{-1} R_* C_o x_0, R_* C_o x_0) \qquad (15.48) \\ &= \gamma^2 ((\gamma^2 I - TT^* + \gamma^2 FF^*)^{-1} C_o x_0, C_o x_0). \end{aligned}$$

Notice that if $\gamma^2 I - TT^* + \gamma^2 FF^*$ is not positive, then $\|R_* T\| > \gamma$. In this case, $d(x_0)$ is infinite. So, $\|R_* T\|$ is the infimum over the set of all $\gamma > 0$ such that the optimal cost $d(x_0)$ is finite.

To complete the proof it remains to compute the optimal pair $\{\hat{w}, \hat{u}\}$ which achieves the optimal cost $d(x_0)$. As before, let $\hat{v}$ be the optimal input which achieves the cost $\xi(x_0, w)$ in (15.43). Applying Lemma 15.2.2 with $\Upsilon = R_* T$ to (15.45), shows that the optimal input $\hat{w}$ which attains the optimal cost $d(x_0) = \sup_w \xi(x_0, w)$ is given by

$$
\begin{aligned}
\hat{w} &= T^* R_* (\gamma^2 I - R_* T T^* R_*)^{-1} R_* C_o x_0 \qquad\qquad (15.49) \\
&= T^* (\gamma^2 I - T T^* + \gamma^2 F F^*)^{-1} C_o x_0 .
\end{aligned}
$$

Moreover, Lemma 15.2.2 shows that this $\hat{w}$ is the only disturbance which attains the optimal cost $d(x_0)$. By setting $\hat{u} = \hat{v}$ and $w = \hat{w}$ in (15.46), we arrive at an optimal pair $\{\hat{w}, \hat{u}\}$ which achieves the optimal cost $d(x_0)$. Moreover, we claim that

$$
\hat{u} = -\gamma^2 F^* (\gamma^2 I - T T^* + \gamma^2 F F^*)^{-1} C_o x_0 . \qquad\qquad (15.50)
$$

Using (15.49) in (15.46) with $\hat{u} = \hat{v}$, we obtain

$$
\begin{aligned}
\hat{u} &= -F^* (I + F F^*)^{-1} (C_o x_0 + T \hat{w}) \\
&= -F^* (I + F F^*)^{-1} \left(I + T T^* (\gamma^2 I - T T^* + \gamma^2 F F^*)^{-1} \right) C_o x_0 \\
&= -F^* (I + F F^*)^{-1} \left(\gamma^2 I - T T^* + \gamma^2 F F^* + T T^* \right) (\gamma^2 I - T T^* + \gamma^2 F F^*)^{-1} C_o x_0 \\
&= -\gamma^2 F^* (\gamma^2 I - T T^* + \gamma^2 F F^*)^{-1} C_o x_0 . \qquad\qquad (15.51)
\end{aligned}
$$

This yields (15.50). We claim that the optimal output $\hat{z} = C_o x_0 + F \hat{u} + T \hat{w} + D \hat{u}$ is given by

$$
\hat{z} = \gamma^2 (\gamma^2 I - T T^* + \gamma^2 F F^*)^{-1} C_o x_0 + D \hat{u} . \qquad\qquad (15.52)
$$

Let $\hat{g} = C_o x_0 + F \hat{u} + T \hat{w}$. Using (15.50) and (15.49), we obtain

$$
\begin{aligned}
\hat{g} &= C_o x_0 + T \hat{w} + F \hat{u} \\
&= C_o x_0 + (T T^* - \gamma^2 F F^*)(\gamma^2 I - T T^* + \gamma^2 F F^*)^{-1} C_o x_0 \\
&= \left(\gamma^2 I - T T^* + \gamma^2 F F^* + T T^* - \gamma^2 F F^* \right) (\gamma^2 I - T T^* + \gamma^2 F F^*)^{-1} C_o x_0 \\
&= \gamma^2 (\gamma^2 I - T T^* + \gamma^2 F F^*)^{-1} C_o x_0 .
\end{aligned}
$$

Since $\hat{z} = \hat{g} + D \hat{u}$, equation (15.52) holds. Because $F^* D = 0$, equations (15.52) and (15.50) show that $\hat{u} = -F^* \hat{g} = -F^* \hat{z}$. Equations (15.52) and (15.49) imply that $\hat{w} = \gamma^{-2} T^* \hat{g} = \gamma^{-2} T^* \hat{z}$. Finally, it is noted that $\{\hat{w}, \hat{u}\}$ in (15.49) and (15.50) is the only pair which attains the optimal cost $d(x_0)$. If $\{w_o, u_o\}$ is another pair which attains the optimal cost $d(x_0)$, then w_o must be the unique vector which attains the optimal cost $d(x_0) = \sup_w \xi(x_0, w)$. Hence, $w_o = \hat{w}$ is given by (15.49). Because there is only one vector which attains the cost $\xi(x_0, \hat{w})$, the calculation in (15.51) shows that $u_o = \hat{u}$ is given by (15.50). Therefore, the optimal pair $\{\hat{w}, \hat{u}\}$ which attains $d(x_0)$ is unique. ∎

The proof of the previous theorem readily yields the following result.

Corollary 15.2.4 *Let C_o be an operator from $\mathcal{X}$ into $\mathcal{H}$ while T is an operator from $\mathcal{K}$ into $\mathcal{H}$ and F is an operator from $\mathcal{F}$ to $\mathcal{H}$. Let $d(x_0)$ be the optimal cost in the optimization problem (15.42). Then*

$$
\| (I + F F^*)^{-1/2} T \| = \inf \{ \gamma : d(x_0) < \infty \} . \qquad\qquad (15.53)
$$

Corollary 15.2.5 *Assume that $\gamma^2 I - T^*T + \gamma^2 FF^*$ is strictly positive. Then the pair $\{\hat{w}, \hat{u}\}$ which uniquely attains the optimal cost $d(x_0)$ in the optimization problem (15.42) is given by solving the following block matrix system of equations*

$$\begin{bmatrix} I + F^*F & F^*T \\ -T^*F & \gamma^2 I - T^*T \end{bmatrix} \begin{bmatrix} \hat{u} \\ \hat{w} \end{bmatrix} = \begin{bmatrix} -F^*C_o x_0 \\ T^*C_o x_0 \end{bmatrix}. \tag{15.54}$$

Moreover, the 2×2 block matrix in (15.54) is an invertible operator on $\mathcal{F} \oplus \mathcal{K}$.

PROOF. According to Theorem 15.2.3, the optimal pair $\{\hat{w}, \hat{u}\}$ which achieves the optimal cost $d(x_0)$ is uniquely given by $\hat{u} = -F^*\hat{g}$ and $\hat{w} = \gamma^{-2}T^*\hat{g}$ where $\hat{g} = C_o x_0 + T\hat{w} + F\hat{u}$. Hence,

$$(I + F^*F)\hat{u} = -F^*T\hat{w} - F^*C_o x_0 \qquad \text{and} \qquad (\gamma^2 I - T^*T)\hat{w} = T^*F\hat{u} + T^*C_o x_0.$$

By rearranging these equations, we arrive at the system of equations in (15.54). Let X be an invertible operator on $\mathcal{F}$. Then recall that the block matrix

$$\begin{bmatrix} X & Y \\ W & Z \end{bmatrix} \quad \text{on} \quad \begin{bmatrix} \mathcal{F} \\ \mathcal{K} \end{bmatrix}$$

is invertible if and only if the Schur complement $Z - WX^{-1}Y$ is invertible. The Schur complement for the 2×2 block matrix in (15.54) is given by

$$\gamma^2 I - T^*T + T^*F(I + F^*F)^{-1}F^*T = \gamma^2 I - T^*\left(I - (I + FF^*)^{-1}FF^*\right)T$$
$$= \gamma^2 I - T^*(I + FF^*)^{-1}\left(I + FF^* - FF^*\right)T = \gamma^2 I - T^*(I + FF^*)^{-1}T$$
$$= \gamma^2 I - T^*R_* R_* T.$$

Because $\gamma^2 I - T^*T + \gamma^2 FF^*$ is strictly positive, the norm of R_*T is strictly bounded by γ. So, the Schur complement $\gamma^2 I - (R_*T)^*R_*T$ is invertible. Therefore, the 2×2 block matrix in (15.54) is invertible. ∎

A connection to game theory. Let us conclude this section by making a simple connection with game theory. To this end, let $q(w, u)$ be a real valued function of u in $\mathcal{F}$ and w in $\mathcal{K}$. Then,

$$\sup_{w} \inf_{u} q(w, u) \leq \inf_{u} \sup_{w} q(w, u) \tag{15.55}$$

where the infimum is taken over all u in $\mathcal{F}$ and the supremum is taken over all w in $\mathcal{K}$. We say that q defines a *game* if we have equality in (15.55). To motivate this terminology consider a contest with two players a and b. Player a is trying to maximize the cost function q by choosing a strategy from w in $\mathcal{K}$, while player b is trying to minimize the cost function q by choosing a strategy from u in $\mathcal{F}$. If q defines a game, then it does not matter which player a or b goes first. We conclude this section with the following result.

Theorem 15.2.6 *Let T be an operator from $\mathcal{K}$ into $\mathcal{H}$ while F is an operator from $\mathcal{F}$ into $\mathcal{H}$ and C_o is an operator from $\mathcal{X}$ into $\mathcal{H}$. Assume that $\|T\| < \gamma$ and let q be the function defined by*

$$q(w, u) = \|C_o x_0 + Tw + Fu\|^2 + \|u\|^2 - \gamma^2 \|w\|^2. \tag{15.56}$$

Then q defines a game and the optimal cost is given by

$$\inf_u \sup_w q(w, u) = \sup_w \inf_u q(w, u) = \gamma^2 (C_o^*(\gamma^2 I - TT^* + \gamma^2 FF^*)^{-1} C_o x_0, x_0). \tag{15.57}$$

Moreover, the optimal pair $\{\hat{w}, \hat{u}\}$ which attains this optimal cost is uniquely given by $\hat{w} = \gamma^{-2} T^* \hat{g}$ and $\hat{u} = -F^* \hat{g}$ where $\hat{g} = C_o x + T\hat{w} + F\hat{u}$.

PROOF. Let D_* be the positive square root of $(\gamma^2 I - TT^*)^{-1}$. By combining Lemma 15.2.2 with Lemma 15.2.1, we have

$$\begin{aligned}
\inf_u \sup_w q(w, u) &= \inf_u \sup_w \{\|C_o x_0 + Tw + Fu\|^2 - \gamma^2 \|w\|^2 + \|u\|^2\} \\
&= \inf_u \{\gamma^2 \|D_* C_o x + D_* Fu\|^2 + \|u\|^2\} \\
&= ((I + \gamma^2 D_* FF^* D_*)^{-1} \gamma D_* C_o x_0, \gamma D_* C_o x_0) \\
&= \gamma^2 ((\gamma^2 I - TT^* + \gamma^2 FF^*)^{-1} C_o x_0, C_o x_0).
\end{aligned}$$

On the other hand, Theorem 15.2.3 shows that

$$\sup_w \inf_u q(w, u) = d(x_0) = \gamma^2 ((\gamma^2 I - TT^* + \gamma^2 FF^*)^{-1} C_o x_0, C_o x_0).$$

Therefore, (15.57) holds and q defines a game. ∎

Exercise 33 Let ϕ be the function defined in (15.34). Consider the feedback controller $u = \hat{v} = -F^* z$. Then show that $\|\phi(x_0, w, \hat{v})\| = \|R_*(C_o x_0 + Tw)\|$. In particular, this implies that $\|\phi(0, w, \hat{v})\| \le \|R_* T\| \|w\|$.

Exercise 34 Consider the system $\{A, B, C, D, E\}$ in (15.1). Assume that the Riccati differential equation

$$\dot{R} + A^* R + RA + \gamma^{-2} REE^* R + C^* C = 0 \qquad (R(t_1) = 0)$$

has a solution on the interval $[t_0, t_1]$ for some $\gamma > 0$. Then show that

$$\sup_w \inf_u \left(\|z\|^2 - \gamma^2 \|w\|^2\right) = \inf_u \sup_w \left(\|z\|^2 - \gamma^2 \|w\|^2\right).$$

The supremum is taken over all w in $L^2([t_0, t_1], \mathcal{W})$ while the infimum is taken over all u in $L^2([t_0, t_1], \mathcal{U})$.

15.3   The Riccati differential equation and norms

Now let us return to the system $\{A, B, C, D, E\}$ in (15.1). Recall that D is an isometry satisfying $D^* C = 0$. In this section, we will show that the Riccati differential equation in (15.8) has a solution over the interval $[t_0, t_1]$ if and only if a certain operator is strictly positive. To this end, let T be the operator from $\mathcal{K} = L^2([t_0, t_1], \mathcal{W})$ into $\mathcal{H} = L^2([t_0, t_1], \mathcal{Z})$ and F the operator from $\mathcal{F} = L^2([t_0, t_1], \mathcal{U})$ into $L^2([t_0, t_1], \mathcal{Z})$ defined by

$$(Tw)(t) = \int_{t_0}^t Ce^{A(t-\tau)} Ew(\tau)\, d\tau \qquad \text{and} \qquad (Fu)(t) = \int_{t_0}^t Ce^{A(t-\tau)} Bu(\tau)\, d\tau. \tag{15.58}$$

In this setting C_o is the observability operator from $\mathcal{X}$ into $L^2([t_0, t_1], \mathcal{Z})$ defined by $(C_o x_0)(t) = Ce^{A(t-t_0)} x_0$ where x_0 is in $\mathcal{X}$.

Theorem 15.3.1 *Consider the system in (15.1) and let T be the operator from $L^2([t_0, t_1], \mathcal{W})$ into $L^2([t_0, t_1], \mathcal{Z})$ and F the operator from $L^2([t_0, t_1], \mathcal{U})$ into $L^2([t_0, t_1], \mathcal{Z})$ defined in (15.58). Then the following statements are equivalent.*

(i) The norm $\|(I + FF^*)^{-1/2}T\| < \gamma$.

(ii) The operator $\gamma^2 I - TT^* + \gamma^2 FF^*$ is strictly positive.

(iii) The Riccati differential equation in (15.8) has a solution P on the interval $[t_0, t_1]$.

(iv) The optimal cost $d(x_0)$ defined in (15.5) is finite for all initial states x_0 in $\mathcal{X}$.

In this case, $d(x_0)$ is uniquely attained and given by $d(x_0) = (P(t_0)x_0, x_0)$.

PROOF. Recall that Parts (*i*) and (*ii*) are equivalent. According to Theorem 15.1.1, Parts (*iii*) and (*iv*) are equivalent. Obviously, the optimization problem in (15.5) is a special case of the optimization problem in (15.42). So, if Part (*ii*) holds, then equation (15.47) in Theorem 15.2.3 shows that $d(x_0)$ is finite for all x_0 in $\mathcal{X}$. Hence, Part (*ii*) implies Parts (*iii*) and (*iv*).

Now assume that the Riccati differential equation in (15.8) has a solution P on the interval $[t_0, t_1]$, that is, Part (*iii*) holds. Recall that if a nonlinear differential equation $\dot{q} = f(q, \eta)$ has a solution on the interval $[t_0, t_1]$ where f is a continuous function and η is a parameter, then $\dot{q} = f(q, \eta - \epsilon)$ also has a solution on the interval $[t_0, t_1]$ for all ϵ in some neighborhood of the origin; see [28]. Hence, the Riccati differential equation in (15.8) also has a solution on the interval $[t_0, t_1]$ when γ is replaced by $\gamma - \epsilon$ for some $\epsilon > 0$. Theorem 15.1.1 shows that for this $\gamma - \epsilon$ the corresponding optimal cost $d(x_0)$ is finite. Corollary 15.2.4 implies that $\|R_*T\| \leq \gamma - \epsilon$ where $R_* = (I + FF^*)^{-1/2}$. Therefore, $\|R_*T\| < \gamma$ and Part (*iii*) implies Part (*i*). ∎

The following consequence of the previous theorem provides a method to compute the norm of the infinite dimensional operator $(I + FF^*)^{-1/2}T$.

Corollary 15.3.2 *Let T be the operator from $L^2([t_0, t_1], \mathcal{W})$ into $L^2([t_0, t_1], \mathcal{Z})$ and F be the operator from $L^2([t_0, t_1], \mathcal{U})$ into $L^2([t_0, t_1], \mathcal{Z})$ defined in (15.58). Then $\|(I + FF^*)^{-1/2}T\|$ is the infimum of the set of all positive numbers γ such that the Riccati differential equation in (15.8) has a solution on the interval $[t_0, t_1]$.*

Remark 15.3.1 Let T be the operator from $L^2([t_0, t_1], \mathcal{W})$ into $L^2([t_0, t_1], \mathcal{Z})$ and F be the operator from $L^2([t_0, t_1], \mathcal{U})$ into $L^2([t_0, t_1], \mathcal{Z})$ defined in (15.58). Assume that the Riccati differential equation in (15.8) has a solution over the interval $[t_0, t_1]$. Then $\gamma^2 I - TT^* + \gamma^2 FF^*$ is strictly positive. By combining Theorems 15.1.1 and 15.2.3, we obtain

$$P(t_0) = \gamma^2 C_o^*(\gamma^2 I - TT^* + \gamma^2 FF^*)^{-1}C_o. \tag{15.59}$$

Recall that the pair $\{C, A\}$ is observable if and only if the operator C_o is one to one. Hence, equation (15.59) shows that the pair $\{C, A\}$ is observable if and only if $P(t_0)$ is strictly positive. Obviously, this Riccati differential equation has a solution over the interval $[t', t_1]$ for any $t_0 \leq t' \leq t_1$. So, by replacing t_0 by t in (15.59), we see that $P(t)$ is strictly positive for any $t_0 \leq t < t_1$ if and only if the pair $\{C, A\}$ is observable. In particular, $P(t)$ is strictly positive for any $t_0 \leq t < t_1$ if and only if $P(t)$ is strictly positive for all $t_0 \leq t < t_1$. Likewise $\Omega(\tau) = P(t_1 - \tau)$ is strictly positive for any $0 < \tau \leq t_1 - t_0$ if and only if $\Omega(\tau)$ is strictly positive for all $0 < \tau \leq t_1 - t_0$.

15.4   The infimal achievable gain

As before, consider the system $\{A, B, C, D, E\}$ in (15.1). The output z of this system is given by

$$z = C_o x_0 + T w + (F + D) u \,.$$

Here we consider the problem of finding a feedback controller to minimize the effect of the disturbance w on the output z. In general, a linear feedback controller is described by

$$u = K(x \oplus w) \tag{15.60}$$

where K is an operator mapping $L^2([t_0, t_1], \mathcal{X}) \oplus L^2([t_0, t_1], \mathcal{W})$ into $L^2([t_0, t_1], \mathcal{U})$. Assume that the initial state $x_0 = 0$. Then, the closed loop system corresponding to the feedback controller (15.60) applied to system (15.1) satisfies

$$x = L E w + L B K (x \oplus w) \qquad \text{and} \qquad z = T w + (F + D) K (x \oplus w) \tag{15.61}$$

where L is the operator on $L^2([t_0, t_1], \mathcal{X})$ defined by

$$L f(t) = \int_{t_0}^{t} e^{A(t - \sigma)} f(\sigma) \, d\sigma \,.$$

We say that the feedback map K is *admissible*, if for each w in $L^2([0, t_1], \mathcal{W})$, there is a unique element x of $L^2([t_0, t_1], \mathcal{X})$ which satisfies the first equation in (15.61). If K is admissible, then with $x_0 = 0$, there is an operator Z from $L^2([t_0, t_1], \mathcal{W})$ into $L^2([t_0, t_1], \mathcal{U})$ such that the feedback controller in (15.60) can be described by

$$u = Z w \,. \tag{15.62}$$

Thus, the closed loop system (with $x_0 = 0$) satisfies

$$z = T_Z w \tag{15.63}$$

where T_Z is the operator from $L^2([t_0, t_1], \mathcal{W})$ into $L^2([t_0, t_1], \mathcal{Z})$ given by $T_Z = T + (F + D)Z$. We refer to T_Z as the input output map for the closed loop system.

Consider now the problem of finding an admissible controller to minimize the effect of the disturbance w on the output z. One can approach this problem by considering the problem of finding an operator Z to minimize the norm of T_Z. We refer to $\|T_Z\|$ as the *gain* of the closed loop system. This leads to the following optimization problem:

$$d_\infty = \inf_Z \|T_Z\| \,. \tag{15.64}$$

We say that d_∞ is the infimal closed loop gain achievable by an admissible linear controller.

Consider any operator Z. By consulting (15.37) and noting that $T_Z w = \phi(0, w, Zw)$, we see that

$$
\begin{aligned}
\|R_* T\| &= \sup_{\|w\| \le 1} \inf_u \|\phi(0, w, u)\| \\
&\le \sup_{\|w\| \le 1} \|\phi(0, w, Zw)\| = \sup_{\|w\| \le 1} \|T_Z w\|^2 \\
&= \|T_Z\|
\end{aligned}
$$

where $R_* = (I + FF^*)^{-1/2}$. Hence,

$$
\|R_* T\| \le \inf_Z \|T_Z\| \le d_\infty .
$$

We claim that $\|R_* T\| = d_\infty$. In other words, one can find an admissible controller such that $\|T_Z\| \approx \|R_* T\|$.

If $\gamma > \|R_* T\|$, then Theorem 15.3.1 shows that the Riccati differential equation in (15.8) has a solution over the interval $[t_0, t_1]$. Now consider the corresponding central controller as given by $u = -B^* P x$. This is an admissible controller and the input output operator $\hat{T}_\gamma$ from $L^2([t_1, t_2], \mathcal{W})$ into $L^2([t_1, t_2], \mathcal{Z})$ corresponding to this controller is given by (15.26). In other words, if the initial condition $x(t_0) = 0$, then $z = \hat{T}_\gamma w$. Corollary 15.1.2, shows that the norm of $\hat{T}_\gamma$ is strictly bounded by γ. So, letting γ approach $\|R_* T\|$, we see that $d_\infty = \|R_* T\|$. In other words, d_∞ is the infimum of the set all $\gamma > 0$ such that the Riccati differential equation in (15.8) has a solution on the interval $[t_0, t_1]$. Summing up the previous analysis yields the following result.

Theorem 15.4.1 *Consider the system in (15.1) and let T be the operator from $L^2([t_0, t_1], \mathcal{W})$ into $L^2([t_0, t_1], \mathcal{Z})$ and F be the operator from $L^2([t_0, t_1], \mathcal{U})$ into $L^2([t_0, t_1], \mathcal{Z})$ defined in (15.58). Then, the infimal closed loop gain d_∞ achievable by an admissible linear controller is given by*

$$
d_\infty = \|(I + FF^*)^{-1/2} T\| .
$$

Moreover, d_∞ equals the infimum of the set all $\gamma > 0$ such that the Riccati differential equation in (15.8) has a solution on the interval $[t_0, t_1]$. In particular, if $\gamma > d_\infty$, then the central controller $u = -B^* P x$ yields a closed loop system whose gain $\|\hat{T}_\gamma\|$ is strictly less than γ.

15.5   A two point boundary value problem

In this section we derive a two point boundary value problem associated with the optimization problem in (15.5). This will provide us with a natural derivation of the Riccati differential equation in (15.8). As before, let T be the operator from $L^2([t_0, t_1], \mathcal{W})$ into $L^2([t_0, t_1], \mathcal{Z})$ and F be the operator from $L^2([t_0, t_1], \mathcal{U})$ into $L^2([t_0, t_1], \mathcal{Z})$ defined in (15.58). Moreover, assume that $\gamma^2 I - TT^* + \gamma^2 FF^*$ is strictly positive. According to Theorem 15.2.3, the optimal cost $d(x_0)$ in the optimization problem (15.5) is uniquely attained by the pair $\{\hat{w}, \hat{u}\}$ where $\hat{u} = -F^* \hat{z}$ and $\hat{w} = \gamma^{-2} T^* \hat{z}$. The optimal output $\hat{z}$ is given by

$$
\dot{\hat{x}} = A\hat{x} + E\hat{w} + B\hat{u} \qquad \text{and} \qquad \hat{z} = C\hat{x} + D\hat{u} \tag{15.65}
$$

where $\hat{x}(t_0) = x_0$. Since T^*D and F^*D are both zero, the optimal pair $\{\hat{w}, \hat{u}\}$ is given by $\hat{u} = -F^*C\hat{x}$ and $\hat{w} = \gamma^{-2}T^*C\hat{x}$.

To compute the optimal pair we need the adjoints of T and F. A simple calculation shows that T^* is the operator from $L^2([t_0, t_1], \mathcal{Z})$ into $L^2([t_0, t_1], \mathcal{W})$ and F^* is the operator from $L^2([t_0, t_1], \mathcal{Z})$ into $L^2([t_0, t_1], \mathcal{U})$ defined by

$$(T^*g)(t) = \int_t^{t_1} E^*e^{-A^*(t-\tau)}C^*g(\tau)\,d\tau \quad \text{and} \quad (F^*g)(t) = \int_t^{t_1} B^*e^{-A^*(t-\tau)}C^*g(\tau)\,d\tau \quad (15.66)$$

where g is in $L^2([t_0, t_1], \mathcal{Z})$. Let λ be the function defined by

$$\lambda(t) = \int_t^{t_1} e^{-A^*(t-\tau)}C^*g(\tau)\,d\tau. \quad (15.67)$$

Obviously, $E^*\lambda = T^*g$ and $B^*\lambda = F^*g$. Using Leibnitz's rule, we obtain following state space representation of λ

$$\dot{\lambda} = -A^*\lambda - C^*g \quad \text{with} \quad \lambda(t_1) = 0. \quad (15.68)$$

To compute λ for a specified g, one must integrate the previous differential equation backward in time from t_1. Finally, we call λ the *adjoint state*.

Recall that the optimal pair is given by $\hat{w} = \gamma^{-2}T^*C\hat{x}$ and $\hat{u} = -F^*C\hat{x}$. So, by using $g = C\hat{x}$ in (15.68), we obtain $\dot{\lambda} = -A^*\lambda - C^*C\hat{x}$. Moreover, $\hat{w} = \gamma^{-2}E^*\lambda$ and $\hat{u} = -B^*\lambda$. Combining these equations with $\dot{\hat{x}} = A\hat{x} + E\hat{w} + B\hat{u}$, we see that $\hat{x} \oplus \lambda$ solves the following two point boundary value problem

$$\begin{bmatrix} \dot{\hat{x}} \\ \dot{\lambda} \end{bmatrix} = \begin{bmatrix} A & \gamma^{-2}EE^* - BB^* \\ -C^*C & -A^* \end{bmatrix} \begin{bmatrix} \hat{x} \\ \lambda \end{bmatrix} \quad \text{with} \quad \begin{bmatrix} \hat{x}(t_0) \\ \lambda(t_1) \end{bmatrix} = \begin{bmatrix} x_0 \\ 0 \end{bmatrix}. \quad (15.69)$$

Summing up the previous analysis we obtain the following result. If $\gamma^2 I - TT^* + \gamma^2 FF^*$ is strictly positive, then the corresponding two point boundary value problem in (15.69) has a solution for every x_0 in $\mathcal{X}$. Furthermore, the optimal cost $d(x_0)$ in (15.5) is attained for every initial state x_0. Moreover, the optimal pair which achieves this optimal cost is given by $\hat{w} = \gamma^{-2}E^*\lambda$ and $\hat{u} = -B^*\lambda$.

We now claim that, if $\gamma^2 I - TT^* + \gamma^2 FF^*$ is strictly positive, then for every x_0 in $\mathcal{X}$ and for every t' in the interval $[t_0, t_1)$, the two point boundary value problem

$$\begin{bmatrix} \dot{\hat{x}} \\ \dot{\lambda} \end{bmatrix} = \begin{bmatrix} A & \gamma^{-2}EE^* - BB^* \\ -C^*C & -A^* \end{bmatrix} \begin{bmatrix} \hat{x} \\ \lambda \end{bmatrix} \quad \text{with} \quad \begin{bmatrix} \hat{x}(t') \\ \lambda(t_1) \end{bmatrix} = \begin{bmatrix} x_0 \\ 0 \end{bmatrix} \quad (15.70)$$

has a solution $\hat{x} \oplus \lambda$ on $[t', t_1]$.

To see this, let T' be the operator from $L^2([t', t_1], \mathcal{W})$ into $L^2([t', t_1], \mathcal{Z})$ and F' the operator from $L^2([t', t_1], \mathcal{U})$ into $L^2([t', t_1], \mathcal{Z})$ defined by

$$(T'w)(t) = \int_{t'}^t Ce^{A(t-\tau)}Ew(\tau)\,d\tau \quad \text{and} \quad (F'u)(t) = \int_{t'}^t Ce^{A(t-\tau)}Bu(\tau)\,d\tau. \quad (15.71)$$

If $\epsilon I \leq \gamma^2 I - TT^* + \gamma^2 FF^*$ for some $\epsilon > 0$, then we claim that $\epsilon I \leq \gamma^2 I - T'T'^* + \gamma^2 F'F'^*$. To see this let $\mathcal{B}$ be the set of all unit vectors g in $L^2([t_0, t_1], \mathcal{Z})$ satisfying $g(\sigma) = 0$ for $t_0 + t_1 - t' \leq \sigma \leq t_1$. Using this notation, we obtain

$$
\begin{aligned}
\epsilon \ &\leq\ \inf\{\gamma^2\|g\|^2 - \|T^*g\|^2 + \gamma^2\|F^*g\|^2 : g \in L^2([t_0, t_1], \mathcal{Z}) \text{ and } \|g\| = 1\} \\
&\leq\ \inf\{\int_{t_0}^{t_0+t_1-t'} \left(\gamma^2\|g(\sigma)\|^2 - \|(T^*g)(\sigma)\|^2 + \gamma^2\|(F^*g)(\sigma)\|^2\right) d\sigma : g \in \mathcal{B}\} \quad (15.72) \\
&=\ \inf\{\gamma^2\|g\|^2 - \|T'^*g\|^2 + \gamma^2\|F'^*g\|^2 : g \in L^2([t', t_1], \mathcal{Z}) \text{ and } \|g\| = 1\}.
\end{aligned}
$$

Hence, $\gamma^2 I - T'T'^* + \gamma^2 F'F'^*$ is strictly positive. By using this fact, in our previous analysis with t' and T' and F' replacing t_0 and T and F respectively, we see that the two point boundary value problem (15.70) has a solution for every x_0 in $\mathcal{X}$ and for every t' in the interval $[t_0, t_1)$.

15.5.1   The Riccati differential equation

As before, let T and F be the operators defined in (15.58) acting between the appropriate $L^2([t_0, t_1], \cdot)$ spaces. Assume that $\gamma^2 I - TT^* + \gamma^2 FF^*$ is strictly positive. Theorem 15.3.1 states that the Riccati differential equation in (15.8) has a solution P on the interval $[t_0, t_1]$. Here we use the two point boundary value problem in (15.69) to derive the Riccati differential equation in (15.8). In particular, we show that the solution $\hat{x} \oplus \lambda$ to this two point boundary value problem satisfies $\lambda(t) = P(t)\hat{x}(t)$. As in the linear quadratic regulator problem, we will also obtain an explicit formula for P in terms of the elements of the state transition matrix for the corresponding Hamiltonian matrix. In this setting the *Hamiltonian matrix* associated with the optimization problem in (15.5) is given by

$$
H = \begin{bmatrix} A & \gamma^{-2}EE^* - BB^* \\ -C^*C & -A^* \end{bmatrix}. \tag{15.73}
$$

Notice that H is simply the state space matrix for the system in (15.69).

Since $\gamma^2 I - TT^* + \gamma^2 FF^*$ is strictly positive, the two point boundary value problem in (15.70) has a solution $\hat{x} \oplus \lambda$ for every x_0 in $\mathcal{X}$ and t' in $[t_0, t_1)$. Hence, using Lemma 13.6.1, this is equivalent to the existence of a solution P to the Riccati differential equation in (15.8) on the interval $[t_0, t_1]$. Moreover, it follows from Remark 13.6.1 that $\lambda(t) = P(t)\hat{x}(t)$ and

$$
P(t) = \Phi_{21}(t - t_1)\Phi_{11}(t - t_1)^{-1} \qquad (\text{for } t_0 \leq t \leq t_1) \tag{15.74}
$$

where Φ_{11} and Φ_{21} are obtained from the following matrix partition of e^{Ht}:

$$
e^{Ht} = \begin{bmatrix} \Phi_{11}(t) & \Phi_{12}(t) \\ \Phi_{21}(t) & \Phi_{22}(t) \end{bmatrix} \text{ on } \begin{bmatrix} \mathcal{X} \\ \mathcal{X} \end{bmatrix}. \tag{15.75}
$$

So, one can obtain P by either using (15.74) or solving the Riccati differential equation backwards in time with $P(t_1) = 0$.

Combining the results in this section with Theorem 15.3.1 yields the following result.

Theorem 15.5.1 *let T be the operator from $L^2([t_0, t_1], \mathcal{W})$ into $L^2([t_0, t_1], \mathcal{Z})$ and F be the operator from $L^2([t_0, t_1], \mathcal{U})$ into $L^2([t_0, t_1], \mathcal{Z})$ defined in (15.58). Let $d(x_0)$ be the optimal cost for the optimization problem in (15.5). Then the following statements are equivalent.*

(i) The optimal cost $d(x_0)$ is finite for every $x_0 \in \mathcal{X}$.

(ii) The optimal cost $d(x_0)$ is uniquely attained for every $x_0 \in \mathcal{X}$.

(iii) The Riccati differential equation in (15.8) has a solution on the interval $[t_0, t_1]$.

(iv) The two point boundary value problem in (15.70) has a solution for every x_0 in $\mathcal{X}$ and t' in the interval $[t_0, t_1)$.

(v) The operator $\gamma^2 I - TT^* + \gamma^2 FF^*$ is strictly positive.

15.6   The infinite horizon problem

As before, consider the system $\{A, B, C, D, E\}$ where D is an isometry satisfying $D^*C = 0$. This section is concerned with the following infinite horizon version of the optimization problem in (15.5). For each initial state x_0 in $\mathcal{X}$, find the optimal cost $d(x_0)$ and an optimal disturbance pair $\{\hat{w}, \hat{u}\}$ which solves the optimization problem:

$$d(x_0) = \sup_w \inf_u \int_0^\infty \left(\|z(\sigma)\|^2 - \gamma^2 \|w(\sigma)\|^2 \right) d\sigma$$
$$\text{subject to} \quad \dot{x} = Ax + Ew + Bu \text{ and } z = Cx + Du \text{ and } x(0) = x_0. \quad (15.76)$$

The infimum is taken over all u in $L^2([0, \infty), \mathcal{U})$ while the supremum is taken over all w in $L^2([0, \infty), \mathcal{W})$. To be more explicit, let $\xi(x_0, w)$ be the infimal cost defined by

$$\xi(x_0, w) = \inf_u \int_0^\infty \left(\|z(\sigma)\|^2 - \gamma^2 \|w(\sigma)\|^2 \right) d\sigma \qquad (15.77)$$

where the infimum is taken over all u in $L^2([0, \infty), \mathcal{U})$. Then $d(x_0) = \sup_w \xi(x_0, w)$.

As in Remark 15.1.3, let $\Omega(\tau) = P(t_1 - \tau)$ where P is the solution to the Riccati differential equation in (15.8). Remark 15.1.3 shows that Ω satisfies the Riccati differential equation in (15.23). Recalling (15.24), we have

$$(\Omega(\tau)x_0, x_0) = \rho(x_0, 0, \tau) := \sup_w \inf_u \int_0^\tau \left(\|z(\sigma)\|^2 - \gamma^2 \|w(\sigma)^2\| \right) d\sigma. \quad (15.78)$$

Proceeding as in Remark 15.1.1, one can readily show that

$$\rho(x_0, 0, \tau) \le \sup_w \inf_u \int_0^\infty \left(\|z(\sigma)\|^2 - \gamma^2 \|w(\sigma)^2\| \right) d\sigma = d(x_0). \quad (15.79)$$

Hence,

$$(\Omega(\tau)x_0, x_0) \le d(x_0) \qquad (15.80)$$

for all τ. Recall that Ω is a *uniformly bounded solution* to the Riccati differential equation in (15.23) if Ω is a solution to (15.23) for all $\tau \geq 0$ and $\|\Omega(\tau)\| \leq m$ for some finite scalar m. If Ω is a uniformly bounded solution to this Riccati differential equation, then according to Lemma 13.6.3, the solution Ω converges to a limit Ω_∞ as τ approaches infinity. In addition, $Q = \Omega_\infty$ is the minimal positive solution to the following algebraic Riccati equation

$$A^*Q + QA + \gamma^{-2}QEE^*Q - QBB^*Q + C^*C = 0. \tag{15.81}$$

This sets the stage for the following result.

Lemma 15.6.1 *Let $d(x_0)$ be the optimal cost for the infinite horizon optimization problem in (15.76). Then the following statements are equivalent.*

(i) *The optimal cost $d(x_0)$ is finite for all x_0 in $\mathcal{X}$.*

(ii) *There exists a uniformly bounded solution Ω to the Riccati differential equation in (15.23).*

(iii) *There exists a positive solution Q to the algebraic Riccati equation in (15.81).*

In this case, the solution Ω converges to a limit Ω_∞, that is,

$$\Omega_\infty = \lim_{\tau \to \infty} \Omega(\tau). \tag{15.82}$$

and the optimal cost is given by

$$d(x_0) = (\Omega_\infty x_0, x_0). \tag{15.83}$$

In addition, the limit Ω_∞ is the minimal positive solution to the algebraic Riccati equation (15.81). Finally, if $\{C, A\}$ is observable, then Q is strictly positive.

PROOF. The proof is similar to the proof of Lemma 14.7.1. We first demonstrate that (i) implies (ii). So, suppose that the optimal cost $d(x_0)$ is finite for all x_0 in $\mathcal{X}$. Let Ω be the solution to the Riccati differential equation in (15.23) which is defined on some interval $[0, t_1)$. Lemma 13.6.2 implies that $\Omega(\tau)$ is self-adjoint. It now follows from (15.80) that

$$\|\Omega(\tau)\| \leq \operatorname{trace} \Omega(\tau) \leq \sum d(\varphi_j) < \infty$$

where $\{\varphi_j\}$ is any orthonormal basis for $\mathcal{X}$. Therefore, Ω can be extended to a uniformly bounded solution of (15.23) and Part (ii) holds.

The equivalence of Parts (ii) and (iii) follow from Lemma 13.6.3. This lemma also states that the solution Ω converges to a limit Ω_∞. Moreover, the limit Ω_∞ is the minimal positive solution to the algebraic Riccati equation (15.81).

We now demonstrate that (iii) implies (i). So, suppose that the algebraic Riccati equation in (15.81) admits a positive solution Q. Using the algebraic Riccati equation in (15.81), along with $\dot{x} = Ax + Ew + Bu$ and $z = Cx + Du$, we obtain

$$
\begin{aligned}
\frac{d}{dt}(Qx, x) &= 2\Re(Q\dot{x}, x) \\
&= 2\Re(QAx, x) + 2\Re(QEw, x) + 2\Re(QBu, x) \\
&= ((QA + A^*Q)x, x) + 2\Re(w, E^*Qx) + 2\Re(u, B^*Qx) \\
&= -\|Cx\|^2 - \gamma^{-2}\|E^*Qx\|^2 + \|B^*Qx\|^2 + 2\Re(w, E^*Qx) + 2\Re(u, B^*Qx) \\
&= -\|z\|^2 + \gamma^2\|w\|^2 - \|\gamma^{-1}E^*Qx - \gamma w\|^2 + \|u + B^*Qx\|^2 . \quad (15.84)
\end{aligned}
$$

Integrating from 0 to t and using $x(0) = x_0$ results in

$$
\begin{aligned}
\int_0^t \left(\|z(\sigma)\|^2 - \gamma^2\|w(\sigma)\|^2\right) d\sigma &= (Qx_0, x_0) - (Qx(t), x(t)) + \int_0^t \|u(\sigma) + B^*Qx(\sigma)\|^2 d\sigma \\
&\quad - \int_0^t \|\gamma^{-1}E^*Qx(\sigma) - \gamma w(\sigma)\|^2 d\sigma . \quad (15.85)
\end{aligned}
$$

Letting t approach infinity and using the fact that Q is positive, we obtain

$$
\int_0^\infty \left(\|z(\sigma)\|^2 - \gamma^2\|w(\sigma)\|^2\right) d\sigma \le (Qx_0, x_0) + \int_0^\infty \|u(\sigma) + B^*Qx(\sigma)\|^2 d\sigma . \quad (15.86)
$$

If we let $u = -B^*Qx$, then

$$
\int_0^\infty \left(\|z(\sigma)\|^2 - \gamma^2\|w(\sigma)\|^2\right) d\sigma \le (Qx_0, x_0) . \quad (15.87)
$$

Since w is in $L^2([0, \infty), \mathcal{W})$, it follows that z is in $L^2([0, \infty), \mathcal{Z})$. Recalling that $\|z\|^2 = \|Cx\|^2 + \|u\|^2$, we obtain that u is in $L^2([0, \infty), \mathcal{U})$. Inequality (15.87) now implies that $\xi(x_0, w) \le (Qx_0, x_0)$. Hence,

$$
d(x_0) \le (Qx_0, x_0)
$$

and Part (i) holds.

Recall that Ω_∞ is also a positive solution to the algebraic Riccati equation in (15.81). So, we obtain $d(x_0) \le (\Omega_\infty x_0, x_0)$. Combining this with $d(x_0) \ge (\Omega_\infty x_0, x_0)$, yields the equality $d(x_0) = (\Omega_\infty x_0, x_0)$. ∎

Remark 15.6.1 It follows from the previous lemma that, whenever the algebraic Riccati equation in (15.81) admits a positive solution, then is has an *minimal positive solution* Ω_∞, that is, Ω_∞ is a positive solution and $\Omega_\infty \le Q$ for any other positive solution Q. Moreover Ω_∞ is the limit of the corresponding Riccati differential equation and the optimal cost in (15.76) is finite and given by $d(x_0) = (\Omega_\infty x_0, x_0)$.

Remark 15.6.2 Suppose that Q is a positive solution of the algebraic Riccati equation in (15.81) and consider the feedback controller

$$
u = -B^*Qx . \quad (15.88)
$$

In the proof of the previous lemma, it is shown that, for any initial state x_0 in $\mathcal{X}$ and for any disturbance input in $L^2([0, \infty), \mathcal{W})$, we must have

$$\int_0^\infty (\|z(\sigma)\|^2 - \gamma^2 \|w(\sigma)\|^2) \, d\sigma \leq (Qx_0, x_0) \, .$$

In particular, if $Q = \Omega_\infty$, then

$$\int_0^\infty (\|z(\sigma)\|^2 - \gamma^2 \|w(\sigma)\|^2) \, d\sigma \leq d(x_0) \, .$$

The closed loop system corresponding to the above feedback controller is given by

$$\dot{x} = (A - BB^*Q)x + Ew \qquad \text{and} \qquad z = (C - DB^*Q)x \, . \tag{15.89}$$

We now claim that, if the pair $\{C, A\}$ is detectable and Q is a positive solution to the algebraic Riccati equation in (15.81), then $A - BB^*Q$ is stable. To see this, notice that by rearranging terms in (15.81), we obtain

$$(A - BB^*Q)^*Q + Q(A - BB^*Q) + QBB^*Q + \gamma^{-2}QEE^*Q + C^*C = 0 \, . \tag{15.90}$$

Let $A_c = A - BB^*Q$ and $\tilde{C} = [C \quad B^*Q \quad \gamma^{-1}E^*Q]^{tr}$. Then the pair $\{\tilde{C}, A_c\}$ satisfies the Lyapunov equation $A_c^*Q + QA_c + \tilde{C}^*\tilde{C} = 0$. To show that A_c is stable it is sufficient to verify that the pair $\{\tilde{C}, A_c\}$ is detectable; see Lemma 12.7.3. We now claim that the kernel of

$$J(\lambda) = \begin{bmatrix} A_c - \lambda I \\ \tilde{C} \end{bmatrix} \tag{15.91}$$

is zero for all complex numbers λ in the closed right half plane. Then by the PBH test $\{\tilde{C}, A_c\}$ is detectable; see Lemma 9.1.2. If f is any vector in the kernel of $J(\lambda)$, then $\tilde{C} = 0$. In particular, $B^*Qf = 0$ and $Cf = 0$. Thus, $(A - \lambda I)f = A_c f = 0$. Since $\{C, A\}$ is detectable and $\Re\lambda \geq 0$, the PBH test guarantees that the kernel of $[A - \lambda I \quad C]^{tr}$ is zero. In other words, f must be zero which proves our claim. Therefore, A_c is stable.

15.6.1   The stabilizing solution

We say that Q is a *stabilizing solution* to the algebraic Riccati equation in (15.81) if Q is a solution to (15.81) and $A + \gamma^{-2}EE^*Q - BB^*Q$ is stable. The algebraic Riccati equation in (15.81) can have many different solutions. However, if there exists a stabilizing solution Q, then Q is self-adjoint and is the only stabilizing solution; see Chapter 13. One can use the Hamiltonian techniques in Chapter 13 to determine if there exists a stabilizing solution. Suppose that the Hamiltonian matrix H in (15.73) has no eigenvalues on the imaginary axis. Then, according to Lemma 13.1.3, there exists operators X and Y on $\mathcal{X}$ satisfying

$$H\begin{bmatrix} X \\ Y \end{bmatrix} = \begin{bmatrix} A & \gamma^{-2}EE^* - BB^* \\ -C^*C & -A^* \end{bmatrix} \begin{bmatrix} X \\ Y \end{bmatrix} = \begin{bmatrix} X \\ Y \end{bmatrix} \Lambda \tag{15.92}$$

where Λ is a stable operator on $\mathcal{X}$, and the operator $[X^*, Y^*]^*$ from $\mathcal{X}$ into $\mathcal{X} \oplus \mathcal{X}$ is one to one. Then the algebraic Riccati equation in (15.81) admits a stabilizing solution if and only if X is invertible. In this case, the unique stabilizing solution is given by $Q = YX^{-1}$; see Theorem 13.1.5. In our control problem, we are interested in positive stabilizing solutions. So, there exists a positive stabilizing solution to the algebraic Riccati equation in (15.81) if and only if the Hamiltonian matrix H has no eigenvalues on the imaginary axis and X is invertible with YX^{-1} positive. This readily yields the following result.

Theorem 15.6.2 *Let H be the Hamiltonian matrix in (15.73) determined by the operators A, B, C and E acting between the appropriate spaces. Then the algebraic Riccati equation in (15.81) admits a positive stabilizing solution if and only if the following three conditions hold.*

(i) *The Hamiltonian matrix H has no eigenvalues on the imaginary axis.*

(ii) *If $[X^*, Y^*]^*$ from $\mathcal{X}$ into $\mathcal{X} \oplus \mathcal{X}$ is any one to one operator satisfying (15.92) where Λ on $\mathcal{X}$ is stable, then X is invertible.*

(iii) *The operator YX^{-1} is positive.*

In this case, YX^{-1} is the unique positive stabilizing solution to the algebraic Riccati equation.

Example 15.6.1 If the Hamiltonian matrix has no eigenvalues on the imaginary axis, then it does not necessarily follow that the stabilizing solution is positive. For example, consider the system $\{1, 1, 1, 1, 1\}$ in (15.1). Then the eigenvalues for the Hamiltonian H corresponding to this system are given by $\pm\sqrt{2 - 1/\gamma^2}$. In this case, H has no eigenvalues on the imaginary axis if and only if $\gamma > 1/\sqrt{2}$. Now assume that $\gamma > 1/\sqrt{2}$. Then $\lambda = -\sqrt{2 - 1/\gamma^2}$ is the stable eigenvalue with eigenvector $[\lambda + 1, -1]^{tr}$. The stabilizing solution is given by $Q = -1/(\lambda + 1)$ when $\gamma \neq 1$. Hence, the algebraic Riccati equation admits a positive stabilizing solution if and only if $\gamma > 1$. In other words, if $1/\sqrt{2} < \gamma < 1$, then the stabilizing solution is not positive. If $\gamma = 1$, then there is no stabilizing solution.

Theorem 15.6.3 *Let $d(x_0)$ be the optimal cost for the infinite horizon optimization problem in (15.76) and suppose that $\{A, B\}$ is stabilizable and $\{C, A\}$ is detectable. Then the following statements are equivalent.*

(i) *There exists a positive stabilizing solution Q to the algebraic Riccati equation in (15.81).*

(ii) *The Riccati differential equation in (15.23) admits a uniformly bounded solution Ω and $A + \gamma^{-2}EE^*\Omega_\infty - BB^*\Omega_\infty$ is stable where $\Omega_\infty = \lim_{\tau \to \infty} \Omega(\tau)$.*

(iii) *For every x_0 in $\mathcal{X}$, there exists an optimal pair $\{\hat{w}, \hat{u}\}$ which attains the optimal cost $d(x_0)$.*

(iv) *For the Hamiltonian matrix H in (15.73), conditions (i),(ii) and (iii) in Theorem 15.6.2 hold.*

In this case, the unique stabilizing solution to the algebraic Riccati equation in (15.81) is given by $Q = \Omega_\infty = YX^{-1}$.

PROOF. Assume that Q is a positive stabilizing solution to the algebraic Riccati equation in (15.81). According to Lemma 15.6.1, there exists a uniformly bounded solution Ω to the Riccati differential equation in (15.23). Moreover, $\Omega_\infty = \lim_{\tau \to \infty} \Omega(\tau)$ exists, $d(x_0) = (\Omega_\infty x_0, x_0)$ and $\Omega_\infty \le Q$. Hence, $d(x_0) \le (Qx_0, x_0)$. We claim that the optimal cost $d(x_0) = (Qx_0, x_0)$, that is, $\Omega_\infty = Q$. Moreover, an optimal pair $\{\hat{w}, \hat{u}\}$ which attains this cost is given by $\hat{w} = \gamma^{-2}E^*Q\hat{x}$ and $\hat{u} = -B^*Q\hat{x}$ where the optimal state $\hat{x}$ and output $\hat{z}$ are determined by

$$\begin{aligned} \dot{\hat{x}} &= (A + \gamma^{-2}EE^*Q - BB^*Q)\hat{x} \qquad (\hat{x}(0) = x_0) \\ \hat{z} &= (C - DB^*Q)\hat{x}. \end{aligned} \qquad (15.93)$$

Because Q is a stabilizing solution, $\hat{x}$ is in $L^2([0, \infty), \mathcal{X})$ and $\lim_{t \to \infty} \hat{x}(t) = 0$. Hence, $\hat{w}$ and $\hat{u}$ are in the appropriate L^2 spaces. By letting t approach infinity and setting $u = \hat{u}$ and $w = \hat{w}$ in (15.85), it follows that $(Qx_0, x_0) = \|\hat{z}\|^2 - \gamma^2\|\hat{w}\|^2$.

To show that $(Qx_0, x_0) = d(x_0)$ it remains to show that $\hat{u}$ is the optimal input which attains the cost $\xi(x_0, \hat{w})$. The proof is similar to the proof of Theorem 15.1.1. To achieve our objective, notice that for the disturbance $\hat{w} = \gamma^{-2}E^*Q\hat{x}$, we obtain

$$\dot{x} = Ax + E\hat{w} + Bu.$$

Now let us introduce the new state $\tilde{x} = x - \hat{x}$ and set $\tilde{u} = u + B^*Qx$. Then subtracting the differential equation in (15.93) from the previous state equation, yields

$$\dot{\tilde{x}} = (A - BB^*Q)\tilde{x} + B\tilde{u} \qquad \text{with} \qquad \tilde{x}(0) = 0. \qquad (15.94)$$

Using $\hat{w} = \gamma^{-2}E^*Q\hat{x}$ in (15.85) with $\tilde{x} = x - \hat{x}$, gives

$$\begin{aligned} \int_0^t \left(\|z(\sigma)\|^2 - \gamma^2\|\hat{w}(\sigma)\|^2 \right) d\sigma &= (Qx_0, x_0) - (Qx(t), x(t)) - \int_0^t \|\gamma^{-1}E^*Q\tilde{x}(\sigma)\|^2 d\sigma \\ &+ \int_0^t \|\tilde{u}(\sigma)\|^2 d\sigma. \end{aligned} \qquad (15.95)$$

At this point there are several technical issues to consider. Obviously, $\xi(x_0, \hat{w}) \le d(x_0)$ is finite. Let $\mathcal{F}$ be the subset of $L^2([0, \infty), \mathcal{U})$ consisting of the set of all inputs u such that Cx is in $L^2([0, \infty), \mathcal{Z})$. Then

$$\xi(x_0, \hat{w}) = \inf \left\{ \|Cx\|^2 + \|u\|^2 - \gamma^2\|\hat{w}\|^2 : u \in \mathcal{F} \right\}.$$

If Cx is in $L^2([0, \infty), \mathcal{Z})$, then as a consequence of the detectability of $\{C, A\}$, the state trajectory x is in $L^2([0, \infty), \mathcal{X})$; see Lemma 12.8.2. So, if u is in $\mathcal{F}$, then $\tilde{u} = u + B^*Qx$ is in $L^2([0, \infty), \mathcal{U})$. On the other hand, if $\tilde{u}$ is in $L^2([0, \infty), \mathcal{U})$, then the stability of $A - BB^*Q$ and equation (15.94) implies that $\tilde{x}$ is in $L^2([0, \infty), \mathcal{X})$. Obviously, $\hat{x}$ is in $L^2([0, \infty), \mathcal{X})$. Thus, $x = \hat{x} + \tilde{x}$ is in $L^2([0, \infty), \mathcal{X})$. This readily implies that $u = \tilde{u} - B^*Qx$ is in $\mathcal{F}$. In other words, $\mathcal{F}$ equals the set of all functions of the form $\tilde{u} - B^*Qx$ where $\tilde{u}$ is in $L^2([0, \infty), \mathcal{U})$. So, if $\tilde{u}$ is in $L^2([0, \infty), \mathcal{U})$, then x is in $L^2([0, \infty), \mathcal{X})$. Lemma 12.8.3 shows that $x(t)$ approaches zero as t tends to infinity. Using this in (15.95) yields

$$\begin{aligned} \int_0^\infty \left(\|z(\sigma)\|^2 - \gamma^2\|\hat{w}(\sigma)\|^2 \right) d\sigma &= (Qx_0, x_0) - \int_0^\infty \|\gamma^{-1}E^*Q\tilde{x}(\sigma)\|^2 d\sigma \\ &+ \int_0^\infty \|\tilde{u}(\sigma)\|^2 d\sigma. \end{aligned} \qquad (15.96)$$

Hence, the cost $\xi(x_0, \hat{w})$ in (15.77) is given by

$$\xi(x_0, \hat{w}) = (Qx_0, x_0) - \sup_{\tilde{u} \in L^2} \int_0^\infty \left(\|\gamma^{-1} E^* Q \tilde{x}(\sigma)\|^2 - \|\tilde{u}(\sigma)\|^2 \right) d\sigma . \tag{15.97}$$

The algebraic Riccati equation in (15.81) can be rewritten as (15.90). Notice that this is precisely the algebraic Riccati equation one obtains by replacing C with $[C \ \ \gamma^{-1} E^* Q]^{tr}$ and E with B in Theorem 14.7.4. So, by employing Theorem 14.7.4 and (15.94), it follows that

$$\sup_{\tilde{u} \in L^2} \int_0^\infty \left(\|C\tilde{x}(\sigma)\|^2 + \|\gamma^{-1} E^* Q \tilde{x}(\sigma)\|^2 - \|\tilde{u}(\sigma)\|^2 \right) d\sigma = 0$$

and this supremum is uniquely attained with $\tilde{u} = 0$. Noting that, for any $\tilde{u}$,

$$\int_0^\infty \left(\|\gamma^{-1} E^* Q \tilde{x}(\sigma)\|^2 - \|\tilde{u}(\sigma)\|^2 \right) d\sigma \le \int_0^\infty \left(\|C\tilde{x}(\sigma)\|^2 + \|\gamma^{-1} E^* Q \tilde{x}(\sigma)\|^2 - \|\tilde{u}(\sigma)\|^2 \right) d\sigma$$

it follows that the supremum in (15.97) is zero and this is uniquely achieved with $\tilde{u}$ equal to zero. In this case, (15.94) implies that $\tilde{x} = 0$, that is, $\hat{x} = x$. Thus, $u = -B^* Q x = -B^* Q \hat{x}$. This readily implies that $\xi(x_0, \hat{w}) = (Qx_0, x_0)$. Therefore, $(Qx_0, x_0) = d(x_0)$. Since $d(x_0) = (\Omega_\infty x_0, x_0)$, we must have $Q = \Omega_\infty$. In other words, Parts (*ii*) and (*iii*) hold.

Obviously, Part (*ii*) implies Part (*i*). We now demonstrate that Part (*iii*) implies Part (*ii*). So assume that $\{w_o, u_o\}$ is an optimal pair which attains the optimal cost $d(x_0)$. In particular, $d(x_0)$ is finite and there exists a uniformly bounded solution Ω to the Riccati differential equation in (15.23). Furthermore, $\Omega_\infty = \lim_{\tau \to \infty} \Omega(\tau)$ is a positive solution to the algebraic Riccati equation in (15.81) and $d(x_0) = (\Omega_\infty x_0, x_0)$. Because the pair $\{C, A\}$ is detectable, $A - BB^* \Omega_\infty$ is stable. If we set $u = -B^* \Omega_\infty x$, then the state trajectory x corresponding to the pair $\{w_o, u\}$ is given by $\dot{x} = (A - BB^* \Omega_\infty)x + E w_o$. The output $z = (C - DB^* \Omega_\infty)x$. Since w_o is in $L^2([0, \infty), \mathcal{W})$, the state x is in $L^2([0, \infty), \mathcal{X})$, and thus, u is in $L^2([0, \infty), \mathcal{U})$. Moreover, $x(t)$ converges to zero as t tends to infinity; see Lemma 12.8.3. So, by letting t approach infinity in (15.85) with $Q = \Omega_\infty$, we obtain

$$
\begin{aligned}
(\Omega_\infty x_0, x_0) &= \xi(x_0, w_o) \le \int_0^\infty \left(\|z(\sigma)\|^2 - \gamma^2 \|w_o(\sigma)\|^2 \right) d\sigma \\
&= (\Omega_\infty x_0, x_0) - \int_0^\infty \|\gamma^{-1} E^* \Omega_\infty x(\sigma) - \gamma w_o(\sigma)\|^2 \, d\sigma .
\end{aligned} \tag{15.98}
$$

This readily implies that $w_o = \gamma^{-2} E^* \Omega_\infty x$ and the infimum $\xi(x_0, w_o)$ is achieved with u. Hence, $\{w_o, u\}$ is an optimal pair which attains the optimal cost $d(x_0)$. Combining this with $u = -B^* \Omega_\infty x$, we see that the optimal state x is given by $x = \hat{x}$ where $\hat{x}$ is the solution to the following differential equation

$$\dot{\hat{x}} = (A + \gamma^{-2} E E^* \Omega_\infty - BB^* \Omega_\infty)\hat{x} \qquad (\hat{x}(0) = x_0) .$$

Since $\hat{x}(t)$ converges to zero, $A + \gamma^{-2} E E^* \Omega_\infty - BB^* \Omega_\infty$ is stable. In other words, Ω_∞ is the positive stabilizing solution to the algebraic Riccati equation in (15.81), that is, Part (*i*) holds.

Notice that $w_o = \hat{w} = \gamma^{-2} E^* Q \hat{x}$ where $\hat{w}$ is our previous optimal disturbance defined in (15.93) with $\Omega_\infty = Q$. Since $d(x_0) = \xi(x_0, \hat{w})$ and we have shown that $\xi(x_0, \hat{w})$ is uniquely attained with $\tilde{u} = 0$, it follows that the optimal cost $d(x_0)$ is uniquely achieved with $u_o = \hat{u} = -B^* \Omega_\infty \hat{x}$ and $w_o = \hat{w} = \gamma^{-2} E^* Q \hat{x}$. ∎

The previous proof readily yields the following result.

Theorem 15.6.4 *Consider the system $\{A, B, C, D, E\}$ where $\{A, B\}$ is stabilizable and $\{C, A\}$ is detectable. Assume that the algebraic Riccati equation in (15.81) has a positive stabilizing solution Q. Then the optimal cost $d(x_0) = (Qx_0, x_0)$. Moreover, this optimal cost is uniquely achieved by the disturbance/control input pair $\hat{w} = \gamma^{-2} E^* Q \hat{x}$ and $\hat{u} = -B^* Q \hat{x}$ where the optimal state trajectory $\hat{x}$ is determined by*

$$\dot{\hat{x}} = (A + \gamma^{-2} E E^* Q - B B^* Q) \hat{x} \qquad with \qquad \hat{x}(0) = x_0. \tag{15.99}$$

15.6.2 The scalar valued case

Consider the system $\{A$ on $\mathcal{X}, B, C, D, E\}$ where $\{A, B\}$ is stabilizable, $\{C, A\}$ is detectable and n is the dimension of $\mathcal{X}$. Let $\mathbf{F}$ be the transfer function for $\{A, B, C, 0\}$ and $\mathbf{G}$ be the transfer function for $\{A, E, C, 0\}$. Let d be the characteristic polynomial for A and H the Hamiltonian matrix in (15.73). By replacing $\mathbf{T}$ by $\gamma^{-1} \mathbf{G}$ in Theorem 13.5.1, we see that the characteristic polynomial for H is given by

$$\det[sI - H] = (-1)^n d d^\sharp \det[I - \gamma^{-2} \mathbf{G} \mathbf{G}^\sharp + \mathbf{F} \mathbf{F}^\sharp]. \tag{15.100}$$

Now assume that $\{A, B, C, D, E\}$ is a single input single output system consisting of matrices with real entries. Moreover, assume that $\mathbf{G} = p_1/d$ and $\mathbf{F} = p_2/d$ where p_1 and p_2 are polynomials. Then (15.100) reduces to

$$(-1)^n \det[sI - H] = d^\sharp d + p_2^\sharp p_2 - \gamma^{-2} p_1^\sharp p_1. \tag{15.101}$$

So, the eigenvalues of H are precisely the zeros of $d^\sharp d + p_2^\sharp p_2 - \gamma^{-2} p_1^\sharp p_1$. Notice that one can apply the root locus techniques in Section 14.8 to the polynomial in (15.101). (To do this simply replace $d^\sharp d$ with $d^\sharp d + p_2^\sharp p_2$ and p with p_1 in Section 14.8.) In other words, we can graph the zeros of $d^\sharp d + p_2^\sharp p_2 - \gamma^{-2} p_1^\sharp p_1$ as γ varies from infinity to zero. Then the eigenvalues of H are contained in this root locus. Recall that the root locus of $d^\sharp d + p_2^\sharp p_2 - \gamma^{-2} p_1^\sharp p_1$ always has an asymptote on the imaginary axis. Let γ_o be the largest value of γ such that $d^\sharp d + p_2^\sharp p_2 - \gamma^{-2} p_1^\sharp p_1$ has roots on the imaginary axis. Then the root locus has a branch on the imaginary axis for $0 < \gamma \leq \gamma_o$, or equivalently, the Hamiltonian matrix H has an eigenvalue on the imaginary axis for all $0 < \gamma \leq \gamma_o$. Hence, the algebraic Riccati equation in (15.81) does not have a stabilizing solution for all $0 < \gamma \leq \gamma_o$. In other words, if the algebraic Riccati equation admits a stabilizing solution, then $\gamma > \gamma_o$.

Assume that Q is a stabilizing solution for the algebraic Riccati equation for some $\gamma > 0$. Obviously, $\gamma > \gamma_o$. Let Δ be the characteristic polynomial for $A + \gamma^{-2} E E^* Q - B B^* Q$. By replacing $\mathbf{T}$ by $\gamma^{-1} \mathbf{G}$ in Corollary 13.5.2, we obtain

$$\Delta^\sharp \Delta = d^\sharp d + p_2^\sharp p_2 - \gamma^{-2} p_1^\sharp p_1. \tag{15.102}$$

Since Q is a stabilizing solution, all the roots of Δ are in the open left half complex plane. Therefore, all the roots of $\Delta^\sharp$ live in the open right half plane. In particular, Δ and $\Delta^\sharp$ have no common roots and their product $\Delta^\sharp\Delta$ has no roots on the imaginary axis. According to (15.102) the roots of Δ are the left half plane roots of $d^\sharp d + p_2^\sharp p_2 - \gamma^{-2} p_1^\sharp p_1$. In other words, the roots of Δ are contained in the left half plane root locus of $d^\sharp d + p_2^\sharp p_2 - \gamma^{-2} p_1^\sharp p_1$. Finally, it is noted that if $\gamma > \gamma_o$, then it does necessarily follow that there exists a stabilizing solution or that the stabilizing solution is positive; see Example 15.6.1. However, if there exists a positive solution to the algebraic Riccati equation for some $\gamma = \gamma_1$, then the algebraic Riccati has a positive stabilizing solution for all $\gamma > \gamma_1$; see Section 15.8.

Exercise 35 Assume that the Hamiltonian H in (15.73) has eigenvalues on the imaginary axis for some $\gamma = \gamma_o > 0$. Then show that H has eigenvalues on the imaginary axis for all $0 < \gamma \leq \gamma_o$.

Exercise 36 Consider the system $\{A, B, C, D, E\}$ where $\{A, B\}$ is stabilizable and $\{C, A\}$ is detectable. Assume that Q is a positive solution to the algebraic Riccati equation in (15.81). Assume that $z = \phi(x_0, u, w)$ is in $L^2([0, \infty), \mathcal{Z})$ where u is in $L^2([0, \infty), \mathcal{U})$ and w is in $L^2([0, \infty), \mathcal{W})$. Then show that the corresponding state x is in $L^2([0, \infty), \mathcal{X})$ and $\lim_{t \to \infty} x(t) = 0$. Moreover,

$$
\int_0^\infty \left(\|z(\sigma)\|^2 - \gamma^2 \|w(\sigma)\|^2 \right) d\sigma = (Q x_0, x_0) + \int_0^\infty \|u(\sigma) + B^* Q x(\sigma)\|^2 d\sigma
$$
$$
- \int_0^\infty \|\gamma^{-1} E^* Q x(\sigma) - \gamma w(\sigma)\|^2 d\sigma . \tag{15.103}
$$

15.7 The central controller

In this section we present the central solution for the infinite horizon problem. As before, consider the system $\{A, B, C, D, E\}$ where $\{A, B\}$ is stabilizable and $\{C, A\}$ is detectable. Assume that the algebraic Riccati equation in (15.81) has a positive stabilizing solution Q for some specified $\gamma > 0$. Then the feedback controller $u = -B^* Q x$ is called the *central infinite horizon controller* corresponding to the weight γ. The closed loop system corresponding to this feedback controller is described by

$$
\dot{x} = (A - BB^* Q)x + Ew
$$
$$
z = (C - DB^* Q)x . \tag{15.104}
$$

Because the pair $\{C, A\}$ is detectable, $A - BB^* Q$ is stable; see Remark 15.6.2. The transfer function

$$
\mathbf{G}_\gamma(s) = (C - DB^* Q)(sI - A + BB^* Q)^{-1} E \tag{15.105}
$$

is referred to as the *central* transfer function for γ. Obviously, $\mathbf{G}_\gamma$ is stable. Now let $\hat{W}_\gamma$ be the operator from $L^2([0, \infty), \mathcal{W})$ into $L^2([0, \infty), \mathcal{Z})$ defined by

$$
(\hat{W}_\gamma w)(t) = \int_0^t (C - DB^* Q) e^{(A - BB^* Q)(t - \tau)} E w(\tau) \, d\tau \qquad (w \in L^2([0, \infty), \mathcal{W})) . \tag{15.106}
$$

Obviously, $\hat{W}_\gamma$ is the input output operator for the system in (15.104). Moreover, $\mathcal{L}(\hat{W}_\gamma w) = G_\gamma \mathcal{L}(w)$ where $\mathcal{L}$ denotes the Laplace transform. Finally, $\|\hat{W}_\gamma\| = \|G_\gamma\|_\infty$.

The following transfer function plays a fundamental role in studying the central solution

$$\Theta_\gamma(s) := \gamma I - \gamma^{-1} E^* Q (sI - A + BB^*Q)^{-1} E. \tag{15.107}$$

According to Proposition 6.4.1, the inverse of Θ_γ is given by

$$\Theta_\gamma(s)^{-1} = \gamma^{-1} I + \gamma^{-3} E^* Q (sI - A - \gamma^{-2} EE^*Q + BB^*Q)^{-1} E. \tag{15.108}$$

Since Q is a positive stabilizing solution both Θ_γ and its inverse are stable transfer functions, that is, Θ_γ is an invertible outer function. The following result uses Θ_γ to show that the H^∞ norm of G_γ is strictly bounded by γ.

Proposition 15.7.1 *Consider the system* $\{A, B, C, D, E\}$ *where* $\{A, B\}$ *is stabilizable and* $\{C, A\}$ *is detectable. Assume that the algebraic Riccati equation in (15.81) has a positive stabilizing solution* Q *for some specified* $\gamma > 0$. *Then*

$$\gamma^2 I - G_\gamma^\sharp G_\gamma = \Theta_\gamma^\sharp \Theta_\gamma. \tag{15.109}$$

In particular, $\|G_\gamma\|_\infty < \gamma$. *Finally, for any initial state* $x(0) = x_0$, *the closed loop system in (15.104) satisfies* $\|z\|^2 \leq \gamma^2 \|w\|^2 + (Qx_0, x_0)$.

PROOF. Using $D^*D = I$ and $D^*C = 0$, we obtain

$$(C - DB^*Q)^*(C - DB^*Q) = QBB^*Q + C^*C.$$

Substituting this into the algebraic Riccati equation in (15.90), it follows that this Riccati equation can be written as

$$(A - BB^*Q^*)^*Q + Q(A - BB^*Q) + \gamma^{-2} QEE^*Q + (C - DB^*Q)^*(C - DB^*Q) = 0. \tag{15.110}$$

This is precisely the algebraic Riccati equation one obtains by replacing C and A with respectively $C - DB^*Q$ and $A - BB^*Q$ in Proposition 14.7.5. So, by employing Proposition 14.7.5 with $G = G_\gamma$, we obtain (15.109). Since Θ_γ is an invertible outer function, it follows that $\|G_\gamma\|_\infty < \gamma$. If $u = -B^*Qx$, then equation (15.85) shows that $\|z\|^2 - \gamma^2 \|w\|^2 \leq (Qx_0, x_0)$ for any initial state x_0. ∎

Exercise 37 Assume that the algebraic Riccati equation in (15.81) has an invertible self-adjoint solution Q, and set $R = Q^{-1}$. Show that R satisfies the following algebraic Riccati equation

$$RA^* + AR + RC^*CR + \gamma^{-2} EE^* - BB^* = 0. \tag{15.111}$$

Let F be the transfer function for $\{A, B, C, 0\}$ and G be the transfer function for $\{A, E, C, 0\}$. Let Λ be the transfer function for $\{A, RC^*, -\gamma C, \gamma I\}$. Then show that the following factorization holds

$$\gamma^2 I - GG^\sharp + \gamma^2 FF^\sharp = \Lambda\Lambda^\sharp. \tag{15.112}$$

15.8   An operator perspective

In this section we provide an operator perspective on the infinite horizon optimization problem in (15.76). As before, let $\{A, B, C, D, E\}$ be the system in (15.1) where the pair $\{C, A\}$ is detectable and $\{A, B\}$ is stabilizable. Because the pair $\{A, B\}$ is stabilizable, there exists an operator K from $\mathcal{X}$ into $\mathcal{U}$ such that $A - BK$ is stable. Now let v be the input defined by $v = u + Kx$. Then the system in (15.1) can be rewritten as

$$\dot{x} = (A - BK)x + Ew + Bv \qquad \text{and} \qquad z = Cx + D(v - Kx). \qquad (15.113)$$

Using this notation we see that for all w in $L^2([0, \infty), \mathcal{W})$

$$J(x_0, w, u) = \|z\|^2 - \gamma^2 \|w\|^2 = \|Cx\|^2 + \|v - Kx\|^2 - \gamma^2 \|w\|^2.$$

We claim that $J(x_0, w, u)$ is finite if and only if $u = v - Kx$ where v is in $L^2([0, \infty), \mathcal{U})$. In this case, x is in $L^2([0, \infty), \mathcal{X})$. If v is in $L^2([0, \infty), \mathcal{U})$, then the stability of $A - BK$ along with (15.113) shows that the state x is in $L^2([0, \infty), \mathcal{X})$. Hence, $u = v - Kx$ is in $L^2([0, \infty), \mathcal{U})$, and thus, $J(x_0, w, u)$ is finite. On the other hand, if $J(x_0, w, u)$ is finite for some specified u in $L^2([0, \infty), \mathcal{U})$, then Cx must be in $L^2([0, \infty), \mathcal{Z})$. Because the pair $\{C, A\}$ is detectable, Lemma 12.8.2 along the state space representation in (15.1), shows that the state x is in $L^2([0, \infty), \mathcal{X})$. So, $v = u + Kx$ is in $L^2([0, \infty), \mathcal{U})$. Therefore, $u = v + Kx$ where v is in $L^2([0, \infty), \mathcal{U})$, which proves our claim.

Clearly, the cost $\xi(x_0, w)$ is obtained by taking the infimum over the set of all u such that $J(x_0, w, u)$ is finite. In other words,

$$
\begin{aligned}
\xi(x_0, w) &= \inf\left\{ \|Cx\|^2 + \|v - Kx\|^2 - \gamma^2 \|w\|^2 : v \in L^2([0, \infty), \mathcal{U}) \right\} \\
&= \inf\left\{ \|Cx + D(v - Kx)\|^2 - \gamma^2 \|w\|^2 : v \in L^2([0, \infty), \mathcal{U}) \right\}. \quad (15.114)
\end{aligned}
$$

To turn this into a least squares optimization problem, let L be the operator on $L^2([0, \infty), \mathcal{X})$ defined by

$$(Lf)(t) = \int_0^t e^{(A-BK)(t-\tau)} f(\tau) \, d\tau \qquad (f \in L^2([0, \infty), \mathcal{X})). \qquad (15.115)$$

Now let Φ_0 be the operator from $\mathcal{X}$ into $L^2([0, \infty), \mathcal{X})$ defined by $\Phi_0 x_0 = e^{(A-BK)t} x_0$ where x_0 is in $\mathcal{X}$. Obviously, the state x in (15.113) is given by $x = \Phi_0 x_0 + LBv + LEw$. Moreover, $u = v - Kx = (I - KLB)v - KLEw - K\Phi_0 x_0$. Now let Γ be the operator defined by

$$\Gamma = \begin{bmatrix} I - KLB \\ -CLB \end{bmatrix} : L^2([0, \infty), \mathcal{U}) \to \begin{bmatrix} L^2([0, \infty), \mathcal{U}) \\ L^2([0, \infty), \mathcal{Z}) \end{bmatrix}. \qquad (15.116)$$

Let Φ be the operator from $\mathcal{X}$ into $L^2([0, \infty), \mathcal{U}) \oplus L^2([0, \infty), \mathcal{Z})$ defined by $\Phi x_0 = K\Phi_0 x_0 \oplus C\Phi_0 x_0$. Finally, let W be the operator defined by

$$W = \begin{bmatrix} KLE \\ CLE \end{bmatrix} : L^2([0, \infty), \mathcal{W}) \to \begin{bmatrix} L^2([0, \infty), \mathcal{U}) \\ L^2([0, \infty), \mathcal{Z}) \end{bmatrix}. \qquad (15.117)$$

Because $A - BK$ is stable, the operators Γ, Φ and W are all bounded linear operators. Notice that $Cx = C\Phi_0 x_0 + CLEw + CLBv$. Using these operators we see that the optimization problem in (15.114) is equivalent to the following optimization problem

$$\xi(x_0, w) = \inf\left\{ \|\Phi x_0 + Ww - \Gamma v\|^2 - \gamma^2 \|w\|^2 : v \in L^2([0, \infty), \mathcal{U}) \right\}. \qquad (15.118)$$

We claim that the operator Γ is bounded below, that is, $\|\Gamma v\| \geq \epsilon \|v\|$ for all v in $L^2([0, \infty), \mathcal{U})$ and some scalar $\epsilon > 0$. Once this is established, then the optimization problem in (15.114) reduces to a simple least squares optimization problem. According to Theorem 16.2.4 in the Appendix, the cost $\xi(x_0, w)$ is uniquely obtained by the control input

$$\hat{v} = (\Gamma^*\Gamma)^{-1}\Gamma^*(\Phi x_0 + Ww). \tag{15.119}$$

To show that Γ is bounded below, notice that if $\Gamma v = g \oplus y$, then $g \oplus y$ is the output for the following state space system

$$\begin{aligned} \dot{x} &= (A - BK)x + Bv \qquad (x(0) = 0) \\ g &= -Kx + v \quad \text{and} \quad y = -Cx. \end{aligned} \tag{15.120}$$

Now assume that $\{v_n\}$ is a sequence of unit vectors such that $g_n \oplus y_n = \Gamma v_n$ approaches zero as n tends to infinity. Let x_n be the state corresponding to v_n, that is, $\dot{x}_n = (A - BK)x_n + Bv_n$. Then $g_n = -Kx_n + v_n$ and $y_n = -Cx_n$ both converge to zero as n tends to infinity. This readily implies that $\dot{x}_n = Ax_n + Bg_n$. Because the pair $\{C, A\}$ is detectable, there exists an operator K_o from $\mathcal{Z}$ into $\mathcal{X}$ such that $A + K_oC$ is stable. Thus,

$$\dot{x}_n = (A + K_oC)x_n - K_oCx_n + Bg_n = (A + K_oC)x_n + K_oy_n + Bg_n.$$

Because $A + K_oC$ is stable and $g_n \oplus y_n$ converges to zero, it follows that x_n converges to zero in the $L^2([0, \infty), \mathcal{X})$ topology. Thus, $v_n = g_n + Kx_n$ also converges to zero. This contradicts the fact that v_n is a unit vector for all n. Therefore, Γ is bounded below.

Because Γ is bounded below the range of Γ is closed. Hence, the orthogonal projection onto the range of Γ is given by $\Gamma(\Gamma^*\Gamma)^{-1}\Gamma^*$. So, if $\mathcal{H} = (\text{ran }\Gamma)^\perp$, then the orthogonal projection onto $\mathcal{H}$ is computed by $P_\mathcal{H} = I - \Gamma(\Gamma^*\Gamma)^{-1}\Gamma^*$. This readily implies that the optimal cost is given by

$$\xi(x_0, w) = \|P_\mathcal{H}\Phi x_0 + P_\mathcal{H}Ww\|^2 - \gamma^2\|w\|^2.$$

Therefore, the optimal cost $d(x_0)$ in the infinite horizon optimization problem (15.76) can be expressed as

$$d(x_0) = \sup\left\{\|P_\mathcal{H}\Phi x_0 + P_\mathcal{H}Ww\|^2 - \gamma^2\|w\|^2 : w \in L^2([0, \infty), \mathcal{W})\right\}.$$

This is precisely the optimization problem in (15.39) with $h = P_\mathcal{H}\Phi x_0$ and $\Upsilon = P_\mathcal{H}W$. So, if $d(x_0)$ is finite, then the norm of $P_\mathcal{H}W$ is bounded by γ. On the other hand, if the norm of $P_\mathcal{H}W$ is strictly bounded by γ, then $d(x_0)$ is finite and there exists a unique optimal disturbance $\hat{w}$ which attains the optimal cost $d(x_0)$. In fact, $\hat{w} = (\gamma^2 I - W^*P_\mathcal{H}W)^{-1}W^*P_\mathcal{H}\Phi x_0$ and

$$d(x_0) = \gamma^2((\gamma^2 I - P_\mathcal{H}WW^*|\mathcal{H})^{-1}P_\mathcal{H}\Phi x_0, P_\mathcal{H}\Phi x_0); \tag{15.121}$$

see Lemma 15.2.2. Because there exists a unique optimal $\hat{v}$ which attains the cost $\xi(x_0, \hat{w})$, the pair $\{\hat{w}, \hat{v}\}$ uniquely attains the optimal cost $d(x_0)$. In particular, $\|P_\mathcal{H}W\|$ is the infimum over the set of all γ such that $d(x_0)$ is finite. This also shows that the norm of $P_\mathcal{H}W$ is independent of the choice of the stabilizing feedback gain K.

Recall that $d(x_0)$ is finite if and only if the algebraic Riccati equation in (15.81) admits
a positive solution; see Lemma 15.6.1. Therefore, $\|P_{\mathcal{H}}W\|$ equals the infimum over the set
of all γ such that the algebraic Riccati equation admits a positive solution. If the norm
of $P_{\mathcal{H}}W$ is strictly bounded by γ, then the optimal cost $d(x_0)$ is uniquely obtained by an
optimal pair $\{\hat{w}, \hat{u}\}$, and thus, there exists a positive stabilizing solution Q to the algebraic
Riccati equation in (15.81); see Theorem 15.6.3. In other words, if the algebraic Riccati
equation in (15.81) admits a positive solution for some specified $\gamma = \gamma_1$, then this algebraic
Riccati equation admits a positive stabilizing solution for all $\gamma > \gamma_1$. Hence, $\|P_{\mathcal{H}}W\|$ equals
the infimum over the set of all γ such that the algebraic Riccati equation admits a positive
stabilizing solution. On the other hand, if the algebraic Riccati equation does not admit a
positive stabilizing solution for some $\gamma = \gamma_1$, then there is no positive stabilizing solution for
all $0 \leq \gamma \leq \gamma_1$. This proves part of the following result.

Proposition 15.8.1 *Consider the system* $\{A, B, C, D, E\}$ *where* $\{A, B\}$ *is stabilizable and*
$\{C, A\}$ *is detectable. Let K be an operator from $\mathcal{X}$ into $\mathcal{U}$ such that $A - BK$ is stable. Let*
Γ *be the operator defined in (15.116) and $P_{\mathcal{H}}$ the orthogonal projection onto* $(\operatorname{ran}\Gamma)^{\perp}$. *Then*
the following statements are equivalent.

(i) *The norm of $P_{\mathcal{H}}W$ is strictly bounded by γ.*

(ii) *The algebraic Riccati equation in (15.81) admits a positive stabilizing solution Q.*

(iii) *For the Hamiltonian matrix H in (15.73), the conditions (i),(ii) and (iii) in Theorem*
15.6.2 hold.

In this case, the optimal cost $d(x_0) = (Qx_0, x_0)$ is uniquely achieved and

$$Q = YX^{-1} = \gamma^2 \Phi^* P_{\mathcal{H}}(\gamma^2 I - P_{\mathcal{H}}WW^*|\mathcal{H})^{-1}P_{\mathcal{H}}\Phi. \tag{15.122}$$

PROOF. To complete the proof it remains to show that Part (ii) implies Part (i). If the
algebraic Riccati equation in (15.81) admits a positive stabilizing solution Q, then we claim
that the norm of $P_{\mathcal{H}}W$ is strictly bounded by γ. Recall that if Q is a positive stabilizing
solution to the algebraic Riccati equation, then $A - BB^*Q$ is stable. Because $\|P_{\mathcal{H}}W\|$ is the
infimum over the set of all γ such that $d(x_0)$ is finite, the norm of $P_{\mathcal{H}}W$ is in independent
of the choice of K. So, without loss of generality we can choose $K = B^*Q$. Notice that the
algebraic Riccati equation can be rewritten as

$$(A - BB^*Q)^*Q + Q(A - BB^*Q) + \gamma^{-2}QEE^*Q + \tilde{C}^*\tilde{C}$$

where $\tilde{C} = [B^*Q\, C]^{tr}$. By replacing C by $\tilde{C}$ in Proposition 14.7.5, we see that the H^∞ norm
of

$$\mathbf{W} = \begin{bmatrix} B^*Q(sI - A + BB^*Q)^{-1}E \\ C(sI - A + BB^*Q)^{-1}E \end{bmatrix}$$

is strictly bounded by γ. Notice that $\mathbf{W}$ is the Laplace transform of W with $K = B^*Q$,
that is, $\mathcal{L}(Ww) = \mathbf{W}\mathcal{L}(w)$. Thus, the operator norm of W equals the H^∞ norm of $\mathbf{W}$. This
readily implies that the norm of W is strictly bounded by γ. In particular, the norm of
$P_{\mathcal{H}}W$ is strictly bounded by γ. Therefore, the algebraic Riccati equation in (15.81) admits

a positive stabilizing solution if and only if the norm of $P_{\mathcal{H}}W$ is strictly bounded by γ. The formula in (15.122) follows from (15.121) along with $d(x_0) = (Qx_0, x_0)$ when Q is the stabilizing solution.   ∎

Now consider the system $\{A, B, C, D, E\}$ where A is stable. Let T from $L^2([0, \infty), \mathcal{W})$ into $L^2([0, \infty), \mathcal{Z})$ and F from $L^2([0, \infty), \mathcal{U})$ into $L^2([0, \infty), \mathcal{Z})$ be the operators defined by

$$(Tw)(t) = \int_0^t C e^{A(t-\tau)} E w(\tau)\, d\tau \quad \text{and} \quad (Fu)(t) = \int_0^t C e^{A(t-\tau)} B u(\tau)\, d\tau . \qquad (15.123)$$

Let C_o be the operator from $\mathcal{X}$ into $L^2([0, \infty), \mathcal{Z})$ defined by $C_o x_0 = C e^{At} x_0$ where x_0 is in $\mathcal{X}$. Because A is stable T, F and C_o are all bounded linear operators. In this case, we can set K to be zero. Then the operators Γ and W in (15.116) and (15.117) reduce to

$$\Gamma = \begin{bmatrix} I \\ -F \end{bmatrix} \quad \text{and} \quad W = \begin{bmatrix} 0 \\ T \end{bmatrix} . \qquad (15.124)$$

The orthogonal projection onto $\mathcal{H} = (\operatorname{ran} \Gamma)^{\perp}$ is given by $P_{\mathcal{H}} = I - \Gamma(\Gamma^*\Gamma)^{-1}\Gamma^*$. Let R_* be the positive square root of $(I + FF^*)^{-1}$. Then for h in $L^2([0, \infty), \mathcal{W})$, we have

$$
\begin{aligned}
\|P_{\mathcal{H}}Wh\|^2 &= \|Wh\|^2 - \|\Gamma(\Gamma^*\Gamma)^{-1}\Gamma^*Wh\|^2 \\
&= \|Th\|^2 - (\Gamma(\Gamma^*\Gamma)^{-1}\Gamma^*(0 \oplus Th), (0 \oplus Th)) \\
&= \|Th\|^2 - (F(I + F^*F)^{-1}F^*Th, Th) \\
&= \|Th\|^2 - ((I + FF^*)^{-1}FF^*Th, Th) \\
&= ((I - (I + FF^*)^{-1}FF^*)Th, Th) \\
&= ((I + FF^*)^{-1}(I + FF^* - FF^*)Th, Th) \\
&= ((I + FF^*)^{-1}Th, Th) = \|R_*Th\|^2 .
\end{aligned}
$$

Hence, $\|P_{\mathcal{H}}Wh\| = \|R_*Th\|$ for all h in $L^2([0, \infty), \mathcal{W})$. This readily implies that $\|P_{\mathcal{H}}W\| < \gamma$ if and only if $\gamma^2 I - TT^* + \gamma^2 FF^*$ is strictly positive. In particular, Proposition 15.8.1 shows that the algebraic Riccati equation in (15.81) admits a stabilizing solution if and only if $\gamma^2 I - TT^* + \gamma^2 FF^*$ is strictly positive. If Q is the stabilizing solution, then Theorem 15.2.3 and Proposition 15.8.1 yield

$$(Qx_0, x_0) = d(x_0) = \gamma^2 (C_o^*(\gamma^2 I - TT^* + \gamma^2 FF^*)^{-1}C_o x_0, x_0) .$$

This readily implies that $Q = \gamma^2 C_o^*(\gamma^2 I - TT^* + \gamma^2 FF^*)^{-1}C_o$. Summing up this analysis proves the following result.

Corollary 15.8.2 *Consider the stable system $\{A, B, C, D, E\}$. Let T and F be the operators defined in (15.123) acting between the appropriate $L^2([0, \infty), \cdot)$ spaces. Then the following statements are equivalent.*

(i) The the operator $\gamma^2 I - TT^* + \gamma^2 FF^*$ is strictly positive.

(ii) The algebraic Riccati equation in (15.81) admits a positive stabilizing solution Q.

In this case, $Q = \gamma^2 C_o^*(\gamma^2 I - TT^* + \gamma^2 FF^*)^{-1}C_o$.

Let $\rho(x_0, \gamma)$ be the optimal cost in the infinite horizon optimization problem in (15.76). Notice that the cost function $q(w, u, \gamma) = \|z\|^2 - \gamma^2\|w\|^2$ is decreasing in γ, that is, if $0 < \gamma_1 \leq \gamma_2$, then $q(w, u, \gamma_1) \geq q(w, u, \gamma_2)$. Hence, $\rho(x_0, \gamma)$ is also decreasing. So, if Q_γ denotes the stabilizing solution to the algebraic Riccati equation for the parameter γ, then Q_γ is a decreasing set of positive operators, that is, $Q_{\gamma_1} \geq Q_{\gamma_2}$ when $0 < \gamma_1 \leq \gamma_2$. Now let $\gamma_{opt} = \|P_\mathcal{H}W\|$. Then γ_{opt} is the infimum over the set of all γ such that the algebraic Riccati equation admits a positive stabilizing solution. One can use Theorem 15.6.3 to compute γ_{opt}. If $\gamma > \gamma_{opt}$ and γ converges to γ_{opt}, then the positive stabilizing solution Q_γ increases with decreasing γ. Moreover, if $\gamma < \gamma_{opt}$, then at least one of the three conditions in Theorem 15.6.2 will fail. Typically what happens is that for $\gamma = \gamma_{opt}$, the operator X in condition (ii) is singular. Let $\gamma_o > 0$ be the largest scalar such that the Hamiltonian matrix has eigenvalues on the imaginary axis. Then for $\gamma_o < \gamma < \gamma_{opt}$ the stabilizing solution may exist, however it is not positive. Finally, for $0 < \gamma \leq \gamma_o$ the Hamiltonian matrix has eigenvalues on the imaginary axis.

Exercise 38 Consider the stable system $\{A, B, C, D, E\}$. If the algebraic Riccati equation in (15.81) admits a positive stabilizing solution, then the Hamiltonian matrix H in (15.73) has no eigenvalues on the imaginary axis. However, even for a stable system the converse is not necessarily true. In other words, if the Hamiltonian H has no eigenvalues on the imaginary axis and A is stable, then it does not necessarily follow that algebraic Riccati equation admits a positive stabilizing solution. For a counter example, let $\{A, [B, E], C, 0\}$ be any minimal realization of

$$\left[\ 3(1-s)/(s+1)^2 \quad 3/(s+1)\ \right] = C(sI - A)^{-1}\left[\ B \quad E\ \right].$$

Here B and E can be viewed as column vectors in $\mathbb{C}^2$. Then compute the stabilizing solution Q for the corresponding algebraic Riccati equation and show that this Q is not positive.

15.9   A tradeoff between norms

In this section, we derive a bound to demonstrate a tradeoff between minimizing the operator norm and the L^2 norm of a closed loop input output system.

15.9.1   The L^2 norm of an operator

Let M be an operator on a finite dimensional space $\mathcal{W}$. The trace of M is defined by

$$\text{trace}(M) = \sum_{i=1}^{n}(M\phi_i, \phi_i) \tag{15.125}$$

where $\{\phi_i\}_1^n$ is an orthonormal basis for $\mathcal{W}$. The trace of M is independent of the orthonormal basis chosen for $\mathcal{W}$. To see this, let $\{\psi_i\}_1^n$ be another orthonormal basis for $\mathcal{W}$. So, if f and g are vectors in $\mathcal{W}$, then we have $(f, g) = \Sigma(f, \psi_j)(\psi_j, g)$. This fact readily implies that

$$\sum_{i=1}^{n}(M\phi_i, \phi_i) = \sum_{i=1}^{n}\sum_{j=1}^{n}(M\phi_i, \psi_j)(\psi_j, \phi_i) = \sum_{j=1}^{n}\sum_{i=1}^{n}(\psi_j, \phi_i)(\phi_i, M^*\psi_j)$$

$$= \sum_{j=1}^{n} (\psi_j, M^* \psi_j) = \sum_{j=1}^{n} (M\psi_j, \psi_j).$$

Therefore, the trace of M is independent of the choice of the orthonormal basis.

Now let M be an operator mapping $\mathcal{W}$ into $\mathcal{Z}$ where $\mathcal{W}$ and $\mathcal{Z}$ are finite dimensional spaces. Let N be a linear operator mapping $\mathcal{Z}$ into $\mathcal{W}$. Then it is an easy exercise to verify that $\text{trace}(NM) = \text{trace}(MN)$. Using the trace, one defines the Hilbert-Schmidt inner product on the set of all linear operators mapping $\mathcal{W}$ into $\mathcal{Z}$ by

$$(M, R)_{HS} = \text{trace}(MR^*) = \text{trace}(R^*M) = \sum_{i=1}^{n} (M\phi_i, R\phi_i) \qquad (15.126)$$

where M and R are linear operators mapping $\mathcal{W}$ into $\mathcal{Z}$ and $\{\phi_i\}_1^n$ is an orthonormal basis for $\mathcal{W}$. The Hilbert-Schmidt inner product is independent of the choice of the orthonormal basis. The Hilbert-Schmidt norm of M is defined by

$$\|M\|_{HS}^2 = (M, M)_{HS} = \text{trace}(MM^*) = \text{trace}(M^*M) = \sum_{i=1}^{n} \|M\phi_i\|^2 \qquad (15.127)$$

where $\{\phi_i\}_1^n$ is an orthonormal basis for $\mathcal{W}$. The Hilbert-Schmidt norm, denoted by $\|\cdot\|_{HS}$, is independent of the choice of the orthonormal basis. It should be clear from (15.127) that $\|M\|_{HS} = \|M^*\|_{HS}$. Obviously, the set of all linear operators mapping $\mathcal{W}$ into $\mathcal{Z}$ is a Hilbert space under the Hilbert-Schmidt inner product. We denote this Hilbert space by $\mathcal{H}(\mathcal{W}, \mathcal{Z})$. One can define Hilbert-Schmidt norms on infinite dimensional spaces; see [59]. However, the infinite dimensional setting is not needed in our work.

Now consider the Hilbert space $L^2([0, t_1], \mathcal{H}(\mathcal{W}, \mathcal{Z}))$ under the inner product

$$(M, R)_2 = \int_0^{t_1} \text{trace}(M(\sigma)R(\sigma)^*) \, d\sigma \qquad (M, R \in L^2([0, t_1], \mathcal{H}(\mathcal{W}, \mathcal{Z}))). \qquad (15.128)$$

The norm induced by the above inner product is called an L^2 norm. For a given operator M in the above Hilbert space, the L^2 norm of M is denoted by $\|M\|_2$ and is given by

$$\|M\|_2^2 = \int_0^{t_1} \text{trace}(M(\sigma)M(\sigma)^*) \, d\sigma = \int_0^{t_1} \text{trace}(M(\sigma)^*M(\sigma)) \, d\sigma. \qquad (15.129)$$

15.9.2 The L^2 norm of a system

In this section we introduce the L^2 norm of a linear system. Consider the linear system

$$\dot{x} = Ax + Ew \qquad \text{and} \qquad z = Cx \qquad (15.130)$$

where A is an operator on $\mathcal{X}$ while E maps $\mathcal{W}$ into $\mathcal{X}$ and C maps $\mathcal{X}$ into $\mathcal{Z}$. As before, the spaces $\mathcal{W}$, $\mathcal{X}$ and $\mathcal{Z}$ are all finite dimensional. This is precisely the system in (15.1) with B and D equal to zero. Let us consider the behavior of this system over some time interval

$[0, t_1]$ where $t_1 > 0$. Consider any initial state x_0 in $\mathcal{X}$ and any input w in $L^2([0, t_1], \mathcal{W})$. Then with initial condition $x(0) = x_0$, the output z is in $L^2([0, t_1], \mathcal{Z})$ and is given by

$$z = C_o x_0 + Tw \tag{15.131}$$

where C_o is the observability operator from $\mathcal{X}$ into $L^2([0, t_1], \mathcal{Z})$ defined by

$$(C_o x_0)(t) = Ce^{At} x_0 \qquad (x_0 \in \mathcal{X}). \tag{15.132}$$

The linear operator T maps $L^2([0, t_1], \mathcal{W})$ into $L^2([0, t_1], \mathcal{Z})$ and is defined by

$$(Tw)(t) = \int_0^t G(t-\tau)w(\tau)\,d\tau \qquad (w \in L^2([0, t_1], \mathcal{W})) \tag{15.133}$$

where

$$G(t) = Ce^{At}E. \tag{15.134}$$

Assume that the initial state x_0 equals zero. Then $z = Tw$ and the behavior of system (15.130) is completely described by T. We refer to T as the input output map for system (15.130).

Let ϕ be any vector in $\mathcal{W}$. Then the *impulse response corresponding* to ϕ for T is the output $z = Tw$ obtained by choosing the input $w(t) = \delta(t)\phi$ where δ is the delta function. Recall that if f is any continuous function, then $\int_a^b \delta(\sigma)f(\sigma)\,d\sigma = f(0)$ when $a \le 0 \le b$. This readily implies that the impulse response corresponding to ϕ is given by

$$T(\delta\phi)(t) = G(t)\phi. \tag{15.135}$$

So, G is simply referred to as the *impulse response* for T or the impulse response for system (15.1). Recalling the definition of the observability operator C_o, we see that

$$T(\delta\phi) = C_o E\phi, \tag{15.136}$$

that is, the zero initial state response of system (15.130) to input $w = \delta\phi$ is the same as the response of the system to $w = 0$ and initial state $x_0 = E\phi$.

By a slight abuse of notation we define the L^2 norm of T or the L^2 norm of system (15.130) by

$$\|T\|_2^2 = \int_0^{t_1} \text{trace}(G(\sigma)G(\sigma)^*)\,d\sigma = \int_0^{t_1} \text{trace}(G(\sigma)^*G(\sigma))\,d\sigma. \tag{15.137}$$

Recalling the definition of G, we obtain that

$$\|T\|_2^2 = \int_0^{t_1} \text{trace}(Ce^{A\sigma}EE^*e^{A^*\sigma}C^*)d\sigma = \int_0^{t_1} \text{trace}(E^*e^{A^*\sigma}C^*Ce^{A\sigma}E)d\sigma. \tag{15.138}$$

Notice that the L^2 norm of T is precisely the L^2 norm of the impulse response for T. In other words, the L^2 norm of T is simply the $L^2([0, t_1], \mathcal{H}(\mathcal{W}, \mathcal{Z}))$ norm of G where G is the impulse response for the system in (15.130) with $x(0) = 0$. Consider any orthonormal basis $\{\phi_i\}^n$ for $\mathcal{W}$. Then

$$
\begin{aligned}
\|T\|_2^2 &= \int_0^{t_1} \operatorname{trace}(G(\sigma)^* G(\sigma)) \, d\sigma = \int_0^{t_1} \sum_{i=1}^n \|G(\sigma)\phi_i\|^2 \, d\sigma \\
&= \int_0^{t_1} \sum_{i=1}^n \|T(\delta\phi_i)(\sigma)\|^2 \, d\sigma = \sum_{i=1}^n \int_0^{t_1} \|T(\delta\phi_i)(\sigma)\|^2 \, d\sigma \\
&= \sum_{i=1}^n \|T(\delta\phi_i)\|^2 ,
\end{aligned}
$$

that is,

$$
\|T\|_2^2 = \sum_{i=1}^n \|T(\delta\phi_i)\|^2 . \tag{15.139}
$$

Remark 15.9.1 Let T be the input output operator from $L^2([0,t_1],\mathcal{W})$ into $L^2([0,t_1],\mathcal{Z})$ defined in (15.133). By combining (15.138) with Lemmas 4.3.1 and 5.3.1, we see that the L^2 norm of T is given by

$$
\|T\|_2^2 = \operatorname{trace}(C X(t_1) C^*) = \operatorname{trace}(E^* Y(t_1) E) . \tag{15.140}
$$

Here X and Y are the solutions to the Lyapunov differential equations

$$
\dot{X} = AX + XA^* + EE^* \quad \text{and} \quad \dot{Y} = A^*Y + YA + C^*C \tag{15.141}
$$

subject to the initial conditions $X(0) = 0$ and $Y(0) = 0$.

Moreover, if $t_1 = \infty$, and A is stable, then the L^2 norm of T is given by $\|T\|_2^2 = \operatorname{trace}(CXC^*) = \operatorname{trace}(E^* Y E)$, where X and Y are the solutions to the Lyapunov equations

$$
AX + XA^* + EE^* = 0 \quad \text{and} \quad A^*Y + YA + C^*C = 0 . \tag{15.142}
$$

15.9.3 The L^2 optimal cost

Now let us return to the linear system in (15.1), that is,

$$
\dot{x} = Ax + Ew + Bu \quad \text{and} \quad z = Cx + Du \tag{15.143}
$$

where D is an isometry satisfying $D^*C = 0$. Recall that w is the disturbance input, and u is the control input. We wish to design a feedback controller for u which the minimizes the effect of the disturbance w on the output z. Recall that T is the operator from $L^2([0,t_1],\mathcal{W})$ into $L^2([0,t_1],\mathcal{Z})$ defined in (15.133) and C_o is the observability operator mapping $\mathcal{X}$ into $L^2([0,t_1],\mathcal{Z})$ defined in (15.132). Finally, let F be the operator mapping $L^2([0,t_1],\mathcal{U})$ into $L^2([0,t_1],\mathcal{Z})$ defined by

$$
(Fu)(t) = \int_0^t C e^{A(t-\tau)} Bu(\tau) \, d\tau \qquad (u \in L^2([0,t_1],\mathcal{U})) . \tag{15.144}
$$

Then, with initial condition $x(0) = x_0$, we have

$$z = C_o x_0 + Tw + Fu + Du. \tag{15.145}$$

Consider now zero initial state, that is, $x_0 = 0$. Then, recalling Section 15.4, any admissible linear feedback controller can described by

$$u = Zw \tag{15.146}$$

where Z is an operator from $L^2([t_0, t_1], \mathcal{W})$ into $L^2([t_0, t_1], \mathcal{U})$. The closed loop system (with $x_0 = 0$) corresponding to the feedback controller (15.146) applied to system (15.143) satisfies

$$z = T_Z w \tag{15.147}$$

where T_Z is the operator from $L^2([t_0, t_1], \mathcal{W})$ into $L^2([t_0, t_1], \mathcal{Z})$ given by $T_Z = T + (F + D)Z$. We refer to T_Z as the input output map for the closed loop system. This leads to the design problem of finding an operator Z to minimize the effect of the disturbance w on the output z. So, the idea is to compute an operator Z to minimize the norm of T_Z. However, this minimization problem clearly depends upon which norm of T_Z we use. In this section we concentrate on minimizing the L^2 norm of T_Z defined by

$$\|T_Z\|_2^2 = \sum_{i=1}^{n} \|T_Z(\delta\phi_i)\|^2$$

where $\{\phi_i\}^n$ is an orthonormal basis for $\mathcal{W}$. This leads to the following optimization problem: Find an operator Z to solve the minimization problem

$$d_2 = \inf_Z \|T_Z\|_2. \tag{15.148}$$

We refer to d_2 as the optimal L^2 norm associated with the original system (15.143).

To obtain a solution to this L^2 optimization problem, consider any feedback operator Z. Choose any vector ϕ in $\mathcal{W}$ and let $w = \delta\phi$. Using (15.145) and (15.136), the output $z = T_Z(\delta\phi)$ is given by

$$z = T(\delta\phi) + Fu + Du = C_o E\phi + Fu + Du$$

where $u = Zw$. Using the fact that D is an isometry and $D^*C = 0$, we obtain that

$$\|T_Z(\delta\phi)\|^2 = \|z\|^2 = \|C_o E\phi + Fu\|^2 + \|u\|^2. \tag{15.149}$$

Hence, $\|T_Z(\delta\phi)\|^2$ is the cost in the linear quadratic regulator problem discussed in Section 12.1 with $x_0 = E\phi$. Recalling Theorem 12.1.1, we have

$$\|T_Z(\delta\phi)\|^2 \geq (P_2(0)E\phi, E\phi) = (E^* P_2(0)E\phi, \phi) \tag{15.150}$$

where P_2 is the solution to the Riccati differential equation

$$\dot{P_2} + A^* P_2 + P_2 A - P_2 BB^* P_2 + C^* C = 0 \qquad (P_2(t_1) = 0). \tag{15.151}$$

Moreover, equality is achieved if and only if $u(t) = \hat{u}_2(t) = -B^*P_2(t)\hat{x}_2(t)$ where the optimal state trajectory $\hat{x}_2$ is given by $\dot{\hat{x}}_2 = (A - BB^*P_2)\hat{x}_2$ with $\hat{x}_2(0) = E\phi$. Recall that this Riccati differential equation does not have a finite escape time. Also, Remark 12.2.1 shows that $P_2(0) = C_o^*(I + FF^*)^{-1}C_o$. Noticing that $\hat{x}_2 = x$ where

$$\dot{x} = Ax + B\hat{u}_2 + E(\delta\phi) \qquad \text{with} \quad x(0) = 0\,,$$

we see that $\hat{x}_2$ is the state response of system (15.143) when $u = \hat{u}$, $w = \delta\phi$ and $x_0 = 0$. Thus, $\hat{u}_2 = \hat{Z}_2 w$ where the feedback operator $\hat{Z}_2$ corresponds to the admissible controller given by

$$u(t) = -B^*P_2(t)x(t)\,. \tag{15.152}$$

Thus, for any vector ϕ in $\mathcal{W}$, we have

$$(E^*P_2(0)E\phi,\ \phi) = \|T_{\hat{Z}_2}(\delta\phi)\|^2 \leq \|T_Z(\delta\phi)\|^2\,.$$

Consider now any orthonormal basis $\{\phi_i\}_1^n$ in $\mathcal{W}$. Then, for any operator Z, we have

$$\|T_Z\|_2^2 = \sum_{i=1}^n \|T_Z(\delta\phi_i)\|^2 \geq \sum_{i=1}^n (E^*P_2(0)E\phi_i, \phi_i) = \text{trace}(E^*P_2(0)E)\,,$$

while

$$\|T_{\hat{Z}_2}\|_2^2 = \sum_{i=1}^n \|T_{\hat{Z}_2}(\delta\phi_i)\|^2 = \sum_{i=1}^n (E^*P_2(0)E\phi_i, \phi_i) = \text{trace}(E^*P_2(0)E)\,.$$

Thus,

$$\text{trace}(E^*P_2(0)E) = \|T_{\hat{Z}_2}\|_2^2 \leq \|T_Z\|_2^2$$

and the optimal solution to the L^2 optimization problem in (15.148) is given by $\hat{Z}_2$ and the optimal L^2 cost is given by

$$d_2^2 = \text{trace}(E^*P_2(0)E) = \text{trace}(E^*C_o^*(I + FF^*)^{-1}C_oE)\,. \tag{15.153}$$

Summing up this analysis yields the following result.

Theorem 15.9.1 *Consider the system* $\{A, B, C, D, E\}$ *and let* P_2 *be the solution to the Riccati differential equation in (15.151). Then the optimal cost* d_2 *for the* L^2 *optimization problem in (15.148) is given by* $d_2^2 = \text{trace}(E^*P_2(0)E)$. *Moreover, this optimal cost is obtained by the optimal controller* $u = -B^*P_2(t)x$.

15.9.4 A tradeoff between d_∞ and d_2

Consider system (15.143) with $x(0) = 0$ and subject to an admissible linear controller. Then, the controller can be described by $u = Zw$ where Z is an operator from $L^2([0, t_1], \mathcal{W})$ into $L^2([0, t_1], \mathcal{U})$. Then, as before, the resulting closed loop system satisfies $z = T_Z w$ where T_Z is the operator from $L^2([0, t_1], \mathcal{W})$ into $L^2([0, t_1], \mathcal{Z})$ defined by $T_Z = T + FZ + DZ$. Recall from Section 15.3 that the problem of finding an admissible controller for u to minimize the operator norm from the disturbance input w to the output z, leads to the following operator optimization problem:

$$d_\infty = \inf_Z \|T_Z\|. \tag{15.154}$$

Here d_∞ is the optimal cost with respect to the operator norm. According to Theorem 15.4.1, the optimal cost $d_\infty = \|(I + FF^*)^{-1/2}T\|$. Moreover, d_∞ is the infimum over the set of all $\gamma > 0$ such that the Riccati differential equation in (15.8) has a solution over the interval $[0, t_1]$.

Now let ϱ be any scalar satisfying $\varrho > 1$. Let P be the solution to the Riccati differential equation in (15.8) with $\gamma = \varrho d_\infty$. Let $u = -B^*Px$ be the central controller corresponding to the weight $\gamma = \varrho d_\infty$. The state x and output z corresponding to this central controller is given by

$$\dot{x} = (A - BB^*P)x + Ew \quad \text{and} \quad z = (C - DB^*P)x. \tag{15.155}$$

Furthermore, the input output operator $\hat{T}_\gamma$ from $L^2([0, t_1], \mathcal{W})$ into $L^2([0, t_1], \mathcal{Z})$ associated with the central controller is given by

$$(\hat{T}_\gamma w)(t) = \int_0^t (C - DB^*P(t))\Psi(t, \tau)Ew(\tau)\,d\tau \qquad (w \in L^2([0, t_1], \mathcal{W})) \tag{15.156}$$

where Ψ is the state transition matrix for $A - BB^*P$. According to Theorem 15.4.1, the norm of $\hat{T}_\gamma$ is strictly bounded by $\gamma = \varrho d_\infty$. The following result provides a bound on the L^2 norm of $\hat{T}_\gamma$.

Theorem 15.9.2 *Let ϱ be a scalar satisfying $\varrho > 1$ and set $\gamma = \varrho d_\infty$. Let P be the solution to the Riccati differential equation in (15.8) on the interval $[0, t_1]$. Let $u = -B^*Px$ be the central controller for $\{A, B, C, D, E\}$. Then*

$$\|\hat{T}_\gamma\| < \varrho d_\infty \quad \text{and} \quad \|\hat{T}_\gamma\|_2 \le \frac{\varrho d_2}{\sqrt{\varrho^2 - 1}}. \tag{15.157}$$

In particular, if $\varrho = \sqrt{2}$, then

$$\|\hat{T}_\gamma\| < \sqrt{2}d_\infty \quad \text{and} \quad \|\hat{T}_\gamma\|_2 \le \sqrt{2}d_2. \tag{15.158}$$

PROOF. Corollary 15.1.2 guarantees that $\|\hat{T}_\gamma\| < \varrho d_\infty$. To obtain an upper bound on the L^2 norm of $\hat{T}_\gamma$, notice that for $u = -B^*Px$ and ϕ in $\mathcal{W}$ we have

$$
\begin{aligned}
\|\hat{T}_\gamma(\delta\phi)\|^2 &= \|T(\delta\phi) + Fu\|^2 + \|u\|^2 \\
&= \|C_oE\phi + T0 + Fu\|^2 + \|u\|^2 - \gamma^2\|0\|^2 \le d(E\phi) \\
&= \gamma^2(E^*C_o^*(\gamma^2I - TT^* + \gamma^2FF^*)^{-1}C_oE\phi, \phi).
\end{aligned}
$$

The inequality follows from the last statement in Corollary 15.1.2 with $x_0 = E\phi$. The last equality is a consequence of (15.47) in Theorem 15.2.3. This readily implies that

$$\|\hat{T}_\gamma\|_2 \le \gamma^2 \text{trace}(E^*C_o^*(\gamma^2I - TT^* + \gamma^2FF^*)^{-1}C_oE). \tag{15.159}$$

We claim that

$$\gamma^2(\gamma^2 I - TT^* + \gamma^2 FF^*)^{-1} \le \frac{\varrho^2(I + FF^*)^{-1}}{\varrho^2 - 1}. \tag{15.160}$$

Notice that if (15.160) holds, then (15.159) and (15.153) imply that $\|\hat{T}_\gamma\|_2^2 \le \varrho^2 d_2^2/(\varrho^2 - 1)$ which finishes the proof. So, to complete the proof let us verify that (15.160) holds when $\gamma = \varrho d_\infty$. To this end, recall that d_∞ equals the norm of $N = (I + FF^*)^{-1/2}T$. Using $NN^* \le d_\infty^2 I$ we readily see that $d_\infty^2 I - TT^* + d_\infty^2 FF^*$ is positive. Hence, $-d_\infty^2(I + FF^*) \le -TT^*$. By adding $\varrho^2 d_\infty^2(I + FF^*)$ to both sides, we obtain

$$(\varrho^2 - 1)d_\infty^2(I + FF^*) \le (\varrho^2 d_\infty^2 I - TT^* + \varrho^2 d_\infty^2 FF^*).$$

Recall that if M and R are two strictly positive operators on $\mathcal{H}$ satisfying $M \le R$, then $R^{-1} \le M^{-1}$; see Lemma 14.4.3. Using this and $\gamma = \varrho d_\infty$ in the previous expression, yields

$$(\gamma^2 I - TT^* + \gamma^2 FF^*)^{-1} \le \frac{(I + FF^*)^{-1}}{(\varrho^2 - 1)d_\infty^2}.$$

Multiplying both sided by $\varrho^2 d_\infty^2$ gives (15.160). ■

Finally, it is noted that Theorem 15.9.2 also shows that the central solution converges to the optimal L^2 solution of the optimization problem in (15.148) as ϱ tends to infinity.

Remark 15.9.2 Let ϱ be a scalar satisfying $\varrho > 1$ and set $\gamma = \varrho d_\infty$. Let P be the solution to the Riccati differential equation in (15.8) on the interval $[0, t_1]$. Let P_2 be the solution to the linear quadratic Riccati differential equation in (15.151). By employing Equation (15.160) along with $P_2(0) = C_o^*(I + FF^*)^{-1}C_o$ and $P(0) = \gamma^2 C_o^*(\gamma^2 I - TT^* + \gamma^2 FF^*)^{-1}C_o$, we obtain

$$P(0) \le \varrho^2 P_2(0)/(\varrho^2 - 1). \tag{15.161}$$

15.10 A tradeoff between the H^2 and H^∞ norms.

In this section, we present an infinite horizon version of Theorem 15.9.2. Consider system (15.143) where $\{A, B\}$ is stabilizable and $\{C, A\}$ is detectable. Suppose $x(0) = 0$. Then an admissible linear controller can be described by $u = Zw$. As before, the resulting closed loop system satisfies $z = T_Z w$ where T_Z is the linear map defined by $T_Z w = \phi(0, w, Zw)$. The L^2 norm of T_Z is defined by

$$\|T_Z\|_2^2 = \sum_{i=1}^n \|T_Z(\delta\phi_i)\|^2 = \sum_{i=1}^n \int_0^\infty \|T_Z(\delta\phi_i)(t)\|^2 dt$$

where $\{\phi_i\}^n$ is an orthonormal basis for $\mathcal{W}$. The optimal norms ρ_2 and ρ_∞ are defined by

$$\rho_2 = \inf_Z \|T_Z\|_2 \quad \text{and} \quad \rho_\infty = \inf_Z \|T_Z\|. \tag{15.162}$$

Let $\mathbf{V}$ be a stable strictly proper rational transfer function with values in $\mathcal{L}(\mathcal{W}, \mathcal{Z})$. Then the L^2 norm of $\mathbf{V}$ is defined by $\|\mathbf{V}\|_2 = \|V\|_2$ where V is the inverse Laplace transform of

V and $\|V\|_2$ is the $L^2([0,\infty), \mathcal{H}(\mathcal{W}, \mathcal{Z}))$ norm of V. Notice that if J is the operator from $L^2([0,\infty), \mathcal{W})$ into $L^2([0,\infty), \mathcal{Z})$ defined by

$$(Jf)(t) = \int_0^t V(t-\tau)u(\tau)\,d\tau \qquad (u \in L^2([0,\infty), \mathcal{W})),$$

then $\|J\|_2 = \|\mathbf{V}\|_2$. This sets the stage for the following result.

Theorem 15.10.1 *Consider the system* $\{A, B, C, D, E\}$ *where* $\{A, B\}$ *is stabilizable and* $\{C, A\}$ *is detectable. Let* ϱ *be a scalar satisfying* $\varrho > 1$ *and set* $\gamma = \varrho\rho_\infty$. *If* $\mathbf{G}_\gamma$ *is the central feedback transfer function, then*

$$\|\mathbf{G}_\gamma\|_\infty < \varrho\rho_\infty \qquad and \qquad \|\mathbf{G}_\gamma\|_2 \le \frac{\varrho\rho_2}{\sqrt{\varrho^2 - 1}}. \tag{15.163}$$

In particular, if $\varrho = \sqrt{2}$, *then*

$$\|\mathbf{G}_\gamma\|_\infty < \sqrt{2}\rho_\infty \qquad and \qquad \|\mathbf{G}_\gamma\|_2 \le \sqrt{2}\rho_2. \tag{15.164}$$

PROOF. The inequality $\|\mathbf{G}_\gamma\|_\infty < \gamma$ follows Proposition 15.7.1. Let ϕ be any vector in $\mathcal{W}$ and consider the following optimization problem

$$\rho_2(\phi)^2 = \inf_u \left\{ \|Cx\|^2 + \|u\|^2 : \dot{x} = Ax + Bu \text{ and } x(0) = E\phi \right\}. \tag{15.165}$$

Notice that this is precisely the infinite horizon linear quadratic regulator problem solved in Theorem 12.8.1 with $x_0 = E\phi$. The optimal cost $\rho_2(\phi)^2 = (Q_2 E\phi, E\phi)$ where Q_2 is the unique positive solution to the algebraic Riccati equation

$$A^*Q_2 + Q_2 A - Q_2 BB^* Q_2 + C^* C = 0. \tag{15.166}$$

Notice that for any Z, we have $\|Cx\|^2 + \|u\|^2 = \|\phi(0, \delta\phi, Z\delta\phi)\|_2^2$ where δ is the delta function and $u = Z\delta\phi$. This implies that $\rho_2^2 = \sum_1^n \rho_2^2(\phi_j)$ where $\{\phi_j\}_1^n$ is an orthonormal basis for $\mathcal{W}$. Therefore, $\rho_2 = \text{trace } E^* Q_2 E$.

Now let $d_\infty(t_1) = d_\infty$ be the optimal norm defined in (15.64) where Z is an operator from $L^2([0,t_1], \mathcal{W})$ into $L^2([0,t_1], \mathcal{U})$. It is easy to show that $d_\infty(t_1)$ is increasing, that is, if $t \le \sigma$, then $d_\infty(t) \le d_\infty(\sigma)$. In particular, $d_\infty(t) \le \rho_\infty$ for all t. Recall that $d_\infty(t_1)$ equals the infimum over the set of all γ such that the Riccati differential equation in (15.8) has a solution over the interval $[0, t_1]$. In particular, ρ_∞ is greater than or equal to the infimum over the set of all γ such that there exists a uniformly bounded solution to the Riccati differential equation in (15.8). However, if there exists a uniformly bounded solution to this Riccati differential equation for some specified γ, then the central controller provides a causal feedback operator satisfying $\rho_\infty \le \|\mathbf{G}_\gamma\|_\infty < \gamma$. Therefore, ρ_∞ equals the infimum over the set of all γ such that the Riccati differential equation in (15.8) admits a uniformly bounded solution. Moreover, $d_\infty(t_1)$ converges to ρ_∞ as t_1 tends to infinity.

Assume that the algebraic Riccati equation in (15.81) admits a positive stabilizing solution Q, and let $u = -B^* Qx$ be the central controller. Let $\tilde{C} = C - DB^* Q$ and $A_c =$

$A - BB^*Q$. Because A_c is stable, the form of the algebraic Riccati equation in (15.110) along with (15.104) yields

$$\|\mathbf{G}_\gamma\|_2^2 = \text{trace} \int_0^\infty E^* e^{A_c^* \sigma} \left(\tilde{C}^* \tilde{C} \right) e^{A_c \sigma} E \, d\sigma$$

$$\leq \text{trace} \int_0^\infty E^* e^{A_c^* \sigma} \left(\gamma^{-2} Q E E^* Q + \tilde{C}^* \tilde{C} \right) e^{A_c \sigma} E \, d\sigma = \text{trace} \, E^* Q E \,.$$

The equality is a consequence of (15.167) in Exercise 39. Hence, $\|\mathbf{G}_\gamma\|_2^2 \leq \text{trace} \, E^* Q E$. Recall that $\gamma = \varrho \rho_\infty$ where $\varrho > 1$. Notice that $P_2(0) = \Omega_2(t_1)$ in (15.151) and $P(0) = \Omega(t_1)$ in (15.8) both depend upon the final time t_1. Moreover, by letting t_1 approach infinity, it follows that $\Omega_2(t_1)$ approaches Q_2 and $\Omega(\tau_1)$ approaches Q as t_1 tends to infinity. Recall that $d_\infty(t_1)$ approaches ρ_∞ as t_1 tends to infinity. So, by passing limits in (15.161), we arrive at

$$\|\mathbf{G}_\gamma\|_2^2 \leq \text{trace} \, E^* Q E \leq \varrho^2 \text{trace} \, E^* Q_2 E / (\varrho^2 - 1) = \varrho^2 \rho_2 / (\varrho^2 - 1) \,.$$

This proves the second inequality in (15.163).   ∎

Exercise 39 Assume that Q is a self-adjoint solution to the algebraic Riccati equation in (15.81) and $A_c = A - BB^*Q$ is stable. Then show that

$$Q = \int_0^\infty e^{A_c^* \sigma} \left(\gamma^{-2} Q E E^* Q + C^* C + Q B B^* Q \right) e^{A_c \sigma} \, d\sigma \,. \tag{15.167}$$

15.11 Notes

This chapter concentrates on the full information H^∞ control problem, and is only a brief introduction to H^∞ control theory. The main purpose was to show how elementary techniques from operator theory can be used to gain some further insight into a fundamental H^∞ control problem. There are many different ways to solve the finite horizon optimization problem in (15.5) and derive the corresponding two point boundary value problem. For example one can use the calculus of variations and game theory; see Green-Limebeer [57], Limebeer-Anderson-Khargonekar-Green [83] and Basar-Bernhard [13]. The derivation of the Riccati differential equation from the two point boundary value problem is classical. The papers of Peterson [96] and Khargonekar-Peterson-Rotea [76] initiated the study of the full information H^∞ control problem. The full information control problem in the infinite horizon case and many other H^∞ control problems were solved in Doyle-Glover-Khargonekar-Francis [38]. For a historical account of the full information H^∞ control problem see Section 6.4 in Green-Limebeer [57]. The H^∞ filtering problem is the dual of the full information H^∞ control problem. By combining the full information control and filtering problems, one can solve a general H^∞ control problem. This and many other results in H^∞ control theory are presented in Green-Limebeer [57] and Zhou-Doyle-Glover [131]. For some further results in H^∞ control theory see Basar-Bernhard [13], Burl [23], Chui-Chen [27], Doyle-Francis-Tannenbaum [37], Helton [63], Francis [44] and Mustafa-Glover [92].

The central solution presented in Sections 15.1 and 15.7 is a causal controller whose norm is bounded by γ. One can show that the set of all causal controllers whose norm is bounded by γ is parameterized by the unit ball in some Banach space; see Green-Limebeer [57] and Zhou-Doyle-Glover [131]. The central solution is simply the solution corresponding to zero in the parameterization of all solutions. The central solution is also the maximal entropy solution studied in Mustafa-Glover [92]. The trade off between d_2 and d_∞ presented in Section 15.9 was taken from Rotea-Frazho [107]. This result is a special case of the fact that the central solution in the commutant lifting theorem satisfies a similar bound; see Foias-Frazho-Gohberg-Kaashoek [41]. Kaftal-Larson-Weiss [66] were the first to show that the central solution in the Nehari interpolation problem satisfies the corresponding bound in (15.157) between the H^2 and H^∞ norm. This result was extended to the general setting of the commutant lifting theorem in Foias-Frazho [40] and Foias-Frazho-Li [42]. Finally, it is noted that the set of all solutions in the commutant lifting theorem can be parameterized by the closed unit ball in some $H^\infty(\mathcal{F}, \mathcal{G})$ space; see Foias-Frazho-Gohberg-Kaashoek [41]. This is a operator theoretic generalization of the parameterization of all controllers in H^∞ theory.

Glover [52] was the first to use state space techniques to solve the rational Nehari interpolation problem. For a nice reference on how state space techniques can be used to solve a large number of interpolation problems for rational functions with applications to H^∞ control theory see Ball-Gohberg-Rodman [10]. Sarason [112] used operator techniques to solve some H^∞ interpolation problems. Using dilation theory Sz.-Nagy-Foias [119] developed an operator interpolation result known as the Sz.-Nagy-Foias commutant lifting theorem. The commutant lifting theorem can be used to solve many H^∞ interpolations problems. For further results on the commutant lifting theorem and its applications see Foias-Frazho [39], Foias-Frazho-Gohberg-Kaashoek [41] and Rosenblum-Rovnyak [104].

Chapter 16

Appendix: Least Squares

In this chapter we introduce and use the *Projection Theorem* to solve some basic least squares optimization problems. This naturally leads to the pseudo inverse of an operator. Then we will develop the singular value decomposition for finite rank operators. Some examples from linear systems will be given.

16.1 The Projection Theorem

To establish some notation, let $\mathcal{H}$ be a Hilbert space. Then two vectors f and g in $\mathcal{H}$ are *orthogonal*, denoted by $f \perp g$, if $(f, g) = 0$. We say that a vector f is orthogonal to a subset $\mathcal{M}$ of $\mathcal{H}$, denoted by $f \perp \mathcal{M}$, if f is orthogonal to every vector g in $\mathcal{M}$. Two subsets $\mathcal{M}$ and $\mathcal{N}$ of $\mathcal{H}$ are orthogonal, denoted by $\mathcal{M} \perp \mathcal{N}$, if every vector f in $\mathcal{M}$ is orthogonal to every vector g in $\mathcal{N}$. We say that $\mathcal{M}$ is a *subspace* of $\mathcal{H}$ if $\mathcal{M}$ is a closed linear space contained in $\mathcal{H}$. (A subset of $\mathcal{H}$ is closed if it contains all its limit points.) The subspace $\mathcal{M}$ can be zero $\{0\}$ or the whole space $\mathcal{H}$. Finally, if $\mathcal{M}$ is any subset of $\mathcal{H}$, then the orthogonal complement of $\mathcal{M}$ is the subspace of $\mathcal{H}$ defined by $\mathcal{M}^{\perp} := \{f \in \mathcal{H} : f \perp \mathcal{M}\}$.

Let h be a vector in $\mathcal{H}$ and $\mathcal{M}$ a subspace of $\mathcal{H}$. A basic least squares optimization problem is to find an element $\hat{h}$ in $\mathcal{M}$, which is closer to h than any other element of $\mathcal{M}$. This naturally leads to the following optimization problem:

$$d(h, \mathcal{M}) := \inf\{\|h - g\| : g \in \mathcal{M}\}. \tag{16.1}$$

The distance from h to the subspace $\mathcal{M}$ is defined as $d(h, \mathcal{M})$. By a slight abuse of terminology we sometimes abbreviate the above optimization problem as $d(h, \mathcal{M}) = \inf \|h - \mathcal{M}\|$. Because $\mathcal{M}$ is closed, it follows that the distance from h to $\mathcal{M}$ is zero if and only if h is a vector in $\mathcal{M}$. The following theorem, known as the Projection Theorem, states that there is a unique vector $\hat{h}$ in $\mathcal{M}$ which achieves the minimum, that is, $d(h, \mathcal{M}) = \|h - \hat{h}\|$. The Projection Theorem plays a fundamental role in operator theory.

Theorem 16.1.1 (Projection Theorem.) *Let $\mathcal{M}$ be a subspace of a Hilbert space $\mathcal{H}$. Then for every h in $\mathcal{H}$, there exists a unique vector $\hat{h}$ in $\mathcal{M}$ solving the following optimization problem:*

$$\|h - \hat{h}\| = \inf\{\|h - g\| : g \in \mathcal{M}\}. \tag{16.2}$$

Moreover, $\hat{h}$ is the only vector in $\mathcal{M}$ such that $h - \hat{h}$ is orthogonal to $\mathcal{M}$, that is, if f is any vector in $\mathcal{M}$ and $h - f \perp \mathcal{M}$, then $f = \hat{h}$ is the unique solution to the optimization problem in (16.2). Finally, if $\tilde{h} = h - \hat{h}$, then the distance $d(h, \mathcal{M}) = \|\tilde{h}\|$ is given by

$$d(h, \mathcal{M})^2 = \|\tilde{h}\|^2 = \|h - \hat{h}\|^2 = \|h\|^2 - \|\hat{h}\|^2. \tag{16.3}$$

If $\hat{h}$ is the unique solution to the optimization problem in (16.2), then we say that $\hat{h}$ is the *orthogonal projection* of h onto the subspace $\mathcal{M}$. The orthogonal projection onto $\mathcal{M}$ is denoted by $P_{\mathcal{M}}$, that is, $\hat{h} = P_{\mathcal{M}}h$. In other words, $P_{\mathcal{M}}h$ is the unique vector in $\mathcal{M}$ which comes closest to h. Obviously, h is in $\mathcal{M}$ if and only if $h = P_{\mathcal{M}}h$. Finally, it is noted that the range of $P_{\mathcal{M}}$ equals $\mathcal{M}$.

We will not present a proof of the Projection Theorem. However, we will establish a few important facts concerning this theorem. In many applications, one computes the orthogonal projection $\hat{h} = P_{\mathcal{M}}h$ by finding the unique vector $\hat{h}$ in $\mathcal{M}$ such that $h - \hat{h}$ is orthogonal to $\mathcal{M}$. So, let us directly show that if $\hat{h}$ is in $\mathcal{M}$ and $h - \hat{h}$ is orthogonal to $\mathcal{M}$, then $\hat{h}$ is the unique solution to the optimization problem in (16.2), and thus $\hat{h} = P_{\mathcal{M}}h$. To see this, recall that if x and y are any vectors in $\mathcal{H}$, then

$$\|x + y\|^2 = \|x\|^2 + 2\Re(x, y) + \|y\|^2.$$

Let g be any vector in $\mathcal{M}$. Then, using $x = h - \hat{h}$ and $y = \hat{h} - g$, we have

$$\begin{aligned}
\|h - g\|^2 &= \|h - \hat{h} + \hat{h} - g\|^2 = \|h - \hat{h}\|^2 + 2\Re(h - \hat{h}, \hat{h} - g) + \|\hat{h} - g\|^2 \\
&= \|h - \hat{h}\|^2 + \|\hat{h} - g\|^2. \tag{16.4}
\end{aligned}$$

Notice that $(h - \hat{h}, \hat{h} - g) = 0$ because $\hat{h} - g$ is in the linear space $\mathcal{M}$ and $h - \hat{h}$ is orthogonal to $\mathcal{M}$. Equation (16.4), yields

$$\|h - g\|^2 = \|h - \hat{h}\|^2 + \|\hat{h} - g\|^2 \geq \|h - \hat{h}\|^2. \tag{16.5}$$

This readily implies that

$$\|h - \hat{h}\|^2 \leq \inf\{\|h - g\|^2 : g \in \mathcal{M}\} = d(h, \mathcal{M})^2.$$

Because $\hat{h}$ is in $\mathcal{M}$, it follows that we have equality, that is, $\|h - \hat{h}\| = d(h, \mathcal{M})$. Equation (16.5) also shows that $\hat{h}$ is the only solution to the optimization problem in (16.2). If g is another solution to this optimization problem, then $\|h - g\| = \|h - \hat{h}\|$. By consulting (16.5), this implies that $\|\hat{h} - g\|^2 = 0$. Hence, $\hat{h} = g$ which proves our claim.

To establish (16.3), recall that $h - \hat{h}$ is orthogonal to $\mathcal{M}$. Since $\hat{h}$ is in $\mathcal{M}$, it follows that $(h - \hat{h}, \hat{h})$ is zero. Using this we have,

$$\begin{aligned}
\|h\|^2 &= \|h - \hat{h} + \hat{h}\|^2 = \|h - \hat{h}\|^2 + 2\Re(h - \hat{h}, \hat{h}) + \|\hat{h}\|^2 \\
&= \|h - \hat{h}\|^2 + \|\hat{h}\|^2.
\end{aligned}$$

Hence, $\|h - \hat{h}\|^2 = \|h\|^2 - \|\hat{h}\|^2$. This result and $\tilde{h} := h - \hat{h}$ yields (16.3).

As before, let $\hat{h}$ be the orthogonal projection of h onto the subspace $\mathcal{M}$. Observe that $h = \hat{h} + \tilde{h}$ where $\tilde{h} = h - \hat{h}$. According to the Projection Theorem $\tilde{h}$ is orthogonal to $\mathcal{M}$, that is, $\tilde{h}$ is a vector in $\mathcal{M}^\perp$. Therefore, every vector h in $\mathcal{H}$ admits a unique orthogonal decomposition of the form $h = \hat{h} + \tilde{h}$ where $\hat{h}$ is in $\mathcal{M}$ and $\tilde{h}$ is in $\mathcal{M}^\perp$. In fact, $\hat{h}$ is the orthogonal projection of h onto $\mathcal{M}$ and $\tilde{h}$ is the orthogonal projection of h onto $\mathcal{M}^\perp$. Moreover, $\|h\|^2 = \|\hat{h}\|^2 + \|\tilde{h}\|^2$. Motivated by this decomposition, we introduce the notation $\mathcal{H} = \mathcal{M} \oplus \mathcal{N}$. This means that $\mathcal{M}$ and $\mathcal{N}$ are two orthogonal spaces which span $\mathcal{H}$, that is, every vector h in $\mathcal{H}$ admits a unique orthogonal decomposition of the form $h = \hat{h} + \tilde{h}$ where $\hat{h}$ is in $\mathcal{M}$, while $\tilde{h}$ is in $\mathcal{N}$ and the subspace $\mathcal{M}$ is orthogonal to $\mathcal{N}$. If $\mathcal{H} = \mathcal{M} \oplus \mathcal{N}$, then obviously $\mathcal{N} = \mathcal{M}^\perp$. If $\mathcal{H} = \oplus_1^n \mathcal{M}_j$, then $\mathcal{M}_1, \mathcal{M}_2, \cdots, \mathcal{M}_n$ are n pairwise orthogonal subspaces which span $\mathcal{H}$. If $\mathcal{M}$ and $\mathcal{R}$ are two subspaces satisfying $\mathcal{M} \subset \mathcal{R}$, then $\mathcal{R} \ominus \mathcal{M}$ denotes the orthogonal complement of $\mathcal{M}$ in $\mathcal{R}$, that is, $\mathcal{R} \ominus \mathcal{M} = \{g \in \mathcal{R} : g \perp \mathcal{M}\}$. Obviously, $\mathcal{M}^\perp = \mathcal{H} \ominus \mathcal{M}$.

Recall that $P_\mathcal{M}$ is the orthogonal projection onto the subspace $\mathcal{M}$, that is, $\hat{h} = P_\mathcal{M} h$ where $\hat{h}$ is the unique solution to the optimization problem in (16.2). We claim that $P_\mathcal{M}$ is a positive operator on $\mathcal{H}$ satisfying $P_\mathcal{M} = P_\mathcal{M}^2 = P_\mathcal{M}^*$. Moreover, the range of $P_\mathcal{M}$ equals $\mathcal{M}$ and $0 \leq P_\mathcal{M} \leq I$. To verify this, first notice that $P_\mathcal{M}$ is a mapping from $\mathcal{H}$ into $\mathcal{H}$ whose range equals $\mathcal{M}$. If h is in $\mathcal{M}$, then obviously, $P_\mathcal{M} h = h$. Hence, $P_\mathcal{M}^2 = P_\mathcal{M}$. Now let us show that $P_\mathcal{M}$ is a linear map, that is, $P_\mathcal{M}(\alpha f + \beta h) = \alpha P_\mathcal{M} f + \beta P_\mathcal{M} h$ for all vectors f, h in $\mathcal{H}$ and scalars α, β. To this end, let $\hat{f} = P_\mathcal{M} f$ and $\hat{h} = P_\mathcal{M} h$. Clearly, the vector $\alpha \hat{h} + \beta \hat{f}$ is in the subspace $\mathcal{M}$. Because both $f - \hat{f}$ and $h - \hat{h}$ are orthogonal to $\mathcal{M}$, it follows that $\alpha f + \beta h - (\alpha \hat{f} + \beta \hat{h})$ is also orthogonal to $\mathcal{M}$. By the Projection Theorem $\alpha \hat{h} + \beta \hat{f}$ must be the orthogonal projection of $\alpha h + \beta f$ onto the subspace $\mathcal{M}$. Therefore, $P_\mathcal{M}(\alpha h + \beta f) = \alpha P_\mathcal{M} h + \beta P_\mathcal{M} f$. In other words, $P_\mathcal{M}$ is a linear map. Recall that any vector h in $\mathcal{H}$ admits an orthogonal decomposition of the form $h = \hat{h} + \tilde{h}$ where $\hat{h} = P_\mathcal{M} h$ and $\tilde{h}$ is in $\mathcal{M}^\perp$. Using this, we obtain

$$\|P_\mathcal{M} h\|^2 = \|\hat{h}\|^2 \leq \|\hat{h}\|^2 + \|\tilde{h}\|^2 = \|h\|^2 .$$

So, $\|P_\mathcal{M}\| \leq 1$. Therefore, the orthogonal projection is a bounded operator. Finally, notice that for any h in $\mathcal{H}$, we have

$$(P_\mathcal{M} h, h) = (\hat{h}, \hat{h} + \tilde{h}) = (\hat{h}, \hat{h}) \leq (h, h) .$$

This readily implies that $0 \leq P_\mathcal{M} \leq I$. In particular, $P_\mathcal{M}$ is a self-adjoint operator. Therefore, the orthogonal projection is an operator satisfying $P_\mathcal{M} = P_\mathcal{M}^2 = P_\mathcal{M}^*$.

An operator P on $\mathcal{H}$ is an *orthogonal projection* if $P = P_\mathcal{M}$ where $P_\mathcal{M}$ is an orthogonal projection onto some subspace $\mathcal{M}$. For example, if $P_\mathcal{M}$ is an orthogonal projection onto $\mathcal{M}$, then it is easy to verify that $I - P_\mathcal{M}$ is the orthogonal projection onto $\mathcal{M}^\perp$, that is, $I - P_\mathcal{M} = P_{\mathcal{M}^\perp}$. We claim that an operator P on $\mathcal{H}$ is an orthogonal projection if and only if $P = P^2 = P^*$. In this case, $P = P_\mathcal{M}$ where $\mathcal{M} = \operatorname{ran} P := P\mathcal{H}$. (The range of an operator is denoted by ran.) To prove this fact it remains to show that if $P = P^2 = P^*$, then the range of P is closed and $P = P_\mathcal{M}$ where $\mathcal{M} = \operatorname{ran} P$. Notice that for any h in $\mathcal{H}$, we have $\|Ph\|^2 = (P^*Ph, h) = (Ph, h) \leq \|Ph\| \|h\|$. This implies that $\|Ph\| \leq \|h\|$. Thus, P is a contraction, that is, $\|P\| \leq 1$. To show that the range of P is closed, let $\{h_n\}_0^\infty$ be any convergent sequence of vectors in the range of P with limit h. We need to show that h is in the range of P. Since P is a bounded operator, it follows that the sequence $\{Ph_n\}_0^\infty$ converges to Ph. Because $P^2 = P$ and h_n is in the range of P, it follows that $Ph_n = h_n$ for all n. Hence the sequence $\{Ph_n\}_0^\infty$ converges to h and therefore, $Ph = h$. This implies that h is in the range of P. Finally, using $P = P^2 = P^*$, a simple calculation shows that $h - Ph$ is orthogonal to $P\mathcal{H}$. By the Projection Theorem, $\hat{h} = Ph$ is the orthogonal projection onto the range of P, that is, $P = P_\mathcal{M}$ where $\mathcal{M} = \operatorname{ran} P$. This completes the proof.

16.2   A general least squares optimization problem

This section presents some elementary methods to compute the orthogonal projection. Then these results are used to solve a least squares optimization problem. Some of these results play a fundamental role in the theory of controllability and observability for linear systems. Throughout the rest of this chapter all operator are bounded operators.

First, let us establish some terminology. Consider any Hilbert spaces $\mathcal{U}$ and $\mathcal{Y}$ and let T be an operator from $\mathcal{U}$ into $\mathcal{Y}$. Recall that the *kernel* (or null space) of T is denoted by $\ker T$, that is,

$$\ker T = \{u \in \mathcal{U} : Tu = 0\} .\tag{16.6}$$

Notice that the kernel of T is closed. The Projection Theorem shows that $\mathcal{U}$ admits an orthogonal decomposition of the form

$$\mathcal{U} = \ker T \oplus (\ker T)^\perp .\tag{16.7}$$

The closure of a set $\mathcal{M}$ in a Hilbert space, denoted by $\overline{\mathcal{M}}$, is the smallest closed subset of $\mathcal{H}$ which contains $\mathcal{M}$. It consists of all the elements of $\mathcal{M}$ along with all the limit points of $\mathcal{M}$. It should be clear that $\left(\overline{\mathcal{M}}\right)^\perp = \mathcal{M}^\perp$. Recall that the *range* of an operator T mapping $\mathcal{U}$ into $\mathcal{Y}$ is denoted by $\operatorname{ran} T$, that is,

$$\operatorname{ran} T = T\mathcal{U} = \{Tu : u \in \mathcal{U}\} .\tag{16.8}$$

Since $(\overline{\operatorname{ran} T})^\perp = (\operatorname{ran} T)^\perp$, the Projection Theorem shows that $\mathcal{Y}$ admits an orthogonal decomposition of the form

$$\mathcal{Y} = \overline{\operatorname{ran} T} \oplus (\operatorname{ran} T)^\perp .\tag{16.9}$$

We are now ready to present the following fundamental result for operators.

Lemma 16.2.1 *Let T be an operator mapping $\mathcal{U}$ into $\mathcal{Y}$. Then*

$$(\ker T)^\perp = \overline{\operatorname{ran} T^*} \qquad and \qquad \ker T = (\operatorname{ran} T^*)^\perp \tag{16.10a}$$
$$(\ker T^*)^\perp = \overline{\operatorname{ran} T} \qquad and \qquad \ker T^* = (\operatorname{ran} T)^\perp . \tag{16.10b}$$

In particular, T maps $(\ker T)^\perp$ into $(\ker T^*)^\perp$, and T^* maps $(\ker T^*)^\perp$ into $(\ker T)^\perp$.

PROOF. We first show that the second equation in (16.10a) holds, that is, $\ker T = (\operatorname{ran} T^*)^\perp$. Let u be in $\ker T$. Then for all y in $\mathcal{Y}$, we have $0 = (Tu, y) = (u, T^*y)$. Thus, u is orthogonal to T^*y for all y in $\mathcal{Y}$. This shows that u is in $(\operatorname{ran} T^*)^\perp$. Therefore, $\ker T \subset (\operatorname{ran} T^*)^\perp$. On the other hand, if u is in $(\operatorname{ran} T^*)^\perp$, then $0 = (u, T^*y) = (Tu, y)$ for all y in $\mathcal{Y}$. By setting $y = Tu$, we obtain $0 = (Tu, Tu)$. So, $Tu = 0$ and u is in $\ker T$. In other words, $(\operatorname{ran} T^*)^\perp \subset \ker T$. Combining this with $\ker T \subset (\operatorname{ran} T^*)^\perp$ yields $\ker T = (\operatorname{ran} T^*)^\perp$.

Notice that $\mathcal{X}^{\perp\perp} = \overline{\mathcal{X}}$ for any linear space $\mathcal{X}$. By taking the orthogonal complement of $\ker T = (\operatorname{ran} T^*)^\perp$, we obtain the first equation in (16.10a). The equations in (16.10b) follow by substituting T^* for T in the first two equations and using $T^{**} = T$.   ∎

We say that an operator T has *closed range* if $\operatorname{ran} T$ is closed, that is, $\overline{\operatorname{ran} T} = \operatorname{ran} T$. In many of our applications the rank of T is finite. If the rank of T is finite, then its range is closed. For example, if T is a matrix from $\mathbb{C}^m$ to $\mathbb{C}^n$, then its range is closed. If T is an operator from $\mathbb{C}^m$ to any space $\mathcal{Y}$, its range is closed. The range of an operator from any space $\mathcal{U}$ to $\mathbb{C}^n$ is closed. In general, the range of an operator on an infinite dimensional space is not closed. For example, let T be the diagonal operator on l^2 defined by $T = \operatorname{diag}(1, 1/2, 1/3, \cdots)$. Recall that l^2 is the Hilbert space consisting of all square summable vectors of the form $[x_1, x_2, x_3, \cdots]^{tr}$, that is, x_j is in $\mathbb{C}$ for all integers $j \geq 1$ and $\sum_1^\infty |x_j|^2$ is finite. In this case, the closure of the range of T equals l^2. Notice that $y = [1, 1/2, 1/3, \cdots]^{tr}$ is not in the range of T. If $y = Tu$, then u is given by $u = [1, 1, 1, \cdots]^{tr}$. However, this vector u is not in l^2. Therefore, the range of T is not closed.

Recall that an operator T is *one to one* if $Tu = 0$ implies that $u = 0$. Clearly, T is one to one if and only if the kernel of T is zero. Moreover, T is *onto* a linear space $\mathcal{H}$ if $\operatorname{ran} T = \mathcal{H}$. If T maps $\mathcal{U}$ into $\mathcal{Y}$, then T is *onto* if the range of T equals the whole space $\mathcal{Y}$, that is, $\operatorname{ran} T = \mathcal{Y}$. An operator T from $\mathcal{U}$ into $\mathcal{Y}$ is *invertible*, if there exists an operator S from $\mathcal{Y}$ into $\mathcal{U}$ satisfying $ST = I$ and $TS = I$. In this case, S is called the inverse of T and is denoted by $S = T^{-1}$. If T is invertible, then T is one to one and onto. Moreover, if T is bounded, one to one and onto, then T is invertible and the inverse of T is also a bounded operator. If T is one to one and onto, then it is easy to show that there exists a unique linear map S from $\mathcal{Y}$ into $\mathcal{U}$ satisfying $ST = I$ and $TS = I$. However, it is not obvious that S is bounded. Fortunately, the open mapping theorem in operator theory shows that T has a bounded inverse if and only if T is one to one and onto; see Conway [30], Halmos [59] and Taylor-Lay [117]. Finally, it is noted that an operator T is invertible if and only if its adjoint T^* is invertible. In this case, $(T^*)^{-1} = (T^{-1})^*$.

As before, let T be an operator mapping $\mathcal{U}$ into $\mathcal{Y}$. (All operators in this chapter are bounded.) According to (16.7) and (16.9), the operator T admits a matrix representation of the form:

$$T = \begin{bmatrix} T_{11} & 0 \\ 0 & 0 \end{bmatrix} : \begin{bmatrix} (\ker T)^\perp \\ \ker T \end{bmatrix} \to \begin{bmatrix} \overline{\operatorname{ran} T} \\ (\operatorname{ran} T)^\perp \end{bmatrix}. \tag{16.11}$$

Here T_{11} is the operator from $(\ker T)^\perp$ into $\overline{\operatorname{ran} T}$ defined by $T_{11} = T|(\ker T)^\perp$. (The notation $|\mathcal{V}$ means restricted to the subspace $\mathcal{V}$.) Notice that T and T_{11} have the same range. In particular, the range of T is closed if and only if the range of T_{11} is closed. Obviously, T_{11} is one to one and onto $\operatorname{ran} T$. So, the range of T_{11} is closed if and only if T_{11} is one to one and onto $\overline{\operatorname{ran} T}$, or equivalently, T_{11} is invertible. In other words, the range of T is closed if and only if T_{11} is invertible. By consulting (16.10) and (16.11), we see that the adjoint T^* of T is given by

$$T^* = \begin{bmatrix} T_{11}^* & 0 \\ 0 & 0 \end{bmatrix} : \begin{bmatrix} (\ker T^*)^\perp \\ \ker T^* \end{bmatrix} \to \begin{bmatrix} \overline{\operatorname{ran} T^*} \\ (\operatorname{ran} T^*)^\perp \end{bmatrix}. \tag{16.12}$$

Here T_{11}^* is the operator from $(\ker T^*)^\perp$ into $\overline{\operatorname{ran} T^*}$ defined by $T_{11}^* = T^*|(\ker T^*)^\perp$. Since T_{11}^* is one to one, it follows that the range of T^* is closed if and only if T_{11}^* is invertible.

However, T_{11}^* is invertible if and only if T_{11} is invertible. Therefore, the range of T is closed if and only if the range of T^* is closed. This proves part of the following result.

Lemma 16.2.2 *Let T be an operator mapping $\mathcal{U}$ into $\mathcal{Y}$. Then the range of T is closed if and only if the range of T^* is closed. In this case, T^*T maps $(\ker T)^\perp$ one to one and onto $(\ker T)^\perp$, and TT^* maps $\operatorname{ran} T$ one to one and onto $\operatorname{ran} T$. Moreover, when the range of T is closed, $T^*T|(\ker T)^\perp$ is an invertible operator on $(\ker T)^\perp$, and $TT^*|\operatorname{ran} T$ is an invertible operator on $\operatorname{ran} T$.*

PROOF. By consulting (16.11) and (16.12), we see that T^*T and TT^* admit matrix representations of the form

$$T^*T = \begin{bmatrix} T_{11}^*T_{11} & 0 \\ 0 & 0 \end{bmatrix} \text{ on } \begin{bmatrix} (\ker T)^\perp \\ \ker T \end{bmatrix}$$

$$TT^* = \begin{bmatrix} T_{11}T_{11}^* & 0 \\ 0 & 0 \end{bmatrix} \text{ on } \begin{bmatrix} (\ker T^*)^\perp \\ \ker T^* \end{bmatrix}.$$

(16.13)

This readily implies that $T^*T|(\ker T)^\perp = T_{11}^*T_{11}$ is an operator on the subspace $(\ker T)^\perp$, while $TT^*|(\ker T^*)^\perp = T_{11}T_{11}^*$ is an operator on $(\ker T^*)^\perp$. Now assume that the range of T is closed. Then T_{11} is invertible. So, $T_{11}^*T_{11}$ and $T_{11}T_{11}^*$ are both invertible operators. Hence, $T^*T|(\ker T)^\perp$ is an invertible operator on $(\ker T)^\perp$, and $TT^*|(\ker T^*)^\perp$ is an invertible operator on $(\ker T^*)^\perp$. Finally, using $\operatorname{ran} T = (\ker T^*)^\perp$ completes the proof. ∎

As before, let T be an operator mapping $\mathcal{U}$ into $\mathcal{Y}$. Recall that T is invertible if and only if $\ker T = \{0\}$ and $\operatorname{ran} T = \mathcal{Y}$. If T is invertible, then T is bounded below, that is, there exists a scalar $\delta > 0$ such that $\|Tu\| \geq \delta\|u\|$ for all u in $\mathcal{U}$. To verify this, let S be the inverse of T. Then $\|u\| = \|STu\| \leq \|S\|\,\|Tu\|$. So, if $\delta = 1/\|S\|$, then $\|Tu\| \geq \delta\|u\|$ for all u in $\mathcal{U}$, and thus, T is bounded below.

If T is bounded below, then the kernel of T is zero and the range of T is closed. If T is bounded below, then obviously, the kernel of T is zero. To show that the range of T is closed, assume that $\{Tu_n\}_1^\infty$ is a sequence of vectors which converge to some vector y in $\mathcal{Y}$. Then

$$\delta\|u_n - u_m\| \leq \|Tu_n - Tu_m\| \leq \|Tu_n - y\| + \|y - Tu_m\| \to 0$$

as n and m tend to infinity. So, $\|u_n - u_m\|$ approaches zero as n and m tend to infinity. In other words, $\{u_n\}_1^\infty$ is a Cauchy sequence in the Hilbert space $\mathcal{U}$. Hence, the sequence $\{u_n\}_1^\infty$ converges to some vector u. This readily implies that

$$\|Tu - y\| \leq \|Tu - Tu_n\| + \|Tu_n - y\| \leq \|T\|\,\|u - u_n\| + \|Tu_n - y\| \to 0$$

as n tends to infinity. Thus, $\|Tu - y\| = 0$, or equivalently, $y = Tu$ is in the range of T. Therefore, the range of T is closed when T is bounded below.

Let T be an operator mapping $\mathcal{U}$ into $\mathcal{Y}$. Then T is invertible if and only if T is bounded below and the range of T is dense in $\mathcal{Y}$, that is, $\overline{\operatorname{ran} T} = \mathcal{Y}$. If T is invertible, then we have all ready shown that T is bounded below. Obviously, the range of T is dense in $\mathcal{Y}$ when T is invertible. In fact, in this case, $\operatorname{ran} T = \mathcal{Y}$. Now assume that T is bounded below and the range of T is dense in $\mathcal{Y}$. Since T is bounded below, the kernel of T is zero. Because T is bounded below, the range of T is closed. Thus, $\mathcal{Y} = \overline{\operatorname{ran} T} = \operatorname{ran} T$. Therefore, T is one to one and onto. So, T is invertible.

As before, let T_{11} be the operator mapping $(\ker T)^{\perp}$ into $\overline{\operatorname{ran} T}$ defined by $T_{11} = T|(\ker T)^{\perp}$. Obviously, T_{11} is one to one and the range of T_{11} is dense in $\overline{\operatorname{ran} T}$. So, T_{11} is invertible if and only if the range of T_{11} is closed, or equivalently, T_{11} is bounded below. Equation (16.11) shows that T and T_{11} have the same range. Thus, the range of T is closed if and only if T_{11} is invertible, or equivalently, $T|(\ker T)^{\perp}$ is bounded below. Therefore, the range of T is closed if and only if there exists a scalar $\delta > 0$ such that $\|Tu\| \geq \delta\|u\|$ for all u in $(\ker T)^{\perp}$.

Recall that T^*T is positive. To see this simply observe that $(T^*Tu, u) = \|Tu\|^2 \geq 0$ for all u in $\mathcal{U}$. This equation also shows that T is bounded below if and only if T^*T is strictly positive. In particular, if the kernel of T is zero, then the range of T is closed if and only if T^*T is strictly positive. Moreover, if the range of T is closed, then the kernel of T is zero if and only if T^*T is invertible. Clearly, T^*T maps $(\ker T)^{\perp}$ into $(\ker T)^{\perp}$. So, the operator $T^*T|(\ker T)^{\perp}$ on $(\ker T)^{\perp}$ is strictly positive if and only if $T|(\ker T)^{\perp}$ is bounded below, or equivalently, the range of T is closed. Summing up this analysis, we obtain the following useful result.

Remark 16.2.1 Let T be any operator mapping $\mathcal{U}$ into $\mathcal{Y}$. Then T^*T is positive. Moreover, if T has finite rank or the range of T is closed, then the following statements are equivalent:

(i) The operator T^*T is invertible.

(ii) $\ker T = \{0\}$, or equivalently, $\operatorname{ran} T^* = \mathcal{U}$.

(iii) The operator T^*T is strictly positive.

By replacing T with T^* we see that TT^* is positive, and the following statements are equivalent when the range of T is closed:

(i) The operator TT^* is invertible.

(ii) $\ker T^* = \{0\}$, or equivalently, $\operatorname{ran} T = \mathcal{Y}$.

(iii) The operator TT^* is strictly positive.

16.2.1 Computation of orthogonal projections

Let T be an operator mapping $\mathcal{U}$ into $\mathcal{Y}$. Let y be a vector in $\mathcal{Y}$. The equation $y = Tu$ has a solution if and only if y is in the range of T. If $y = Tu$ does not have a solution, then it makes sense to look for a vector $\hat{u}$ in $\mathcal{U}$ such that $T\hat{u}$ is closer to y than any other element in $T\mathcal{U}$, that is, find a vector $\hat{u}$ in $\mathcal{U}$ which makes $\|y - Tu\|$ as small as possible. This naturally leads to the optimization problem $\inf \|y - T\mathcal{U}\|$. To solve this problem, let $\mathcal{R}$ be the closure of the range of T and note that the distance from y to $\mathcal{R}$ is given by

$$d(y, \mathcal{R}) = \inf \left\{ \|y - Tu\| : u \in \mathcal{U} \right\}. \tag{16.14}$$

Let $P_\mathcal{R}$ be the orthogonal projection onto $\mathcal{R}$ and set $\hat{y} = P_\mathcal{R}y$. Then, according to the Projection theorem, $d(y, \mathcal{R}) = \|y - \hat{y}\|$; hence

$$\|y - \hat{y}\| = \inf\{\|y - Tu\| : u \in \mathcal{U}\}. \tag{16.15}$$

In general $\hat{y}$ may not be in the range of T, that is, $\hat{y} = Tu$ may not have a solution. However, if the range of T is closed, then $\hat{y}$ is in the range of T and there exists a vector $\hat{u}$ in $\mathcal{U}$ such that $\hat{y} = T\hat{u}$. In this case, $d(y, \mathcal{R}) = \|y - T\hat{u}\|$ and $T\hat{u}$ is the unique vector in $T\mathcal{U}$ which is closer to y than any other element in $T\mathcal{U}$. Furthermore, the vector $\hat{y} = T\hat{u}$ achieves the minimum in the optimization problems (16.14) and (16.15). If the range of T is not closed, then the minimum may not be achieved. In this section, we will develop several techniques to compute the orthogonal projection $P_\mathcal{R}$ and solve these optimization problems when the range of T is closed.

Let T be an operator from $\mathcal{U}$ into $\mathcal{Y}$ with closed range $\mathcal{R}$. Let y be a vector in $\mathcal{Y}$ and set $\hat{y} = P_\mathcal{R}y$. Then there exists a unique vector $\hat{u}$ in $(\ker T)^\perp$ satisfying $T\hat{u} = \hat{y}$. Because the range of T is closed, $\hat{y}$ is in the range of T and thus, the equation $Tu = \hat{y}$ has a solution. Let $u = \hat{u} + \tilde{u}$ be the orthogonal decomposition of any u solution where $\hat{u}$ is in $(\ker T)^\perp$ and $\tilde{u}$ is in $\ker T$. Then using $T\tilde{u} = 0$, we obtain $T\hat{u} = \hat{y}$. Moreover, $\hat{u}$ is the only vector in $(\ker T)^\perp$ satisfying $T\hat{u} = \hat{y}$. If $Tv = \hat{y}$ where v is in $(\ker T)^\perp$, then $T(\hat{u} - v) = \hat{y} - \hat{y} = 0$. So, $\hat{u} - v$ is in the kernel of T. Since $(\ker T)^\perp$ is a linear space, $\hat{u} - v$ is also in $(\ker T)^\perp$. Because $\ker T \cap (\ker T)^\perp = \{0\}$, the vector $\hat{u} - v = 0$, or equivalently, $\hat{u} = v$. Therefore, there is a unique vector $\hat{u}$ in $(\ker T)^\perp$ solving the equation $T\hat{u} = \hat{y} = P_\mathcal{R}y$.

The *restricted inverse* T^{-r} of T is the operator from $\mathcal{Y}$ to $\mathcal{U}$ defined by $T^{-r}y = \hat{u}$ where $\hat{u}$ is the unique element in $(\ker T)^\perp$ such that $T\hat{u} = \hat{y} = P_\mathcal{R}y$. Obviously, if T is invertible, then $T^{-r} = T^{-1}$. If $\hat{u} = T^{-r}y$, then $\hat{y} = T\hat{u}$ is the solution to the optimization problems in (16.14) and (16.15). So, computing the restricted inverse plays a fundamental role in solving the optimization problems in (16.14) and (16.15). Recall that T admits a matrix representation of the form (16.11) where T_{11} is the operator from $(\ker T)^\perp$ onto $\operatorname{ran} T$ defined by $T_{11} = T|(\ker T)^\perp$. (Because the range of T is closed $\overline{\operatorname{ran} T} = \operatorname{ran} T$.) Recall that the range of T is closed if and only if T_{11} is invertible. Therefore, the operator T^{-r} admits a matrix representation of the form:

$$T^{-r} = \begin{bmatrix} T_{11}^{-1} & 0 \\ 0 & 0 \end{bmatrix} : \begin{bmatrix} \operatorname{ran} T \\ (\operatorname{ran} T)^\perp \end{bmatrix} \to \begin{bmatrix} (\ker T)^\perp \\ \ker T \end{bmatrix}. \tag{16.16}$$

In particular, $T^{-r} = T_{11}^{-1}P_\mathcal{R}$. In other words, T^{-r} admits a matrix representation of the form

$$T^{-r} = \begin{bmatrix} I \\ 0 \end{bmatrix} T_{11}^{-1} \begin{bmatrix} I & 0 \end{bmatrix}.$$

Notice that $TT^{-r}y = T\hat{u} = P_\mathcal{R}y$. Since this holds for all y in $\mathcal{Y}$, we have $TT^{-r} = P_\mathcal{R}$. This also follows from the fact that

$$TT^{-r} = \begin{bmatrix} T_{11} & 0 \\ 0 & 0 \end{bmatrix} \begin{bmatrix} T_{11}^{-1} & 0 \\ 0 & 0 \end{bmatrix} = \begin{bmatrix} I & 0 \\ 0 & 0 \end{bmatrix} = P_\mathcal{R}.$$

A similar calculation also shows that $T^{-r}T$ is the orthogonal projection onto $(\ker T)^\perp$.

Notice that T^*T and TT^* are positive operators. Shortly, we will demonstrate how to obtain $P_{\mathcal{R}}$ and T^{-r} by computing the restricted inverse of T^*T or TT^*. In particular, if $\mathcal{U}$ or $\mathcal{Y}$ equals $\mathbb{C}^n$, then T^*T or TT^*, respectively, admits a matrix representation. In this case, one can use standard matrix computational techniques to compute the restricted inverse of T^*T or TT^*, respectively, and thus, readily obtain $P_{\mathcal{R}}$ or T^{-r}. Let us begin with the computation of $P_{\mathcal{R}}$.

Lemma 16.2.3 *Let T be an operator from $\mathcal{U}$ into $\mathcal{Y}$ with closed range $\mathcal{R}$. Then the orthogonal projection onto $\mathcal{R}$ is given by*

$$P_{\mathcal{R}} = T(T^*T)^{-r}T^* = TT^*(TT^*)^{-r} = (TT^*)^{-r}TT^*. \tag{16.17}$$

Moreover, the distance from any vector y in $\mathcal{Y}$ to $\mathcal{R}$ is given by

$$d(y, \mathcal{R})^2 = \|y - P_{\mathcal{R}}y\|^2 = \|y\|^2 - \left((T^*T)^{-r}T^*y, T^*y\right). \tag{16.18}$$

In particular, if T is one to one, then T^*T is invertible and

$$P_{\mathcal{R}} = T(T^*T)^{-1}T^* \tag{16.19}$$
$$\|y - P_{\mathcal{R}}y\|^2 = \|y\|^2 - \left((T^*T)^{-1}T^*y, T^*y\right). \tag{16.20}$$

PROOF. Consider any y in $\mathcal{Y}$ and let $\hat{y} = P_{\mathcal{R}}y$. By the Projection Theorem, the vector $y - \hat{y}$ is orthogonal to $T\mathcal{U}$. Thus,

$$0 = (y - \hat{y}, Tu) = (T^*y - T^*\hat{y}, u) \qquad \text{(for all } u \in \mathcal{U}\text{).} \tag{16.21}$$

So, $T^*y - T^*\hat{y}$ is orthogonal to the entire space $\mathcal{U}$. Hence,

$$T^*y = T^*\hat{y}. \tag{16.22}$$

Because the range of T is closed, $\hat{y} = P_{\mathcal{R}}y$ is in the range of T. So, there exists a unique vector $\hat{u}$ in $(\ker T)^\perp$ such that $\hat{y} = T\hat{u}$. Also, as a result of (16.22), this vector satisfies

$$T^*T\hat{u} = T^*y. \tag{16.23}$$

Recall that T^*T maps $(\ker T)^\perp$ one to one and onto $(\ker T)^\perp = \operatorname{ran} T^*$; see Lemma 16.2.2. So, the unique vector $\hat{u}$ in $(\ker T)^\perp$ satisfying (16.23) is given by $\hat{u} = (T^*T)^{-r}T^*y$. Therefore, the restricted inverse of T is given by

$$T^{-r} = (T^*T)^{-r}T^*. \tag{16.24}$$

This implies that $\hat{y} = P_{\mathcal{R}}y = T\hat{u} = T(T^*T)^{-r}T^*y$. Since this holds for all y in $\mathcal{Y}$, we obtain the first equality in (16.17).

Equation (16.18) follows from equation (16.3) in the Projection Theorem, that is,

$$\|y - \hat{y}\|^2 = \|y\|^2 - (P_{\mathcal{R}}y, y) = \|y\|^2 - \left((T^*T)^{-r}T^*y, T^*y\right). \tag{16.25}$$

To prove the second equality in (16.17), let $y = \hat{y} + \tilde{y}$ be the orthogonal decomposition of y where $\hat{y} = P_{\mathcal{R}}y$ and $\tilde{y} = (I - P_{\mathcal{R}})y$ is the orthogonal projection of y onto $(\operatorname{ran} T)^\perp = \mathcal{R}^\perp$.

Because $\ker TT^* = (\operatorname{ran} T)^\perp$, we have $(TT^*)^{-r}\tilde{y} = 0$. By employing the definition of the pseudoinverse, $TT^*(TT^*)^{-r}y = TT^*(TT^*)^{-r}\hat{y} = \hat{y} = P_{\mathcal{R}}y$. Since this holds for all y, we obtain $TT^*(TT^*)^{-r} = P_{\mathcal{R}}$. A similar argument yields the last equality in (16.17). Finally, it is noted that equation (16.17) also follows from (16.13) along with the fact that T_{11} is invertible.   ■

The previous lemma readily proves part of the following result.

Theorem 16.2.4 *Let T be an operator from $\mathcal{U}$ to $\mathcal{Y}$ with closed range. Then the following holds.*

(i) *The restricted inverse T^{-r} of T is given by*

$$T^{-r} = (T^*T)^{-r}T^* = T^*(TT^*)^{-r}. \tag{16.26}$$

(ii) *If T is one to one, then $T^{-r} = (T^*T)^{-1}T^*$.*

(iii) *If T is onto, then $T^{-r} = T^*(TT^*)^{-1}$.*

(iv) *If y is a specified vector in $\mathcal{Y}$, then $\hat{u} = T^{-r}y$, is a solution to the least squares optimization in (16.14), that is,*

$$\|y - T\hat{u}\| = \inf\{\|y - Tu\| \, : \, u \in \mathcal{U}\} = d(y, T\mathcal{U}). \tag{16.27}$$

(v) *If T is one to one, then there is a unique solution to the optimization problem in (16.27) and this solution is given by $\hat{u} = (T^*T)^{-1}T^*y$.*

PROOF. The first equality in (16.26) follows from (16.24). To show that the second equality in (16.26) holds, let y be a vector in $\mathcal{Y}$. According to the previous lemma, $u_1 = (T^*T)^{-r}T^*y$ and $u_2 = T^*(TT^*)^{-r}y$ are two vectors satisfying $P_{\mathcal{R}}y = Tu_1 = Tu_2$. Clearly, u_1 is in $(\ker T)^\perp$. Since $\operatorname{ran} T^* = (\ker T)^\perp$, it follows that u_2 is also in $(\ker T)^\perp$. In particular, $u_1 - u_2$ is in $(\ker T)^\perp$. Because $T(u_1 - u_2) = 0$, the vector $u_1 - u_2$ is also in the kernel of T. Hence, $u_1 - u_2 = 0$, or equivalently, $u_1 = u_2$. Therefore, (16.26) holds. Part (ii) follows from the fact that T^*T is invertible if and only if T is one to one. To prove part (iii) simply notice that TT^* is invertible if and only if T is onto. Equation (16.27) in Part (iv) follows from the Projection Theorem and the fact that $T\hat{u} = P_{\mathcal{R}}y$. To verify that Part (v) holds, recall that $\hat{y} = P_{\mathcal{R}}y$ is the only vector in $\mathcal{Y}$ satisfying $\|y - \hat{y}\| = d(y, T\mathcal{U})$. Since T is one to one, the equation $Tu = P_{\mathcal{R}}y$ has a unique solution. So, there is only one vector u in $\mathcal{U}$ satisfying $\|y - Tu\| = d(y, T\mathcal{U})$ and this u is the unique solution to $Tu = P_{\mathcal{R}}y$. Hence, there exists a unique solution to the optimization problem in (16.27). Because T is one to one, the vector $\hat{u} = (T^*T)^{-1}T^*y$ is well defined and satisfies $T\hat{u} = P_{\mathcal{R}}y$. Therefore, $\hat{u}$ is the unique solution to this optimization problem. Finally, it is noted that Part (i) also follows from the matrix representations in (16.13) and (16.16).   ■

16.3 The Gram matrix

In this section, we use the previous least squares optimization theory to solve a classical optimization problem via the Gram matrix. Suppose that $\{\psi_1, \psi_2, \cdots, \psi_n\}$ is a finite set of vectors in a Hilbert space $\mathcal{Y}$ and $\mathcal{R}$ is the space spanned by these vectors. A classical least squares optimization problem is to compute the orthogonal projection $\hat{y} = P_{\mathcal{R}} y$ for some fixed y in $\mathcal{Y}$. In other words, find an element $\hat{y}$ of $\mathcal{R}$ to solve the following optimization problem

$$\|y - \hat{y}\| = \inf \left\{ \|y - \sum_{i=1}^{n} \alpha_i \psi_i\| : \alpha_i \in \mathbb{C} \right\}. \tag{16.28}$$

To solve this problem, let T be the operator mapping $\mathbb{C}^n$ into $\mathcal{Y}$ defined by

$$T\alpha = \sum_{i=1}^{n} \alpha_i \psi_i = \begin{bmatrix} \psi_1 & \psi_2 & \cdots & \psi_n \end{bmatrix} \alpha \tag{16.29}$$

where $\alpha = [\alpha_1, \alpha_2, \cdots, \alpha_n]^{tr}$ is in $\mathbb{C}^n$ and ^{tr} denotes transpose. Clearly, T is a finite rank operator whose range is $\mathcal{R}$. Moreover, the optimization problem in (16.28) is a special case of the optimization problem in (16.15), that is, $\|y - \hat{y}\| = \inf\{\|y - T\mathbb{C}^n\|\}$. Notice that the vectors $\{\psi_i\}_1^n$ are linearly independent if and only if T is one to one, or equivalently, T^*T is strictly positive. Since the range of T is $\mathcal{R}$, the solution to the least squares problem in (16.28) is given by (16.17), that is,

$$\hat{y} = P_{\mathcal{R}} y = T(T^*T)^{-r}T^*y.$$

We can also express this as $\hat{y} = T\hat{\alpha}$ where $\hat{\alpha} = T^{-r}\hat{y}$, that is,

$$\hat{y} = P_{\mathcal{R}} y = \sum_{i=1}^{n} \hat{\alpha}_i \psi_i \quad \text{where} \quad \begin{bmatrix} \hat{\alpha}_1 & \hat{\alpha}_2 & \cdots & \hat{\alpha}_n \end{bmatrix}^{tr} = \hat{\alpha} := (T^*T)^{-r}T^*y. \tag{16.30}$$

To obtain $\hat{\alpha}$, we need to compute the adjoint T^* of T. To this end, let f be any vector in $\mathcal{Y}$, then for each α in $\mathbb{C}^n$, we have

$$(\alpha, T^*f) = (T\alpha, f) = (\sum_{i=1}^{n} \alpha_i \psi_i, f) = \sum_{i=1}^{n} \alpha_i (\psi_i, f) = \sum_{i=1}^{n} \alpha_i \overline{(f, \psi_i)}. \tag{16.31}$$

Therefore, the adjoint T^* of T is given by

$$T^*f = \begin{bmatrix} (f, \psi_1) \\ (f, \psi_2) \\ \vdots \\ (f, \psi_n) \end{bmatrix}. \tag{16.32}$$

We now define γ to be the "cross covariance" vector in $\mathbb{C}^n$ given by

$$\gamma = \begin{bmatrix} (y, \psi_1) & (y, \psi_2) & \cdots & (y, \psi_n) \end{bmatrix}^{tr} = T^*y. \tag{16.33}$$

We now compute T^*T. Clearly, T^*T maps $\mathbb{C}^n$ into $\mathbb{C}^n$. Thus, T^*T has a matrix representation. To compute the matrix representation for T^*T, let $\{e_1, e_2, \cdots, e_n\}$ be the standard orthonormal basis for $\mathbb{C}^n$, that is, the i-th component of e_i is one, while all the other components are zero. Notice that the ij-th entry $(T^*T)_{ij}$ of T^*T is given by

$$(T^*T)_{ij} = (T^*Te_j, e_i) = (Te_j, Te_i) = (\psi_j, \psi_i). \tag{16.34}$$

Therefore, the matrix representation for T^*T is precisely the Gram matrix G given by

$$G = \begin{bmatrix} (\psi_1, \psi_1) & (\psi_2, \psi_1) & \cdots & (\psi_n, \psi_1) \\ (\psi_1, \psi_2) & (\psi_2, \psi_2) & \cdots & (\psi_n, \psi_2) \\ \vdots & \vdots & & \vdots \\ (\psi_1, \psi_n) & (\psi_2, \psi_n) & \cdots & (\psi_n, \psi_n) \end{bmatrix}. \tag{16.35}$$

The matrix G in (16.35) is referred to as the *Gram matrix generated* by $\{\psi_j\}_1^n$.

It now follows from (16.30) and (16.33) that

$$\hat{\alpha} = (T^*T)^{-r}T^*y = G^{-r}\gamma. \tag{16.36}$$

Combining (16.30) and (16.36), the optimal solution $\hat{y}$ to the least squares optimization problem (16.28) is given by

$$\hat{y} = P_{\mathcal{R}}y = \sum_{i=1}^{n} \hat{\alpha}_i\psi_i \quad \text{where} \quad \begin{bmatrix} \hat{\alpha}_1 & \hat{\alpha}_2 & \cdots & \hat{\alpha}_n \end{bmatrix}^{tr} = G^{-r}\gamma. \tag{16.37}$$

Finally, notice that $\hat{y}$ can also be expressed as

$$\hat{y} = P_{\mathcal{R}}y = \begin{bmatrix} \psi_1 & \psi_2 & \cdots & \psi_n \end{bmatrix} G^{-r} \begin{bmatrix} (y, \psi_1) & (y, \psi_2) & \cdots & (y, \psi_n) \end{bmatrix}^{tr}. \tag{16.38}$$

Remark 16.3.1 Let G be the Gram matrix in (16.35) generated by any set of vectors $\{\psi_i\}_1^n$ in a Hilbert space $\mathcal{Y}$. Then G is positive. Moreover, G is strictly positive if and only if the set of vectors $\{\psi_i\}_1^n$ are linearly independent. To see this, let T be the operator from $\mathbb{C}^n$ to $\mathcal{Y}$ defined in (16.29). Since $G = T^*T$, it follows that G is positive. The kernel of T is zero if and only if $\{\psi_i\}_1^n$ is linearly independent. Therefore, the Gram matrix G is strictly positive if and only if the set of vectors $\{\psi_i\}_1^n$ are linearly independent. Finally, in this case, the optimal solution $\hat{y}$ to the least squares optimization problem in (16.28) is given by (16.37) where G^{-1} replaces G^{-r}, or equivalently,

$$\hat{y} = P_{\mathcal{R}}y = \begin{bmatrix} \psi_1 & \psi_2 & \cdots & \psi_n \end{bmatrix} G^{-1} \begin{bmatrix} (y, \psi_1) & (y, \psi_2) & \cdots & (y, \psi_n) \end{bmatrix}^{tr}. \tag{16.39}$$

Remark 16.3.2 Suppose that $\{\psi_1, \psi_2, \cdots, \psi_n\}$ forms an orthonormal set in a Hilbert space $\mathcal{Y}$, that is, $(\psi_i, \psi_j) = 0$ if $i \neq j$ and $\|\psi_i\| = 1$. Then, clearly, $G = I$ and $\{\psi_1, \psi_2, \cdots, \psi_n\}$ is linearly independent. Moreover, if $\mathcal{R}$ is the space spanned by $\{\psi_1, \psi_2, \cdots, \psi_n\}$, then formulas (16.33) and (16.37) show that for $y \in \mathcal{Y}$,

$$P_{\mathcal{R}}y = \sum_{i=1}^{n}(y, \psi_i)\psi_i \quad \text{and} \quad \|P_{\mathcal{R}}y\|^2 = \sum_{i=1}^{n} |(y, \psi_i)|^2. \tag{16.40}$$

The last equality follows from

$$\|P_{\mathcal{R}}y\|^2 = (P_{\mathcal{R}}y, y) = (\sum_{i=1}^{n}(y, \psi_i)\psi_i, y) = \sum_{i=1}^{n}|(y, \psi_i)|^2.$$

This is the classical Fourier representation for the orthogonal projection in terms of an orthonormal basis. In particular, if y is in $\mathcal{R}$, then $P_{\mathcal{R}}y = y$ and (16.40) reduces to

$$y = \sum_{i=1}^{n}(y, \psi_i)\psi_i \quad \text{and} \quad \|y\|^2 = \sum_{i=1}^{n}|(y, \psi_i)|^2 \quad (\text{if } y \in \mathcal{R}). \tag{16.41}$$

Example 16.3.1 Consider the problem of approximating the exponential function, $y(t) = e^t$, by a polynomial of degree at most two in the $L^2[0, 1]$ norm. To be precise, we wish to find the optimal polynomial $\hat{y} = \hat{\alpha}_0 + \hat{\alpha}_1 t + \hat{\alpha}_2 t^2$ to solve the optimization problem

$$\|e^t - \hat{y}\|^2 = \inf\left\{\int_0^1 |e^t - \alpha_0 - \alpha_1 t - \alpha_2 t^2|^2 dt : \alpha_i \in \mathbb{C}\right\}. \tag{16.42}$$

To obtain a solution to this problem, let $\psi_1 = 1$, $\psi_2 = t$ and $\psi_3 = t^2$. Clearly, ψ_1, ψ_2 and ψ_3 are linearly independent. Therefore, the Gram matrix G corresponding to these vectors is strictly positive. Moreover, the optimal polynomial $\hat{y}$ is given by (16.39). In this case, the entries of the Gram matrix are of the form

$$(\psi_j, \psi_i) = \int_0^1 t^{i-1} t^{j-1} \, dt = \frac{1}{i+j-1}.$$

So, the Gram matrix G is given by

$$G = \begin{bmatrix} 1 & 1/2 & 1/3 \\ 1/2 & 1/3 & 1/4 \\ 1/3 & 1/4 & 1/5 \end{bmatrix}. \tag{16.43}$$

Furthermore, the cross covariance vector γ is given by

$$\gamma = \begin{bmatrix} (e^t, 1) \\ (e^t, t) \\ (e^t, t^2) \end{bmatrix} = \begin{bmatrix} e^1 - 1 \\ 1 \\ e^1 - 2 \end{bmatrix}. \tag{16.44}$$

So, by combining (16.43) and (16.44), we see that the optimal polynomial $\hat{y}$ of degree at most 2 approximating e^t in the $L^2[0, 1]$ norm is

$$\hat{y} = \begin{bmatrix} 1 & t & t^2 \end{bmatrix} G^{-1}\gamma \approx 1.013 + .8511t + .8392t^2. \tag{16.45}$$

Finally, notice that the optimal polynomial $\hat{y}$ in (16.45) does not equal $1 + t/1! + t^2/2!$ which comes from the Taylor series expansion of e^t.

16.4   An application to curve fitting

In this section, we will use the Vandermonde matrix to solve a classical least squares polynomial fit problem. To this end, let $\deg p$ denote the degree of a polynomial p. Now let $\{\lambda_1, \lambda_2, \cdots, \lambda_m\}$ be a set of distinct complex numbers and $\{y_1, y_2, \cdots, y_m\}$ be a set of complex numbers. Our problem is to find a polynomial $\hat{p}$ of a complex variable λ of degree at most $n - 1$ to solve the following classical polynomial curve fitting problem:

$$d^2 := \inf \left\{ \sum_{i=1}^{m} |y_i - p(\lambda_i)|^2 : p \text{ is a polynomial and } \deg p \leq n - 1 \right\}. \tag{16.46}$$

Here d is the error for the polynomial fit. Moreover, we say that $\hat{p}$ is a solution to this curve fitting problem if $\hat{p}$ is a polynomial of degree at most $n - 1$ and

$$d^2 = \sum_{i=1}^{m} |y_i - \hat{p}(\lambda_i)|^2. \tag{16.47}$$

Without loss of generality, we assume that $n < m$. If $n \geq m$, then we can find a polynomial p of degree at most $m - 1 \leq n - 1$ such that $p(\lambda_i) = y_i$ for all $i = 1, 2, \cdots, m$. In fact, one such polynomial is obtained by the classical Lagrange interpolation formula

$$p(\lambda) = \sum_{i=1}^{m} y_i p_i(\lambda) \qquad \text{where} \qquad p_i(\lambda) = \prod_{\substack{j=1 \\ i \neq j}}^{m} \frac{(\lambda - \lambda_j)}{(\lambda_i - \lambda_j)}. \tag{16.48}$$

(Notice that $p_i(\lambda_i) = 1$ and $p_i(\lambda_j) = 0$ for $j \neq i$.) Moreover, this is the only polynomial of degree at most $m - 1$ satisfying the interpolation conditions $p_i(\lambda_i) = y_i$ for $i = 1, 2, \cdots, m$. To see this, assume that q is another polynomial of degree of at most $m - 1$ satisfying $q_i(\lambda_i) = y_i$ for $i = 1, 2, \cdots, m$. Then $p - q$ is a polynomial of degree at most $m - 1$ with m roots. Hence, $p - q$ must be zero and this proves our claim.

Recall that a Vandermonde matrix is a matrix of the form:

$$V = \begin{bmatrix} 1 & \lambda_1 & \lambda_1^2 & \cdots & \lambda_1^{n-1} \\ 1 & \lambda_2 & \lambda_2^2 & \cdots & \lambda_2^{n-1} \\ 1 & \lambda_3 & \lambda_3^2 & \cdots & \lambda_3^{n-1} \\ \vdots & \vdots & \vdots & & \vdots \\ 1 & \lambda_m & \lambda_m^2 & \cdots & \lambda_m^{n-1} \end{bmatrix} \tag{16.49}$$

where $\{\lambda_1, \lambda_2, \cdots, \lambda_m\}$ are complex numbers. Notice that, if p is a polynomial of the form $p(\lambda) = \sum_{i=0}^{n-1} \alpha_i \lambda^i$, then V can be used to evaluate p at the points $\{\lambda_1, \lambda_2, \cdots, \lambda_m\}$ in the following fashion:

$$V \begin{bmatrix} \alpha_o \\ \alpha_1 \\ \vdots \\ \alpha_{n-1} \end{bmatrix} = \begin{bmatrix} p(\lambda_1) \\ p(\lambda_2) \\ \vdots \\ p(\lambda_m) \end{bmatrix}. \tag{16.50}$$

We claim that if the $\{\lambda_i\}_1^m$ are distinct and $n \leq m$, then V is one to one. If $V\alpha = 0$ for some α in $\mathbb{C}^n$, then (16.50) shows that $p(\lambda_j) = 0$ for $j = 1, 2, \cdots, m$. However, p is a polynomial of degree at most $n - 1$, with m roots. Since $n - 1 < m$, it follows that $p(\lambda) = 0$ for all λ and thus $\alpha = 0$. Hence, V is one to one when $n \leq m$.

If the $\{\lambda_i\}_1^m$ are distinct and $n = m$, then the Vandermonde matrix is invertible. In this case, (16.50) shows that the unique polynomial of degree at most $m - 1$ satisfying the interpolation conditions $p_i(\lambda_i) = y_i$ for $i = 1, 2, \cdots, m$ is given by

$$p(\lambda) = \begin{bmatrix} 1 & \lambda & \cdots & \lambda^{m-1} \end{bmatrix} V^{-1} \begin{bmatrix} y_1 & y_2 & \cdots & y_m \end{bmatrix}^{tr}.$$

This is precisely the polynomial obtained by the Lagrange interpolation formula (16.48).

As before, consider the polynomial interpolation problem in (16.46) where $n < m$. Using (16.49) and (16.50), it follows that this interpolation problem is equivalent to the following least squares optimization problem:

$$\|y - V\hat{\alpha}\| = d := \inf\{\|y - V\alpha\| : \alpha \in \mathbb{C}^n\} \tag{16.51}$$

where $y = \begin{bmatrix} y_1 & y_2 & \cdots & y_m \end{bmatrix}^{tr}$ and $\|\cdot\|$ is the standard norm on $\mathbb{C}^m$. If $\hat{\alpha}$ is the solution to this optimization problem, then the corresponding optimal polynomial given by

$$\hat{p}(\lambda) = \sum_{i=0}^{n-1} \hat{\alpha}_i \lambda^i \quad \text{where} \quad \hat{\alpha} = \begin{bmatrix} \hat{\alpha}_0 & \hat{\alpha}_1 & \cdots \hat{\alpha}_{n-1} \end{bmatrix}^{tr}$$

is the solution to the least squares optimization in (16.51). According to Part (v) of Theorem 16.2.4, the optimal solution $\hat{\alpha}$ is unique and is given by $\hat{\alpha} = (V^*V)^{-1}V^*y$. Therefore, the polynomial $\hat{p}$ which solves the least squares problem in (16.46) is given by

$$\hat{p}(\lambda) = \sum_{i=0}^{n-1} \hat{\alpha}_i \lambda^i = \begin{bmatrix} 1 & \lambda & \cdots & \lambda^{n-1} \end{bmatrix} (V^*V)^{-1}V^*y.$$

The error d is computed by

$$d^2 = \|y\|^2 - ((V^*V)^{-1}V^*y, V^*y).$$

Moreover, $d = d(y, \operatorname{ran} V)$, where $d(y, \operatorname{ran} V)$ is the distance from y to the range of V.

Remark 16.4.1 The above analysis shows that a square Vandermonde matrix generated by the scalars $\{\lambda_i\}_1^m$ is nonsingular if and only if $\{\lambda_i\}_1^m$ are distinct.

16.5 Minimum norm problems

This section is devoted to minimum norm problems. As before, let T be an operator from $\mathcal{U}$ to $\mathcal{Y}$ with finite rank or closed range. Recall that the restricted inverse T^{-r} of T is the operator mapping $\mathcal{Y}$ into $\mathcal{U}$ defined by $\hat{u} = T^{-r}y$ where $\hat{u}$ is the unique vector in $(\ker T)^{\perp}$ satisfying $T\hat{u} = P_{\mathcal{R}}y$ and $P_{\mathcal{R}}$ is the orthogonal projection onto the range $\mathcal{R}$ of T. This sets the stage for the following *minimum norm optimization problem* :

Given a specified vector y in $\mathcal{Y}$, find an optimal $\hat{u}$ in $\mathcal{U}$ such that

$$\|\hat{u}\| = \inf\left\{\|u\| : u \in \mathcal{U} \text{ and } Tu = P_{\mathcal{R}}y\right\} . \tag{16.52}$$

In other words, given a specified vector y in $\mathcal{Y}$ find an optimal $\hat{u}$ in $\mathcal{U}$ satisfying

$$\|\hat{u}\| = \inf\left\{\|u\| : u \in \mathcal{U} \text{ and } \|y - Tu\| = d(y, \mathcal{R})\right\} . \tag{16.53}$$

Here $d(y, \mathcal{R})$ is the distance from y to the range of T in the $\mathcal{Y}$ norm.

Theorem 16.5.1 *Let T be an operator from $\mathcal{U}$ to $\mathcal{Y}$ with closed range $\mathcal{R}$ and let y be in $\mathcal{Y}$. Then there exists a unique solution $\hat{u}$ in $\mathcal{U}$ to the minimum norm problem (16.52). Moreover, this solution is given by $\hat{u} = T^{-r}y$. In other words, the optimal $\hat{u}$ is the unique element in $(\ker T)^{\perp}$ satisfying $T\hat{u} = P_{\mathcal{R}}y$.*

PROOF. Since $\hat{y} = P_{\mathcal{R}}y$ is in $\operatorname{ran}T$, it follows that there exists an element u in $\mathcal{U}$ satisfying $\hat{y} = Tu$. By the Projection Theorem u admits a unique decomposition of the form $u = \hat{u} + \tilde{u}$ where $\hat{u}$ is in $(\ker T)^{\perp}$ and $\tilde{u}$ is in $\ker T$. We claim that $\hat{u}$ is unique, that is, there is only one $\hat{u}$ in $(\ker T)^{\perp}$ satisfying $\hat{y} = T\hat{u}$. To see this, assume that $\hat{y} = T\hat{u} = T\hat{v}$ where both $\hat{u}$ and $\hat{v}$ are in $(\ker T)^{\perp}$. Then $T(\hat{u} - \hat{v}) = \hat{y} - \hat{y} = 0$. Thus, $\hat{u} - \hat{v}$ is in $\ker T$. However, $(\ker T)^{\perp}$ is a linear space. So, $\hat{u} - \hat{v}$ is also in $(\ker T)^{\perp}$. This implies that $\hat{u} - \hat{v} \in (\ker T)^{\perp} \cap \ker T = \{0\}$. Hence, $\hat{u} - \hat{v} = 0$, or equivalently, $\hat{u} = \hat{v}$. So, there is only one $\hat{u}$ in $(\ker T)^{\perp}$ satisfying $T\hat{u} = \hat{y}$.

The previous analysis shows that the set of all solutions u to $Tu = \hat{y}$ (or equivalently $\|y - Tu\| = d(y, T\mathcal{U})$) is given by $u = \hat{u} + \tilde{u}$ where $\hat{u}$ is the unique vector in $(\ker T)^{\perp}$ satisfying $T\hat{u} = \hat{y}$ and $\tilde{u}$ is any vector in $\ker T$. Since $\ker T$ is orthogonal to $(\ker T)^{\perp}$

$$\|u\|^2 = \|\hat{u} + \tilde{u}\|^2 = \|\hat{u}\|^2 + 2\Re(\hat{u}, \tilde{u}) + \|\tilde{u}\|^2 = \|\hat{u}\|^2 + \|\tilde{u}\|^2 \geq \|\hat{u}\|^2 .$$

Notice that we have equality $\|u\|^2 = \|\hat{u}\|^2$ if and only if $\|\tilde{u}\| = 0$, or equivalently, $u = \hat{u}$. Therefore, the solution $\hat{u}$ to the minimum norm optimization problem in (16.52) is unique and given by the unique element $\hat{u}$ in $(\ker T)^{\perp}$ satisfying $T\hat{u} = \hat{y} = P_{\mathcal{R}}y$. ∎

The following two special cases of the minimum norm problem plays a fundamental role in many applications.

Case 1.   If the kernel of T is zero, then the infimum in (16.52) and (16.53) is not needed. In this case, there is only one solution to the equation $Tu = P_{\mathcal{R}}y$, and thus, the optimal $\hat{u}$ is the only vector in $\mathcal{U}$ satisfying $T\hat{u} = P_{\mathcal{R}}y$. In other words, if T is one to one, then the optimization problems in (16.52) and (16.53) reduce to

$$\|y - T\hat{u}\| = \inf\{\|y - Tu\| : u \in \mathcal{U}\} . \tag{16.54}$$

Case 2.   On the other hand, if T is onto $\mathcal{Y}$, then the optimization problems in (16.52) and (16.53) are equivalent to finding the optimal $\hat{u}$ in $\mathcal{U}$ satisfying

$$\|\hat{u}\| = \inf\{\|u\| : u \in \mathcal{U} \text{ and } Tu = y\} . \tag{16.55}$$

By combining Theorems 16.2.4 and 16.5.1, we obtain the following result which is useful in the computation of the optimal $\hat{u}$.

Corollary 16.5.2 *Let T be an operator from $\mathcal{U}$ into $\mathcal{Y}$ with closed range $\mathcal{R}$ and let y be a vector in $\mathcal{Y}$. Then the following holds.*

(i) The unique solution $\hat{u}$ to the minimum norm optimization problem (16.52) is given by

$$\hat{u} = (T^*T)^{-r}T^*y = T^*(TT^*)^{-r}y \qquad and \qquad \|\hat{u}\|^2 = ((TT^*)^{-r}y, y). \qquad (16.56)$$

(ii) If T is one to one, then T^*T is invertible and $\hat{u} = (T^*T)^{-1}T^*y$. Moreover, this $\hat{u}$ is the unique solution to the optimization problem (16.54) in Case 1.

(iii) If T is onto, then TT^* is invertible and $\hat{u} = T^*(TT^*)^{-1}y$. Furthermore, this $\hat{u}$ is the unique solution to the optimization problem (16.55) in Case 2.

PROOF. It remains to verify the last equation in (16.56). This follows from

$$\begin{aligned}\|\hat{u}\|^2 &= (\hat{u}, \hat{u}) = (T^*(TT^*)^{-r}y, T^*(TT^*)^{-r}y) \\ &= (TT^*(TT^*)^{-r}y, (TT^*)^{-r}y) = ((TT^*)^{-r}y, y)\end{aligned}$$

which completes the proof. ∎

Example 16.5.1 Let T be the column matrix from $\mathbb{C}^1$ to $\mathbb{C}^3$ defined by

$$T = \begin{bmatrix} 1 \\ 1 \\ 1 \end{bmatrix} \qquad and \qquad y = \begin{bmatrix} 1 \\ 2 \\ 0 \end{bmatrix}.$$

Then find the scalar $\hat{u}$ to solve the minimum norm optimization problem in (16.52). Because $\ker T = \{0\}$, this optimization problem corresponds to Case 1 above.

SOLUTION. Part (ii) of the previous theorem shows that the optimal $\hat{u} = (T^*T)^{-1}T^*y$. Clearly, $T^*T = 3$ and $T^*y = 3$. Hence, $\hat{u} = (T^*T)^{-1}T^*y = 1$ is the optimal solution. Notice that the second formula $\hat{u} = T^*(TT^*)^{-r}y$ in (16.56) is much harder to use. It requires the computation of the restricted inverse of the nonsingular 3×3 matrix TT^*.

Example 16.5.2 Let T be the operator from $\mathbb{C}^3$ to $\mathbb{C}^1$ defined by

$$T = \begin{bmatrix} 1 & 1 & 1 \end{bmatrix} \qquad and \qquad y = 2.$$

Then find the optimal $\hat{u}$ to solve the following minimum norm optimization problem:

$$\|\hat{u}\| = \inf\{ \|u\| : u \in \mathbb{C}^3 \text{ and } Tu = 2 \}.$$

Notice that in this case the range of T is $\mathbb{C}^1$. This corresponds to Case 2 above.

SOLUTION. According to Part (iii) of the previous theorem, the optimal solution is given by $\hat{u} = T^*(TT^*)^{-1}y$. Since $TT^* = 3$, it follows that

$$\hat{u} = T^*(TT^*)^{-1}y = \begin{bmatrix} 1 \\ 1 \\ 1 \end{bmatrix} 3^{-1}2 = \frac{2}{3}\begin{bmatrix} 1 \\ 1 \\ 1 \end{bmatrix}.$$

Notice that the first formula $\hat{u} = (T^*T)^{-r}T^*y$ in (16.56) is much harder to use. It requires the computation of the restricted inverse of the nonsingular 3×3 matrix T^*T.

Example 16.5.3 Find a function $\hat{u}$ in $L^2[0,1]$ such that

$$\|\hat{u}\|^2 = \inf \int_0^1 |u(t)|^2 \, dt \qquad \text{subject to} \qquad \int_0^1 e^t u(t) \, dt = 2 \,.$$

SOLUTION. Let T be the operator mapping $L^2[0,1]$ into $\mathbb{C}^1$ defined by

$$Tu = \int_0^1 e^t u(t) \, dt \qquad (u \in L^2[0,1]) \,.$$

Clearly, T is onto and this corresponds to Case 2 above. So, $y = 2$ is in the range of T and this optimization problem makes sense. We claim that the adjoint T^* of T is the operator mapping $\mathbb{C}^1$ to $L^2[0,1]$ defined by

$$(T^*\gamma)(t) = e^t \gamma \qquad (\gamma \in \mathbb{C}^1) \,.$$

To see this notice that

$$(u, T^*\gamma)_{L^2} = (Tu, \gamma)_{\mathbb{C}^1} = \int_0^1 e^t u(t) \, dt \, \bar{\gamma} = \int_0^1 u(t) \overline{e^t \gamma} \, dt = (u, e^t \gamma)_{L^2} \,.$$

(Here $(f,g)_{\mathcal{H}}$ denotes the inner product on the Hilbert space $\mathcal{H}$.) Therefore, $(T^*\gamma)(t) = e^t \gamma$. By equation (16.56) in Corollary 16.5.2, the optimal solution $\hat{u} = T^*(TT^*)^{-1} 2$. To find $\hat{u}$ notice that the operator TT^* on $\mathbb{C}^1$ is given by

$$TT^* = \int_0^1 e^t e^t \, dt = \frac{e^2 - 1}{2} \,.$$

Hence, the optimal solution

$$\hat{u}(t) = T^*(TT^*)^{-1} 2 = \frac{4e^t}{e^2 - 1} \,.$$

Finally, recall that the optimal solution is also given by $\hat{u} = (T^*T)^{-r} T^* y$. To use this formula one would have to compute the restricted inverse of the operator T^*T on $L^2[0,1]$. This is obviously much harder to do.

Example 16.5.4 Find the optimal function $\hat{u} = \begin{bmatrix} \hat{u}_1 & \hat{u}_2 \end{bmatrix}^{tr}$ in $L^2([0,1], \mathbb{C}^2)$ to solve the following optimization problem:

$$\|\hat{u}\|^2 = \inf \int_0^1 (|u_1(t)|^2 + |u_2(t)|^2) \, dt \qquad \text{subject to} \qquad \int_0^1 (u_1(t) + t u_2(t)) \, dt = 4 \,.$$

SOLUTION. Let T be the operator from $L^2([0,1], \mathbb{C}^2)$ into $\mathbb{C}^1$ defined by

$$T \begin{bmatrix} u_1 & u_2 \end{bmatrix}^{tr} = \int_0^1 (u_1(t) + t u_2(t)) \, dt \,.$$

Clearly, T is onto and this corresponds to Case 2 above. A simple computation shows that T^* is the operator from $\mathbb{C}^1$ to $L^2([0,1], \mathbb{C}^2)$ given by

$$(T^*\gamma)(t) = \begin{bmatrix} 1 \\ t \end{bmatrix} \gamma \qquad (\gamma \in \mathbb{C}^1).$$

Because T is onto, the optimal $\hat{u} = T^*(TT^*)^{-1}y$ where $y = 4$. Observe that

$$TT^* = \int_0^1 \begin{bmatrix} 1 & t \end{bmatrix} \begin{bmatrix} 1 \\ t \end{bmatrix} dt = \frac{4}{3}.$$

So, the optimal function $\hat{u}$ is given by

$$\hat{u}(t) = T^*(TT^*)^{-1}4 = 3\begin{bmatrix} 1 \\ t \end{bmatrix}.$$

Exercise 40 Find the optimal function $\hat{u} = \begin{bmatrix} \hat{u}_1 & \hat{u}_2 \end{bmatrix}^{tr}$ to solve

$$\|\hat{u}\|^2 = \inf \int_0^1 (|u_1(t)|^2 + |u_2(t)|^2)\, dt \quad \text{subject to} \quad \int_0^1 \begin{bmatrix} 2t & 1 \\ 0 & 2 \end{bmatrix} \begin{bmatrix} u_1(t) \\ u_2(t) \end{bmatrix} dt = \begin{bmatrix} 1 \\ 2 \end{bmatrix}.$$

HINT. Let T be the operator from $L^2([0,1], \mathbb{C}^2)$ into $\mathbb{C}^2$ defined by

$$T\begin{bmatrix} u_1 \\ u_2 \end{bmatrix} = \int_0^1 \begin{bmatrix} 2t & 1 \\ 0 & 2 \end{bmatrix} \begin{bmatrix} u_1(t) \\ u_2(t) \end{bmatrix} dt.$$

Show that T is onto and the adjoint T^* of T is the operator from $\mathbb{C}^2$ into $L^2([0,1], \mathbb{C}^2)$ given by

$$T^*\begin{bmatrix} \gamma_1 \\ \gamma_2 \end{bmatrix} = \begin{bmatrix} 2t & 0 \\ 1 & 2 \end{bmatrix} \begin{bmatrix} \gamma_1 \\ \gamma_2 \end{bmatrix} \qquad (\begin{bmatrix} \gamma_1 & \gamma_2 \end{bmatrix}^{tr} \in \mathbb{C}^2).$$

16.6   The singular value decomposition

This section, presents a brief description of the singular value decomposition for finite rank operators and its relationship to the restricted inverse. First, we introduce the polar decomposition of an operator.

Polar decomposition.   Recall that an isometry U is a linear operator mapping $\mathcal{U}$ into $\mathcal{Y}$ satisfying $U^*U = I$, or equivalently, $\|Uu\| = \|u\|$ for all u in $\mathcal{U}$. If Q is a positive operator on a Hilbert space $\mathcal{H}$, then Q has a positive root, that is, there exists a unique positive operator S on $\mathcal{H}$ satisfying $S^2 = Q$. In this case, S is denoted by $Q^{1/2}$. In fact, there exists a sequence of polynomials $\{p_n\}_1^\infty$ such that $\{p_n(Q)h\}_1^\infty$ converges to $Q^{1/2}h$ for every h in $\mathcal{H}$; see Halmos [59]. Now let T be any operator mapping $\mathcal{U}$ into $\mathcal{Y}$. The decomposition

$$T = WR \tag{16.57}$$

where R is a positive operator and W is an isometry from the closure of the range of R into $\mathcal{Y}$, is known as the *polar decomposition* of T. Notice that the polar decomposition of an operator generalizes the result that any complex number γ has a polar representation of the form $\gamma = e^{i\omega}r$ where $r = |\gamma|$ and ω is in $[0, 2\pi)$. We claim that any operator T admits a polar decomposition. To demonstrate this, let u be in $\mathcal{U}$. Then

$$\|Tu\|^2 = (T^*Tu, u) = \|(T^*T)^{1/2}u\|^2 = \|Ru\|^2 \tag{16.58}$$

where R is the positive square root of T^*T. We now show that as a consequence of (16.58), we can define a linear map W from $\operatorname{ran} R$ to $\mathcal{Y}$ by $WRu = Tu$ where u is in $\mathcal{U}$. Clearly, $WRu = Tu$ defines a relation W from $\operatorname{ran} R$ to $\mathcal{Y}$. This relation is a function if and only if Wv has only one value for every v in $\operatorname{ran} R$, or equivalently, if $WRu_1 = Tu_1$ and $WRu_2 = Tu_2$ and $Ru_1 = Ru_2$, then $Tu_1 = Tu_2$. If $Ru_1 = Ru_2$, then (16.58) shows that

$$\|Tu_1 - Tu_2\| = \|T(u_1 - u_2)\| = \|R(u_1 - u_2)\| = \|Ru_1 - Ru_2\| = 0.$$

Hence, $Tu_1 = Tu_2$ and W is a well defined function from $\operatorname{ran} R$ into $\mathcal{Y}$. The linearity of W follows from the linearity of R and T. Utilizing (16.58) we obtain $\|WRu\| = \|Tu\| = \|Ru\|$. This shows that W is an isometry. Finally, the isometry W has a unique extension by continuity to the closure of $\operatorname{ran} R$ and we also denote this isometry by W. Hence, $T = WR$ is the polar decomposition of T.

Singular value decomposition.   Let T be an operator of finite rank n mapping $\mathcal{U}$ into $\mathcal{Y}$. Then a singular value decomposition of T is a factorization of the form

$$T = U\Lambda V^* \tag{16.59}$$

where U is an isometry mapping $\mathbb{C}^n$ into $\mathcal{Y}$ and V is an isometry from $\mathbb{C}^n$ into $\mathcal{U}$ while Λ is a diagonal matrix with nonzero positive diagonal elements arranged in decreasing order, that is,

$$\Lambda = \operatorname{diag}(\sigma_1, \sigma_2, \cdots, \sigma_n) \quad \text{with} \quad \sigma_1 \geq \sigma_2 \geq \cdots \geq \sigma_n > 0. \tag{16.60}$$

The numbers $\{\sigma_1, \sigma_2, \cdots, \sigma_n\}$, are referred to as the *singular values* of T. In this presentation, we do not consider zero a singular value for T.

   To obtain a singular value decomposition of T, let $T = WR$ be the polar decomposition of T where R is the positive square root of T^*T and W is an isometry. Since R is a positive operator of finite rank n, it admits a spectral decomposition of the form $R = V\Lambda V^*$, where V is an isometry from $\mathbb{C}^n$ into $\mathcal{U}$ and Λ is given by (16.60) where $\sigma_1 \geq \sigma_2 \geq \cdots \geq \sigma_n > 0$ are the nonzero eigenvalues of R. Substituting $R = V\Lambda V^*$ into the polar decomposition of T, we arrive at $T = WV\Lambda V^*$. By letting $U = WV$, we obtain the singular value decomposition $T = U\Lambda V^*$.

   Let $U\Lambda V^*$ be a singular value decomposition of T. Because U is an isometry, $T^*T = V\Lambda^2 V^*$; this is precisely a spectral decomposition of T^*T. Hence, the squares of the nonzero singular values $\{\sigma_1^2, \sigma_2^2, \cdots, \sigma_n^2\}$ of T are precisely the nonzero eigenvalues of T^*T. Since $T^* = V\Lambda U^*$ is the singular value decomposition of T^*, it follows that T and T^* have the same nonzero singular values. Moreover, using the fact that V is an isometry, $TT^* = U\Lambda^2 U^*$ is the spectral decomposition of TT^*. In particular, T^*T and TT^* have the same nonzero eigenvalues which are the squares of the nonzero singular values of T, or equivalently, T^*. We are now ready to prove the following result.

Theorem 16.6.1 *Let T be a finite rank operator with singular value decomposition $T = U\Lambda V^*$. Then the following statements hold.*

(i) The orthogonal projection onto the range of T is UU^*

(ii) The orthogonal projection onto $(\operatorname{ran} T)^\perp = \ker T^* = I - UU^*$

(iii) The orthogonal projection onto $(\ker T)^\perp$ is VV^*

(iv) The orthogonal projection onto $\ker T = (\operatorname{ran} T^*)^\perp = I - VV^*$

(v) The restricted inverse T^{-r} of T is given by $T^{-r} = V\Lambda^{-1}U^*$.

PROOF. Because V is an isometry, it has zero kernel, and hence, V^* is onto. Since Λ is invertible, it now follows that T and U have the same range which we denote by $\mathcal{R}$. By applying Lemma 16.2.3 with U replacing T, we have $P_\mathcal{R} = U(U^*U)^{-1}U^* = UU^*$. So, UU^* is the orthogonal projection onto $\operatorname{ran} U = \operatorname{ran} T$. This proves (i).

Part (ii) follows from the fact that P is the orthogonal projection onto any subspace $\mathcal{H}$ if and only if $I - P$ is the orthogonal projection onto $\mathcal{H}^\perp$.

Statements (iii) and (iv) follow from Parts (i) and (ii) by observing that $V^*\Lambda U$ is the singular value decomposition of T^*.

To complete the proof it remains to verify Part (v). To this end, recall that the restricted inverse of T is defined by $T^{-r}y = \hat{u}$, where $\hat{u}$ is the unique element of $(\ker T)^\perp$ satisfying $T\hat{u} = P_\mathcal{R}y$. By using the singular value decomposition of T and Part (i), we see that

$$U\Lambda V^*\hat{u} = T\hat{u} = P_\mathcal{R}y = UU^*y. \tag{16.61}$$

Since $U^*U = I$, it follows that $\Lambda V^*\hat{u} = U^*y$, or equivalently, $V^*\hat{u} = \Lambda^{-1}U^*y$. Multiplying by V gives $VV^*\hat{u} = V\Lambda^{-1}U^*y$. However, VV^* is the orthogonal projection onto $(\ker T)^\perp$. Therefore, $VV^*\hat{u} = \hat{u}$ and $T^{-r} = V\Lambda^{-1}U^*$. ∎

When T is a matrix, there are fast and efficient algorithms to compute its singular value decomposition. In this case, the restricted inverse $T^{-r} = V\Lambda^{-1}U^*$ is particularly easy to compute. In many of our applications we will have to compute the restricted inverse of T^*T. In this case, T^*T is a positive operator and its singular value decomposition is precisely the spectral decomposition of T^*T, that is, $T^*T = V\Lambda^2V^*$ where V is an isometry. Hence, $(T^*T)^{-r} = V\Lambda^{-2}V^*$.

16.7 Schmidt pairs

A classical way of obtaining a singular value decomposition for a finite rank operator is through Schmidt pairs. To see this, consider a finite rank operator T from $\mathcal{U}$ into $\mathcal{Y}$. Let $\{u_i\}_1^n$ be an orthonormal basis of eigenvectors corresponding to the nonzero eigenvalues $\sigma_1^2 \geq \sigma_2^2 \geq \cdots \geq \sigma_n^2 > 0$ of T^*T, that is, $T^*Tu_i = \sigma_i^2u_i$ where $(u_i, u_j) = \delta_{ij}$. (Recall that $\delta_{ij} = 1$ if $i = j$ and zero otherwise.) The sequence $\{u_i\}_1^n$ forms an orthonormal basis for $(\ker T^*T)^\perp = (\ker T)^\perp$. Now let $\{y_i\}_1^n$ be the vectors defined by $Tu_i = \sigma_iy_i$, for $i = 1, 2, \cdots, n$. The pair $\{u_i, y_i\}$ is known as a *Schmidt pair* for T corresponding to the singular value σ_i. By multiplying by T^* we see that $\sigma_i^2 u_i = T^* T u_i = T^* \sigma_i y_i$. Therefore, $T^* y_i = \sigma_i u_i$ for $i = 1, 2, \cdots, n$. In particular, $\{u_i, y_i\}$ is a Schmidt pair for T corresponding to σ_i if and only if $\{y_i, u_i\}$ is a Schmidt pair for T^* corresponding to σ_i.

We claim that $\{y_i\}_1^n$ forms an orthonormal basis for $(\ker T^*)^\perp = \operatorname{ran} T$. The orthogonality follows from

$$(y_i, y_j) = \frac{(T u_i, T u_j)}{\sigma_i \sigma_j} = \frac{(T^* T u_i, u_j)}{\sigma_i \sigma_j} = \frac{\sigma_i}{\sigma_j}(u_i, u_j) = \delta_{ij}. \qquad (16.62)$$

Since $\{u_i\}_1^n$ is a basis for $(\ker T)^\perp$ and $T u_i = \sigma_i y_i$, it now follows that $\{y_i\}_1^n$ is an orthonormal basis for $\operatorname{ran} T = (\ker T^*)^\perp = (\ker TT^*)^\perp$. Because $T^* y_i = \sigma_i u_i$ we also see that $TT^* y_i = \sigma_i^2 y_i$. So, T^*T and TT^* have the same nonzero eigenvalues $\{\sigma_1^2, \sigma_2^2, \cdots, \sigma_n^2\}$. Summing up the previous analysis, we see that the Schmidt pairs $\{u_i, y_i\}_1^n$ corresponding to the singular values $\{\sigma_i\}_1^n$, satisfy the following relationship

$$T u_i = \sigma_i y_i \qquad \text{and} \qquad T^* y_i = \sigma_i u_i \qquad (\text{for } i = 1, \cdots, n). \qquad (16.63)$$

Moreover, $\{u_i\}_1^n$ are orthonormal eigenvectors for T^*T corresponding to the nonzero eigenvalues $\sigma_1^2 \geq \sigma_2^2 \geq \cdots \geq \sigma_n^2$ of T^*T, while the vectors $\{y_i\}_1^n$ are orthonormal eigenvectors for TT^* corresponding to the nonzero eigenvalues $\sigma_1^2 \geq \sigma_2^2 \geq \cdots, \geq \sigma_n^2$ of TT^*.

Since $\{u_i\}_1^n$ is a basis for $(\ker T)^\perp$, Remark 16.3.2 shows that the orthogonal projection $P_{\mathcal{H}}$ onto $\mathcal{H} = (\ker T)^\perp$ is given by

$$P_{\mathcal{H}} u = \sum_{i=1}^n (u, u_i) u_i \qquad (u \in \mathcal{U}). \qquad (16.64)$$

Using this expression for $P_{\mathcal{H}}$ along with $T u_i = \sigma_i y_i$, we have

$$T u = T P_{\mathcal{H}} u = T \sum_{i=1}^n (u, u_i) u_i = \sum_{i=1}^n \sigma_i y_i (u, u_i) = \begin{bmatrix} y_1 & y_2 & \cdots & y_n \end{bmatrix} \Lambda \begin{bmatrix} (u, u_1) \\ (u, u_2) \\ \vdots \\ (u, u_n) \end{bmatrix} \qquad (16.65)$$

where $\Lambda = \operatorname{diag}(\sigma_1, \sigma_2, \cdots, \sigma_n)$.

To obtain a singular value decomposition, let V be the isometry mapping $\mathbb{C}^n$ into $\mathcal{U}$ defined by

$$V\alpha = \begin{bmatrix} u_1 & u_2 & \cdots & u_n \end{bmatrix} \alpha = \sum_{i=1}^n \alpha_i u_i \qquad (16.66)$$

where $\alpha = [\alpha_1, \alpha_2, \cdots, \alpha_n]^{tr}$ is a vector in $\mathbb{C}^n$. Notice that V is an isometry because $\{u_i\}_1^n$ is an orthonormal set. Recall that $V^* f = [(f, u_1), \cdots, (f, u_n)]^{tr}$; see (16.29) and (16.32) with T replaced by V. Let U be the isometry mapping $\mathbb{C}^n$ into $\mathcal{Y}$ defined by

$$U\alpha = \begin{bmatrix} y_1 & y_2 & \cdots & y_n \end{bmatrix} \alpha = \sum_{i=1}^n \alpha_i y_i. \qquad (16.67)$$

Now the singular value decomposition $T = U\Lambda V^*$ follows from (16.65). In other words, one can obtain the singular value decomposition $T = U\Lambda V^*$ by computing the orthonormal eigenvectors $\{u_i\}_1^n$ for the nonzero eigenvalues $\sigma_1^2 \geq \sigma_2^2 \geq \cdots \geq \sigma_n^2 > 0$ of T^*T, that is $T^*Tu_i = \sigma_i^2 u_i$. Then the scalars $\sigma_1, \sigma_2, \cdots, \sigma_n$ are the singular values of T. Next compute the orthonormal eigenvectors $\{y_i\}_1^n$ for TT^* by $Tu_i = \sigma_i y_i$. Then the singular value decomposition of T is

$$Tu = U\Lambda V^* u = \sum_{i=1}^n \sigma_i (u, u_i) y_i \qquad (u \in \mathcal{U}) \tag{16.68}$$

where U and V are the isometries in (16.66) and (16.67) formed by the Schmidt pairs $\{u_i\}_1^n$ and $\{y_i\}_1^n$, respectively. It is noted that in some applications it may be easier to compute the orthonormal eigenvectors y_i for TT^* first. For example, if T is a matrix with more columns than rows. Then compute the vectors $u_i = T^*y_i/\sigma_i$ in the Schmidt pair $\{u_i, y_i\}$. Finally, it is noted that the restricted inverse T^{-r} of T is given by

$$T^{-r}y = V\Lambda^{-1}U^*y = \sum_{i=1}^n \frac{(y, y_i)}{\sigma_i} u_i \qquad (y \in \mathcal{Y}). \tag{16.69}$$

The last equality follows from the fact that U^*y equals $[\ (y, y_1)\ \ (y, y_2)\ \cdots\ (y, y_n)\ ]^{tr}$.

Remark 16.7.1 Let T be any finite rank operator mapping $\mathcal{U}$ into $\mathcal{Y}$. We can also use the singular value decomposition of T to obtain another proof of equation (16.26) in Theorem 16.2.4, that is,

$$T^{-r} = (T^*T)^{-r}T^* = T^*(TT^*)^{-r}. \tag{16.70}$$

To verify this result, let $T = U\Lambda V^*$ be the singular value decomposition of T. Since U is an isometry, the singular value decomposition of T^*T is $V\Lambda^2 V^*$. Therefore, the restricted inverse of T^*T is $V\Lambda^{-2}V^*$. Using this and $T^* = V\Lambda U^*$ we have

$$T^{-r} = V\Lambda^{-1}U^* = V\Lambda^{-2}\Lambda U^* = V\Lambda^{-2}V^*V\Lambda U^* = (T^*T)^{-r}T^*.$$

To obtain the second equality in (16.70), notice that the singular value decomposition of TT^* is $U\Lambda^2 U^*$. So, using $(TT^*)^{-r} = U\Lambda^{-2}U^*$ we have

$$T^{-r} = V\Lambda^{-1}U^* = V\Lambda\Lambda^{-2}U^* = V\Lambda U^*U\Lambda^{-2}U^* = T^*(TT^*)^{-r}.$$

This completes the proof of (16.70).

Now, let T be a finite rank operator and $T = U\Lambda V^*$ be a singular value decomposition of T. Because U and V are isometries, it follows that $\|T\| = \|\Lambda\|$. Therefore, $\|T\| = \sigma_1$, the largest singular value of T, or equivalently, $\|T\|^2 = \sigma_1^2$ where σ_1^2 is the largest eigenvalue of T^*T. We say that u attains the norm of T if u is a nonzero vector in $\mathcal{U}$ satisfying $\|Tu\| = \|T\|\|u\|$. Obviously, the vector u_1 corresponding to the largest singular value of T attains the norm of T. In other words, the norm of T is attained by the eigenvectors corresponding to the largest eigenvalue of T^*T. In many applications, it is easier to compute the orthogonal eigenvectors $\{y_i\}_1^n$ for TT^* rather than the orthogonal eigenvectors $\{u_i\}_1^n$ for T^*T. In this case, the vector $u_1 = T^*y_1/\sigma_1$ attains the norm of T, where y_1 is an eigenvector corresponding to the largest eigenvalue σ_1^2 of TT^*.

Consider the following classical problem in linear algebra: Given a finite rank operator T and a positive integer k, construct an operator T_k of rank less than or equal to k which comes closest to T in the operator norm. The following classical result uses the singular value decomposition to solve this problem.

Theorem 16.7.1 *Let T mapping $\mathcal{U}$ into $\mathcal{Y}$ be a finite rank operator and $T = U\Lambda V^*$ be its singular value decomposition where $\sigma_1 \geq \sigma_2 \geq \cdots \geq \sigma_n > 0$ are the singular values of T and $\Lambda = diag\,(\sigma_1, \sigma_2, \cdots, \sigma_n)$. If $k < n$, then*

$$\sigma_{k+1} = \inf\{\|T - Z\| : Z \in \mathcal{L}(\mathcal{U}, \mathcal{Y}) \text{ and } rank\,Z \leq k\}. \tag{16.71}$$

Moreover, an operator T_k of rank k which solves the optimization problem in (16.71) is given by

$$T_k = U\,diag\,(\sigma_1, \sigma_2, \cdots, \sigma_k, 0, \cdots, 0)V^*, \tag{16.72}$$

that is, $\|T - T_k\| = \sigma_{k+1}$.

PROOF. Let $\{u_i, y_i\}_1^n$ be a Schmidt pair for T, that is, $\{u_i\}_1^n$ and $\{y_i\}_1^n$ are orthonormal sets satisfying $Tu_i = \sigma_i y_i$ for all i. Let Z be any operator whose rank is less than or equal to $k < n$ and let $\mathcal{H}$ be the $k + 1$ dimensional space spanned by $\{u_1, u_2, \cdots, u_{k+1}\}$. Since $\dim \ker(Z)^{\perp} \leq k < \dim \mathcal{H}$, it follows that there exists a unit vector u in $(\ker Z) \cap \mathcal{H}$. (If $\mathcal{M}$ is any subspace of $\mathcal{U}$ where $\dim \mathcal{M} \leq k$, then there exists a unit vector u in $\mathcal{H}$ which is orthogonal to $\mathcal{M}$. To see this, let $P_{\mathcal{H}}$ be the orthogonal projection onto $\mathcal{H}$ and Q the operator from $\mathcal{M}$ into $\mathcal{H}$ defined by $Qf = P_{\mathcal{H}}f$. Clearly, Q is not onto, and thus, there exists a unit vector u in $\mathcal{H}$ orthogonal to the range of Q. Hence, $0 = (u, P_{\mathcal{H}}f) = (P_{\mathcal{H}}u, f) = (u, f)$ for all $f \in \mathcal{M}$. Therefore, u is orthogonal to $\mathcal{M}$.) According to Remark 16.3.2, for u in $\mathcal{H}$ we have

$$u = P_{\mathcal{H}}u = \sum_{i=1}^{k+1}(u, u_i)u_i. \tag{16.73}$$

Using the fact that $\{y_i\}_1^n$ is orthonormal with (16.73) and Bessel's equality ($\|\sum \alpha_i \phi_i\|^2 = \sum |\alpha_i|^2$ for any orthonormal set $\{\phi_i\}$ and scalars $\{\alpha_i\}$), we arrive at

$$\|T - Z\|^2 \geq \|(T - Z)u\|^2 = \|Tu\|^2 = \|T\sum_{i=1}^{k+1}(u, u_i)u_i\|^2 = \|\sum_{i=1}^{k+1}(u, u_i)\sigma_i y_i\|^2$$

$$= \sum_{i=1}^{k+1}\sigma_i^2|(u, u_i)|^2 \geq \sigma_{k+1}^2 \sum_{i=1}^{k+1}|(u, u_i)|^2 = \sigma_{k+1}^2\|u\|^2 = \sigma_{k+1}^2.$$

The equality $\|u\|^2 = \sum_{i=1}^{k+1}|(u, u_i)|^2$ follows from (16.41) in Remark 16.3.2. Thus,

$$\sigma_{k+1} \leq \inf\{\|T - Z\| : Z \in \mathcal{L}(\mathcal{U}, \mathcal{Y}) \text{ and } rank\,Z \leq k\}.$$

To obtain equality, notice that the rank of the operator T_k in (16.72) is precisely k. Because U and V are isometries, it follows that the norm of $T - T_k$ equals the norm of the diagonal operator, $diag\,(0, 0, \cdots, 0, \sigma_{k+1}, \cdots, \sigma_n)$. Therefore, $\|T - T_k\| = \sigma_{k+1}$. ∎

Remark 16.7.2 Let T be any finite rank operator mapping $\mathcal{U}$ into $\mathcal{Y}$. Let $\{u_i, y_i\}_1^n$ be the Schmidt pairs corresponding to the singular values $\sigma_1 \geq \sigma_2 \geq \cdots \geq \sigma_n > 0$ of T. Then the operator T_k of rank k in (16.72) satisfying $\|T - T_k\| = \sigma_{k+1}$ can also be written as

$$T_k u = \sum_{i=1}^{k}\sigma_i(u, u_i)y_i \qquad (u \in \mathcal{U}).$$

This follows from (16.68) along with the definition of the isometries U and V in (16.66) and (16.67).

16.8   A control example

In this section we obtain a singular value decomposition for a certain operator T which arises in the analysis of control systems. Suppose G is a continuous function on $[0, t_1]$ with values in $\mathcal{L}(\mathcal{U}, \mathcal{Y})$ where $\mathcal{U}$ and $\mathcal{Y}$ are finite dimensional Hilbert spaces and t_1 is positive. Consider the operator T mapping $L^2([0, t_1], \mathcal{U})$ into $\mathcal{Y}$ defined by the convolution integral:

$$Tu = \int_0^{t_1} G(t_1 - \tau)u(\tau)\, d\tau \qquad (u \in L^2([0, t_1], \mathcal{U}))\,. \tag{16.74}$$

We shall obtain expressions for a singular value decomposition of T and the norm of T. To this end, notice that the adjoint T^* of T is the linear operator mapping $\mathcal{Y}$ into $L^2([0, t_1], \mathcal{U})$ defined by

$$(T^*g)(t) = G(t_1 - t)^*g \qquad (g \in \mathcal{Y})\,. \tag{16.75}$$

To verify this notice that for any u in $L^2([0, t_1], \mathcal{U})$ and g in $\mathcal{Y}$, we have

$$
\begin{aligned}
(Tu, g)_{\mathcal{Y}} &= (\int_0^{t_1} G(t_1 - \tau)u(\tau)\, d\tau,\, g)_{\mathcal{Y}} = \int_0^{t_1} (G(t_1 - \tau)u(\tau),\, g)_{\mathcal{Y}}\, d\tau \\
&= \int_0^{t_1} (u(\tau),\, G(t_1 - \tau)^*g)_{\mathcal{U}}\, d\tau = (u, T^*g)_{L^2}\,.
\end{aligned}
$$

(Here $(f, g)_{\mathcal{H}}$ denotes the inner product on the Hilbert space $\mathcal{H}$.)   Hence, (16.75) holds. Combining equations (16.74) and (16.75), we see that TT^* is the positive operator on $\mathcal{Y}$ given by

$$TT^* = \int_0^{t_1} G(t_1 - \tau)G(t_1 - \tau)^*\, d\tau = \int_0^{t_1} G(\sigma)G(\sigma)^*\, d\sigma\,. \tag{16.76}$$

Let $\{y_1, y_2, \cdots, y_n\}$ be an orthonormal set of eigenvectors corresponding to the nonzero eigenvalues $\sigma_1^2 \geq \sigma_2^2 \geq \cdots \geq \sigma_n^2 > 0$ of TT^*, that is,

$$TT^*y_i = \sigma_i^2 y_i \qquad (\text{for } i = 1, 2, \cdots, n)\,. \tag{16.77}$$

Notice that one can readily compute the eigenvectors y_i and the eigenvalues σ_i^2 because in general $\mathcal{Y} = \mathbb{C}^p$ and thus TT^* is a positive matrix. Now let $\{u_1, u_2, \cdots, u_n\}$ be the orthonormal set defined by

$$u_i = T^*y_i/\sigma_i \qquad (\text{for } i = 1, 2, \cdots, n)\,. \tag{16.78}$$

Then using the formula for T^* in (16.75), we see that

$$u_i(t) = G(t_1 - t)^*y_i/\sigma_i \qquad (\text{for } i = 1, 2, \cdots, n)\,. \tag{16.79}$$

Moreover, $\{u_i, y_i\}_1^n$ forms the Schmidt pairs for the operator T, that is, $\{u_i\}_1^n$ and $\{y_i\}_1^n$ are orthonormal sets satisfying

$$Tu_i = \sigma_i y_i \quad \text{and} \quad T^*y_i = \sigma_i u_i \qquad (\text{for } i = 1, 2, \cdots, n) \tag{16.80}$$

where $\sigma_1 \geq \sigma_2 \geq \cdots \geq \sigma_n > 0$ are the (nonzero) singular values of T. So, according to (16.68), a singular value decomposition of T is given by

$$Tu = \sum_{i=1}^{n} \sigma_i (u, u_i) y_i \qquad (u \in L^2([0, t_1], \mathcal{U})). \tag{16.81}$$

An operator T_k of rank $k < n$ which comes closest to T in the operator norm, over the class of all operators mapping $L^2([0, t_1], \mathcal{U})$ into $\mathcal{Y}$ whose rank is less than or equal to k is given by (see Theorem 16.7.1)

$$T_k u = \sum_{i=1}^{k} \sigma_i (u, u_i) y_i \qquad (u \in L^2([0, t_1], \mathcal{U})). \tag{16.82}$$

Finally, the norm of T is given by

$$\|T\| = \sigma_1 = \lambda_{\max}(TT^*)^{1/2} \tag{16.83}$$

where $\lambda_{\max}(R)$ denotes the maximum eigenvalue of a self-adjoint operator R. Furthermore, a vector u_1 in $L^2([0, t_1], \mathcal{U})$ which attains the norm of T is given by $u_1 = T^* y_1 / \sigma_1$, that is,

$$u_1(t) = G(t_1 - t)^* y_1 / \sigma_1 \tag{16.84}$$

where y_1 is an eigenvector corresponding to the maximum eigenvalue σ_1^2 of TT^*, or equivalently, $\{u_1, y_1\}$ is the Schmidt pair corresponding to the largest singular value σ_1 of T. In many cases $\mathcal{Y} = \mathbb{C}^p$, so (16.83) and (16.84) give us a simple procedure to compute the norm of T and a vector u_1 which attains the norm of T.

16.8.1 State space

Consider the state space system

$$\dot{x} = Ax + Bu \qquad \text{and} \qquad y = Cx \tag{16.85}$$

where the initial state $x(0)$ is zero, A is an operator on a finite dimensional space $\mathcal{X}$, while B maps $\mathcal{U}$ into $\mathcal{X}$ and C maps $\mathcal{X}$ into $\mathcal{Y}$. Here both $\mathcal{U}$ and $\mathcal{Y}$ are finite dimensional spaces. For any positive t_1, the output $y(t_1)$ of system (16.85) is given by $y(t_1) = Tu$ where T is defined by the convolution integral in (16.74) with $G(t) = Ce^{At}B$. In this case, equation (16.76) reduces to

$$TT^* = CQ(t_1)C^* \tag{16.86}$$

where $Q(t_1)$ is the positive operator on $\mathcal{X}$ defined by

$$Q(t_1) = \int_0^{t_1} e^{A\sigma} BB^* e^{A^*\sigma} \, d\sigma = \int_0^{t_1} e^{A(t_1 - \tau)} BB^* e^{A^*(t_1 - \tau)} \, d\tau. \tag{16.87}$$

Notice that $Q(t_1)$ is the unique solution at t_1 to the following differential equation:

$$\dot{Q} = AQ + QA^* + BB^* \qquad \text{with} \qquad Q(0) = 0. \tag{16.88}$$

This follows by applying Leibnitz's rule,

$$\frac{d}{dt} \int_0^t f(t,\tau)\, d\tau = f(t,t) + \int_0^t \frac{\partial f}{\partial t}(t,\tau)\, d\tau$$

to $Q(t) = \int_0^t f(t,\tau)\, d\tau$ where $f(t,\tau) = e^{A(t-\tau)} B B^* e^{A^*(t-\tau)}$. It now follows from (16.86) and (16.83) that

$$\|T\| = [\lambda_{\max}(CQ(t_1)C^*)]^{1/2}\,. \tag{16.89}$$

Also, a vector u_1 in $L^2([0,t_1], \mathcal{U})$ which attains the norm of T is given by

$$u_1(t) = B^* e^{A^*(t_1-t)} C^* y_1 / \sigma_1 \tag{16.90}$$

where y_1 is an eigenvector corresponding to the maximum eigenvalue σ_1^2 of $CQ(t_1)C^*$. So, the norm of T and the singular value decomposition of T can readily be computed by solving a linear matrix differential equation in (16.88). Then $\|T\|^2$ is the largest eigenvalue of $CQ(t_1)C^*$.

A minimum control norm problem. Consider now the following optimization problem associated with the state space system in (16.85). Given a vector y in $\mathcal{Y}$, find an input $\hat{u} \in L^2([0,t_1], \mathcal{U})$ with the smallest possible norm which drives the output $y(t_1) = Tu$ at time t_1 as close as possible to y. In mathematical terminology this is equivalent to finding a control $\hat{u}$ to solve the following minimum norm optimization problem:

$$\|\hat{u}\| = \inf\left\{\|u\| : Tu = P_\mathcal{R} y \text{ and } u \in L^2([0,t_1], \mathcal{U}))\right\} \tag{16.91}$$

where $P_\mathcal{R}$ is the orthogonal projection onto the range of T. So, according to Theorem 16.5.1, the solution $\hat{u}$ to this minimum norm control problem is unique and given is by $\hat{u} = T^{-r} y$. By using the singular value decomposition to compute T^{-r}, we see that the unique optimal control solving (16.91) is

$$\hat{u}(t) = T^{-r} y = \sum_{i=1}^n \frac{(y, y_i)}{\sigma_i} u_i(t)\,. \tag{16.92}$$

By consulting Corollary 16.5.2, the optimal control $\hat{u}$ is also given by $\hat{u} = T^*(TT^*)^{-r} y$. Since T^* is given by (16.75) with $G(t) = Ce^{At}B$ and $TT^* = CQ(t_1)C^*$, the optimal $\hat{u}$ can be computed by

$$\hat{u}(t) = B^* e^{A^*(t_1-t)} C^* (CQ(t_1)C^*)^{-r} y\,. \tag{16.93}$$

These optimization problems play a role in controllability of linear systems.

16.8.2   The L^2-L^∞ gain

Let $L^\infty([0,\infty), \mathcal{Y})$ denote the set of all Lebesgue measurable functions over $[0,\infty)$ with values in $\mathcal{Y}$ such that

$$\|y\|_\infty = \operatorname{ess\,sup}\{\|y(t)\| : t \geq 0\} < \infty\,. \tag{16.94}$$

As before, $\mathcal{U}$ and $\mathcal{Y}$ are finite dimensional vector spaces. Moreover, assume that G is a continuous (or Lebesgue measurable) function on $[0, \infty)$ with values in $\mathcal{L}(\mathcal{U}, \mathcal{Y})$ such that

$$W_\infty := \int_0^\infty G(\sigma)G(\sigma)^* \, d\sigma \tag{16.95}$$

is a bounded operator on $\mathcal{Y}$. Obviously, W_∞ is positive. For example, if G is a continuous function satisfying $\|G(t)\| \leq me^{-\alpha t}$ for some positive m and $\alpha > 0$, then the operator W_∞ is positive and bounded. Now let M be the linear map from $L^2([0, \infty), \mathcal{U})$ into $L^\infty([0, \infty), \mathcal{Y})$ defined by

$$(Mu)(t) = \int_0^t G(t - \tau)u(\tau) \, d\tau \qquad (u \in L^2([0, \infty), \mathcal{U})\,).$$

In this setting, the norm of M is defined by

$$\|M\| := \sup\{\|Mu\|_\infty : u \in L^2([0, \infty), \mathcal{U}) \text{ and } \|u\| \leq 1\}. \tag{16.96}$$

We claim that $\|M\|^2 = \lambda_{\max}(W_\infty)$.

If $y = Mu$, then for any $t_1 \geq 0$, we have $y(t_1) = Tu$ where T is the operator from $L^2([0, t_1], \mathcal{U})$ into $\mathcal{Y}$ defined in (16.74). Using (16.76), we obtain

$$TT^* = W(t_1) := \int_0^{t_1} G(\sigma)G(\sigma)^* \, d\sigma \leq W_\infty\,.$$

So, using $\|y(t_1)\| = \|Tu\|$, we have

$$\|y(t_1)\|^2 \leq \|T\|^2 \int_0^{t_1} \|u(t)\|^2 \, dt \leq \lambda_{\max}(TT^*) \int_0^\infty \|u(t)\|^2 \, dt$$

$$\leq \lambda_{\max}(W_\infty)\|u\|^2\,.$$

Since the above holds for all $t_1 \geq 0$, it follows that $\|y\|_\infty^2 \leq \lambda_{\max}(W_\infty)\|u\|^2$. Therefore, $\|M\|^2 \leq \lambda_{\max}(W_\infty)$.

It remains to show that $\|M\|^2 = \lambda_{\max}(W_\infty)$. To this end, consider any t_1 and let u_1 be the unit vector in (16.84) which attains the norm of T in (16.74), that is, $\|u_1\| = 1$ and $\|Tu_1\| = \|T\| = [\lambda_{\max}(W(t_1))]^{1/2}$. Now let u be the unit vector in $L^2([0, \infty), \mathcal{U})$ defined by

$$u(t) = \begin{cases} u_1(t) & \text{if } t \leq t_1 \\ 0 & \text{if } t > t_1\,. \end{cases}$$

Setting $y = Mu$, we have $y(t_1) = Tu_1$ and

$$\|M\| \geq \|y\|_\infty \geq \|y(t_1)\| = \|Tu_1\| = [\lambda_{\max}(W(t_1))]^{1/2}\,.$$

Because this holds for all $t_1 \geq 0$ and $W(t_1)$ approaches W_∞ as t_1 approaches infinity, we obtain $\|M\| \geq [\lambda_{\max}(W_\infty)]^{1/2}$. Since we have also shown that this inequality holds in the other direction, we must have equality, that is, $\|M\| = [\lambda_{\max}(W_\infty)]^{1/2}$.

Consider now the finite dimensional system (16.85) where A is stable. In this setting, $G(t) = Ce^{At}B$ for all $t \geq 0$, and thus, W_∞ is a well defined bounded operator on $\mathcal{Y}$. In this case, $W_\infty = CQC^*$ where Q is the positive operator on $\mathcal{X}$ defined by

$$Q = \int_0^\infty e^{A\sigma}BB^*e^{A^*\sigma} \, d\sigma\,.$$

By consulting Lemma 3.1.2, we see that $W_\infty = CQC^*$ where Q is the unique solution to the Lyapunov equation

$$AQ + QA^* + BB^* = 0. \tag{16.97}$$

So, if G is the impulse response for a stable system $\{A, B, C, 0\}$, then $\|M\|^2 = \lambda_{max}(CQC^*)$.

Exercise 41 Let A be a stable operator on a finite dimensional space $\mathcal{X}$ and C an operator mapping $\mathcal{X}$ into $\mathcal{Y}$. Let T be the operator mapping $\mathcal{X}$ into $L^2([0,\infty), \mathcal{Y})$ defined by

$$Tx = Ce^{At}x \qquad (x \in \mathcal{X}).$$

Show that $T^*T = P$ where P is the solution to the Lyapunov equation

$$A^*P + PA + C^*C = 0.$$

Let $\{x_i\}_1^n$ be the orthogonal eigenvectors corresponding to the nonzero eigenvalues $\sigma_1^2 \geq \sigma_2^2 \geq \cdots \geq \sigma_n^2 > 0$ of P. Then show that the singular value decomposition of T is given by

$$Tx = \sum_{i=1}^n \sigma_i(x, x_i) y_i(t) \qquad (x \in \mathcal{X})$$

where $\sigma_i y_i(t) = Ce^{At}x_i$ for $i = 1, \cdots, n$. Show that $\|T\|^2 = \lambda_{max}(P)$.

16.9 Notes

The results in Sections 16.2 to 16.5 are classical results in Hilbert space; see Akhiezer-Glazman [2], Balakrishnan [8], Conway [30], Gohberg-Goldberg [53], Halmos [59], Luenberger [85], and Taylor-Lay [117]. For a solution to the least square polynomial fit problem based on the Gram matrix see Gohberg-Goldberg [53]. Our approach to the singular value decomposition is standard. For a nice presentation of the singular value decomposition for compact operators see Chapter VI in Gohberg-Goldberg-Kaashoek [54]. The results in the L^2-L^∞ gain section were taken from Wilson [126] and Zhu-Corless-Skelton [31].

Bibliography

[1] Ackermann, J., Der entwurf linearer regelungssysteme im zustandsraum, *Regulungestechnik und Prozessedatenverarbeitung*, **7** (1972) pp. 297-300.

[2] Akhiezer, N.I. and I.M. Glazman, *Theory of Linear Operators in Hilbert Space*, Dover Publishing, New York, 1993.

[3] Arnold, W.F. and A.J. Laub, Generalized eigenproblem algorithms and software for algebraic Riccati equations, *Proceedings of the IEEE*, **72** (1984) pp. 1746-1754.

[4] Anderson, B.D.O. and J.B. Moore, *Linear Optimal Control*, Prentice Hall, Englewood Cliffs, New Jersey, 1971.

[5] Anderson, B.D.O. and J.B. Moore, *Optimal Filtering*, Prentice Hall, Englewood Cliffs, New Jersey, 1979.

[6] Anderson, B.D.O. and J.B. Moore, *Optimal Control: Linear Quadratic Methods*, Prentice Hall, Englewood Cliffs, New Jersey, 1990.

[7] Athans, M. and P.L. Falb, *Optimal Control*, McGraw-Hill, New York, 1966.

[8] Balakrishnan, A.V., *Applied Functional Analysis*, Springer-Verlag, New York, 1976.

[9] Ball, J.A. and A.C.M. Ran, Optimal Hankel norm model reductions and Wiener-Hopf Factorizations I: The canonical case, *SIAM J. Control and Optimization*, **25** (1987) pp. 362-382.

[10] Ball, J. A., Gohberg, I. and L. Rodman, *Interpolation for Rational Matrix Functions*, Operator Theory: Advances and Applications, **45**, Birkhäuser Verlag, Basel, 1990.

[11] Barnett, S., *Polynomials and Linear Control Systems*, Marcel Dekker, New York, 1983.

[12] Bart, H., Gohberg, I. and M. A. Kaashoek, *Minimal Factorization of Matrix and Operator Functions*, Operator Theory: Advances and Applications, **1**, Birkhäuser Verlag, Basel, 1979.

[13] Basar, T. and P. Bernhard, *Optimal Control and Related Minimax Design Problems: A Dynamic Game Approach*, Systems and Control: Foundations and Applications, Birkhäuser Verlag, Boston, 1991.

[14] Belevitch, V., *Classical Network Theory*, Holden-Day, San Francisco, 1968.

[15] Berkovitz, L.D., *Optimal Control Theory*, Springer-Verlag, New York, 1974.

[16] Boyd, S., El Ghaoui, L., Feron, E. and V. Balakrishnan, *Linear Martix Inequalities in Systems and Control*, SIAM, Philadelphia, 1994.

[17] de Branges, L., Factorization and invariant subspaces, *J. Mathematical Analysis and Applications*, **29** (1970) pp. 163-200.

[18] de Branges, L., Pertubation Theory, *J. Mathematical Analysis and Applications*, **57** (1977) pp. 393-415.

[19] de Branges, L. and J. Rovnyak, *Square Summable Power Series*, Holt Rinehart and Winston, New York, 1966.

[20] Brasch, F.M. and J.B. Pearson, Pole placement using dynamic compensators, *IEEE Trans. Automatic Contrtrol*, **15** (1970) pp. 34-43.

[21] Brockett, R.W., *Finite Dimensional Linear Systems*, John Wiley and Sons, New York, 1970.

[22] Bryson, A.E. and Y.C. Ho, *Applied Optimal Control: Optimization, Estimation and Control*, Hemisphere Publishing, New York, 1981.

[23] Burl, B., *Linear Optimal Control: H_2 and H_∞ Methods*, Addison Wesley Longman, Reading Massachusetts, 1998.

[24] Caines, P. E., *Linear Stochastic Systems*, John Wiley and Sons, Montreal, 1988.

[25] Callier, F.M. and C.A. Desoer, *Multivariable Feedback Systems*, Springer-Verlag, New York, 1982.

[26] Chen, C.T., *Linear Systems Theory and Design*, Oxford University Press, New York, 1998.

[27] Chui, C.K. and G. Chen, *Signal Processing and Systems Theory*, Springer-Verlag, Berlin, 1992.

[28] Coddington, E. A. and N. Levinson, *Theory of Ordinary Differential Equations*, McGraw-Hill, New York, 1955.

[29] Conway, J. B., *Functions of One Complex Variable*, Springer-Verlag, New York, 1978.

[30] Conway, J. B., *A Course in Functional Analysis*, Springer-Verlag, Berlin, 1985.

[31] Corless, M., Zhu, G., and Skelton, R., Improved Robustness Bounds Using Covariance Matrices, *Proceedings of the 28th IEEE Conference on Decision and Control*, Tampa, Florida, 1989.

[32] Damen, A.A.H., Van den Hof, P.M.J. and A.K. Hajdasinski, Approximate realization based upon an alternative to the Hankel matrix: the Page matrix, *Systems and Control Letters*, **2** (1982) pp. 202-208.

[33] DeCarlo, R.A., *Linear Systems*, Prentice Hall, Englewood Cliffs, New Jersey, 1989.

[34] Delchamps, D.F., *State Space and Input-Output Linear Systems*, Springer-Verlag, New York, 1988.

[35] Desoer, C.A. and M. Vidyasagar, *Feedback Systems: Input-Output Properties*, Academic Press, New York, 1975.

[36] Dorato, P., Abdallah, C. and V. Cerone, *Linear-Quadratic Control an Introduction*, Prentice Hall, New Jersey, 1995.

[37] Doyle, J. C., Francis, B. A. and A. Tannenbaum, *Feedback Control Theory*, MacMillan, New York, 1991.

[38] Doyle, J. C., Glover, K., Khargonekar, P.P. and B.A. Francis, State-space solutions to standard H_2 and H_∞ control problems, *IEEE Trans. Automatic Contrtrol*, **34** (1988) pp. 831-847.

[39] Foias, C. and A. E. Frazho, *The Commutant Lifting Approach to Interpolation Problems*, Operator Theory Advances and Applications, **44**, Birkhäuser Verlag, Basel, 1990.

[40] Foias, C. and A. E. Frazho, Commutant lifting and simultaneous H^∞ and L^2 suboptimization, *SIAM J. Math. Anal.*, **23** (1992) pp. 984-994.

[41] Foias, C., Frazho, A.E, Gohberg, I. and M.A. Kaashoek, *Metric Constrained Interpolation, Communtant Lifting and Systems*, Operator Theory: Advances and Applications, **100**, Birkhäuser Verlag, Basel, 1998.

[42] Foias, C., Frazho, A. E. and W. S. Li, The exact H^2 estimate for the central H^∞ interpolant, in: *New aspects in interpolation and completion theories* (Ed. I Gohberg) Operator Theory Advances and Applications, **64**, Birkhäuser Verlag, Basel (1993) pp.119-156.

[43] Foias, C., Ozbay, H. and A.R. Tannenbaum, *Robust Control of Infinite Dimensional Systems*, Springer-Verlag, London, 1996.

[44] Francis, B. A., *A Course in H_∞ Control Theory*, Lecture Notes in Control and Information Sciences, Springer-Verlag, New York, 1987.

[45] Frazho, A.E., A shift operator approach to bilinear systems theory, *SIAM J. Control*, **18** (1980) pp. 640-658.

[46] Frazho, A. E. and M. A. Rotea, A remark on mixed L^2/L^∞ bounds, *Integral Equations and Operator Theory*, **15** (1992) pp. 343-348.

[47] Fuhrmann, P.A., *Linear Systems and Operators in Hilbert Space*, McGraw-Hill, New York, 1981.

[48] Gantmacher, F.R., *The Theory of Matrices*, Vol. I, Chelsea Publishing Co., New York, 1977.

[49] Gantmacher, F.R., *The Theory of Matrices*, Vol. II, Chelsea Publishing Co., New York, 1960.

[50] Garnett, J.B., *Bounded Analytic Functions*, Academic Press, New York, 1981.

[51] Gilbert, E., Controllability and observability in multivariable control systems, *SIAM J. Control*, **1** (1963) pp. 128-151.

[52] Glover, K., All optimal Hankel-norm approximations of linear multivariable systems and their error bounds, *International J. Control*, **39** (1984) pp. 1115-1193.

[53] Gohberg, I. and S. Goldberg, *Basic Operator Theory*, Birkhäuser, Basel, 1981.

[54] Gohberg, I., Goldberg, S. and M.A. Kaashoek, *Classes of linear operators*, Vol. I, **49**, Birkhäuser Verlag, Basel, 1990.

[55] Gohberg, I., Goldberg, S. and M.A. Kaashoek, *Classes of linear operators*, Vol. II, **63**, Birkhäuser Verlag, Basel, 1993.

[56] Golub, G.H. and C.F. Van Loan *Matrix Computations*, The Johns Hopkins University Press, Baltimore, 1983.

[57] Green, M. and D. Limebeer, *Linear Robust Control*, Prentice Hall, New Jersey, 1995.

[58] Halmos, P. R., *Finite-Dimensional Vector Spaces*, Springer-Verlag, New York, 1974.

[59] Halmos, P. R., *A Hilbert Space Problem Book*, Springer-Verlag, New York, 1982.

[60] Hanselman, D., and Littlefield, B. R., *Mastering MATLAB 6*, Prentice Hall, New Jersey, 2000.

[61] Hautus, M.L.J., Controllability and observability conditions of linear autonomous systems, *Ned. Akad. Wetenschappen, Proceedings Ser. A*, **72** (1969) pp. 443-448.

[62] Helton, J.W., Discrete time systems, operator models and scattering theory, *J. Functional Analysis*, **16** (1974) pp. 15-38.

[63] Helton, J.W., *Operator Theory, Analytic Functions, Matrices, and Electrical Engineering*, CBMS Regional Conference Series in Math., **68**, American Mathematical Society, Providence Rhode Island, 1987.

[64] Hoffman, K., *Banach Spaces of Analytic Functions*, Dover Publications, New York, 1988.

[65] Horn, R.A. and C.R. Johnson, *Martix Analysis*, Cambridge University Press, Cambridge, 1985.

[66] Kaftal, V., Larson, D. and G. Weiss, Quasitriangular subalgebras of semifinite Von Neumann algebras are closed, *J. Functional Analysis*, **107** (1992) pp. 387-401.

[67] Kailath, T., *Lectures on Linear Least-Squares Estimation*, CISM Courses and Lectures, **140**, Springer-Verlag, New York, 1978.

[68] Kailath, T., *Linear Systems*, Englewood Cliffs: Prentice Hall, New Jersey, 1980.

[69] Kalman, R.E., A new approach to linear filtering and prediction problems, *Trans. ASME J. Basic Engineering*, **82** (1960) pp. 34-45.

[70] Kalman, R.E., Contribution to the theory of optimal control, *Bol. Soc. Matem. Mex.*, **5** (1960) pp. 102-119.

[71] Kalman, R.E., Mathematical description of linear dynamical systems, *SIAM J. Control*, **1** (1963) pp 152-192.

[72] Kalman, R.E., Irreducible realizations and the degree of a rational matrix, *SIAM J. Applied Math.*, **13** (1965) pp. 520-544.

[73] Kalman, R.E., Falb, P.L. and M.A. Arbib, *Topics in Mathematical System Theory*, McGraw-Hill, New York, 1969.

[74] Khalil, H.K., *Nonlinear Systems*, Prentice Hall, Englewood Cliffs, New Jersey, 1996.

[75] Khargoneker, P.P. and K.M. Nagpal, Filtering and smoothing in an H_∞ setting, *IEEE Trans. Automatic Control*, **36** (1991) pp. 152-166.

[76] Khargoneker, P.P., Peterson, I.R. and M.A. Rotea, H_∞-optimal control with state feedback, *IEEE Trans. Automatic Control*, **33** (1988) pp. 786-788.

[77] Kwakernaak, H. and R. Sivan, *Linear Optimal Control Systems*, Wiley Interscience, New York, 1972.

[78] Lancaster, P. and L. Rodman, *Algebraic Riccati Equations*, Clarendon Press, Oxford, 1995.

[79] Laub, A.J., A Schur method for solving the algebraic Riccati equation, *IEEE Trans. Automatic Control*, **24** (1979) pp. 913-921.

[80] Lee, E.B. and L. Markus, *Foundations of Optimal Control Theory*, John Wiley and Sons, New York, 1967.

[81] Leitmann, G., *The Calculus of Variations and Optimal Control*, Plenum Publishing, New York, 1981.

[82] Lewis, F. L. and V. L. Syrmos, *Optimal Control*, John Wiley and Sons, New York, 1995.

[83] Limebeer, D.J.N., Anderson, B.D.O., Khargonekar, P.P. and M. Green, A game theoretic approach to H_∞ control for time varying systems, *SIAM J. Control and Optimization*, **30** (1992) pp. 262-283.

[84] Luenberger, D.G., Observing the state of a linear system, *IEEE Transactions on Military Eletronics*, **8** (1964) pp. 74-80.

[85] Luenberger, D.G., *Optimization by Vector Space Methods*, John Wiley and Sons, New York, 1969.

[86] Luenberger, D.G., An Introduction to observers, *IEEE Trans. Automatic Contrtrol*, **16** (1971) pp. 596-602.

[87] Mageirou, E.F., Values and strategies for infinite time linear quadratic games, *IEEE Trans. Automatic Contrtrol*, **21** (1976) pp. 547-550.

[88] Mageirou, E.F. and Y.C. Ho, Decentralized stabilization via game theoretic methods, *Automatica*, **13** (1977) pp. 393-399.

[89] Meirovitch, L., *Elements of Vibration Analysis*, McGraw Hill, New York, 1986.

[90] Meirovitch, L., *Dynamics and Control of Structures*, John Wiley and Sons, New York, 1990.

[91] Moore, B.C., Principle component analysis in linear systems: Controllability, observability, and model reduction, *IEEE Trans. Automatic Contrtrol*, **26** (1981) pp. 17-32.

[92] Mustafa, D. and K. Glover, *Minimum Entropy H_∞ Control*, Lecture notes in Control and Information Sciences, *Springer-Verlag*, New York, 1990.

[93] Naylor, A.W. and G.R. Sell, *Linear Operator Theory in Engineering and Science*, Springer-Verlag, New York, 1982.

[94] Nikolskii, N.K., *Treatise on the Shift Operator*, Springer-Verlag, New York, 1986.

[95] Ogata, K., *Modern Control Engineering*, Prentice Hall, Englewood Cliffs, New Jersey, 1970.

[96] Peterson, I.R., Disturbance attenuation and H_∞ optimization: A design method based on the algrbraic Riccati equation, *IEEE Trans. Automatic Contrtrol*, **32** (1987) pp. 427-429.

[97] Peterson, I.R. and C.V. Hollot, A Riccati equation approach to the stabilization of uncertain systems, *Automatica*, **22** (1986) pp. 397-411.

[98] Polderman, J.W. and J.C. Willems, *Introduction to Mathematical Systems Theory: A Behavioral Approach*, Springer-Verlag, New York, 1997.

[99] Popov, V.M., *Hyperstability of Control Systems*, Springer, Berlin, 1973.

[100] Porter, W.A., *Modern Foundations of Systems Engineering*, The Macmillan Company, New York, 1966.

[101] Porter, W.A., A basic optimization problem in linear systems, *Mathematical Systems Theory*, **5** (1971) pp. 20-44.

[102] Ravi, R., Nagpal, K.M. and P.P. Khargonekar, H_∞ control of linear time varying systems: A state-space approach, *SIAM J. Control and Optimization*, **29** (1991) pp. 1394-1413.

[103] Riesz, F. and B. Sz.-Nagy, *Functional Ananysis*, Dover Publications, New York, 1990.

[104] Rosenblum, M. and J. Rovnyak, *Hardy Classes and Operator Theory*, Oxford University Press, New York, 1985.

[105] Rosenbrock, H.H., *State Space and Multivariable Theory*, Wiley Interscience, New York, 1970.

[106] Rota, G.C., On models for linear operators, *Comm. Pure Applied Math.*, **13** (1960) pp. 468-472.

[107] Rotea, M. and A. E. Frazho, Bounds on solutions to H^∞ algebraic Riccati equations and properties of the central solution, *Systems and Control Letters*, **19** (1992) pp.341-352.

[108] Royden, H.L., *Real Analysis*, Prentice Hall, New Jersey, 1988.

[109] Rugh, W. J., *Nonlinear System Theory, The Volterra Wiener Approach*, The Johns Hopkins University Press, Baltimore, 1981.

[110] Rugh, W. J., *Linear System Theory*, Prentice Hall, New Jersey, 1993.

[111] Sarason, D., On spectral sets having connected complement, *Acta Sci. Math.*, **26** (1965) pp. 289-299.

[112] Sarason, D., Generalized interpolation in H^∞, *Trans. American Math. Soc.*, **127** (1967) pp. 179-203.

[113] Sarason, D., *Sub-Hardy Hilbert Spaces in the Unit Disc*, The University of Arkansas Lecture notes in the Mathematical Sciences, **10**, John Wiley and Sons, New York, 1994.

[114] Skelton, R.E., *Dynamic Systems Control*, John Wiley and Sons, New York, 1988.

[115] Skelton, R.E., Iwasaki, T. and K.M. Grigoriadis, *A Unified Algebraic Approach to Control Design*, Taylor and Francis, London, 1997.

[116] Sontag, E.D., *Mathematical Control Theory*, Springer-Verlag, 1990.

[117] Taylor, A.E. and D.C. Lay, *Introduction to Functional Analysis*, John Wiley and Sons, New York, 1980.

[118] Tadmor, G., Worst-case design in the time domain: The maximum principle and the standard H_∞ problem, *Mathematics of Control, Signals and Systems*, **3** (1990) pp. 301-325.

[119] Sz.-Nagy, B. and C. Foias, Dilatation des commutants d'opérateurs, *C.R. Acad. Sci. Paris, Serie A*, **266** (1968) pp.493-495.

[120] Sz.-Nagy, B. and C. Foias, *Harmonic Analysis of Operators on Hilbert Space*, North-Holland Publishing Co., Amsterdam, 1970.

[121] Van Dooren, P., A generalized eigenvalue approach for solving algebraic Riccati equations, *SIAM J. Sci. Stat. Comput.*, **2** (1981) pp. 121-135.

[122] Vardulakis, A.I.G., *Linear Multivariable Control*, John Wiley, Chichester, 1991.

[123] Vidyasagar, M., *Control System Synthesis: A Factorization Approach*, The MIT Press, Cambridge, Massachusetts, 1985.

[124] Vidyasagar, M., *Nonlinear Systems Analysis*, Prentice Hall, Englewood Cliffs, New Jersey, 1993.

[125] Willems, J.C., Least squares stationary optimal control and the algebraic Riccati equation, *IEEE Trans. Automatic Contrtrol*, **16** (1971) pp. 621-634.

[126] Wilson, D.A., Convolution and Hankel operator norms for linear systems, *IEEE Trans. Automatic Contrtrol*, **34** (1989) pp. 94-97.

[127] Wolovich, W.A., *Linear Multivariable Systems*, Springer-Verlag, New York, 1974.

[128] Wong, W.S., *Operator Theoretic Methods in Nonlinear Systems*, Phd. Thesis, Harvard University, Cambridge, Massachusetts, 1980.

[129] Wonham, W.M., *Linear Multivariable Control: A Geometric Approach*, Springer-Verlag, New York, 1985.

[130] Zadeh, L.A. and C.A. Desoer, *Linear System Theory*, McGraw-Hill, New York, 1963.

[131] Zhou, K., Doyle J. C., and K. Glover, *Robust and Optimal Control*, Prentice Hall, New Jersey, 1996.

Index

$A(G)$, 106
A_G, 85
$B(G)$, 106
B_G, 85
$C(G)$, 106
C_G, 85
$D(G)$, 106
D_G, 85
$H^\infty(\cdot,\cdot)$, 25
H^∞ norm, 25
$L^2([a,b],\mathcal{U})$, 2
$L^2[a,b]$, 2
$P_\mathcal{M}$, 300
$\bigvee$, 2
$\mathbb{C}^n$, 2
ker, 302
$\lambda_{\max}$, 26
ran , 302
$\mathcal{H}(G)$, 106
$\mathcal{H}_G$, 85
$\mathcal{L}(\cdot,\cdot)$, 3
$l_+(\mathcal{Y})$, 84
$l^2_+(\mathcal{U})$, 90
$l_{c+}(\mathcal{U})$, 89
m_A, 77
$\mathcal{H}(\mathcal{W},\mathcal{Z})$, 289
$\mathcal{R}_+(\mathcal{Y})$, 106

Ackermann's Formula, 125
asymptotic estimate, 145

backward shift operator, 84, 106
balanced, 116
boundary problem
    two point, 179

Cauchy sequence, 1
Cauchy-Schwartz, 1

Cayley-Hamilton Theorem, 13
central controller, 262
    infinite horizon, 282
closed range, 303
co-prime polynomials, 87
companion matrix, 16
compression, 67
controllable, 55
    canonical form, 126
    Gramian, 62
    Gramian finite time, 60
    PBH test, 59
    subspace, 56

detectable, 145
dilation, 68
dissipative, 29

eigenvalue
    stable, 122
estimation error
    state, 149
estimator
    state, 145
evaluation operator, 84, 106

feedback
    dynamic output, 156

gain, 119
game theory, 267
Gram matrix, 310

H-infinity analysis problem, 228
Hamiltonian, 179, 205, 273
Hankel matrix, 89
Hilbert space, 2
Hilbert-Schmidt norm, 289

impulse response, 6

increasing
    backwards in time, 221
infinite horizon
    H^∞ analysis, 244
    H^∞ control, 274
    central controller, 282
    linear quadratic, 189
initialization operator, 85, 106
input space, 3
intertwines, 84
invertible, 303

Kalman-Ho, 102
kernel, 302

Laplace transform, 6
linear quadratic regulator, 171
Lyapunov
    equation, 30
    function, 33
    operator, 30

matrix representation, 67, 102
McMillan degree, 84
minimal polynomial, 77
minimal realization, 83

norm
    bounded by γ, 235
    strictly bounded by γ, 235
normal rank, 163

observable, 41
    companion matrix, 135
    Gramian, 50
    Gramian finite time, 47
    pair, 41
    PBH test, 45
    subspace, 43
    unobservable subspace, 42
one to one, 303
onto, 303
operator, 3
    adjoint, 3
    backward shift, 84, 106
    bounded, 3

bounded below, 304
evaluation, 84, 106
initialization, 85, 106
positive, 19
strictly positive, 19
orthogonal complement, 299
orthogonal projection, 300
    operator, 301
outer factor, 198
    finite time, 184
outer function, 198
output space, 3

Parseval's relation, 2
PBH test, 45
    detectable, 147
polar decomposition, 317
pole of order j, 78
polynomial
    minimal, 77
polynomials
    co-prime, 87
positive definite function, 33
Projection Theorem, 299

range, 302
rank
    normal, 163
rational
    proper, 7
    strictly proper, 7
realization, 12, 13, 70
    minimal, 83
    of a sequence, 12, 84
    partial minimal, 96
    partial order of, 96
    restricted backward shift, 85, 106
relative degree, 8
restricted inverse, 306
Riccati
    algebraic equation, 190, 205
    differential equation, 171
    uniformly bounded solution, 245, 275

scalar input, 123
Schmidt pair, 319

Schur complement, 165
similar systems, 84
single input, 123
singular value decomposition, 318
singular values, 318
spectral factor, 184
stabilizable, 119
stabilizing solution, 191, 205
stable
    eigenvalue, 122
    input output, 23
    mechanical system, 21
    operator, 20
    system, 19
    transfer function, 26, 154
state, 3
    adjoint, 178
state estimator, 145
state space, 3
subspace, 299
    co-invariant, 68
    controllable/observable, 73
    controllable/unobservable, 71
    invariant, 68
    reducing, 68
    semi-invariant, 69
    uncontrollable/observable, 74
    uncontrollable/unobservable, 74
system
    closed loop, 141
    open loop, 119

trace, 288
tracking
    linear quadratic, 182, 186
transfer function, 7
    central, 282
    zero of, 163, 164

uncontrollable
    eigenvalue, 58
    eigenvector, 58
    subspace, 57
unobservable
    eigenvalue, 45

eigenvector, 45
subspace, 42

Vandermonde matrix, 312